ISS

The International Space Station

By TD Barnes

Copyright 2024

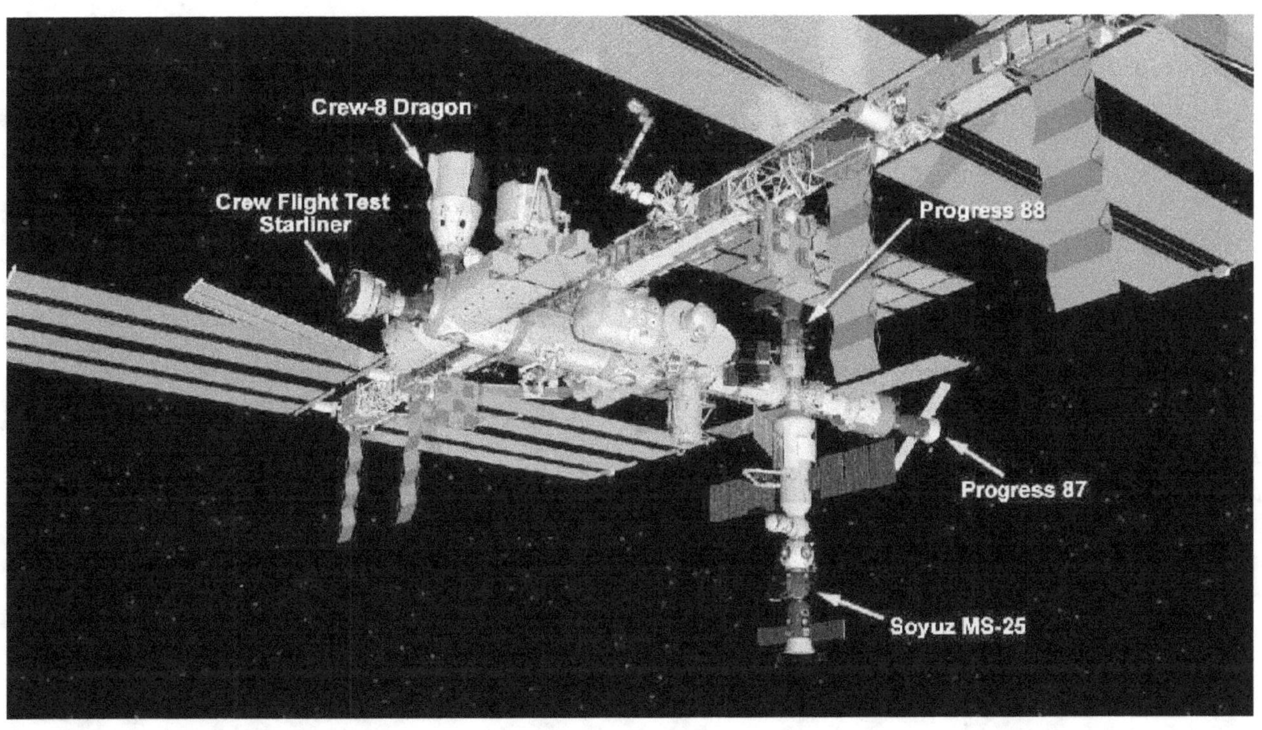

July 12, 2024: International Space Station Configuration. Five spaceships are parked at the space station including Boeing's Starliner spacecraft, the SpaceX Dragon Endeavour spacecraft, the Soyuz MS-25 crew ship, and the Progress 87 and 88 resupply ships.

Contents

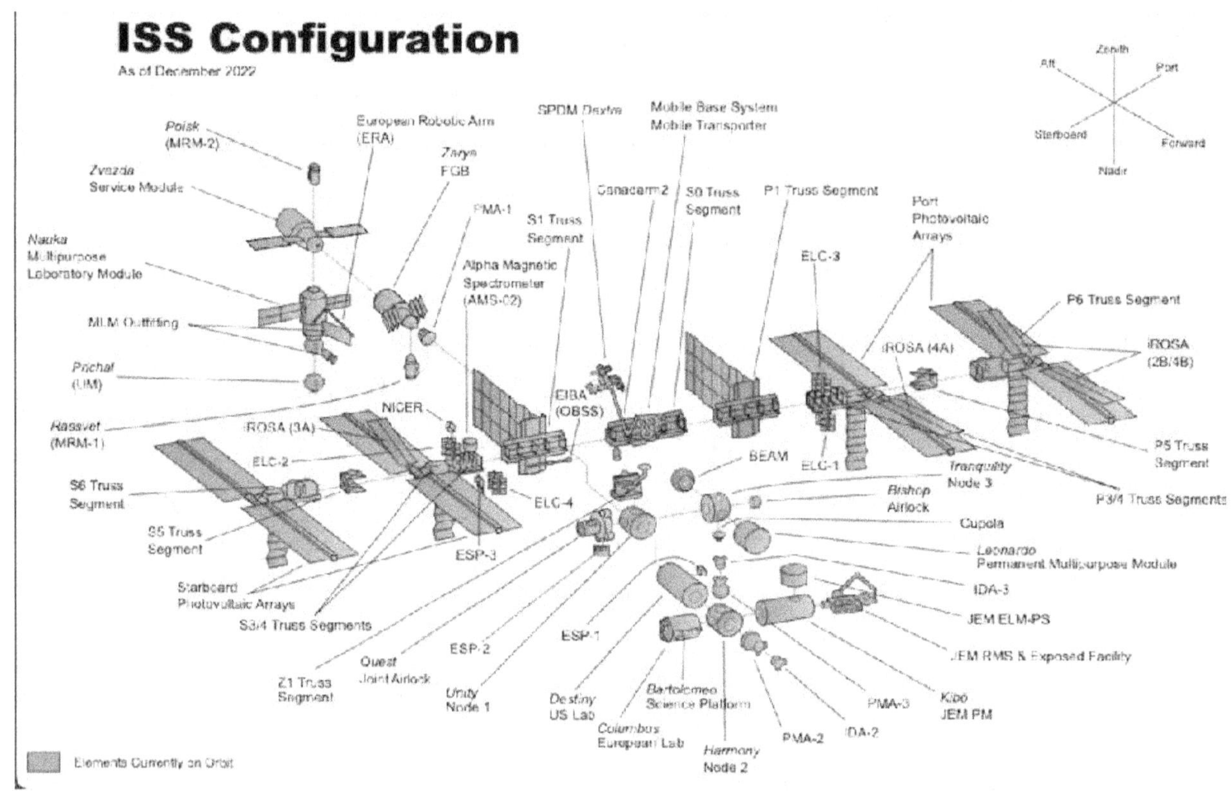

Glossary

ATV (Automated Transfer Vehicle): A European spacecraft designed for cargo resupply missions to the ISS. It delivers supplies, scientific experiments, and waste disposal services, equipped with advanced docking systems.

Atlas V: A family of expendable launch vehicles developed by United Launch Alliance (ULA). It launches various spacecraft, including the Boeing Starliner, into orbit. Known for its reliability, it supports missions to the ISS.

Boe-CFT (Boeing Crew Flight Test): The first crewed mission of Boeing's Starliner capsule. This mission tests the spacecraft's systems and performance in transporting astronauts to and from the ISS.

Boe-OFT (Boeing Orbital Flight Test): The uncrewed test flights of Boeing's Starliner capsule designed to validate the spacecraft's performance and safety before crewed missions.

Capsule: A spacecraft designed for transporting humans or cargo to space and returning them safely to Earth. The ISS uses various capsules, including the Space Shuttle, Boeing Starliner, and SpaceX Dragon.

Cosmonaut: A Russian astronaut. Cosmonauts are trained by the Russian space agency, Roscosmos, to conduct missions on the ISS and other space endeavors.

Docking: The process of bringing two spacecraft together in orbit to attach securely. Accurate docking is essential for transferring crew, cargo, and experiments between spacecraft and the ISS.

Helium Leak: A malfunction involving helium gas escape, which is used in spacecraft systems for pressurization and thermal control. Leaks can impact spacecraft performance and mission success.

International Space Station (ISS): A large, habitable spacecraft orbiting Earth, serving as a laboratory for scientific research and international cooperation in space. It is a joint project involving NASA, Roscosmos, ESA, JAXA, and CSA.

Launch Vehicle: A rocket used to propel spacecraft from Earth into orbit. Examples include the Atlas V, SpaceX Falcon 9, and Soyuz rocket.

Mission Specialist: An astronaut with specific expertise and responsibilities related to the mission's objectives, such as conducting scientific experiments, operating specialized equipment, or performing spacewalks.

Orbital Flight Test (OFT): An uncrewed test flight conducted to assess the performance of a spacecraft in orbit before it carries astronauts. The Starliner OFT missions were crucial for validating the spacecraft's systems.

Pilot: An astronaut responsible for operating the spacecraft during the mission. Pilots assist the commander with navigation, systems management, and other critical tasks.

Space Shuttle: A reusable spacecraft used by NASA from 1981 to 2011. It was designed for transporting astronauts and cargo to and from space, including the ISS. The Shuttle featured wings for atmospheric re-entry and landing, and was capable of carrying large payloads.

Solar Arrays: Large panels on the ISS designed to capture sunlight and convert it into electrical power. They are crucial for maintaining the station's operations and supporting scientific experiments.

Soyuz: A Russian spacecraft used for transporting astronauts to and from the ISS. It has been a reliable vehicle for crewed missions since the early 2000s, known for its robust design and safety features.

Thruster: A propulsion device used on spacecraft for maneuvering and attitude control. Thrusters are essential for docking, orbital adjustments, and maintaining the spacecraft's orientation.

U.S. Orbital Segment (USOS): The portion of the ISS primarily managed by NASA and its partners, including modules and equipment developed by the United States and other international collaborators.

Vehicle: A general term for any spacecraft or launch system used to transport astronauts or cargo to space. This includes shuttles, capsules, and spaceplanes.

Astronaut: An individual trained to travel into space, conduct research, and operate spacecraft. Astronauts are trained by various space agencies, including NASA (United States), ESA (Europe), JAXA (Japan), and CSA (Canada)

Preface

Throughout the history of human spaceflight, several space stations have been launched, operated, and eventually deorbited. These stations played pivotal roles in developing space technology and international cooperation.

The Soviet Union's space program in the 1970s saw the launch of two concurrent space station programs, both named Salyut, but with distinct objectives. The Long Duration Orbital Station (DOS) program was designed for scientific research and long-term habitation, while the Almaz program, a secretive military initiative, focused on space reconnaissance.

Salyut 1, the first of the DOS program, was launched on April 19, 1971. It remained in orbit until October 11, 1971, accumulating 175 days in space. Over its operational period, the station hosted 6 crew members, with 24 days of human occupancy. Its pressurized volume was 100 cubic meters, and its mass was 18,425 kilograms.

DOS-2, the second station in the DOS program, was launched on July 29, 1972. However, it failed to achieve orbit and was never crewed. Its intended mission was to continue the scientific research objectives of its predecessor.

Salyut 2, launched on April 3, 1973, was part of the Almaz program. Like DOS-2, it failed to achieve its mission objectives and was not crewed. It remained in orbit for only 13 days before re-entering.

Kosmos 557, another station from the DOS program, launched on May 11, 1973, and remained operational until May 22, 1973. It was intended to further the DOS program's goals but had a short mission duration of 11 days.

NASA launched Skylab on May 14, 1973, the United States' first space station. Its mission lasted until July 11, 1979, with 2,249 days in orbit. It hosted 9 crew members over 171 days of human occupation. Skylab's pressurized volume was 360 cubic meters, and its mass was 77,088 kilograms.

The Soviet Union's Salyut 3, part of the Almaz program, was launched on June 25, 1974. It operated until January 24, 1975, with 213 days in space. This station had a pressurized volume of 90 cubic meters and a mass of 18,900 kilograms. It hosted 15 crew members over 15 days.

Salyut 4 followed on December 26, 1974, and operated until February 3, 1977. This station accumulated 770 days in space, with 92 days of human occupancy. It had a mass of 18,900 kilograms and a pressurized volume of 90 cubic meters.

Salyut 5, another Almaz program station, launched on June 22, 1976. It was in orbit until August 8, 1977, spending 412 days in space. Salyut 5 had a pressurized volume of 100 cubic meters and a mass of 19,000 kilograms, and it hosted 67 crew days.

Salyut 6, launched on September 29, 1977, was another station from the DOS program. It operated until July 29, 1982, amassing 1,764 days in orbit. This station supported 33 crew members over 683 days of human occupancy. Its pressurized volume was 90 cubic meters, and its mass was 19,000 kilograms.

Salyut 7, launched on April 19, 1982, was the final station in the Salyut program. It remained operational until February 7, 1991, with 3,216 days in space. It hosted 22 crew members over 861 days of human occupancy. Salyut 7 had a pressurized volume of 90 cubic meters and a mass of 19,000 kilograms.

The Mir space station, launched on February 19, 1986, was the largest and most enduring of the Soviet space stations. It continued operations until March 23, 2001, achieving 5,511 days in orbit. Mir supported 125 crew members over 4,594 days of human occupancy. Its pressurized volume was 350 cubic meters, and its mass was 129,700 kilograms.

China's space program introduced the Tiangong-1 on September 29, 2011. It remained in orbit until April 2, 2018, spending 2,377 days in space. Tiangong-1 weighed 8,506 kilograms and had a pressurized volume of 15 cubic meters. During its operational period, it hosted 22 crew members.

Tiangong-2, launched on September 15, 2016, followed its predecessor and operated until July 19, 2019. This station had a similar mass and pressurized volume to Tiangong-1. It supported 29 crew members over 1,037 days of space operation.

As of the current era in space exploration, two fully operational space stations, the International Space Station (ISS) and China's Tiangong Space Station (TSS), represent the pinnacle of human presence in space.

The ISS has been a continuous symbol of international collaboration since October 2000, when Expedition 1 marked the beginning of long-term human occupancy. Over the years, the ISS has hosted astronauts from various space agencies, contributing to a broad range of scientific research and technological advancements. Its role as a hub for space science and international cooperation has been pivotal in our understanding of life and work in space.

China's Tiangong Space Station (TSS) joined the ranks of operational space stations in June 2022 with the arrival of Shenzhou 14. This station reflects China's growing space capabilities and ambitions, serving as a platform for scientific experiments, technology demonstrations, and international collaboration.

The ISS and TSS have collectively marked significant milestones in space station history, including records for the highest number of people aboard a single space station. This record was first achieved on July 23, 2009, during the eleven days when the ISS hosted 13 people. This situation arose from the simultaneous docking of the Space Shuttle Endeavour on its 127th mission, which brought additional astronauts aboard, temporarily increasing the station's crew capacity.

A more recent record for the highest number of people across all space stations was established on May 30, 2023. On this date, the combined total of occupants on the ISS and TSS reached 17. This included 11 astronauts on the ISS and 6 on the TSS, demonstrating the capacity of multiple space stations to accommodate a substantial number of crew members simultaneously.

Orbiting Earth at an average altitude of approximately 410 kilometers (250 miles), the International Space Station (ISS) resides within the realm known as low Earth orbit. This orbital path ensures that the station completes a full revolution around Earth approximately every 90 minutes. Due to atmospheric drag and periodic reboosting maneuvers, the altitude of the ISS could fluctuate by several kilometers over time.

The ISS spans an impressive length of 357 feet (109 meters), nearly equivalent to the length of a football field, including the end zones. The station's living and working space, encompassing approximately 5,600 square feet (520 square meters), offers an environment comparable to that of a six-bedroom house. This space includes six sleeping quarters, two bathrooms, a gym, and a unique 360-degree bay window providing breathtaking views of Earth and space.

The ISS was a testament to human ingenuity and international cooperation, comprising a complex assembly of 16 pressurized modules. These modules, constructed by various space agencies, collectively contribute to the station's diverse functions. Six of these modules are of Russian origin: Zarya, Zvezda, Poisk, Rassvet, Nauka, and Prichal. Each Russian module fulfills a critical role in the ISS's operations.

Zarya, the first module launched, acts as the station's initial power and propulsion component. Zvezda serves as the primary living quarters and command center, integrating essential life support and control systems. Poisk and Rassvet provide additional docking ports and storage capabilities, while Nauka, a multi-purpose laboratory module, supports scientific research and advanced docking functions. Finally, Prichal, the latest addition, extends the ISS's docking capabilities, facilitating the arrival of new modules and spacecraft.

The United States has contributed eight modules to the ISS: BEAM, Leonardo, Harmony, Quest, Tranquility, Unity, Cupola, and Destiny. These modules are essential for various functions, including scientific experiments, crew operations, and maintenance. Destiny, the U.S. laboratory module, was central to the station's research activities, while Harmony and Unity facilitate the connection of other modules and international docking. The Cupola module offers an unparalleled view of space and Earth, enhancing observational and photographic capabilities.

Japan's contribution to the ISS was represented by the Kibō module, which provides additional laboratory space for scientific experiments in a variety of fields, including biology, astronomy, and material science. Similarly, the European Space Agency's Columbus module enhances the ISS's research capabilities, with advanced facilities for conducting experiments that require microgravity conditions.

To date, at least one Russian pressurized module, Pirs, has been deorbited, reflecting the evolving nature of the ISS's configuration as older modules are replaced or removed.

In addition to these permanent modules, the ISS has utilized Multi-Purpose Logistics Modules (MPLMs) during certain Shuttle missions.

Although not permanently docked, MPLMs were attached to either Harmony or Unity for resupply and logistical purposes. These modules facilitated the delivery of cargo and experiments, underscoring the flexibility and adaptability of the ISS in accommodating new supplies and equipment.

The station's pressurized volume was further extended by the spacecraft docked to it. Typically, at least one Soyuz spacecraft was continually docked as a "lifeboat," ensuring that the station's crew has an emergency escape vehicle readily available. These Soyuz spacecraft are replaced every six months as part of a regular crew rotation, maintaining a continuous presence of astronauts and ensuring the station's ongoing operational capability.

The evolution of America's space endeavors could be traced through exploring five fundamental questions: what, why, when, where, and who. At the heart of this exploration was the concept of human habitation in space, which serves as the cornerstone for developing modern space stations. The aspiration to live and travel beyond Earth's atmosphere has its roots in the early 17th century, embodying a vision that has shaped space exploration for centuries.

In the early 1600s, the renowned Renaissance astronomer Johannes Kepler was one of the first to reimagine planets as distant celestial bodies and as distinct worlds separated by immense expanses of space. Kepler's profound insights extended to the possibility that future generations might traverse these interplanetary distances one day. His pioneering ideas laid the groundwork for future space exploration, reflecting an early yet significant understanding of human potential in space travel.

By the 1860s, the concept of space habitation had further evolved through the imaginative narrative of Edward Everett Hale's novella, The Brick Moon. Published in the Atlantic Monthly, Hale's story introduced the revolutionary idea of a man-made satellite, referred to as the "brick moon," orbiting Earth. This fictional creation was depicted as a habitable space station with life support systems and functions resembling a contemporary space station. Hale's work provided a conceptual framework for space habitation, merging imaginative fiction with plausible technological concepts that foreshadowed developing space stations.

The early 20th century saw further advancements in space station theory with the contributions of Konstantin Tsiolkovsky, a prominent Russian scientist. Tsiolkovsky envisioned space stations as self-sustaining environments powered by solar energy, designed to simulate miniature Earths in orbit. His innovative concept included the support of human life through internal vegetation growth, demonstrating a sophisticated understanding of life-support systems in space. Tsiolkovsky's theories were instrumental in shaping developing space stations, reflecting a forward-thinking vision of human life in space.

In 1928, Austrian engineer Herman Noordung made significant strides with his proposal of the "wohnrad," or "living wheel." This rotating structure was designed to simulate gravity through centrifugal force, to be assembled in orbit. Noordung's design was initially tested on Earth and represented a critical step toward creating a viable, habitable space environment. His innovative ideas on simulating gravity and sustaining human life in space provided crucial insights into the practical aspects of space station design, advancing the theoretical framework laid out by his predecessors

By 1952, the notion of space habitation had gained significant momentum, thanks to the collaborative efforts of science writer Willy Ley and artist Chesley Bonestell. Their joint work, The Conquest of Space, vividly illustrated a wheel-shaped space station orbiting Earth, reflecting and expanding upon the ideas initially proposed by Herman Noordung. Ley envisioned this space station as a confined yet functional environment, akin to a submarine but equipped with modern amenities such as air conditioning and artificial gravity. These innovative concepts captured the public's imagination and were prominently featured in Collier's Magazine and the Walt Disney television program, further cementing their influence and inspiring a generation.

The Cold War era introduced a new dimension to space exploration, transforming it from a scientific endeavor into a fierce political and ideological competition. The space race between the United States and the Soviet Union, which began in earnest in the late 1950s, symbolized a broader contest for global supremacy.

The space race was ignited on October 4, 1957, when the Soviet Union launched Sputnik 1, the world's first artificial satellite. This landmark achievement stunned the United States and showcased Soviet technological prowess. The subsequent launch of Sputnik 2, which carried Laika, the first living creature in orbit, further heightened American anxieties. These early successes marked the inception of a new Cold War front, one focused on technological and scientific advancements as symbols of national strength.

In response, the United States rapidly intensified its space efforts, leading to the establishment of NASA in 1958. The American objective was to match and ultimately surpass Soviet achievements. This period saw developing advanced spacecraft and rockets, highlighted by the launch of Explorer 1 in 1958, which discovered the Van Allen radiation belts—a significant American milestone. The competition escalated with the race to the Moon, where the Soviet Union initially took the lead through its Luna program. Luna 2's successful impact on the Moon's surface in 1959 demonstrated Soviet capabilities.

In contrast, the United States 'Apollo program, initiated in 1961 under President John F. Kennedy's ambitious goal to land a man on the Moon by the end of the decade, aimed to showcase American technological superiority. This intense rivalry culminated in the historic Apollo 11 mission, landing astronauts Neil Armstrong and Buzz Aldrin on the lunar surface on July 20, 1969. Armstrong's famous words, "That's one small step for man, one giant leap for mankind," echoed around the world, marking a pinnacle achievement in human history and solidifying the United States' position as a leader in space exploration.

The 1950s marked a pivotal era in the United States' endeavors to develop space stations, laying the groundwork for future advancements in space habitation. One of the earliest and most ambitious proposals emerged from the U.S. Army's Project Horizon, which envisioned a modular orbital station capable of housing crews and serving as a refueling point for spacecraft bound for a lunar base. This project underscored the strategic military and exploratory interests of the era.

Simultaneously, NASA's Manned Spacecraft Center—later renamed the Johnson Space Center in Houston—began to formalize its own requirements for a space station. These early conceptualizations were instrumental in shaping the vision for Skylab, America's first space station. Launched in the 1970s, Skylab marked a significant milestone, providing invaluable insights into living and working in space over extended periods.

The conclusion of the Apollo program heralded a new chapter in space exploration. The landscape began to shift with the introduction of commercial astronauts, diversifying the scope of space missions. Traditionally, astronaut selection and training had been the exclusive purview of government space agencies like NASA and military programs, focusing on missions driven by national interest and scientific advancement. The advent of commercial astronauts opened the doors for private sector involvement, adding a new dimension to human spaceflight.

The International Space Station (ISS) was conceived as a multifaceted testament to human ingenuity and international collaboration in space exploration. Its design envisioned multiple roles: a laboratory for cutting-edge scientific experiments, an observatory for astronomical studies, a factory for manufacturing processes in microgravity, and a hub for essential transportation and maintenance activities. Beyond these roles, the ISS was also seen as a strategic platform in low Earth orbit to support future missions to the Moon, Mars, and beyond. Initial agreements between NASA and Roscosmos, the Russian space agency, outlined these ambitious goals. Still, the ISS has since evolved to embrace commercial, diplomatic, and educational functions as defined by the United States National Space Policy.

The evolution of the ISS concept was deeply rooted in the historical context of the early 1970s when the intense rivalry of the space race began to give way to a more collaborative spirit. This shift in perspective was exemplified by the Apollo-Soyuz Test Project (ASTP) in 1975, a landmark mission that achieved the first successful docking of spacecraft from different nations. The ASTP symbolized a new era of international cooperation in space exploration, fostering goodwill and mutual understanding between the United States and the Soviet Union. The success of this mission was instrumental in paving the way for future joint ventures, ultimately culminating in the creation of the ISS—a symbol of peaceful collaboration and a beacon of shared human achievement in space.

During the late 1970s, several notable concepts emerged in space station development, reflecting both ambition and the complexities of international cooperation. One significant proposal from this period was the International Skylab initiative, which envisioned launching a backup Skylab B space station to facilitate a series of missions involving both Apollo and Soyuz spacecraft. This initiative aimed to build upon the existing Skylab program and extend its capabilities through collaboration with the Soviet Union. Another more ambitious idea was the Skylab-Salyut Space Laboratory, which sought to dock Skylab B with a Soviet Salyut space station, creating a combined research platform. However, as Cold War tensions intensified and budgetary constraints became increasingly stringent, these plans and a proposal to dock the Space Shuttle with a Salyut station were eventually shelved.

In the early 1980s, NASA shifted its focus toward a new initiative: developing a modular space station known as Freedom. This ambitious project was intended to provide a counterpart to the Soviet Union's Salyut and Mir space stations. By 1984, the European Space Agency (ESA) joined the Freedom Project, and by 1987, it had approved the Columbus laboratory as a critical component of the initiative. This collaboration represented a significant step toward establishing an international space infrastructure. In response to a 1982 request from NASA, Japan also contributed by introducing the Japanese Experiment Module (JEM), or Kibō, in 1985, further enhancing the developing framework of the space station.

The formal endorsement of the International Space Station (ISS) project was marked by a pivotal moment on January 25, 1984, when President Ronald Reagan articulated a transformative vision for space exploration in his State of the Union Address. Reagan's statement, "Just as the oceans opened up a new world for clipper ships and Yankee traders, space holds enormous potential for commerce today," signaled the beginning of a mission to establish a permanent international presence in space. The goal was to construct an orbital laboratory within a decade.

The U.S. Congress responded positively to this ambitious vision, allocating the necessary funds to support the ISS project. This critical approval enabled NASA Administrator James Beggs to embark on efforts to secure international partnerships. The global response was swift and enthusiastic, with Canada, Japan, and several European Space Agency member states committing to the project. This marked the beginning of a landmark international collaboration.

From 1984 to 1993, the design and development of the ISS progressed steadily, involving contributions from the United States, Canada, Japan, and Europe. The Columbus program, led by Germany and Italy, emerged as one of ESA's most ambitious space projects. The Columbus module was envisioned as a central element of the Freedom project, with aspirations to evolve into a comprehensive European orbital outpost by the end of the century.

As the early 1990s approached, financial concerns surrounding the Freedom project intensified. The rising costs led to resistance from Congress, which called for increased international participation to alleviate the financial burden and ensure the project's feasibility. Facing these financial pressures, NASA considered canceling the project. Concurrently, the Soviet Union was advancing plans for the Mir-2 space station, with module construction beginning in the mid-1980s. However, the dissolution of the Soviet Union significantly reduced the scope of the Mir-2 project, creating uncertainty about its future.

A pivotal development occurred in September 1993, when American Vice President Al Gore and Russian Prime Minister Viktor Chernomyrdin announced a groundbreaking agreement to launch a new collaborative space station project. This agreement marked the inception of what would eventually become the International Space Station (ISS). It also included provisions for the United States to engage with the Mir program, facilitating dockings with American Space Shuttles through the Shuttle-Mir program.

By the end of 1993, Russia was formally invited to join the ISS collaboration, leading to a transformative shift in the project's trajectory. This integration resulted in an agreement to implement the ISS program in distinct phases, setting the stage for the construction and operation of the ISS as a symbol of international cooperation and scientific advancement in space.

Phase 1: NASA-Mir (1995-1998)

The initial phase of the International Space Station (ISS) program, designated as NASA-Mir, unfolded between 1995 and 1998. This pivotal era marked a significant chapter in space exploration as NASA Space Shuttles undertook the essential task of transporting astronauts and cosmonauts to the Russian Mir Orbital Station. The core objectives of this phase were to adapt and integrate two Russian-built modules for use in U.S. and international scientific experiments and to cultivate a framework of effective collaboration among the participating nations.

Eleven Space Shuttle missions were executed, with ten of these achieving docking with the Mir space station. Each mission was crucial in facilitating personnel exchanges between U.S. and Russian space agencies, strengthening the two nations' collaborative efforts. The exchanges fostered a deeper understanding of joint operations and helped iron out the intricacies of international space cooperation.

Among the key advancements during the NASA-Mir phase were the launches and integration of two new Russian modules, Spektr and Priroda. Adding these modules was instrumental in expanding the operational capabilities of Mir. Spektr, launched in 1995, offered critical space for scientific experiments, particularly in astrophysics and Earth observation. Priroda, launched in 1996, provided further room for U.S. payloads and research, significantly augmenting the station's scientific capacity. These modules not only facilitated the accommodation of seven U.S. astronauts but also played a vital role in advancing the scientific and technological objectives of the space station.

Phase 2: Construction and Expansion (1998-2015)

The successful completion of Phase 1 of the International Space Station (ISS) project set the stage for the pivotal Phase 2, which, with contributions from all participating nations, encompassed the construction and operational establishment of the ISS. This phase was centered on the assembly of the space station and the initiation of its fundamental operations.

The multinational character of the ISS project necessitated the assembly of components manufactured across different countries. In the mid-1990s, significant U.S. components, including the Destiny Laboratory, Unity Module, Integrated Truss Structure, and solar arrays, were fabricated at the Marshall Space Flight Center in Huntsville, Alabama, and the Michoud Assembly Facility in New Orleans. These modules were then transported to the Operations and Checkout Building and the Space Station Processing Facility (SSPF) at the Kennedy Space Center for final assembly and processing in preparation for their launch into space.

The Space Shuttles was crucial in this expansion. Notably, Shuttle missions by Discovery, Atlantis, and Endeavour were instrumental in delivering key components to the ISS. Discovery's missions brought the Destiny Laboratory, to developing the cornerstone of American scientific research on the station. Atlantis followed with the Quest airlock, essential for conducting extravehicular activities (EVAs), while Endeavour delivered the Canadarm2, the primary robotic arm of the ISS, along with additional segments of the Integrated Truss Structure. The Canadarm2, a sophisticated piece of technology, proved indispensable for maneuvering and maintaining the station's various components.

Between 2001 and 2006, the ISS underwent a transformative period of assembly and expansion. This era saw the addition of several critical modules and components that significantly enhanced the station's capabilities. The Destiny Laboratory, a major American contribution, provided dedicated space for scientific experiments, reinforcing the ISS's role as a hub for space research.

The Russian space agency also made substantial contributions with the Zvezda Service Module, crucial for life support and served as the primary living quarters for the crew. This module played an integral role in sustaining long-term missions aboard the ISS.

The ISS's capabilities were notably expanded on February 7, 2001, with the addition of the Destiny Laboratory, which increased the station's laboratory space by 41% and became a cornerstone for U.S. scientific research aboard the ISS. Another significant American addition was the Harmony

Module, which enhanced the station's capacity for international cooperation by acting as a connecting node that linked various research laboratories and facilitated the integration of future modules.

During this period, the European Space Agency's Columbus Laboratory joined the ISS, bringing advanced facilities for scientific research across a range of disciplines. Similarly, with its sophisticated research facilities and critical systems, the Japanese Kibo laboratory added new capabilities and furthered the station's ability to conduct scientific research in microgravity.

Russian modules like Zarya and Zvezda were produced at the Khrunichev State Research and Production Space Center in Moscow. Zvezda, initially manufactured in 1985 for the canceled Mir-2 project, was repurposed as the ISS Service Module.

The European Space Agency's Columbus module was meticulously crafted at the EADS Astrium Space Transportation facilities in Bremen, Germany, with contributions from numerous European contractors. Alongside Columbus, other ESA-built modules such as Harmony, Tranquility, the Leonardo Multi-Purpose Logistics Module (MPLM), and the Cupola were initially constructed at the Thales Alenia Space factory in Turin, Italy. Once completed, these modules were transported by aircraft to the Kennedy Space Center's Space Station Processing Facility (SSPF) for final preparation and launch processing.

The Japanese Experiment Module, known as Kibo, was developed at advanced technology manufacturing facilities in Japan, including the NASDA (now JAXA) Tsukuba Space Center and the Institute of Space and Astronautical Science. After its construction, Kibo was shipped by sea and then flown to the SSPF for final assembly and integration.

The Mobile Servicing System, comprised the Canadarm2 and the Dextre grapple fixture, was a crucial component of the ISS's operational capabilities. Developed at various locations in Canada, including the David Florida Laboratory, and in the United States, the system was created under contract by the Canadian Space Agency. Northrop Grumman constructed the mobile base system that provides the structural framework for the Canadarm2.

The assembly of the ISS, a monumental achievement in space architecture, commenced in November 1998. This process involved deploying Russian modules launched and docked using robotic systems, except for the Rassvet module. Most other modules were delivered by the Space Shuttle and required precise installation by ISS and Shuttle crew members using the Canadarm2 and through extra-vehicular activities (EVAs). By June 5, 2011, these efforts had added 159 components to the station, achieved through more than 1,000 hours of EVA. Of these spacewalks, 127 originated from the station itself, while the remaining 32 were conducted from the airlocks of docked Space Shuttles. Throughout this intricate assembly process, the station's beta angle—its orientation relative to the Sun—had to be continuously monitored and managed.

On November 20, 1998, a significant milestone was achieved with the launch of the Zarya Control Module. This historic event marked the first tangible step toward constructing a permanent human outpost in space. Named "Zarya," meaning "sunrise" in Russian, the module was launched by a Russian Proton rocket from the Baikonur Cosmodrome in Kazakhstan.

Zarya was foundational to the ISS's initial assembly, providing critical functions necessary for the station's operation and future expansion. It offered essential services for subsequent spacecraft, including fuel storage, power supply, and docking capabilities. Zarya also played a significant role in providing propulsion, attitude control, and communications for the ISS, with its electrical power systems supporting the station's early operations.

Despite its crucial contributions, Zarya did not include long-term life support systems, as its primary role was establishing the ISS's fundamental infrastructure. The launch of Zarya represented not only a technological triumph but also the commencement of an international collaborative effort to create a unique platform for scientific research and space exploration.

Two weeks later, on December 4, 1998, the Unity module was launched aboard Space Shuttle flight STS-88. On December 7, 1998, the Space Shuttle Endeavour, on mission STS-88, docked with the ISS. This pivotal mission marked the inaugural assembly of the ISS, as Endeavour delivered and installed the Zarya module, the first major component of the station. The crew, led by Commander Robert Cabana and Pilot Frederick W. Leslie, along with Mission Specialists Jerry L. Ross, James H. Newman, and Sergei Krikalev, conducted crucial setup tasks to prepare the station for future operations. The shuttle undocked on December 13, 1998, completing its mission and leaving the ISS as the foundation for its ongoing assembly.

Unity, a passive module with two Pressurized Mating Adapters (PMAs), was connected to Zarya during extra-vehicular activities (EVAs) conducted by the Shuttle astronauts. One PMA was permanently attached to Zarya, while the other facilitated the docking of future Space Shuttles to the ISS. The precise connection of Unity to Zarya marked a significant advancement in the ISS's construction, setting the stage for further development and expansion of the station.

At this stage in its development, the International Space Station (ISS) remained uncrewed, as the Russian Mir space station continued accommodating crewed missions. The ISS would not see its first human inhabitants until the year 2000. The pivotal moment for the ISS's expansion came on July 12, 2000, with the launch of the Zvezda module into orbit. Zvezda, meticulously preprogrammed to deploy its solar arrays and communications antenna upon reaching space, became the focal point for a critical rendezvous with the Zarya and Unity modules. This docking maneuver was executed through a sophisticated interplay of ground control efforts and the Russian automated rendezvous and docking system.

Upon its successful docking with Zarya and Unity, control of the station was seamlessly transitioned from Zarya's onboard computer to Zvezda's advanced systems. The addition of Zvezda marked a significant enhancement to the ISS's capabilities, incorporating vital amenities for long-term habitation. These included sleeping quarters, a toilet, a kitchen, CO_2 scrubbers, a dehumidifier, oxygen generators, and exercise equipment. Additionally, Zvezda established essential communication links with mission control, including data, voice, and television channels, thus initiating a permanent human presence aboard the ISS.

However, the progress of the ISS's expansion encountered a major setback in 2003 due to the Space Shuttle Columbia disaster. This tragic event grounded the Space Shuttle fleet and halted assembly operations until 2005. It was only with the successful return to flight of Space Shuttle Discovery on mission STS-114 that assembly resumed.

Construction resumption saw significant milestones with the arrival of Space Shuttle Atlantis on STS-115 in 2006, which delivered the second set of solar arrays. Subsequent missions, including STS-116, STS-117, and STS-118, further advanced the station's infrastructure by adding additional truss segments and a third set of solar arrays. These upgrades were crucial in increasing the ISS's power-generating capacity, thus enabling the accommodation of more pressurized modules.

The final years of the decade, from 2007 to 2010, witnessed continued progress with installing solar arrays and docking adapters, essential for future module arrivals. Among the significant additions were the Rassvet module from Russia, the Harmony node, and the Columbus laboratory, representing Europe's contributions to the ISS. Additionally, the first two components of the Japanese Experiment Module, Kibō, were integrated. By March 2009, STS-119 completed the Integrated Truss Structure by installing the fourth and final set of solar arrays. The remaining sections of Kibō were delivered during STS-127 in July 2009, followed by the Russian Poisk module, which docked with the station in May 2010.

The ISS's expansion continued with the arrival of the third node, Tranquility, during STS-130 in February 2010, accompanied by the Cupola module. The penultimate Russian module, Rassvet, was delivered by Space Shuttle Atlantis on STS-132 in May 2010, marking an exchange for the Russian Proton rocket's delivery of the US-funded Zarya module in 1998. The final pressurized module of the United States Orbital Segment (USOS), Leonardo, arrived at the station in February 2011 aboard the final flight of Discovery, STS-133. Later in the year, Endeavour delivered the Alpha Magnetic Spectrometer on STS-134, a key scientific instrument for the station.

By June 2011, the ISS had grown to include 15 pressurized modules and the Integrated Truss

Structure, though two power modules, NEM-1 and NEM-2, remained pending launch. The Russian research module Nauka, which docked in July 2021, was accompanied by the European Robotic Arm, capable of repositioning itself across the Russian modules. The latest addition, the nodal module Prichal, docked in November 2021.

The ISS's mass was a dynamic attribute continually influenced by the ongoing addition of experiments, spare parts, crew personal effects, provisions, and docked spacecraft. As of September 2011, the total launch mass of the modules in orbit was approximately 417,289 kilograms (919,965 pounds). The station's mass was further augmented by hydrogen gas vented from oxygen generators, underscoring this monumental structure's immense scale and complexity and its ongoing operation and maintenance.

Phase 3: Optimization and Innovation (2016-Present)

Phase 3 of the International Space Station (ISS), which began in 2016, represents a significant era of optimization and innovation. This phase focuses on enhancing the station's research capabilities, advancing technological developments, and preparing for future missions beyond Earth's orbit.

The ISS, situated in the relatively stable environment of low Earth orbit, was a pivotal testing ground for spacecraft systems and mission strategies crucial for long-duration space exploration. The unique orbital position of the ISS allows scientists and engineers to test and refine the technologies and techniques necessary for missions to the Moon, Mars, and beyond. By providing a practical setting for astronauts to operate, maintain, and repair spacecraft, the ISS plays a crucial role in reducing the risks associated with deep-space travel and enhancing the capabilities of interplanetary spacecraft. This phase of the ISS's mission was instrumental in developing and validating the skills needed for future explorations far beyond our planet.

One noteworthy complementary initiative to the ISS's research efforts was the MARS-500 project, which aligned closely with its objectives while not conducted aboard the ISS. The MARS-500 project, conducted between 2007 and 2011, was a ground-based simulation designed to study long-term isolation and confinement's psychological and physiological effects. This international research effort, involving Russia, Europe, and China, included significant contributions from the Russian space agency Roscosmos, the European Space Agency (ESA), and the China National Space Administration (CNSA). The project consisted of a series of isolation experiments where crews lived in a specially designed habitat, replicating the conditions of a manned mission to Mars. These experiments provided valuable insights into the challenges of deep-space exploration, including human behavior and health during extended space travel.

Although the MARS-500 project was not conducted on the ISS, it was part of a broader strategy to prepare for future space missions. The ISS itself has been crucial in addressing the unique challenges of space, such as weightlessness and radiation. At the same time, ground-based simulations like MARS-500 have complemented this research by examining human endurance and psychological resilience. In 2011, Sergey Krasnov, head of human spaceflight programs for Roscosmos, proposed that a condensed version of the MARS-500 experiment might be conducted aboard the ISS to investigate these issues further. This proposal underscores the ISS's pivotal role in advancing our understanding of long-duration space missions and preparing for future explorations.

The ISS's historical milestones reflect its significance in space exploration. In November 2000, the ISS welcomed its first resident crew, Expedition 1, aboard Soyuz TM-31. This pioneering crew, consisting of NASA astronaut Bill Shepherd and Russian cosmonauts Yuri Gidzenko and Sergei Krikalev, embarked on a mission to establish a continuous human presence in space. This feat has now persisted for over two decades.

By December 30, 2005, Congress officially designated the U.S. portion of the ISS as the nation's newest national laboratory. This designation aimed to optimize the station's utility for various U.S. government agencies, academic institutions, and private research entities.

The ISS's international collaboration was further enriched with the addition of the European Space Agency's Columbus Laboratory on

February 7, 2008. Shortly after that, on March 11, 2008, the Japanese Experiment Module, Kibo, joined the station. Kibo, which means "hope" in Japanese, represented Japan's first human-rated space facility and marked a significant contribution from the Japan Aerospace Exploration Agency (JAXA) to the ISS program.

On November 2, 2010, the ISS celebrated a decade of continuous human occupancy. By this milestone, the station had hosted 202 visitors from various nations, underscoring its role as a hub for international scientific and collaborative endeavors in space exploration.

In February 2011, NASA took a pivotal step in enhancing the scientific capabilities of the International Space Station (ISS) by issuing a cooperative agreement notice. This announcement aimed to find a management partner for the ISS National Laboratory, formally established later that year. In July 2011, the Center for the Advancement of Science in Space (CASIS) was selected as the official manager of the ISS National Lab, tasked with overseeing its operations and research initiatives.

A landmark achievement for the ISS National Lab came on September 30, 2013, with the launch of the first research flight in the protein crystal growth (PCG) series. This research flight capitalized on the ISS's unique microgravity environment to grow proteins into nearly perfect three-dimensional structures. Cultivating such high-quality crystals in space was invaluable, as it significantly advances drug development and contributes to various scientific fields by providing clearer insights into protein structures.

The ISS's evolution from an ambitious presidential vision to a fully operational space station exemplifies a remarkable journey of international collaboration, technological innovation, and scientific exploration. Its development has been a testament to global partnership, involving the combined efforts of the United States, Russia, Japan, Canada, and Europe, among other partners.

By 2011, the ISS had entered a crucial phase in its development, marking the completion of its core assembly. The addition of the PMA-2 (Pressurized Mating Adapter 2) and the Nauka module, a significant contribution from Russia, represented major milestones in the station's assembly. This period was instrumental in finalizing the station's structure and enhancing its capabilities, setting the stage for its future operations.

2014 and 2015 saw further expansion of the ISS's operational scope. New docking ports were installed, and scientific facilities were upgraded, allowing for greater flexibility in accommodating additional spacecraft and experiments. These enhancements solidified the ISS's role as a premier platform for scientific research and international collaboration.

From 2016 to 2018, the ISS became a hub of groundbreaking scientific research. During this time, the station was the site of numerous pioneering experiments in microgravity, advancements in biological sciences, and innovative technology demonstrations. These efforts allowed scientists to conduct research that would have been impossible under Earth's gravity, driving forward our understanding of various scientific phenomena.

Since 2019, the ISS has continued evolving, supporting diverse international science and technology missions. Upgrades during this period have included the installation of new docking systems and the addition of commercial modules.

The International Space Station (ISS) was a profound symbol of global cooperation and technological innovation. It embodies the collective efforts of a diverse array of international partners, each contributing uniquely to this remarkable orbiting laboratory.

At the heart of this collaborative venture was NASA, the United States' premier space agency, which plays a central role in the ISS project. NASA's contributions include providing essential modules and cutting-edge technology that form the foundation of the station's infrastructure. These components are crucial to the ISS's scientific and operational goals, ensuring that the station remains at the forefront of space exploration and research. NASA's involvement encompasses everything from developing life-support systems to advanced scientific equipment, underscoring its pivotal role in the station's success.

Roscosmos, Russia's space agency, was equally integral to the ISS's operations. Roscosmos was responsible for supplying critical modules and managing crew transport services, essential for the safe transit of astronauts and cosmonauts to and from the station. Beyond transportation, Roscosmos provides vital support services, including maintaining the station's systems and coordinating mission operations. The agency's expertise in these areas ensures the ISS operates smoothly and efficiently.

The European Space Agency (ESA) represents Europe's contribution to the ISS. ESA has enriched the station with the Columbus Laboratory, a state-of-the-art facility dedicated to scientific research. This laboratory supports a wide range of experiments across various disciplines, including biology, materials science, and astronomy. ESA's involvement extends beyond the hardware of the Columbus Laboratory to include an extensive array of scientific equipment and research capabilities, contributing significantly to the ISS's mission of advancing knowledge.

Japan, through the Japan Aerospace Exploration Agency (JAXA), has also made substantial contributions to the ISS. Japan's major contribution was the Kibo Laboratory, a versatile research facility that supports a broad spectrum of scientific experiments. Additionally, JAXA handles cargo resupply missions, ensuring the ISS was continually stocked with necessary supplies and equipment. This role was crucial for maintaining the station's operational capabilities and supporting ongoing research activities.

Canada's contribution, represented by the Canadian Space Agency (CSA), was embodied in the Canadarm2, an advanced robotic system essential for the maintenance and operation of the ISS. This sophisticated arm was used for a variety of tasks, including assembling and repairing the station and conducting scientific experiments in space. The Canadarm2 was a testament to Canada's technological prowess and its critical role in supporting the ISS's mission.

The International Space Station (ISS) stands as a testament to international cooperation and scientific progress, made possible through a series of pivotal support missions that have ensured its continued operation and enhancement.

At the heart of the ISS's construction and ongoing resupply efforts was the Space Shuttle program. This program, a cornerstone of the ISS's development, operated from 1998 to 2011. Throughout 39 missions, beginning with STS-88, the Space Shuttle fleet played an essential role in assembling the ISS, delivering crucial modules, and transporting vital equipment. The shuttles not only helped to build the station's infrastructure but also established the foundation for its future scientific endeavors. The complex assembly involved the integration of various modules, each contributing to the ISS's capabilities as a multi-national laboratory orbiting Earth.

Parallel to this, the Russian space program has supported the ISS. With its impressive record of 83 missions, the Progress spacecraft has been a lifeline for the station, routinely delivering food, fuel, and other essential supplies. This consistent flow of resources has been critical in maintaining the ISS's operational status. The Soyuz spacecraft, equally important, has conducted 83 missions, primarily focused on ferrying astronauts to and from the station. While primarily a crew transport vehicle, Soyuz has also carried cargo, further supporting the ISS's needs.

The advent of commercial spaceflight brought new dimensions to ISS support. SpaceX's Dragon spacecraft, with 37 missions to its name, has become a cornerstone of resupply efforts. Dragon's contributions extend beyond simple deliveries; it has carried a wide array of cargo and scientific experiments, thereby bolstering the station's research capabilities and operational capacity. Northrop Grumman's Cygnus spacecraft has also been a key player, completing 22 missions that ensure the ISS remains stocked with necessary supplies and experimental materials.

Japan and Europe have made significant contributions to the global support network. The Japanese HTV (H-II Transfer Vehicle) has completed 11 missions, providing the ISS essential supplies and experimental payloads. Similarly, the European Automated Transfer Vehicle (ATV) undertook five missions between 2008 and 2015, delivering vital equipment and research materials.

Space - The Last Frontier

The year 1962 was marked by intense developments in space exploration. During this period, the United States and the Soviet Union, locked in the throes of the Nuclear Arms Race, advanced their technological capabilities both on Earth and beyond. This year witnessed a series of groundbreaking achievements that pushed the boundaries of human exploration and technological prowess.

Amid the Cold War, both superpowers were engaged in a fervent race to develop and deploy advanced nuclear technologies. The United States and the Soviet Union not only conducted numerous nuclear tests but also pursued increasingly sophisticated delivery systems and warhead technologies. While the U.S. had made a significant leap with Alan Shepard becoming the first American astronaut to travel into space in 1961, 1962 truly showcased American ambition in space exploration. On February 20, 1962, John Glenn achieved a historic milestone by becoming the first American to orbit the Earth. His mission aboard the Mercury spacecraft Friendship 7 consisted of three orbits that demonstrated the viability of extended space travel and solidified Glenn's place in the annals of space history.

The Soviet Union, meanwhile, continued to make impressive strides in space exploration. Following Yuri Gagarin's historic flight in 1961 as the first human to journey into space, the Soviets further cemented their leadership in the space race by sending Valentina Tereshkova, the first woman to travel beyond Earth's atmosphere, into space in 1963. Not resting on their laurels, the Soviets achieved another significant milestone in August 1962 by simultaneously orbiting two manned spacecraft, Vostok 3 and Vostok 4, around the Earth. This achievement underscored their capability in managing multiple space missions concurrently and advanced their understanding of spaceflight.

The spirit of exploration was not confined to the two superpowers. In 1962, several other nations sought to make their mark in the burgeoning field of space exploration. Countries such as Turkey, Norway, and Lebanon embarked on their inaugural spaceflights, marking their entry into the space arena. The United Kingdom and Canada also made noteworthy advancements, placing their first satellites into orbit and thus contributing to the global space effort.

Rocket technology was crucial in advancing space capabilities during this period. The United States conducted 59 rocket launches in 1962, with a success rate of approximately 85%. This impressive record included the launches of various rockets such as Delta A, B, Scout X-2, Scout X-2M, Scout X-3, and Thor DM-21. Conversely, the Soviet Union carried out 22 rocket launches, achieving 15 successes and experiencing seven failures.

In addition to the high-profile manned missions, significant technological advancements were underway. The United States initiated Project ROVER, a pioneering effort to explore the potential of nuclear thermal rockets for space travel. The project, supported by the US Space Nuclear Propulsion Office, aimed to develop and test nuclear engines capable of heating hydrogen for propulsion in space. The Kiwi phase of the project involved extensive testing of these engines to verify their design and operational potential.

The year also saw notable achievements in aircraft technology. The X-15 program, featuring test flights by pioneering pilots such as Joseph A. Walker, Robert M. White, and Neil A. Armstrong, showcased remarkable advancements in speed and altitude. These flights provided invaluable data that would inform future aerospace endeavors.

NASA's M2-F1 program also made significant progress by testing a lightweight, unpowered lifting body designed to validate the concept of wingless flight. The program included over 400 ground tows and 77 aircraft tow flights, providing critical insights into this innovative approach to atmospheric reentry and landing.

At the Nuclear Rocket Development Station in the Nevada Test Site, NASA's three test stands became operational, facilitating developing the Nuclear Engine for Rocket Vehicle Application (NERVA). This reactor aimed to advance space travel capabilities through nuclear propulsion. Additionally, the Surveyor program made a crucial shift from using the Agena rocket to the Centaur rocket, enhancing its ability demonstrating soft landings on the Moon.

In a strategic shift, the United States redirected its focus from the GAM-87 Skybolt missile program to the UGM-27 Polaris submarine-launched ballistic missile. This decision was driven by technical challenges and cost concerns, highlighting the evolving nature of defense and space strategies during this pivotal era.

The year 1964 emerged as a defining chapter in the relentless space race between the United States and the Soviet Union, embodying the technological and ideological rivalry that marked the Cold War era. This pivotal year witnessed landmark achievements from both superpowers, highlighting their fierce competition and the broader implications of their advancements in space exploration.

On October 12, 1964, the Soviet Union achieved a groundbreaking milestone with the launch of Voskhod 1. This mission marked several significant firsts in human spaceflight. Voskhod 1 became the first spacecraft to carry a multi-person crew into orbit, with cosmonauts Vladimir Komarov, Konstantin Feoktistov, and Boris Yegorov on board. Notably, Voskhod 1 was also the first mission without spacesuits, underscoring the Soviet confidence in their spacecraft's life support systems. This mission was a pioneering step in crewed space exploration, as it featured an engineer and a physician among its crew, reflecting the Soviets' commitment to advancing scientific and biomedical research in space. During their time in orbit, the crew conducted a variety of scientific experiments and biomedical studies, providing invaluable data on the effects of space travel on the human body. Despite the mission's success, it was cut short due to unforeseen issues, and Voskhod 1 completed its mission by safely landing on October 13, having orbited the Earth 16 times. As a symbolic gesture, the spacecraft carried a ribbon from a communist banner associated with the Paris Commune, further emphasizing the ideological undertones of the space race.

In the United States, 1964 marked a crucial advancement for NASA's Gemini program, instrumental in developing the technologies and techniques necessary for future lunar missions. The Gemini 3 mission, launched in March 1964, was the inaugural manned flight of the Gemini program. Astronauts Virgil "Gus" Grissom and John W. Young were aboard the spacecraft, achieving a significant milestone by becoming the first Americans to perform orbital maneuvers. The mission, which lasted four hours, included various experiments and demonstrated new spaceflight techniques, such as spacecraft control and orbital adjustments.

In 1965, the race to space between the United States and the Soviet Union reached new heights, marking a year of extraordinary technological advancements and significant milestones. The era was characterized by intense competition and rapid progress as both superpowers pushed the boundaries of human and technological limits.

In February, the United States achieved a landmark in lunar exploration with the Ranger 8 mission. This spacecraft, part of NASA's ambitious lunar program, sent back over 7,000 high-resolution images of the Moon's surface before its deliberate crash-landing. The detailed photographs provided unprecedented views of the lunar terrain, enhancing scientific understanding of the Moon's surface features. Ranger 8's success underscored American advancements in space technology and set the stage for future lunar missions, demonstrating an ever-increasing capability to explore and document distant celestial bodies.

The Gemini program, a crucial step toward eventual human missions to the Moon, also made significant strides in 1965. March saw the launch of Gemini 3, the first American spaceflight to carry a two-person crew. Astronauts Gus Grissom and John W. Young undertook this historic mission, proving that a pair of astronauts could effectively operate a spacecraft. Their successful flight was pivotal in preparing for more complex missions and laid foundational technologies for future endeavors.

Meanwhile, the Soviet Union marked a historic achievement with the first-ever spacewalk in March 1965. Cosmonaut Alexei Leonov exited the Voskhod 2 spacecraft, spending 12 minutes in the vacuum of space. This pioneering extravehicular activity (EVA) was a groundbreaking event, showcasing the feasibility and potential of human activity outside spacecraft and demonstrating Soviet leadership in pushing the boundaries of space exploration.

In June, the Soviet Luna 5 mission aimed to achieve a soft landing on the Moon. Although the spacecraft ultimately crashed, it was a significant effort in the Soviet lunar exploration program, highlighting their persistent drive towards lunar exploration despite technical setbacks.

Throughout the year, both nations ramped up their satellite activity. The Soviet Union launched five Kosmos satellites, which served various functions, including technology demonstrations and radar calibration for military purposes. Additionally, four Soviet lunar spacecraft were sent into orbit, each contributing to the ongoing quest for lunar exploration. Concurrently, the United States marked a milestone with the launch of Intelsat I, the first commercial communications satellite placed in geosynchronous orbit. This development represented a major advancement in global communications infrastructure, illustrating the expanding role of space technology in everyday life.

The sheer volume of activity in 1965 was remarkable, with 124 orbital launches reflecting the intense pace of space exploration. Notably, the United States Air Force launched 13 CORONA reconnaissance satellites as part of the CIA's National Reconnaissance Office operations. Despite technical issues, the CORONA program delivered valuable reconnaissance data, enhancing U.S. strategic intelligence capabilities.

NASA's Project Gemini further demonstrated its progress with several pivotal missions. Gemini 4, for instance, was notable for carrying out scientific experiments and for astronaut Ed White's first American spacewalk. Gemini 5 set a new record for the longest American crewed spaceflight, lasting a full week. The successful completion of Gemini 6A and 7 missions marked the first-ever space rendezvous, showcasing the feasibility of spacecraft docking and close-approach maneuvers.

The year also saw significant strides towards developing reusable space transportation. The United States conducted two tests of the M2-F2 atmospheric re-entry vehicle, a precursor to future reusable spacecraft. Simultaneously, the Soviet Union advanced the Mikoyan-Gurevich MiG-105, an experimental lifting-body aircraft designed for potential orbital flights, reflecting a shared commitment to enhancing space travel technology.

The X-15 program continued to explore the extremes of high-speed and high-altitude flight. In June, NASA pilot John B. McKay achieved one of the X-15's fastest flights, reaching 938 mph (6,338 km/h) at 29.5 miles (47.5 km). On June 29, Air Force pilot Joe Engle completed his final flight above 50 miles, achieving speeds of 3,432 mph (5,523 km/h) at 53.1 miles (85.5 km). Later in August, Engle reached 3,550 mph (5,713 km/h) at 51.3 miles (82.6 km), and in September, McKay conducted his only spaceflight above 50 miles, reaching speeds of 3,732 mph (6,006 km/h) at 56.0 miles (90.1 km). These flights provided crucial data on high-speed performance and the challenges of operating at the edge of space.

In 1967, the Space Race intensified at the height of Cold War tensions, captivating global attention and underscoring the fierce rivalry between the United States and the Soviet Union. This year was a crucial chapter in the quest for dominance beyond Earth's atmosphere, reflecting this competition's high stakes and profound impacts.

By 1967, space exploration had reached significant milestones. The United States, buoyed by the successes of the Mercury and Gemini programs, embarked on the ambitious Apollo program to land humans on the Moon. This initiative sought to fulfill President John F. Kennedy's challenge to achieve lunar exploration by the decade's end. Meanwhile, the Soviet Union had already achieved groundbreaking milestones with Yuri Gagarin's historic manned spaceflight in 1961 and Valentina Tereshkova's pioneering journey as the first woman in space in 1963.

However, the year was marked by a tragic setback for NASA. On January 27, a routine test of the Apollo 1 spacecraft at Cape Kennedy's Launch Complex 34 ended in disaster. During a countdown simulation, a fire erupted inside the capsule, fueled by the pure oxygen atmosphere and flammable materials. Despite valiant rescue efforts, the fire led to the deaths of astronauts Gus Grissom, Ed White, and Roger Chaffee, who succumbed to asphyxiation from carbon monoxide and other toxic gases. The design of the hatch further impeded their escape. The incident, exacerbated by the test's non-fueled status, revealed gaps in NASA's safety preparations and caused a significant delay in the Apollo program.

In contrast, the Soviet space program made notable advancements. On April 23, 1967, the Soviets launched Soyuz 1, a mission demonstrating the feasibility of manned space rendezvous and docking. Piloted by Cosmonaut Vladimir Komarov, Soyuz 1 was to dock with Soyuz 2.

However, technical failures during reentry led to a fatal crash, marking the first human death in space. The mission underscored the inherent risks of space exploration and the challenges both superpowers face.

The United States, undeterred, pressed on with its lunar ambitions. On October 11, 1968, the Apollo 7 mission, the program's first manned flight, took off with astronauts Wally Schirra, Donn Eisele, and Walter Cunningham. Their successful mission provided a much-needed boost to American space endeavors and restored confidence in the Apollo program, signaling a renewed commitment to lunar exploration.

As 1967 drew to a close, the Space Race continued to escalate. The Soviet Union had made significant strides with the launch of Luna 9 in early 1966, the first spacecraft to land on the Moon and transmit images back to Earth. In response, the United States achieved its own milestone with the successful landing of Surveyor 1 in May 1966, marking the first soft landing of an American spacecraft on the lunar surface.

Throughout 1967, NASA remained focused on advancing space exploration and technology. The Apollo program, central to American space efforts, saw the announcement of its crew in May, including Walter M. Schirra, Jr., Donn F. Eisele, and R. Walter Cunningham. Concurrently, NASA continued its research into supersonic flight, building on the legacy of its predecessor, the National Advisory Committee on Aeronautics (NACA). The X-15 program, which had achieved remarkable speeds, sought to gather more data on sustained supersonic flight, contributing to developing advanced supersonic and hypersonic aircraft.

In the realm of satellite technology, NASA launched several important missions in 1967. The Applications Technology Satellite (ATS-3) and the first Nimbus satellite focused on Earth observation and meteorological research, providing crucial data on weather patterns, ocean currents, and other aspects of Earth science.

A highlight of the year came on October 3, when Major William J. Knight of the U.S. Air Force set a new world airspeed record in the North American X-15A-2. Despite multiple structural failures, Knight reached an unprecedented speed of Mach 6.72 (4,520 mph, 7,274 km/h), the highest ever achieved by an aircraft in the 20th century.

Additional X-15 flights on October 17 and November 15 saw speeds of Mach 5.53 (3,856 mph, 6,206 km/h) and Mach 5.20 (3,570 mph, 5,745 km/h), respectively. Unfortunately, the latter flight ended tragically with the death of Michael J. Adams and the destruction of the X-15-3.

In addition, 1967 saw test flights of the Schweizer X-26 Frigate, a training glider designed for quiet observation and advanced yaw-roll coupling, conducted by DARPA, the U.S. Army, and the U.S. Navy.

In 1968, the Space Race reached a dramatic and pivotal juncture as the United States and the Soviet Union, driven by the fierce Cold War rivalry, competed to assert their supremacy in space exploration. This intense competition spurred both superpowers to push the boundaries of technological and scientific capabilities, achieving milestones that would shape the future of spaceflight.

The United States, still grappling with the tragic loss of the Apollo 1 crew in January 1967, made a significant breakthrough with the launch of Apollo 7. This mission, which took place from October 11 to October 22, 1968, was the first manned flight of the Apollo program since the devastating fire that claimed the lives of astronauts Gus Grissom, Ed White, and Roger Chaffee. Apollo 7, commanded by Wally Schirra and crewed by Donn Eisele and Walter Cunningham, orbited the Earth 163 times, traveling 4,546,918 miles (7,317,556 kilometers). The success of Apollo 7 was a crucial step forward, proving the viability of the Apollo spacecraft and providing essential data for future lunar missions.

The most iconic achievement of the year came with the historic launch of Apollo 8 on December 21, 1968. Commanded by Frank Borman and accompanied by astronauts James Lovell and William Anders, Apollo 8 became the first crewed spacecraft to break free of Earth's orbit and enter lunar orbit. This mission granted the crew the unprecedented opportunity to witness the Moon's far side and capture the iconic "Earthrise" photograph, which portrayed Earth rising above the lunar horizon. The profound significance of this image was further amplified by the crew's reading from the Book of Genesis, adding a deeply symbolic dimension to this groundbreaking journey.

Meanwhile, the Soviet Union made noteworthy advances in space exploration. In September 1968, the Soviets launched Zond 5, to developing the first spacecraft to carry living organisms—tortoises, flies, and plants—on a mission around the Moon before safely returning to Earth. This was followed by the launch of Zond 6 in November, which completed a similar mission but encountered some difficulties upon re-entry. These missions underscored the Soviet Union's progress in both unmanned and manned spaceflight, showcasing their advancements in space technology.

In addition to these achievements, the Soviet Union concentrated on countering U.S. reconnaissance satellites through developing the "Istrebitel Sputnikov" (IS) program. This initiative aimed at deploying military satellites designed to intercept and destroy enemy satellites. Equipped with explosive charges, these was satellites were intended to be guided into orbit and detonated near their targets. By the end of 1968, however, no enemy satellites had been destroyed, highlighting the challenges faced by this ambitious program.

Despite early successes, the Soviet space program encountered significant challenges in the mid-1960s, leading to the launch of the Soyuz program. This initiative focused on developing capabilities for space rendezvous and docking. The program faced setbacks, including the tragic loss of cosmonaut Vladimir Komarov during the Soyuz 1 mission, which underscored the inherent risks of spaceflight. Progress was briefly stalled with the cancellation of Soyuz 2, but Soyuz 3, launched on October 26, 1968, achieved the first Russian space rendezvous with the uncrewed Soyuz 2 spacecraft. Although the planned docking was less successful than hoped, details of the mission remained classified until after the dissolution of the Soviet Union, revealing the complexities of early space docking attempts.

Throughout 1968, several distinguished astronauts and cosmonauts made their mark on space history. Frank Borman, James Lovell, and William Anders, as part of the Apollo 8 mission, achieved remarkable milestones by orbiting the Moon and capturing striking images of Earth. Their journey was a significant precursor to the subsequent Apollo 11 mission, culminating in humanity's first steps on the lunar surface, marking

an enduring legacy in the annals of space exploration.

In 1969, the Space Race reached a pivotal juncture as NASA achieved a historic milestone by landing Apollo 11 on the Moon. This landmark event was the culmination of a decade of relentless effort and innovation, embodying the United States' resolute commitment to reach the Moon and return its astronauts safely to Earth. The mission was a testament to American ingenuity and technical prowess and symbolized triumph in the intense Cold War competition with the Soviet Union.

The most iconic moment of this era occurred on July 20, 1969, when Apollo 11 executed its historic lunar landing. Under NASA's Apollo program, astronauts Neil Armstrong and Edwin "Buzz" Aldrin became the first humans to set foot on the lunar surface. At the same time, Michael Collins orbited above in the Command Module. Armstrong's famous declaration, "That's one small step for man, one giant leap for mankind," was broadcast live to a global audience, underscoring the event's profound scientific and symbolic significance. This monumental achievement highlighted American technological capabilities and served as a powerful symbol of national pride and innovation.

Amidst the celebrations of this historic American accomplishment, the Soviet Union continued to advance its own space exploration efforts. The Soviets made significant strides with several key achievements, including uncrewed lunar missions and the historic spacewalk by cosmonaut Alexei Leonov in 1965. The Soviet Union further cemented its position in space exploration by launching the world's first space station, Salyut 1, in 1971, marking a significant leap in its space program.

During the same period, NASA's scope of exploration extended beyond lunar missions. In 1969, NASA launched Mariner 6 and Mariner 7 spacecraft, which performed flybys of Mars, providing valuable images and data about the Red Planet. Another milestone was achieved with Mariner 9, to develop the first spacecraft to orbit another planet when it entered Mars' orbit in November. This achievement marked a new era in planetary exploration.

NASA's dedication to Earth observation also saw remarkable progress in 1969. The agency's

Nimbus program deployed satellites to study Earth's atmosphere, while the Landsat program initiated efforts to analyze Earth's surface. These programs contributed crucial data for environmental monitoring and resource management.

NASA's research during this period also encompassed diverse areas of aeronautics and propulsion. In March 1969, NASA, in collaboration with the U.S. Air Force, initiated the YF-12 research program. The YF-12A aircraft, with its distinctive design, began a series of high-speed, high-altitude tests. On December 11, 1969, U.S. Air Force pilot Colonel Joseph Rogers and Flight Controls Officer Major Garry Heilbaugh completed the inaugural YF-12 flight. The YF-12A became a cornerstone of the program, undertaking 146 flights from 1969 to 1979, and the surviving aircraft stands as a testament to the program's success.

Simultaneously, NASA and the U.S. Air Force worked on the Martin Marietta X-24A, an aircraft notable for its bulbous shape and three vertical fins. The testing of the NERVA XE engine at the Nevada Test Site coincided with Apollo 11's lunar landing, showcasing NASA's concurrent advancements in space propulsion technologies.

The year 1969 was indeed a watershed moment in the Space Race, highlighting the significance of space exploration for both the United States and the Soviet Union. On July 20th, the United States achieved its ambitious goal with Apollo 11, landing astronauts Neil Armstrong and Edwin "Buzz" Aldrin on the Moon. This success, broadcast live to millions worldwide, represented a monumental achievement in human history.

Meanwhile, the Soviet Union also made noteworthy advancements in its space endeavors. In February 1969, the Soviet Union launched Soyuz 4 and Soyuz 5, performing the first-ever space docking by cosmonauts. Later in the year, the Soyuz 6, 7, and 8 missions achieved another milestone by conducting the first instance of multiple spacecraft flying in close proximity. Despite these successes, the Soviet lunar program faced challenges, including the failure of the N1 rocket during two launches in 1969. Nevertheless, the Soviets attempted 11 sample returns from the Moon, achieving three. Although Luna 15, which flew concurrently with Apollo 11, tragically crashed during its descent, the Soviet Union made significant progress with the first-ever three-craft spaceflight using Soyuz 6, Soyuz 7, and Soyuz 8 in October.

In contrast, the United States achieved a higher success rate in lunar missions in 1969. Six of seven attempts resulted in successful landings, with Apollo missions leading the charge. Apollo 11, in particular, was a historic success, with Neil Armstrong and Buzz Aldrin becoming the first humans to land on the Moon. The subsequent Apollo 12 mission also achieved a successful lunar landing near Surveyor 3, fulfilling President Kennedy's challenge to land a man on the Moon and return him safely to Earth by the end of the 1960s.

In research and development, NASA's exploration of lifting bodies saw significant progress in 1969. The HL-10, the first supersonically flying lifting body, was crucial in advancing this technology. NASA's lifting body program laid the groundwork for developing the Space Shuttle. Additionally, the Martin Marietta X-24A and X-24B were utilized to study aerodynamics and flight characteristics, validating the advantages of the lifting body configuration for hypersonic trans-atmospheric aircraft.

The year also saw the U.S. Air Force launching five CORONA reconnaissance satellites for the CIA. Concurrently, the Soviet Union tested the Kosmos-291, marking the first launch of the 11K69 (Tsyklon-2) booster. While the exact payload of Kosmos-291 remains uncertain, it was speculated to be a dummy satellite killer spacecraft or a failed target, with the official record indicating all 103 launches of the 11K69 rocket were successful by January 2000.

By the mid-1960s, reusable spaceplanes gained considerable traction in the United States. In response, President Richard Nixon established the Space Task Group in 1969, a collaborative effort involving NASA, the Department of Defense, and other key stakeholders. This group evaluated various spacecraft designs and formulated a cohesive national space strategy. The culmination of these efforts was the Space Shuttle program, which aimed to revolutionize space travel with a reusable spacecraft that could carry astronauts and cargo into orbit.

In 1970, the Space Race between the United States and the Soviet Union reached a pivotal juncture, characterized by both remarkable achievements and notable setbacks that underscored the intense rivalry driving each nation's space efforts. This year vividly reflected the Cold War's competitive dynamics as both superpowers pursued their lofty goals with equal fervor, each striving to surpass the other in the realm of space exploration.

For the United States, 1970 was marked by the launch of Apollo 13, a mission that became a profound testament to human ingenuity and resilience despite its failure to achieve its initial objective. Launched in April, Apollo 13 was intended to continue the ambitious lunar exploration program set forth by previous missions. However, the mission was severely compromised when an oxygen tank exploded in the service module. This catastrophic malfunction forced the crew to abort their planned lunar landing and instead focus on a desperate struggle to return safely to Earth. The ensuing rescue operation, characterized by unprecedented teamwork and problem-solving, turned a potentially tragic event into a remarkable success story. The safe return of astronauts Jim Lovell, Fred Haise, and Jack Swigert was a triumph of human spirit and NASA's operational prowess, highlighting the agency's ability to overcome dire challenges through collaboration and innovation.

Meanwhile, the Soviet Union demonstrated significant advancements in space station technology. In June, the launch of Soyuz 9 set a new record for the longest manned spaceflight at the time, with cosmonauts Andriyan Nikolayev and Viktor Patsayev spending 18 days in orbit. This extended mission not only showcased the Soviet Union's progress in sustaining human life in space for prolonged periods but also reinforced its commitment to establishing a permanent human presence beyond Earth. The achievement was a crucial step in the Soviet Union's broader strategy to maintain a competitive edge in long-duration space missions.

Despite these strides, the Soviet lunar ambitions faced considerable obstacles. The Soviet Union's N1 rocket program, designed to achieve manned moon landings, struggled with persistent technical difficulties that ultimately led to its downfall. The repeated failures of the N1 rocket highlighted the formidable challenges inherent in developing new and complex space technologies,

underscoring the risks associated with ambitious space exploration goals.

The competition extended beyond the lunar and space station programs. In February, the Soviet Union achieved a notable milestone by deploying the Terra-3 ground-based laser anti-satellite (ASAT) system. This event marked the world's first successful satellite intercept, illustrating the increasing militarization of space and the growing sophistication of space-based technologies. Additionally, 1970 saw Japan and China enter the space arena as independent launch-capable nations, with the successful launches of their first satellites. Expanding spacefaring nations to five highlighted a broader international engagement in space exploration.

The Soviet Union continued to make strides with its Luna program, notably with the successful landing of the Lunokhod 1 rover on the Moon's surface. This mission marked a significant advancement in lunar exploration by showcasing the capability to conduct remote operations on the Moon. Furthermore, the Luna 16 mission, which returned lunar soil samples to Earth, represented another critical milestone in uncrewed lunar exploration. Additionally, the Venera program achieved a historic first with Venera 7, which landed on Venus and transmitted valuable data about the planet's atmosphere and surface conditions despite succumbing to Venus's harsh environment after just 23 minutes.

However, October brought setbacks to the Soviet space efforts. The launch of Kosmos-373, targeted by other Soviet satellites, ended in an explosion during an attempted intercept, reflecting the inherent risks of space-based defensive technologies. Another failed attempt occurred when a DS-P-1M target satellite launched from Plesetsk exploded, highlighting ongoing challenges in developing reliable space systems.

In 1971, the Space Race between the United States and the Soviet Union remained a focal point of global fascination, with both superpowers achieving notable advancements in their respective space programs amidst their intense rivalry.

On January 31, the United States made headlines with the launch of Apollo 14, a mission that would cement its place in lunar history. Commanded by Alan B. Shepard, with Stuart A. Roosa and Edgar D. Mitchell as the Lunar Module

Pilot and Command Module Pilot, respectively, Apollo 14 was the third mission to land on the Moon. Shepard, a seasoned astronaut from the Mercury program, became renowned not just for his scientific contributions but also for an unconventional moment: hitting a golf ball on the Moon's surface. This act, a whimsical demonstration of human ingenuity, captured the public's imagination and highlighted the adventurous spirit that characterized the era of human space exploration. The mission's success was underscored by the scientific experiments and exploration activities conducted on the lunar surface, contributing valuable data to the ongoing study of the Moon.

Meanwhile, the Soviet Union was making significant strides in space exploration, though not without challenges. On April 23, the Soyuz 10 mission, carrying cosmonauts Vladimir Shatalov, Aleksei Yeliseyev, and Nikolay Rukavishnikov, aimed to dock with the newly launched Salyut 1 space station. The mission, however, was thwarted by a malfunction in the docking mechanism, preventing the cosmonauts from boarding the station. Despite this setback, the Soviet space program's resolve remained unshaken.

The Soviet Union's determination bore fruit on June 6, with the successful docking of Soyuz 11 to the Salyut 1 space station. This achievement marked the world's first space station, a groundbreaking milestone in the quest for space habitation. However, the mission was marred by tragedy. During re-entry, a critical valve failure led to the loss of cabin pressure, resulting in the deaths of Georgy Dobrovolsky, Viktor Patsayev, and Vladislav Volkov. This incident, the only recorded fatality above the Kármán line—the boundary between Earth's atmosphere and space—cast a somber shadow over the mission and the broader space exploration community.

Despite these challenges, the Soviet Union's space station program continued to progress. The launch of Salyut 1 on April 19, 1971, marked the debut of a series of seven Salyut space stations. These stations were designed for various purposes, both civilian and military, and paved the way for future space station projects, including Mir and the International Space Station (ISS). The Salyut program achieved several spaceflight records and contributed significantly to long-term space research, even as it faced its share of setbacks, such as the tragic loss of the Soyuz 11 crew and the failure of the DOS-2 space station, lost in a launch failure on July 29, 1972.

Beyond the realm of space exploration, 1971 was marked by significant geopolitical and technological developments. The ongoing arms race prompted the United States to reassess its military capabilities and the security of its command centers. This period saw increased scrutiny of national defense strategies, particularly in response to the Soviet Union's rapid development of intercontinental ballistic missiles (ICBMs).

On Earth, 1971 witnessed its own share of pivotal events. In the United States, President Richard Nixon's controversial decision to invade Laos to disrupt North Vietnamese supply lines inadvertently activated the US Emergency Broadcast System. Additionally, the United Kingdom made a brief foray into space history by launching its first and only satellite aboard a British rocket. However, this endeavor was short-lived and ultimately canceled, reflecting the shifting priorities and challenges of the era's space exploration efforts.

In 1972, the global arena was charged with a heightened intensity of competition between the United States and the Soviet Union, a period known as the 1972 Missile Race. This year was characterized by a dramatic escalation in the quest to develop and deploy cutting-edge missile technology, primarily focusing on nuclear-armed intercontinental ballistic missiles (ICBMs). These formidable weapons, capable of delivering nuclear warheads across vast distances, posed a substantial threat to international stability and each superpower's national security.

By the early 1970s, the United States had made significant strides in its missile development program, which had begun in the 1950s. A landmark achievement in 1972 was the deployment of the Minuteman III ICBM. This missile represented a considerable advancement over its predecessors, incorporating solid-fuel propulsion and the ability to carry multiple independently targetable reentry vehicles (MIRVs). The Minuteman III's sophisticated design improved accuracy, range, and reliability, allowing a single missile to strike several targets with separate warheads. This capability amplified the missile's destructive potential and introduced a new layer of complexity into the strategic calculations of both superpowers.

In response to the advancements made by the United States, the Soviet Union was equally proactive in its missile development. The SS-18 Satan, introduced in 1972, marked a significant leap in Soviet missile technology. This liquid-fueled ICBM was known for its powerful payload and precision, capable of carrying multiple warheads and equipped with advanced penetration aids to bypass missile defense systems. The deployment of the SS-18 Satan heightened concerns about the shifting balance of power and underscored the Soviet Union's determination to enhance its strategic capabilities.

The 1972 Missile Race extended beyond ICBMs, encompassing advancements in other missile categories as well. Both superpowers intensified their efforts to develop submarine-launched ballistic missiles (SLBMs). These missiles provided a crucial strategic advantage, as submarines equipped with SLBMs could operate stealthily beneath the ocean's surface, making them difficult targets for detection and attack. The introduction of SLBMs added a new dimension to nuclear deterrence, enhancing the survivability and effectiveness of each nation's nuclear arsenal.

Amid this relentless pursuit of strategic advantage, characterized by a continuous quest to surpass each other in missile range, accuracy, payload capacity, and countermeasures, the atmosphere was one of heightened tension and uncertainty. Each superpower endeavored to maintain or gain the upper hand in nuclear capabilities, driving a competitive and precarious arms race.

Despite the pervasive rivalry, 1972 also witnessed a significant diplomatic development aimed at mitigating the risks associated with this arms race. In May, the United States and the Soviet Union signed the Strategic Arms Limitation Treaty (SALT I). This groundbreaking agreement was designed to impose limitations on the number and types of strategic nuclear weapons each side could possess, including both ICBMs and SLBMs. While SALT I did not eliminate the arms race entirely, it marked a crucial step toward arms control and contributed to easing tensions between the two superpowers. The treaty represented the beginning of a new phase in nuclear diplomacy, illustrating the potential for negotiated agreements to manage and reduce the dangers inherent in an ever-expanding arms race.

In 1973, the realm of space exploration continued to evolve as nations pushed the boundaries of scientific understanding and technological innovation. Although the year did not witness an overt "space race," it was marked by significant milestones that advanced humanity's knowledge of space and set the stage for future endeavors.

A notable development in 1973 came from the Soviet Union, which established Civilian Specialist Group 5. This group, including prominent scientists such as Vladimir Aksyonov and Valeri Romanov, was crucial in advancing Soviet space missions. Their work underscored the USSR's ongoing commitment to space exploration and contributed significantly to expanding its space capabilities.

The early 1970s and 1980s were characterized by a concentrated focus on space stations, a trend driven primarily by the United States and the Soviet Union. A landmark event of 1973 was the launch of Skylab, the world's first space station designed for extended human habitation. Skylab's launch on May 14, 1973, was a momentous occasion, achieved using the Saturn V rocket—the same powerful vehicle that had previously transported astronauts to the Moon during the Apollo missions. Skylab was ingeniously adapted from the third stage of a Saturn V rocket, transforming it into a habitable module. This launch signified the beginning of a new era in human space exploration, offering astronauts the opportunity to conduct scientific research in the unique microgravity environment of space while living and working in orbit.

Following Skylab's successful deployment, NASA sent three astronaut crews to the station in 1973 and 1974. The first mission, Skylab 2, launched on May 25, 1973, and was notably referred to as the "Skylab Rescue Mission." This mission, commanded by Charles Conrad, Jr., with astronauts Paul J. Weitz and Joseph P. Kerwin, was tasked with repairing and activating the space station, which had sustained damage during launch. The Skylab 2 crew undertook critical repairs, including deploying a sunshade to manage excessive heat and the repair of damaged solar panels. Their 28-day mission also involved a broad range of scientific experiments, including studies in solar physics, Earth observations, and biomedical research, which provided valuable data across multiple disciplines.

The subsequent missions, Skylab 3 and Skylab 4, were launched in July and November 1973, respectively. Each crew spent approximately two months aboard Skylab, continuing scientific research and contributing to the station's operational efficiency. Skylab 3, launched on July 28, 1973, set a new record with its 59-day mission. Astronauts Alan L. Bean, Owen K. Garriott, and Jack R. Lousma focused on experiments investigating the effects of prolonged space travel on the human body, solar observations, and Earth studies. Their work provided insights into the physiological impacts of spaceflight and contributed to our understanding of solar phenomena and Earth's environment.

Following Skylab 3, Skylab 4, launched on November 16, 1973, set a new record for the longest-duration manned spaceflight at the time, with an impressive 84-day stay. Astronauts Gerald P. Carr, William R. Pogue, and Edward G. Gibson continued their research into solar physics, Earth resources, and human physiology. Their mission also involved essential maintenance and repairs, highlighting the crucial role of human adaptability and ingenuity in sustaining space operations.

Simultaneously, the Soviet Union made notable advancements in space exploration with the launch of Salyut 4 on December 26, 1973. This was the sixth space station in the Soviet series, serving as a precursor to future Soviet space stations and contributing to the nation's ongoing space exploration efforts.

The United States' Skylab program continued to thrive into 1974, providing a platform for ongoing scientific research in microgravity. This period also saw the closure of the Apollo program following Apollo 17's final mission in December. With the cancellation of nuclear rocket and power programs, including the Nuclear Engine for Rocket Vehicle Application (NERVA), and the shutdown of the Plum Brook Station facility, an era of certain space exploration projects came to a close.

On January 1, 1974, a noteworthy development occurred with the formation of the Physician Group within the Soviet space program. Zyyadin Abuzyarov was selected to provide specialized medical support, a critical role ensuring cosmonauts' health and safety during their missions. This initiative reflected the growing recognition of the importance of medical expertise in space exploration, an area crucial for the well-being of astronauts in the harsh environment of space.

During the same period, the Soviet Union made notable advancements with the launch of the Mars 5 space probe and the Soyuz 12 mission, which marked the first crewed flight since the Soyuz 11 tragedy in 1971. Soyuz 13, launched on December 26, 1973, carried cosmonauts Valentin Lebedev and Pyotr Klimuk on the second test of a redesigned Soyuz capsule for astrophysical observations.

Additionally, 1973 saw the signing of the Spacelab Memorandum of Understanding between ESRO (European Space Research Organization) and NASA. The Soviet Union's Space Station Salyut 2, launched earlier in the year, unfortunately, suffered from a loss of attitude control and depressurized within two weeks, leading to its re-entry into the atmosphere on May 28, 1973, without any crew visiting it.

The year 1976 saw the introduction of Air Force Group 6, commonly known as the Space Shuttle Buran crew. This team, which included notable members such as Leonid Ivanov and Leonid Kadenyuk, was trained specifically for the Soviet Buran program. Their preparation aimed to develop reusable spacecraft technology, mirroring the advancements made by the American Space Shuttle program. This initiative highlighted the USSR's ambition to compete in the burgeoning field of reusable space vehicles.

A pivotal moment in international cooperation came on November 25, 1976, with the formation of the Intercosmos Group. This group comprised cosmonauts from Poland, East Germany, and Czechoslovakia, including Mirosław Hermaszewski, Sigmund Jähn, and Vladimír Remek. The creation of this group was a testament to the Cold War-era collaboration in space exploration, showcasing a collective effort to advance human spaceflight and foster international partnerships.

In 1977, the field of space exploration experienced a transformative period, marked by significant advancements and ambitious missions that sought to unveil the cosmos' mysteries.

NASA's Space Shuttle program was central to these developments, a groundbreaking initiative under development since the 1960s. The program's origins could be traced back to the National Advisory Committee for Aeronautics (NACA), NASA's predecessor, which conducted pioneering studies on high-altitude and supersonic flight using the X-15 research aircraft. These early investigations laid the groundwork for"wingless" aircraft and "lifting bodies," leading to developing reusable spaceplanes.

In 1977, the Space Shuttle program reached a pivotal moment in its development, marking a significant step forward in the quest for reusable space transportation. On January 14, NASA received the Space Shuttle Orbiter Enterprise, a crucial milestone in the program's evolution. Unlike the operational shuttles that would follow, the Enterprise was an unpowered prototype designed to test the fundamental aspects of the Shuttle's design and performance. Its first major test flight occurred on February 18, when it was carried aloft by a modified Boeing 747. Although this initial flight did not reach orbit, it was instrumental in assessing the Shuttle's aerodynamic characteristics and handling. The Enterprise's maiden free flight, which took place on August 12, represented a significant achievement. With a human crew onboard, this flight demonstrated the Shuttle's capability to glide and land , marking the first instance of a space shuttle executing a free flight. The successful landing at Edwards Air Force Base was a testament to the Shuttle's design and performance under real-world conditions. Subsequent months saw the Enterprise undertaking additional test flights, further proving its capabilities. Notably, these flights showcased the Shuttle's ability to operate without its tail cone and to land on conventional runways, as opposed to the lakebeds used during earlier tests.

In parallel with these advancements in shuttle technology, NASA embarked on an ambitious endeavor to explore the outer reaches of our solar system. The Voyager program, initiated in 1977, was designed to expand humanity's knowledge of the solar system's distant planets. On August 20, 1977, Voyager 2 was launched from Cape Canaveral, Florida, followed by the launch of Voyager 1 on September 5. Both spacecraft were equipped with sophisticated scientific instruments, including cameras, spectrometers, and magnetometers, intended to capture detailed imagery and gather crucial data about Jupiter, Saturn, Uranus, and Neptune. By March 1979, Voyager 1 had performed its flyby of Jupiter, delivering stunning images of the planet's atmospheric phenomena, including the Great Red Spot, and revealing the presence of active volcanoes on Jupiter's moon Io. Voyager 2, in July 1979, followed with its own flyby, contributing further insights into Jupiter's characteristics.

Meanwhile, the Soviet Union made notable strides in space exploration during the same year. In April 1977, the Soviet Union launched Soyuz 24, its final manned mission to the Salyut 5 space station. The 18-day mission focused on scientific experiments and observations, continuing the Soviet Union's commitment to advancing manned space station technology. Later in August, NASA's Skylab Rescue mission was launched as a contingency plan to support the Skylab 4 crew in case of emergencies, underscoring NASA's dedication to astronaut safety and mission integrity.

In October, the Soviet Union launched the unmanned Venera 11 and Venera 12 spacecraft towards Venus. These missions were designed to gather vital data about Venus' atmosphere and surface, enhancing our understanding of this enigmatic planet. Additionally, 1977 saw the introduction of the second-generation Salyut space stations, Salyut 6 and Salyut 7. These stations featured improved design elements, including a second docking port, allowing for continuous occupation and facilitating crew exchanges and resupply missions through the Progress spacecraft. The Progress, based on the Soyuz design, was crucial in maintaining the operational success of the Salyut stations by delivering supplies, experimental hardware, and repair equipment.

In 1978, NASA embarked on a transformative chapter in space exploration by selecting Group 8 astronauts, a cohort famously known as the "Thirty-Five New Guys" (TFNG). This selection represented a significant departure from the agency's previous astronaut recruitment practices, occurring after a nine-year hiatus following the Apollo missions' conclusion. This new group of astronauts was a testament to NASA's commitment to diversifying its astronaut corps and ushering in a new era of spaceflight.

Among the pioneering members of Group 8 were Judith Resnik and Sally Ride, who made history as the first American women astronauts. Judith Resnik's achievement was particularly notable as she was also the first Jewish American to travel into space. Guion Bluford and Frederick D. Gregory also broke new ground as the first African American astronauts. Additionally, Ellison Onizuka became the first Asian American astronaut, exemplifying NASA's evolving approach to inclusivity and representation within its ranks.

The TFNG cohort comprised a diverse array of specialists, including pilots and mission experts, each selected for their exceptional skills and contributions. The Shuttle Program, a central focus during this period, introduced payload specialists—scientists chosen for specific missions—along with occasional international astronauts, thereby broadening the program's scope and fostering global scientific collaboration.

One of the standout achievements of this group was Sally Ride's historic flight aboard STS-7, which marked her as America's first woman in space. Ride's achievement was soon complemented by Kathryn Sullivan, who became the first American woman to perform an extravehicular activity (EVA). Dr. Norman Thagard, another distinguished member of Group 8, reached a significant milestone as the first American launched aboard a Russian rocket to the Mir space station. Furthermore, Shannon Lucid, also from this group, set new records for American space endurance during her mission aboard Mir, surpassing previous records established during Skylab and Shuttle missions.

As the world continued to marvel at the expanding horizons of space exploration, 1979 emerged as a pivotal year in the annals of human spaceflight. This year was marked by a series of significant events that underscored the extraordinary achievements possible in space and expanded our understanding of the cosmos.

The year began with palpable excitement in both the scientific community and the general public. Space agencies around the globe were rapidly advancing their technological capabilities, driven by a fervent desire to explore the uncharted realms of space. The United States, buoyed by the success of its Apollo missions, had solidified its position as a leader in space exploration. However, the Soviet Union, renowned for its pioneering spirit and ambitious missions, was determined to assert its own dominance in the field.

Amidst this competitive backdrop, the Soviet Union unveiled an ambitious plan that captured global attention. On March 2, 1979, the Soyuz 32 spacecraft was launched from the Baikonur Cosmodrome. This mission began a remarkable journey, as Soyuz 32 carried cosmonauts Anatoliy Filipchenko and Nicolay Rukavishnikov towards the Salyut 6 space station. Launched in 1977, Salyut 6 was a testament to Soviet engineering prowess, orbiting Earth at approximately 220 miles. It was a vital platform for scientific research, technological development, and experimentation.

On March 1, 1978, the USSR's Intercosmos program welcomed a new cohort of international astronauts, enhancing its global collaboration in space. This expansion included astronauts from Bulgaria, Romania, Cuba, Hungary, Mongolia, and other nations, reflecting a shared ambition for scientific discovery that transcended national boundaries. This era of international partnership underscored the program's commitment to cooperative space exploration and the value of diverse contributions worldwide.

Simultaneously, on May 1, 1978, the European Space Agency (ESA) initiated its Spacelab Payload Specialists Group 1. This group included representatives from West Germany, Switzerland, the Netherlands, and Italy, each bringing a wealth of scientific knowledge and expertise. Their involvement in shuttle missions was crucial in advancing research in microgravity and space technology, marking a significant step forward in Europe's contribution to space exploration.

In December 1978, Germany furthered its participation in the Spacelab Payload Specialists Group with Reinhard Furrer and Ernst Messerschmid. Their inclusion underscored the growing importance of European collaboration in shuttle missions and highlighted the role of international cooperation in advancing microgravity research.

In August 1979, the United States Air Force (USAF) Manned Spaceflight Engineer Program introduced its inaugural cohort, known as Group 1.

This pioneering group, comprised of distinguished individuals such as Frank J. Casserino, Michael A. Hamel, and Gary E. Payton, was conceived to integrate military expertise into manned space missions. This initiative underscored a unique fusion of defense and space exploration, reflecting a broader vision of enhancing space capabilities with military precision and knowledge. Among these pioneers, Gary E. Payton achieved a notable milestone by becoming the group's only member to venture into space. Serving as a Payload Specialist on a Department of Defense Shuttle mission, Payton's flight highlighted the program's specialized role in advancing national security through space missions.

Concurrently, the Intercosmos program continued to expand its international outreach. On April 1, 1979, the inclusion of Tuân Pham and Thanh Liem Bui from Vietnam marked a significant step towards fostering global collaboration. This integration of international participants into the program demonstrated a commitment to bringing diverse perspectives and expertise into the realm of space exploration, enhancing the program's global dimension.

As the Soyuz 32 mission approached the Salyut 6 space station, the cosmonauts aboard were enveloped in anticipation and apprehension. The precision required for rendezvous and docking with the space station was immense, demanding flawless execution of maneuvers that had been rigorously rehearsed on Earth. On March 3, 1979, the spacecraft, piloted by Filipchenko and Rukavishnikov, docked with Salyut 6, marking the beginning of an extended stay in space. This achievement was not only a testament to the technical prowess involved but also to the endurance and skill of the crew.

During their 175-day stay, the cosmonauts engaged in a range of scientific experiments. They explored the effects of microgravity on the human body, studied Earth's atmospheric and weather patterns, and investigated the behavior of various materials in space. These experiments contributed valuable insights into fundamental scientific questions and showcased the potential for future space research.

Life aboard Salyut 6 presented its own set of challenges. The confined quarters of the space station posed both physical and psychological hurdles for the crew. Adapting to a new routine, maintaining spacecraft systems, and managing the unique conditions of space required significant resilience and adaptability. Despite these challenges, the mission was deemed a success, culminating in the safe return of Filipchenko and Rukavishnikov to Earth on August 27, 1979. Their successful return highlighted their exceptional skill and the dedication of the Soviet space program.

The Soyuz 32 mission extended its significance beyond the immediate technical achievements. It underscored the Soviet Union's growing capabilities in space exploration and emphasized the critical role of international collaboration in advancing scientific knowledge. This mission set the stage for future long-duration space flights. It paved the way for developing subsequent space stations, such as Mir and the International Space Station (ISS), which would later symbolize global unity and cooperation.

In addition to these space achievements, 1979 was marked by significant geopolitical developments and witnessed an intensification of the missile race between the United States and the Soviet Union. The Soviet Union had deployed a new generation of intermediate-range ballistic missiles, including the SS-20, which posed a considerable threat to NATO countries in Europe. In response, NATO adopted the Double-Track Decision, aiming to address the missile threat through military and diplomatic measures.

Under the Double-Track Decision, NATO deployed 572 new nuclear missiles in Europe, including Pershing II missiles and BGM-109G Ground Launched Cruise Missiles, known as the Gryphon. These missiles were strategically positioned across various European countries, with all Pershing missiles stationed in West Germany. The deployment sparked widespread controversy and concern over the potential for nuclear conflict, leading to significant public protests.

On December 12, 1979, NATO extended an offer to the Warsaw Pact to limit medium-range and intermediate-range ballistic missiles mutually. The Double-Track Decision reflected NATO's commitment to both deterrence and diplomacy, emphasizing a willingness to negotiate with the Soviet Union to reduce or eliminate these weapons on both sides. This approach demonstrated a dedication to fostering stability and reducing the risk of nuclear confrontation through diplomatic engagement.

In 1980, the realm of space exploration was pulsating with ambition and promise, capturing the global imagination as it soared to new heights of achievement and discovery. This pivotal year was marked by groundbreaking missions and significant advancements that expanded the frontiers of human knowledge and deepened our understanding of the cosmos.

On May 29, 1980, NASA unveiled Group 9, a new cohort of astronauts that included both pilots and mission specialists. This diverse group comprised pilots John Blaha, Charles Bolden, and Michael J. Smith, alongside mission specialists Franklin Chang-Diaz, Mary Cleave, and Bonnie Dunbar. Each member of Group 9 brought a unique set of skills essential for NASA's ambitious missions. Notably, Franklin Chang-Diaz made history as the first Hispanic-American astronaut to travel to space. Tragically, Michael J. Smith, who would later become a part of the ill-fated Challenger mission, was among those who lost their lives in the disaster that profoundly impacted the space exploration community.

John Blaha's career extended beyond the Shuttle program, culminating in his journey to the Mir space station, reflecting NASA's commitment to international cooperation. Both Jerry Ross and Franklin Chang-Diaz would set remarkable records for the most crewed spaceflights, each completing seven missions. Charles Bolden also achieved significant milestones, becoming NASA's second astronaut to assume the role of Administrator and the first African-American to hold that position in 2009, following a brief tenure by Frederick Gregory, another African-American Shuttle commander.

In contrast, on July 30, 1980, the Soviet Union introduced the LII–1/IMBP–3/MAP/NPOE–5/AN–2 Cosmonaut Group. This ensemble included specialists from various Soviet organizations, among them Svetlana Savitskaya, who became the second woman to travel into space and the first woman to conduct a spacewalk. Her achievement underscored the Soviet Union's significant contributions to space exploration during this era.

The year 1981 marked a pivotal moment in the history of space exploration with the launch of NASA's Space Shuttle program, an event that would redefine human spaceflight. On April 12, 1981, the Space Shuttle Columbia ascended from Kennedy Space Center in Florida, embarking on its maiden voyage. Commanded by astronaut John W. Young and piloted by Robert L. Crippen, this mission, designated STS-1, was not merely a flight; it was the dawn of a new era in reusable spacecraft.

Columbia's inaugural journey spanned two days, completing 36 orbits around Earth. This landmark mission was meticulously designed to test the shuttle's comprehensive systems, including its launch procedures, in-orbit performance, reentry protocols, and landing capabilities. While STS-1 did not involve scientific experiments, its primary objective was to validate the functionality and reliability of the shuttle's systems, ensuring that they met the rigorous demands of space travel.

The success of STS-1 was instrumental in demonstrating the practicality of reusable spacecraft, offering a glimpse into the future of more frequent and versatile human space exploration. By proving the shuttle's operational capabilities, this mission laid the critical groundwork for the Space Shuttle program, setting the stage for subsequent flights and a host of future accomplishments in space exploration.

As NASA embarked on this new era, the Soviet Union also made significant strides in space exploration. On November 12, 1980, the Soviets launched Venera 13, a spacecraft dedicated to exploring Venus. This mission was part of a series aimed at investigating the neighboring planet. Venera 13 was equipped with an array of scientific instruments, including cameras, spectrometers, and drills, designed to study Venus's atmosphere and surface. After a five-month journey, Venera 13 arrived at Venus and descended through its dense atmosphere. On March 1, 1982, the lander touched down on the planet's surface, withstanding extreme temperatures and pressures. The spacecraft transmitted the first color images of Venus's rocky terrain, providing unprecedented insights into the planet's geology and atmospheric conditions. Venera 13's mission was a notable triumph for Soviet space efforts, significantly enhancing our understanding of Venus, often dubbed Earth's twin due to its similar size and proximity.

Meanwhile, on Earth, the United States advanced its capabilities in space-based observation. On December 4, 1980, NASA launched Landsat 4, part of the Earth Resources Technology Satellite (ERTS) series. Landsat 4 was equipped with sophisticated sensors and cameras designed to capture high-resolution images of Earth's surface, including landmasses, oceans, and natural resources. The satellite revolutionized our ability to monitor Earth's dynamic environment, providing scientists, researchers, and policymakers with valuable data on various aspects of our planet, such as deforestation, urban expansion, agricultural patterns, and environmental changes. Landsat 4's imagery offered a new perspective on Earth's ecosystems, contributing significantly to developing sustainable practices and conservation strategies.

As 1980 drew to a close, the achievements in space exploration stood as a testament to human ingenuity and determination. The launch of the Space Shuttle program signaled a new era in space travel, promising greater opportunities for human exploration beyond Earth. The Soviet Venera 13 mission expanded our knowledge of Venus, while Landsat 4 provided critical insights into our own planet. Collectively, these milestones underscored the relentless spirit of exploration and the pursuit of knowledge that define humanity's quest to understand the universe.

In the history of space exploration, 1981 stands out as a transformative period, marked by pioneering missions and revolutionary achievements that redefined human spaceflight. This year witnessed significant advancements that broadened the scope of our exploration efforts and set new benchmarks for future endeavors.

On November 11, 1981, Columbia took to space again for its STS-2 mission. This flight, carrying astronauts Joseph P. Allen and Philip K. Chapman, aimed to evaluate the shuttle's capabilities further and conduct scientific experiments in the microgravity environment of space. Over five days, the crew engaged in various activities, including deploying and retrieving satellites, testing the shuttle's remote manipulator arm, and conducting experiments in materials processing, Earth observation, and atmospheric studies. STS-2 underscored the shuttle's versatility and its potential for supporting a wide range of scientific investigations, showcasing its capacity for complex missions in orbit.

While the Space Shuttle program made significant strides, other nations also advanced their space exploration efforts. On March 22, 1981, the Soviet Union launched Soyuz 39, a notable joint mission with Bulgaria. This mission marked a significant milestone in international space cooperation, with cosmonaut Georgi Ivanov and Bulgarian astronaut Georgi Ivanov becoming the first international crew to visit the Salyut 6 space station. During their mission, the crew conducted experiments and observations that contributed to our understanding of human adaptation to long-duration spaceflight and fostered international collaboration in space exploration.

Meanwhile, unmanned spacecraft continued their remarkable journeys across the solar system. On August 20, 1981, NASA's Voyager 2 spacecraft achieved a historic milestone with its flyby of Saturn. This encounter provided unprecedented views of Saturn's rings, its complex moon system, and its dynamic atmosphere. The data and images captured by Voyager 2 offered invaluable insights into Saturn's structure and the outer reaches of our solar system, significantly enhancing our understanding of these distant worlds.

In 1982, space exploration thrived with excitement and anticipation, reflecting the dynamic progress of the early space age. The previous decades had set monumental milestones, from Yuri Gagarin's pioneering voyage as the first human in space to the historic Apollo moon landings, representing the pinnacle of human exploration. The 1980s, however, promised to usher in a transformative era, defined by both remarkable achievements and considerable challenges as the boundaries of space exploration continued to expand.

France made notable advances in space exploration during this period. The country launched its inaugural CNES Group 1, which included astronauts Patrick Baudry and Jean-Loup Chrétien. Chrétien achieved historical significance as one of the first French astronauts to travel into space when he flew aboard the Soviet Salyut 7 space station in 1982. Later, Patrick Baudry further enhanced France's role in international space missions by participating in the Space Shuttle STS-51-G mission in 1985.

August 1982 saw the introduction of the USAF Manned Spaceflight Engineer Program's Group 2. This group comprised James B. Armor Jr., Livingston L. Holder Jr., and Maureen C. LaComb, continuing the program's emphasis on integrating military expertise into space missions. Tragically, Charles E. Jones, a member of this group, lost his life in the September 11 attacks aboard American Airlines Flight 11. Of the Group 2 members, William A. Pailes was the sole individual to fly into space, serving as a Payload Specialist on a Department of Defense Shuttle mission.

India's entry into space exploration in 1982 was a significant milestone. On September 11, India selected Ravish Malhotra and Rakesh Sharma for the Intercosmos program. Sharma achieved a historic milestone as the first Indian citizen to journey into space when he flew aboard the Soviet Soyuz T-11 in 1984.

The Space Shuttle program also reached a new landmark with the launch of the Space Shuttle Columbia on January 12. Designated STS-3, this mission marked the third orbital flight of NASA's Space Shuttle program. Commanded by astronaut Jack Lousma and piloted by Gordon Fullerton, STS-3 was an eight-day mission crucial for demonstrating the shuttle's reusability. The mission focused on testing the shuttle's ability to maneuver and land on a conventional runway, a key step toward achieving operational versatility and reliability. STS-3 also featured the debut of the Shuttle Imaging Radar-A (SIR-A) system. This advanced radar technology provided high-resolution imagery of the Earth's surface, facilitating groundbreaking geology, agriculture, and environmental monitoring research.

Meanwhile, the Soviet Union made significant strides with its space endeavors. On March 22, the Soyuz T-5 spacecraft was launched with cosmonauts Anatoli Berezovoy and Valentin Lebedev on board. Their mission, the fifth expedition to the Soviet space station Salyut 7, extended to an unprecedented 211 days in space. This extended mission offered invaluable insights into prolonged space travel's physical and psychological effects, contributing significantly to our understanding of human endurance in space.

Europe also marked a significant achievement with the launch of the European Space Agency's (ESA) first scientific satellite, EXOSAT, on May 26. Designed to probe X-ray emissions from celestial objects such as black holes, neutron stars, and active galactic nuclei, EXOSAT's successful deployment highlighted Europe's growing capabilities in space research. It established the continent's burgeoning role in the international space exploration community.

he year 1983 emerged as a pivotal moment in the ongoing saga of space exploration, characterized by groundbreaking missions, significant scientific advancements, and notable geopolitical developments. Set against the backdrop of the Cold War, space agencies around the globe embarked on ambitious projects, pushing the frontiers of human knowledge and expanding our understanding of the cosmos.

A landmark event unfolded on February 7, 1983, with the launch of the Space Shuttle Challenger on its ninth mission, STS-7. This mission was historic for several reasons. It carried Sally Ride, the first American woman to journey into space, breaking through a significant gender barrier. Accompanied by Commander Robert Crippen and four male crewmates, Ride's participation highlighted the evolving inclusiveness of space exploration. The STS-7 mission was instrumental in showcasing the Space Shuttle program's capabilities. During its six-day mission, the crew deployed two communication satellites: Anik C2 for Telesat Canada and Palapa B1 for Indonesia. These deployments demonstrated the shuttle's growing versatility and laid the groundwork for future advancements in telecommunications.

In the Soviet Union, 1983 marked the continuation of operations aboard the Salyut 7 space station. On June 7, the Soyuz T-9 spacecraft launched with a crew of three cosmonauts: Vladimir Lyakhov, Aleksandr Aleksandrov, and Valery Ryumin. Their mission involved docking with the space station, conducting scientific experiments, and facilitating crew exchanges. The mission encountered significant challenges when the crew discovered malfunctions in the station's systems, including power loss and difficulties with the hatch. Despite these setbacks, the crew's resilience and ingenuity enabled them to stabilize the situation and continue their experiments, underscoring the adaptability required to address unforeseen obstacles in the harsh space environment.

The scientific community also made strides in 1983 with the Infrared Astronomical Satellite (IRAS) launch on August 25. IRAS was designed to survey the entire sky in infrared wavelengths, providing astronomers with unprecedented insights into celestial objects. Over its ten-month mission, IRAS identified numerous new stars, galaxies, and planetary systems, significantly enhancing our understanding of the universe's infrared emissions.

In April 1983, the Soviet Union made a notable advance in its space shuttle program by selecting Ural Sultanov and Magomed Tolboev as the second group of test pilots for the "Buran" project, based at the Gromov Flight Research Institute. This selection underscored the USSR's commitment to developing its space shuttle capabilities, aiming to compete with the United States' achievements.

December 1983 saw Canada join the ranks of spacefaring nations with the National Research Council's selection of its first astronaut group. This pioneering group included Roberta Bondar, Marc Garneau, Steve MacLean, Ken Money, Robert Thirsk, and Bjarni Tryggvason. Their selection set the stage for Canada's involvement in space exploration, with all members, except Ken Money who resigned in 1992, eventually flying aboard U.S. Space Shuttles by 1997. Their contributions marked a significant chapter in Canadian space history.

International collaboration flourished in 1983 as well. On November 28, the European Space Agency (ESA) launched the Spacelab-1 mission in partnership with NASA. Spacelab-1 was a multidisciplinary laboratory module that traveled in the Space Shuttle Columbia cargo bay. This mission enabled scientists from various countries to conduct experiments in microgravity, advancing knowledge in fields such as human physiology, material science, and biology.

However, 1983 also saw moments of tragedy. On September 1, the Soviet Union's space program suffered a severe setback when the Soyuz T-10-1 spacecraft experienced a launch pad abort. Cosmonauts Vladimir Titov and Gennady Strekalov narrowly escaped the explosion, but the spacecraft was destroyed. This incident underscored the inherent risks of space exploration and the continuous need for stringent safety measures.

In the broader geopolitical landscape, 1983 brought significant developments in the Cold War. On September 16, President Ronald Reagan announced that the Global Positioning System (GPS) would be made available for civilian use, marking a pivotal moment in navigation technology with extensive implications across various industries.

The early 1980s also saw a resumption of Soviet anti-satellite weapon tests, reflecting tensions following the end of détente and the ongoing conflict in Afghanistan. In a surprising move, Soviet leader Yuri Andropov declared a unilateral moratorium on anti-satellite (ASAT) tests in 1983, despite opposition from some Soviet military leaders who viewed the ASAT program as crucial for defense. This announcement contrasted with the concerns of Soviet leaders about President Reagan's Strategic Defense Initiative (SDI), which they feared might allow the U.S. to launch missiles without fear of retaliation.

During this period, the rivalry between the United States and the Soviet Union continued to drive advancements in space exploration. Each nation pushed the boundaries of technology and achieved remarkable milestones. A defining moment came when U.S. President Ronald Reagan proposed developing a crewed space station during his State of the Union address, inviting international space agencies to collaborate on this ambitious project. This announcement, though modest at the time, laid the foundation for what would eventually become the International Space Station (ISS), signaling a significant shift toward international cooperation in space.

Around the same time, the Soviet Union strategically enhanced its space capabilities by augmenting its early warning satellite system. The Soviets deployed new satellites into geosynchronous orbit to complement their existing network of highly elliptical orbit (HEO) early warning satellites. This strategic move bolstered the robustness and redundancy of their early warning system, ensuring continuous and reliable surveillance of potential threats.

Grumman Aerospace achieved a notable milestone in aerospace advancements with its X-29A aircraft at Edwards Air Force Base. These experimental aircraft, designed to test forward-swept wing technology, advanced composites, and other aerodynamic innovations, became the first aircraft with forward-swept wings to achieve supersonic flight in level flight. This achievement showcased the remarkable progress in aerospace technology and paved the way for future innovations in aircraft design.

During this period, the United States Army also made significant strides in missile defense technology. The successful launch of the Homing Overlay Experiment (HOE) satellite kill vehicle demonstrated the nation's progress in developing systems capable of intercepting and destroying intercontinental ballistic missiles (ICBMs). The experimental satellite effectively targeted and neutralized an ICBM launched from Vandenberg Air Force Base in California, underscoring advancements in missile defense capabilities.

In 1984, the landscape of space exploration was marked by a series of significant achievements and advancements, each contributing to our understanding of space and showcasing the evolving capabilities of both American and Soviet space programs.

On January 24th, the Space Shuttle Challenger embarked on its tenth mission, designated STS-41-B, with a distinguished crew comprising Commander Vance D. Brand, Pilot Robert L. Gibson, and Mission Specialists Bruce McCandless II, Ronald McNair, and Robert L. Stewart. This mission was notable not only for deploying two communications satellites, Westar 6 and Palapa B2, into orbit but also for its groundbreaking spacewalks. Astronauts McCandless and Stewart conducted the first untethered spacewalks using the Manned Maneuvering Unit (MMU). This device allowed them to float freely in space without physically connecting to the shuttle. These historic spacewalks captured global attention and demonstrated the potential for extravehicular activities beyond traditional tethered operations, opening new avenues for future missions.

In February 1984, the Soviet Union bolstered its space program by selecting Aleksandr Kaleri and Sergei Yemelyanov for the NPOE–6 Cosmonaut Group. This selection was a key development in advancing Soviet space missions, contributing significantly to ongoing exploration and scientific research during intense international competition.

The same month, NASA introduced Group 10, colloquially known as "The Maggots." This diverse group of astronauts included pilots Kenneth Cameron, John Casper, and Frank Culbertson, as well as mission specialists Ellen Baker, Marsha Ivins, and Kathryn Thornton. Among them, William Shepherd would become the commander of the first crew aboard the International Space Station during Expedition 1. James Wetherbee distinguished himself by commanding five spaceflight missions, while Sonny Carter, another member of the group, tragically lost his life in a plane crash in 1991.

In June 1984, the Soviet Union continued its commitment to space exploration by selecting Victor Zabolotski for the third group of test pilots for the "Buran" project at the Gromov Flight Research Institute. This selection underscored the USSR's dedication to advancing its space shuttle program, contributing to the intense global competition in space technology during this era.

Meanwhile, on February 8th, the Soviet Union launched the Soyuz T-10 spacecraft, carrying cosmonauts Leonid Kizim, Vladimir Solovyov, and Oleg Atkov. This mission aimed to repair and resupply the space station Salyut 7, which had experienced a launch pad abort incident the previous year. The crew docked with Salyut 7 and performed essential repairs and maintenance tasks, showcasing their technical expertise and resilience in overcoming space challenges.

Scientific exploration also reached new heights in 1984. On February 24th, NASA's Voyager 2 spacecraft made its closest approach to the planet Uranus, providing unprecedented images and data about this distant gas giant and its moons. Voyager 2's flyby significantly expanded our understanding of the outer solar system and set the stage for future missions to explore these remote regions.

International collaboration in space was exemplified by the joint Soviet-French Vega mission launch on December 15th. The Vega spacecraft, consisting of an orbiter and a lander, was designed to study Venus's atmosphere and surface while also investigating Halley's Comet as it passed through the inner solar system. The mission deployed a lander on Venus, capturing images and conducting experiments while the orbiter studied the planet's atmosphere. Vega's encounter with Halley's Comet provided valuable data and detailed images, highlighting the power of international cooperation in tackling complex scientific challenges and enhancing our knowledge of planetary and cometary science.

The competitive nature of the space race was further highlighted by an intriguing incident in November 1983, when NASA announced that Kathryn D. Sullivan would become the first woman to perform a spacewalk during the STS-41-G mission. This announcement spurred a rapid response from the Soviet Union, which sought to achieve this milestone first. The Soyuz T-12 mission was swiftly assembled, with its crew, including test pilot Volk, preparing to potentially achieve this historic feat. This competition underscored the high stakes and intense rivalry that characterized the space exploration era.

In addition to its space shuttle program, the United States explored innovative aircraft designs in 1984. The X-29 Advanced Technology Demonstrator Aircraft, operational from 1984 to 1992, investigated forward-swept wing designs and other advanced concepts, significantly contributing to aerospace advancements. Concurrently, NASA's Rotor Systems Research Aircraft (RSRA) X-Wing, developed in collaboration with Sikorsky Aircraft, explored the integration of helicopter-like vertical flight stability with the horizontal cruise capabilities of traditional aircraft, further pushing the boundaries of aviation technology.

In 1985, the landscape of space exploration saw a series of significant advancements and tragic events that shaped the future of spaceflight. Among the notable developments was the inclusion of several distinguished astronauts and cosmonauts into various space programs worldwide.

In the United States, NASA's Group 11, selected in June, featured a remarkable lineup of astronauts, including pilots Michael A. Baker and Robert D. Cabana, alongside mission specialists Jerome Apt and Linda Godwin. Tragically, Stephen Thorne, also selected for this group, lost his life in a private airplane crash before he could embark on his first space assignment, a loss that deeply affected the space community.

The following month, on July 19, the NASA Teacher in Space Program made a historic choice by selecting Christa McAuliffe as the primary Payload Specialist for the STS-51-L mission, with Barbara Morgan as her backup. McAuliffe, a civilian teacher chosen to bring educational outreach to space, tragically perished in the Challenger disaster. Her loss was a profound blow, but her legacy endured as Morgan later joined the NASA Astronaut Corps and flew on STS-118 in 2007.

Japan made significant strides in space exploration with the 1985 NASDA Group, announced on August 1. This group included Mamoru Mohri, Chiaki Mukai, and Takao Doi, marking a notable enhancement of Japan's presence in space missions. Their contributions helped establish Japan as a key player in international space exploration.

The same month, the USAF Manned Spaceflight Engineer Program—Group 3, which included astronauts Joseph J. Caretto and Teresa M. Stevens, continued to advance the field by focusing on engineering and technical expertise in spaceflight.

The Soviet Union also marked progress on September 2 by selecting Viktor Afanasyev and Sergei Krikalyov for the GKNII–2/NPOE–7 Cosmonaut Group. Their inclusion underscored the Soviet Union's ongoing commitment to expanding its capabilities in manned spaceflight.

France demonstrated its dedication to space exploration with the CNES Group 2, announced on September 18. This group featured Claudie André-Deshays and Jean-François Clervoy, reflecting France's growing involvement in space research and missions.

Syria's involvement in space exploration was highlighted on September 30 with the selection of Muhammed Ahmed Faris and Munir Habib Habib for the 1985 Intercosmos Group. Their selection marked Syria's entry into international space missions, contributing to the diversity of spacefaring nations.

Indonesia also made strides with the Indonesian Palapa Group, consisting of Taufik Akbar and Pratiwi Sudarmono. Although they prepared for Shuttle missions, the Challenger disaster prevented their flights, a stark reminder of the inherent risks of space exploration.

Dirk D. Frimout from Belgium, who participated in the ATLAS-1 mission on December 27, further marked European contributions, adding to Europe's growing role in space science.

In 1985, other notable developments included India's ISRO Insat Group, featuring Nagapathi Chidambar Bhat and Paramaswaren Radhakrishnan Nair. They were selected for Space Shuttle missions, but the Challenger disaster in 1986 led to the cancellation of Bhat's scheduled flight. Mexico also made its mark with Rodolfo Neri Vela and Ricardo Peralta y Fabi, with Neri Vela achieving a milestone by flying on Shuttle mission STS-61-B in November 1985.

The Soviet Union continued to advance its space capabilities in 1986 by selecting the fourth group of test pilots for the "Buran" project at the Gromov Flight Research Institute. Sergey Tresvyatski and Yuri Schaeffer were key figures in this effort, contributing significantly to developing and testing the Soviet space shuttle program.

However, 1986 was also marred by the Challenger Disaster, a profound tragedy that claimed the lives of all seven crew members, including Christa McAuliffe. The disaster had a lasting impact on NASA and the broader space exploration community. The loss of the Challenger marked a somber moment in the history of spaceflight, leading to a rigorous 32-month hiatus from shuttle flights as NASA undertook a thorough investigation and implemented extensive safety improvements.

Despite this heart-wrenching event, 1986 also witnessed remarkable achievements. The Soviet Union launched the Mir Space Station on February 19, marking the beginning of a pioneering 15-year mission in low Earth orbit. Mir, a modular space station, became a hub for numerous crewed missions and was crucial in advancing scientific research in space.

Voyager 2 made a historic closest approach to Uranus on September 15, providing unprecedented insights into the gas giant's atmosphere, moons, and rings. This encounter expanded humanity's understanding of the outer solar system and showcased the capabilities of interplanetary exploration.

The year also saw significant advancements in modular space station design. The Soviet Union's DOS-7 and DOS-8 modules, originally intended for the next generation of Salyut space stations, were reconfigured into the Mir Core Module and the Zvezda Service Module. These modules played a vital role in constructing the International Space Station (ISS), highlighting a significant advancement in the design of space habitats.

The influence of the Soviet Almaz program was evident as the Almaz TKS spacecraft evolved into the Functional Cargo Block, which served as the foundation for several Mir modules and later contributed to the ISS. The DOS-7 module, after further refinement, became the Mir Core Module, establishing the nucleus of the first modular space station. Mir's design featured upgraded computers, solar arrays, and accommodations for two cosmonauts, with its six docking ports facilitating the integration of various modules and visiting spacecraft.

NASA continued to explore innovative technologies with the X-30, a concept demonstrator aircraft aimed at developing technologies for hypersonic flight and single-stage-to-orbit vehicles. Although the project faced financial constraints and was discontinued in 1994, it represented a significant step in the evolution of aerospace technology. Additionally, NASA's Rotor Systems Research Aircraft (RSRA) underwent taxi and low-altitude flight tests at NASA Armstrong, though this project was also terminated in 1988.

The Soviet Union made notable strides with the Soyuz T-15 mission, which docked with both the Mir and Salyut 7 space stations. The Vega 1 and Vega 2 spacecraft conducted significant studies of Halley's Comet and performed flybys of Venus, while the Phobos 1 and 2 missions aimed to explore Mars' moon Phobos. Despite losing contact with Phobos 1, Phobos 2 reached its destination and began its scientific investigation before communication was lost.

The Proton-K rocket's launch marked the beginning of the Mir space station's assembly in orbit, and the Soviet Union's Granat observatory was launched to study X-ray and gamma-ray sources in the universe, reflecting an ongoing commitment to pushing the boundaries of space exploration.

Despite the challenges faced, NASA's exploration efforts remained resolute. The Space Shuttle Columbia's STS-61C mission included the first Costa Rican astronaut. The Delta rocket launched the Ulysses spacecraft to study the Sun, while the Magellan spacecraft embarked on a mission to explore and map Venus. The Upper Atmosphere Research Satellite (UARS) was also launched to investigate Earth's atmosphere, demonstrating NASA's unwavering commitment to space exploration and scientific discovery.

In 1987, space exploration witnessed transformative advancements that significantly enriched our understanding of the solar system and beyond. Spearheaded by NASA, these developments unfolded through a series of remarkable achievements and technological progress, illustrating the dynamic evolution of space science.

Voyager 2's close encounter with Uranus was a particularly significant milestone of the year. This historic mission provided a wealth of information about the ice giant, allowing the spacecraft to capture detailed images and collect invaluable data about Uranus and its moons. Voyager 2's flyby yielded insights into Uranus' atmospheric composition and magnetic field, offering scientists a clearer understanding of the planet's unique characteristics. The data collected during this mission greatly expanded our knowledge of the solar system's outer reaches.

NASA launched the Galileo spacecraft in tandem with the Voyager 2 mission, marking a pivotal step in exploring Jupiter. Galileo, equipped with an atmospheric probe and sophisticated imaging cameras, embarked on its mission to study the gas giant and its extensive moon system. The spacecraft's observations and data collection significantly advanced our understanding of Jupiter's atmosphere, magnetic field, and its intriguing moons, such as Europa, which was believed to harbor a subsurface ocean. This mission was crucial in deepening our knowledge of Jupiter and its complex system.

A noteworthy event in 1987 occurred on June 5 when the Interdepartmental Qualification Committee (IAC) made a pivotal decision that underscored the significance of their selection. The decision led to the official designation of all Buran test pilots, including prominent figures such as Tresvyatski and Schaeffer, as test cosmonauts. This prestigious qualification recognized their exceptional training and readiness, marking a critical milestone in the Soviet Union's ambitious program to develop and deploy advanced space technologies. It affirmed their capability to pilot and operate the Buran spacecraft, symbolizing Soviet engineering prowess.

The same year, several key astronaut and cosmonaut groups formed, highlighting significant milestones in international space exploration. On January 5, Bulgaria's Shipka Group, including Aleksandr Aleksandrov and Krasimir Stoyanov, was officially announced. These astronauts began rigorous training for future space missions, marking a notable advancement in Bulgaria's space program.

On March 26, the Soviet Union established the TsPK-8/NPOE-8 Cosmonaut Group, an important development for space exploration. This group comprised Valery Korzun, Vladimir Dezhurov, Yuri Gidzenko, Yuri Malenchenko from TsPK, and Sergei Avdeyev from NPOE. They underwent comprehensive training at the Yuri Gagarin Cosmonaut Training Center and the Rocket Space Corporation Energia, preparing for missions aboard the Soyuz spacecraft and future space stations.

On June 5, NASA introduced Group 12, colloquially known as "The GAFFers" or "George Abbey Final Fifteen." This cohort included pilots such as Andrew M. Allen, Kenneth Bowersox, and Curtis Brown, alongside mission specialists like Mae Jemison, who made history as the first African-American woman to travel to space, and Michael Foale, who would later participate in extended missions aboard both the Mir space station and the International Space Station (ISS). Over the subsequent years, individuals such as William Readdy and Kenneth Bowersox would play crucial roles in NASA's leadership and space missions.

In August 1987, Germany announced its astronaut group for the year, which included Renate Brümmer, Hans Schlegel, Gerhard Thiele, Heike Walpot, and Ulrich Walter. This group contributed significantly to European Space Agency (ESA) missions, reflecting the growing international collaboration in space exploration.

NASA also undertook a series of Earth-observation initiatives in 1987. The Earth Radiation Budget Satellite (ERBS), the Upper Atmosphere Research Satellite (UARS), and the Solar Maximum Mission (SMM) were launched to enhance our understanding of Earth's environment.

These satellites provided critical data on the planet's climate, radiation budget, and atmospheric conditions, paving the way for future research on global climate change and atmospheric dynamics.

The year 1987 also marked significant progress in space shuttle technology. NASA's space shuttle program achieved a major milestone with the STS-26 mission, the first shuttle flight since the Challenger disaster. This successful mission symbolized NASA's resilience and commitment to advancing space exploration. The completion of STS-26 demonstrated the agency's renewed focus on safety and innovation in spaceflight.

At the Plum Brook Research Center, efforts were underway to reactivate test facilities for the Strategic Defense Initiative, known as Project TIMBER WIND. This project later evolved into the Space Nuclear Thermal Propulsion (SNTP) program at the Air Force Phillips Laboratory. The aim was to develop advanced rocket propulsion technologies with potential military applications, contributing to the broader space exploration and defense field.

Regarding national security, by December 1987, the United States had strategically positioned nine early warning satellites in High Earth Orbit (HEO) and one geostationary satellite. These satellites were crucial for monitoring potential missile threats and enhancing national defense capabilities during the Cold War era.

Significant events in the Cold War marked the geopolitical landscape of 1987. Mikhail Gorbachev's ascension as the General Secretary of the Communist Party of the USSR introduced a new approach to international relations. In December, Gorbachev and President Reagan signed the INF Treaty, a landmark agreement that eliminated an entire class of nuclear and conventional missiles. This treaty played a significant role in reducing tensions between the United States and the Soviet Union, fostering a more stable international environment.

During this period, NASA's Dryden Flight Research Center was actively engaged in pioneering research on high-angle-of-attack flight and aircraft control. Utilizing a modified F-18 High Alpha Research Vehicle (HARV), the center studied airflow dynamics and control mechanisms under extreme flight conditions. Innovations such as thrust-vectoring systems and movable strakes were developed, enhancing pitch and yaw control and improving maneuverability. These advancements were instrumental in designing safer and more effective fighter aircraft.

In September 1987, the Soviet Union achieved a significant milestone by launching the Soyuz TM-3 mission to the Mir space station. This mission was notable for Svetlana Savitskaya becoming the first woman to perform a spacewalk, highlighting the expanding roles of women in space exploration.

Another groundbreaking event of 1987 was transmitting the first interstellar radio message, "A Message From Earth." Spearheaded by renowned scientist Carl Sagan and a team of researchers, this message was directed toward the M13 globular cluster and aimed to communicate humanity's achievements and characteristics to potential extraterrestrial civilizations.

Throughout 1987, several astronauts and cosmonauts made notable contributions to space exploration. Soviet cosmonaut Aleksandr Viktorenko commanded the Soyuz TM-3 mission, which docked with the Mir space station in July. Alongside him, flight engineer Alexander Serebrov participated in this mission. Earlier in the year, Soyuz TM-2, commanded by Yuri Romanenko and including flight engineer Aleksandr Laveykin, embarked on the first long-duration expedition to Mir, lasting approximately 326 days.

In American spaceflight, Salman bin Abdulaziz Al Saud became the first Arab and Muslim to fly in space as a payload specialist aboard the Space Shuttle Discovery during the STS-51-G mission. This mission, which extended into 1987, was a milestone in international cooperation in space.

American astronaut John M. Lounge participated as a mission specialist in the STS-51-I mission, which also extended into the following year. These missions and the individuals involved underscored space exploration's global and collaborative nature during this transformative period.

In 1988, the landscape of space exploration was profoundly shaped by significant milestones and advancements that highlighted the era's technological and geopolitical dynamism. This pivotal year saw major contributions from various nations, each making strides that would shape the future of space exploration.

The Soviet Union, emerging as a dominant force in space exploration, made a landmark achievement on February 20, 1988, with the launch of the Mir space station. This ambitious project began what would become one of the longest-operating space stations in history, orbiting Earth for over 15 years. Mir was more than a testament to Soviet engineering prowess; it was a crucial platform for long-duration missions and scientific research. The station provided invaluable data on space habitation and human adaptation to long-term spaceflight, offering insights that would pave the way for future space research and exploration. The experiments conducted aboard Mir laid the groundwork for understanding the effects of prolonged exposure to microgravity, which would be essential for future deep-space missions.

In contrast, the United States significantly returned to space exploration with the launch of the Space Shuttle Discovery on September 29, 1988. This mission, designated STS-26, was especially notable as it marked the first shuttle flight following the Challenger disaster of 1986. The successful deployment of the Tracking and Data Relay Satellite-3 (TDRS-3) and the execution of various scientific experiments underscored NASA's resilience and determination to advance its space program despite past setbacks. The STS-26 mission was crucial in restoring confidence in the Space Shuttle program and demonstrated the continued importance of shuttle flights in maintaining and expanding the nation's space capabilities.

The year also saw the entry of Afghanistan into the Soviet space program with the establishment of the OS "Mir" Group on February 12, 1988. Afghan astronauts Mohammad Dauran Ghulam Masum and Abdul Ahad Mohmand were selected to represent their nation in the Mir space station program. This move exemplified the broader international cooperation fostered by space exploration, reflecting the Soviet Union's effort to include diverse participants in its space endeavors. Mohmand's participation was historic as he became the first Afghan to travel into space, marking a significant moment in the internationalization of space exploration.

The Soviet Union launched the Phobos 1 mission on July 7, 1988, with the ambitious goal of exploring Mars and its moons. Although the spacecraft lost communication with Earth just before its scheduled encounter with Mars, bringing an early end to the mission, it represented a significant effort to expand humanity's understanding of the Red Planet and its moons. Despite the setback, the Phobos 1 mission laid the groundwork for future Martian exploration and highlighted the complexities of interplanetary missions.

Later in the year, NASA launched the Space Shuttle Atlantis on November 26, 1988, as part of the STS-27 mission. This flight carried a classified payload for the US Department of Defense, including the Lacrosse 1 radar reconnaissance satellite. The mission underscored the growing intersection between space exploration and national defense, reflecting the strategic importance of space assets during this era. The secrecy surrounding the mission highlighted the sensitive nature of military space operations and the role of space technology in national security.

In addition to these notable missions, the US Air Force made significant strides in enhancing its space capabilities. Activating the Air Force Space Surveillance Element at the Cheyenne Mountain Complex marked a crucial development in establishing effective command and control over space operations. This unit, which would later evolve into the 18th Space Control Squadron of the United States Space Force, played a key role in advancing the Air Force's role in space surveillance and space command.

Throughout 1988, NASA conducted four Space Shuttle missions, demonstrating a sustained commitment to space exploration. Following STS-26, the STS-27 mission on November 30 continued deploying satellites and conducting experiments with military implications. Although STS-29, launched on March 13, 1989, was officially part of the 1989 flight manifest, it was closely associated with the 1988 program. This mission involved deploying a Tracking and Data Relay Satellite (TDRS) and included various scientific experiments. The STS-30 mission, which took off on May 4, 1989, was another significant endeavor featuring the deployment of the Magellan spacecraft. Designed to map Venus's surface using radar, this mission marked a key development in planetary exploration and contributed to the broader understanding of the solar system.

Several notable astronauts and cosmonauts played pivotal roles in the space missions of 1988. Anatoly Solovyev, Viktor Savinykh, and Aleksandr Aleksandrov participated in the Soyuz TM-5 mission, launched on June 7, 1988. Their contributions to the Mir EO-4 mission included a range of scientific experiments aboard the space station, advancing the understanding of long-duration spaceflight. The Soyuz TM-6 mission, launched on August 29, 1988, was particularly noteworthy as it carried Vladimir Lyakhov and Abdul Ahad Mohmand to the Mir space station. Mohmand's participation was historic as he became the first Afghan to travel into space.

Another key figure in this period was Valeri Polyakov, who was part of the Soyuz TM-7 mission launched on April 27, 1989. His work and fellow astronauts' work continued to advance the understanding of long-duration spaceflight, contributing to the knowledge necessary for future deep-space missions. In the US Space Shuttle program, astronauts such as Frederick H. Hauck, Richard O. Covey, John M. Lounge, George D. Nelson, and David C. Hilmers were instrumental in the STS-26 mission, marking the return to flight of the Space Shuttle following the Challenger disaster and including the deployment of the TDRS-3.

Aleksandr Volkov and Sergei Krikalev also made significant contributions during this time as part of the Soyuz TM-7 mission, participating in the crew rotation for the Mir space station. On December 21, 1988, French astronaut Jean-Loup Chrétien made history by becoming the first European to perform an extravehicular activity (EVA) during the Soyuz TM-6 mission.

The STS-27 mission, launched on December 2, 1988, saw the Space Shuttle Atlantis carry a classified payload for the US Department of Defense, including the Lacrosse 1 radar reconnaissance satellite. This mission, executed by a distinguished crew, underscored the critical role of space technology in national security and demonstrated the continued evolution of space exploration in both scientific and strategic realms.

The year 1989 was marked by significant advancements and events in space exploration, reflecting this field's dynamic and evolving nature during international cooperation and intense competition.

On January 25, the Soviet Union's Cosmonaut Group expanded by selecting new astronauts from various institutes. This new cohort included Vladimir Karashtin and Vasili Lukiyanyuk from the Institute of Biomedical Problems (IMBP), Anatoli Polonsky and Valeri Tokarev from the Research Institute of the Central Aerohydrodynamic Institute (GNKII), Aleksandr Yablontsev from the Central Design Bureau of Experimental Machine Building (NPOE), and Sergei Kirchevsky and Gennady Padalka from the Yuri Gagarin Cosmonaut Training Center (TsPK). Their selection highlighted the Soviet Union's continued commitment to advancing its space exploration capabilities, showcasing a broad pool of talent and expertise.

March 22 marked a crucial moment for the Buran project, the Soviet Union's ambitious initiative in reusable spaceflight. The Gromov Flight Research Institute concluded its final round of test pilot selections for the program. Despite experiencing intermittent activity, the Buran project remained a testament to Soviet ambition. Notable participants from the program who ventured into space included Igor Volk and Anatoly Levchenko, illustrating the program's selective and resilient nature. The Buran project, however, would eventually be closed in 1993.

In May, Italy made a notable stride in space exploration by selecting astronauts Franco Malerba, Franco Rossitto, Umberto Guidoni, and Cristiano Batalli Cosmovici. This selection underscored Italy's emerging role in international space missions and its commitment to contributing to global space exploration efforts.

By September 29, NASA made key appointments in its ATLAS program, selecting Charles R. Chappell, Michael Lampton, and Byron K. Lichtenberg as payload specialists. These appointments underscored NASA's dedication to advancing scientific research through specialized payload missions, enhancing the understanding of the cosmos through their targeted efforts.

November 25 saw the United Kingdom's Project Juno significantly contribute to international space cooperation by selecting Helen Sharman and Timothy Mace. Helen Sharman, in particular, would go on to make history as the first British-born person to journey into space when she flew aboard Soyuz TM-12 in May 1991. This milestone highlighted the UK's commitment to space exploration and collaborative efforts with the Soviet Union.

The Soviet Union continued its space endeavors with the launch of the Kosmos 2054 spacecraft in September 1988. This mission focused on testing various space technologies and conducting scientific experiments. Later, in December, the launch of Kosmos 2123, a military reconnaissance satellite, further emphasized the Soviet Union's commitment to enhancing its space-based capabilities.

In 1989, the Space Shuttle program in the United States resumed its activities after reflection. On January 28, the Space Shuttle Discovery launched on mission STS-29 from Kennedy Space Center. This mission, marking a significant return to spaceflight, involved deploying the Tracking and Data Relay Satellite (TDRS-4) and a series of scientific experiments. The crew, including astronauts Steven R. Nagel, Brewster H. Shaw Jr., Guy S. Gardner, and James F. Buchli, completed their mission and returned to Earth on February 8.

The Soviet Union faced a year of both ambition and setbacks. On May 7, the Soviet Union launched Phobos 1, a spacecraft designed to study Mars and its moon, Phobos. However, the mission ended prematurely when contact was lost on September 2, just before its scheduled encounter with Mars.

The European Space Agency (ESA) significantly contributed to astronomical research by launching the Hipparcos satellite on July 23 from French Guiana. This mission was dedicated to precisely measuring star positions and distances within our galaxy, marking a breakthrough in understanding stellar and galactic dynamics.

To rectify earlier setbacks, the Soviet Union launched Phobos 2 on November 8, continuing the study of Mars and Phobos. While Phobos 2 reached Mars, communication was lost on March 27, 1990, leading to another setback in the mission.

NASA's Space Shuttle Atlantis took to the skies on December 2 for mission STS-34, with the primary objective of deploying the Galileo spacecraft. Galileo was tasked with conducting extensive studies of Jupiter and its moons. The mission also featured the first untethered spacewalk since the early days of the Shuttle program, marking a notable achievement.

Throughout the year, the Mir space station remained a focal point for the Soviet space program. On February 6, cosmonauts Aleksandr Volkov and Sergey Krikalev embarked on a 151-day mission, Mir EP-1, which included scientific experiments and equipment testing. Their work contributed valuable data to ongoing space research. On June 28, the Progress M-3 cargo spacecraft docked with Mir, delivering essential supplies. Later, on July 22, the launch of Soyuz TM-8 saw cosmonauts Viktor Afanasyev and Musa Manarov embark on a 175-day mission, Mir EP-2, which included numerous scientific experiments and set a new space endurance record with 168 days in orbit. The Progress M-4 cargo spacecraft arrived at Mir on November 21, further supporting the station's operations with additional supplies.

In addition to these milestones, the year saw several significant astronaut and cosmonaut contributions. The STS-29 mission on March 13 deployed TDRS-4, with the crew engaging in space physiology, materials science, and Earth observations. The STS-30 mission of the Space Shuttle Atlantis on May 4 saw the deployment of the Magellan spacecraft to explore Venus. On August 8, the STS-28 mission of the Space Shuttle Columbia involved a classified operation for the U.S. Department of Defense, deploying military communications and reconnaissance satellites.

The Soyuz TM-8 mission on September 5 saw cosmonauts installing new equipment on Mir, enhancing its capabilities. The STS-34 mission on October 18 involved the deployment of the Galileo probe for Jupiter. On November 22, the STS-33 mission conducted a classified operation for the U.S. Department of Defense, marking a milestone with the first African American astronaut commanding a Space Shuttle.

1989 was also a year of profound geopolitical shifts. The fall of Communist regimes across Eastern Europe began to reshape the global political landscape. The Berlin Wall fell in November, and the Velvet Revolution in Czechoslovakia symbolized the broader movement toward democratic governance. However, the year was also marred by the tragic events of June 4 in Tiananmen Square, Beijing, where the Chinese government's violent suppression of pro-democracy protests resulted in numerous casualties and widespread international condemnation.

In 1990, the realm of space exploration underwent transformative changes, marked by significant scientific advancements and burgeoning international cooperation amidst the backdrop of Cold War tensions. This year saw the realization of several groundbreaking missions that reshaped our understanding of the universe and showcased a new spirit of global collaboration.

A defining achievement of 1990 was the launch of the Hubble Space Telescope, a major joint endeavor between NASA and the European Space Agency. On April 24, 1990, with astronauts F. Story Musgrave and Kathryn D. Sullivan aboard, the Space Shuttle Discovery deployed the Hubble into orbit. The telescope, featuring a 2.4-meter primary mirror, was designed to capture high-resolution images of distant galaxies and stars, aiming to revolutionize our cosmic perspective. However, shortly after its deployment, a mirror's spherical curvature flaw resulted in blurry images, temporarily hindering its mission. This issue was remedied in 1993 during a space shuttle servicing mission, restoring the Hubble's capabilities and heralding a new era of astronomical discovery.

Simultaneously, 1990 witnessed a flurry of other notable space missions. The Magellan probe embarked on a journey to Venus to map its surface using radar technology. This mission was pivotal in enhancing our knowledge of Venus's geological features. The Ulysses spacecraft was launched to study the Sun's polar regions, providing unprecedented data on solar activity and its effects on the solar system. The Galileo spacecraft, dispatched to investigate Jupiter and its moons, promised to expand our understanding of the largest planet in our solar system. The Soviet Union also launched the Phobos 1 and 2 missions , designed to explore Mars and its moons, continuing their robust interplanetary exploration program.

International collaboration was further underscored by the launch of Soyuz TM-10 on August 1, 1990. This mission was historic as it carried American astronaut Norman Thagard, making him the first U.S. astronaut to fly aboard a Soviet spacecraft. This event symbolized a growing spirit of cooperation in space exploration, reflecting the easing of Cold War tensions and a shared commitment to scientific advancement.

In the realm of human spaceflight, 1990 was notable for forming several new astronaut groups and significant missions. On January 17, NASA introduced Group 13, affectionately nicknamed "The Hairballs." This diverse cohort included pilots Kenneth Cockrell and Eileen Collins, who would later make history as the first female shuttle pilot and commander, along with mission specialists Bernard Harris, the first African American to perform a spacewalk, and Ellen Ochoa, the first Hispanic woman in space. Their distinctive patch, featuring a black cat, inspired the group's memorable nickname.

In February, France selected its CNES Group 3, which included astronauts Léopold Eyharts and Jean-Marc Gasparini. This selection marked the final batch of astronauts chosen independently by the French space agency before its integration into the European Space Agency (ESA) Astronaut Corps in 1999, highlighting France's commitment to international collaboration in space.

Germany's astronaut group of 1990, announced on October 8, featured pioneers like Reinhold Ewald and Klaus-Dietrich Flade. This group exemplified Germany's re-emergence as a key player in European space endeavors following its reunification and increasing contributions to space science.

The Soviet space station Mir, operational since 1986, played a central role in space exploration. On February 11, Soyuz TM-9 carried cosmonauts Aleksandr Volkov and Toktar Aubakirov to the Mir space station, where they joined the ongoing Mir EO-6 expedition. Later in the year, Soyuz TM-10 launched with cosmonauts Vladimir Titov and Musa Manarov on August 1. Their mission to Mir extended to a record-breaking 365 days in space, during which they conducted extensive research on the effects of long-duration spaceflight on the human body.

October 6, 1990, was another milestone, as the Space Shuttle Discovery's STS-41 mission included the deployment of the Ulysses spacecraft. During this mission, astronauts Bruce McCandless II and Kathryn D. Sullivan achieved the historic first untethered spacewalk using the Manned Maneuvering Unit (MMU), marking a significant advance in extravehicular activities.

The Soviet Union's space exploration efforts also included the Gamma Astrophysical Observatory and the Foton mission. These missions were dedicated to astrophysical research and the study of microgravity effects on biological and physical systems, contributing to our understanding of the universe and the unique conditions of space.

As the year 1991 began, the world witnessed significant upheaval as the Soviet Union faced dramatic political and economic transformations. The dissolution of the Soviet Union in December 1991, a consequence of sweeping reforms such as perestroika and glasnost initiated by Mikhail Gorbachev, marked a pivotal moment in global politics and had a profound impact on the realm of space exploration. Despite these monumental changes, the Soviet Union's enduring legacy continued to influence international space endeavors.

In this period of transition, Russia, still a constituent part of the Soviet Union, pursued its space ambitions with several notable achievements. On February 19, 1991, the Mir space station, a cornerstone of Soviet space achievements, was launched into orbit. Mir, which means "peace" in Russian, was to serve as a crucial platform for scientific research and experimentation. Its launch represented the zenith of Soviet space technology and an enduring symbol of their commitment to space exploration. The Progress spacecraft, another vital element of the Soviet space program, played an essential role in resupplying Mir with necessary provisions, further showcasing the technical expertise of the Soviet space engineers.

The international space community also reached significant milestones in 1991. On May 18, British astronaut Helen Sharman made history as the first Briton and the first European woman to journey into space. Her mission aboard the Soyuz TM-12 took her to the Mir space station, highlighting the growing global participation in space exploration. Later in the year, on October 2, the Soyuz TM-13 mission launched with a distinguished crew: Toktar Aubakirov and Franz Viehböck, who became the first Kazakh and Austrian astronauts, respectively, to travel to space. Aleksandr Volkov, a seasoned Soviet cosmonaut, also joined this mission, underscoring the continued prominence of Soviet space endeavors even as the geopolitical landscape was shifting.

NASA, too, made strides in its space program during 1991. On September 25, NASA launched the Mars Observer spacecraft to study the Red Planet and its mysteries. Unfortunately, the spacecraft lost contact with mission control before it could complete its journey to Mars, marking a significant setback in the quest to explore the Red Planet. However, the same year brought forth a triumph with the launch of the Galileo spacecraft on October 18. Galileo embarked on an ambitious mission to explore Jupiter and its moons, and its findings would later provide groundbreaking insights into the gas giant's complex system and its surrounding environment.

NASA's Space Shuttle program also saw considerable activity in 1991. The STS-37 mission, conducted from April 5 to 11, deployed the Compton Gamma Ray Observatory, a groundbreaking instrument designed to study gamma rays from distant celestial sources. The following mission, STS-40, took place from June 5 to 14 and was dedicated to life sciences experiments conducted aboard the Spacelab module, which contributed valuable data on the effects of spaceflight on biological systems. Later in the year, the STS-43 mission, held from August 2 to 11, deployed the Tracking and Data Relay Satellite-5 (TDRS-5), enhancing communication capabilities for various spacecraft. The STS-44 mission, which followed, deployed the Defense Support Program satellite, crucial for military reconnaissance. In contrast, the STS-48 mission, conducted from September 12 to 18, launched the Upper Atmosphere Research Satellite (UARS), which focused on studying the Earth's upper atmosphere.

Throughout 1991, NASA also made significant strides in planning for the International Space Station (ISS). The year marked the beginning of formal agreements with Russia, setting the stage for future collaboration. These early agreements laid the crucial groundwork for the ISS, which would evolve into a central hub of international cooperation in space, symbolizing numerous nations' collective efforts to pursue scientific discovery and technological advancement.

In 1992, space exploration experienced a year of remarkable advancements and significant developments, reflecting a period of dynamic progress and heightened international collaboration in space research.

The year began with substantial strides across the global space community. Russia demonstrated its commitment to manned space missions in March by establishing the NPOE-10 Cosmonaut Group. This new cohort included Aleksandr Lazutkin, Sergei Treshchov, and Pavel Vinogradov, marking a significant step as Russia navigated the post-Soviet era and reaffirmed its dedication to space exploration.

On March 31st, NASA's Group 14, colloquially known as "The Hogs," was introduced. This diverse group of astronauts included pilots such as Scott Horowitz, Brent Jett, and Kevin Kregel, alongside international mission specialists such as Marc Garneau and Chris Hadfield from Canada, Maurizio Cheli from Italy, Jean-François Clervoy from France, and Koichi Wakata from Japan. Including these international specialists highlighted NASA's evolving partnerships with global space agencies.

Japan made significant progress with its space program in April by selecting Koichi Wakata for the 1992 NASDA Group. This selection underscored Japan's growing ambition to enhance its role in manned space missions and contribute to international space efforts.

May brought further advancements as Canada announced its CSA Group 2 astronaut team. This group, featuring Dafydd Williams, Julie Payette, and Chris Hadfield, continued Canada's tradition of contributing to space missions, particularly those aboard the US Space Shuttle.

On May 15th, the European Space Agency (ESA) revealed its 1992 astronaut group. This team included Maurizio Cheli from Italy, Jean-François Clervoy from France, Pedro Duque from Spain, Christer Fuglesang from Sweden, Marianne Merchez from Belgium, and Thomas Reiter from Germany. This selection underscored ESA's commitment to fostering European space exploration and enhancing collaborative scientific research.

NASA played a prominent role in advancing space exploration with the launch of the Mars Observer spacecraft. This mission aimed to study Mars's atmosphere and surface, representing a crucial step in humanity's quest to understand the Red Planet. In parallel, NASA was deeply involved in the 1992 Space Exploration Initiative (SEI), an ambitious program to enhance solar system exploration. Despite its high aspirations, the SEI faced challenges, particularly with Space Nuclear Thermal Propulsion (SNTP) technology, which did not significantly outperform earlier technologies such as the Nuclear Engine for Rocket Vehicle Application (NERVA).

Alongside NASA's efforts, the Central Intelligence Agency (CIA) collaborated with the National Reconnaissance Office (NRO) on covert space operations. A notable event in 1992 was the declassification of the CORONA program, a top-secret initiative that had operated from 1960 to 1972. The CORONA program utilized satellites to gather intelligence on foreign territories and had amassed over 860,000 images of Earth's surface, providing invaluable data for military and intelligence purposes.

The year also saw pivotal developments in space exploration with the initial deployment of components for the International Space Station (ISS). This marked the beginning of a new era of international collaboration, as the United States and Russia signed an agreement to jointly pursue space exploration goals, including Mars missions and the construction of the ISS. This agreement symbolized a thaw in Cold War tensions and a commitment to cooperative space research.

The global space community was active throughout 1992, with notable contributions from several countries. Russia continued its space activities by launching the Progress M-14 spacecraft, which resupplied the Mir space station, and the Soyuz TM-15 crewed mission, which docked with Mir. Europe's contributions included the European Space Agency's launch of the Hipparcos spacecraft, designed to measure star positions and movements with unprecedented accuracy. Japan advanced its space capabilities with the launch of the Advanced Earth Observing Satellite (ADEOS) to monitor Earth's atmosphere and environment. China marked a milestone with the launch of its Fengyun series, initiating a new era of meteorological satellite data collection. India achieved significant success with the launch of the INSAT-2D satellite, enhancing its communication and meteorological services.

In the realm of space shuttle missions, NASA conducted several important flights. The STS-42 mission, launched on January 22 aboard the Space Shuttle Discovery, focused on life sciences and materials processing experiments as part of the International Microgravity Laboratory-1 (IML-1). On March 24, the Space Shuttle Atlantis embarked on the STS-45 mission to research Earth's environment and atmosphere. The STS-49 mission, launched on May 7 by the Space Shuttle Endeavour, was notable for its successful repair and rescue of the stranded Intelsat VI F-3 satellite. The year concluded with the STS-50 mission, launched on June 25 by the Space Shuttle Columbia, which marked the first dedicated United States Microgravity Laboratory (USML-1) mission, focusing on materials science, fluid physics, and biotechnology experiments in microgravity.

Politically, 1992 was marked by the dissolution of the Soviet Union. Leaders from Belarus, Russia, and Ukraine declared the end of the USSR, an event rejected by Soviet President Mikhail Gorbachev. This led to the official dissolution of the Soviet Union and the emergence of the Commonwealth of Independent States, signifying the end of the Cold War. Additionally, the nuclear arms race saw its culmination with the US conducting the last nuclear test at the Nevada Test Site under Operation JULIN on September 23, 1992. This test, known as Shot Divider, was part of a series aimed at ensuring the safety of the US's nuclear deterrent forces, amid the onset of negotiations for the Comprehensive Nuclear Test Ban Treaty.

In the domain of human spaceflight, 1992 was notable for several achievements. Gennady Manakov and Gennady Strekalov participated in the Soyuz TM-15 mission, which launched on July 27 and docked with the Mir space station on July 29. This began a significant long-duration stay and scientific research period aboard Mir. The crew of STS-50 aboard the Space Shuttle Columbia, including Richard N. Richards, Kenneth D. Bowersox, Bonnie J. Dunbar, Carl J. Meade, Eugene H. Trinh, Ellen S. Baker, and Lawrence J. DeLucas, conducted a range of experiments in microgravity from June 25 to July 9. Additionally, the STS-46 mission on the Space Shuttle Atlantis, launched on July 31 with a crew including Loren J. Shriver, Andrew M. Allen, Jeffrey A. Hoffman, Franklin R. Chang Díaz, Claude Nicollier, and Marsha S. Ivins, focused on deploying the ESA's EURECA satellite and conducting tethered satellite experiments. Valery Polyakov set a remarkable record by completing the longest continuous spaceflight during the Soyuz TM-14 mission, spending 437 days, 17 hours, and 58 minutes aboard Mir, where he conducted extensive medical and scientific research.

In 1993, the realm of space exploration experienced significant transitions and advancements, set against the backdrop of a shifting geopolitical landscape following the dissolution of the Soviet Union. This pivotal year marked a profound transformation for space programs worldwide, particularly for Russia, which inherited the Soviet space legacy amidst considerable economic and logistical challenges.

As the Soviet Union disintegrated, the Russian Federation emerged as the successor to the Soviet space program, now under the purview of the Russian Federal Space Agency, known as Roscosmos. The economic turmoil that followed the Soviet collapse severely impacted Russia's space ambitions. The country faced diminished import and export capabilities related to space activities and the withdrawal of crucial financial support from former Eastern Bloc nations. These financial constraints placed considerable strain on Russia's ability to maintain its space program, forcing the government to seek external assistance to sustain its capabilities and retain its competitive edge in global space exploration.

Despite these economic hardships, Russia demonstrated remarkable resilience and commitment to space exploration throughout 1993. The Russian space program played a crucial role in supporting the International Space Station (ISS) and launching numerous scientific missions. This ongoing engagement underscored Russia's determination to navigate through the tumultuous period and remain a key player in space exploration. However, not all projects met with success. The Soviet Union had developed an advanced anti-satellite system, IS-MU, designed to engage and maneuver against enemy satellites. Although the system was declared operational in 1991, the cash-strapped Russian government under President Boris Yeltsin decided to terminate the IS-MU project in 1993, reflecting the severe financial constraints affecting Russia's space initiatives during this period.

Meanwhile, NASA continued to make significant strides in space exploration. In 1993, the Space Shuttle Endeavour undertook a notable mission, deploying the fifth Tracking and Data Relay Satellite (TDRS-E) and retrieving the European Retrievable Carrier (EURECA) satellite through a spacewalk. This mission highlighted NASA's unwavering commitment to advancing space technology and maintaining robust satellite communications.

NASA's dedication to the ISS was further exemplified in November 1993 with the launch of Space Station Assembly Flight 3A. This mission was a critical milestone in the ISS construction, reflecting NASA's steadfast commitment to international collaboration in space.

Significant scientific achievements also marked the year for NASA. The launch of the Cosmic Background Explorer (COBE) satellite represented a major advancement in cosmology. COBE was tasked with studying cosmic microwave background radiation, providing crucial evidence that supported the Big Bang theory of the universe's origin. This groundbreaking research significantly expanded our understanding of the universe.

In addition, NASA introduced the Heat Transfer Facility in 1993, an essential asset for testing advanced heat shield materials and technologies designed for spacecraft reentry. This facility played a vital role in ensuring the safety and efficiency of future space missions.

The dissolution of the Soviet Union brought notable changes to the space landscape. As the Russian Federation took over the Soviet space program, Roscosmos continued to operate the Mir space station, which had been launched in 1986. In 1993, several missions were conducted to Mir, including Soyuz TM-16, which delivered a new crew, and Soyuz TM-17, which carried German astronaut Ulf Merbold, marking the first German visit to Mir. These missions underscored Roscosmos's ongoing commitment to space exploration and international cooperation.

Roscosmos also expanded its space activities with the launch of the Spektr-R radio telescope and the Granat X-ray astronomy satellite, both deployed using the Proton-K rocket. These missions contributed to Russia's scientific research and space exploration efforts.

The global space community saw contributions from various nations in 1993. The European Space Agency (ESA) launched the Hipparcos satellite, designed to measure the positions and motions of stars and galaxies with unprecedented accuracy. ESA also initiated the Mars Observer spacecraft mission, although the spacecraft was lost before reaching Mars. Japan demonstrated its space capabilities with the Advanced Earth Observing Satellite (ADEOS) launch and the Solar-A spacecraft, which focused on solar studies. China made its mark by launching its first remote sensing satellite, the FSW-1, while India achieved a significant milestone with the successful launch of its INSAT-2B satellite.

Noteworthy astronauts and cosmonauts played pivotal roles in various missions throughout 1993. Gennady Manakov and Aleksandr Poleshchuk were part of the Soyuz TM-16 mission, which docked with Mir in July, initiating a prolonged stay that involved scientific research and maintenance activities. On the U.S. side, the crew of the Space Shuttle Columbia's STS-50 mission, including Richard N. Richards, Kenneth D. Bowersox, and Ellen L. Ochoa, conducted materials science, fluid physics, and biotechnology experiments.

The Space Shuttle Atlantis's STS-46 mission, featuring astronauts such as Loren J. Shriver and Claude Nicollier, involved deploying ESA's EURECA satellite and conducting experiments with the Tethered Satellite System. During the Soyuz TM-14 mission, Valery Polyakov set a record for the longest continuous stay in space, spending 437 days, 17 hours, and 58 minutes aboard Mir while conducting extensive medical and scientific research.

In 1994, space exploration witnessed a series of transformative events that underscored scientific advancements and international cooperation's burgeoning spirit. This pivotal year marked a significant chapter in the chronicles of space travel as the United States and Russia embarked on groundbreaking missions that would shape the future of their space programs.

On January 7, 1994, the U.S. Space Shuttle Endeavour embarked on its historic STS-60 mission, setting a milestone by carrying Sergei Krikalev, the first Russian cosmonaut to fly aboard an American spacecraft. This mission symbolized the dawn of a new era of collaboration between the United States and Russia, reflecting a shared commitment to advancing space exploration. Just a week later, on January 14, the Russian spacecraft Soyuz TM-17 launched, transporting cosmonauts Valeri Korzun and Aleksandr Kaleri to the Mir space station. This mission was part of Russia's ongoing efforts to maintain and operate Mir, a crucial hub for scientific research and international collaboration in low Earth orbit.

The exploration momentum continued with the U.S. Space Shuttle Columbia launch on March 22, 1994, for mission STS-62. This mission was notable for its array of scientific experiments, including deploying a satellite designed to study Earth's atmosphere and surface. The shuttle's return to Earth further advanced our understanding of these critical environmental systems.

April 26 saw the launch of Soyuz TM-21, which carried cosmonauts Yury Onufrienko and Yuri Usachev, along with U.S. astronaut Shannon Lucid, to the Mir space station. This mission emphasized the growing collaboration between the United States and Russia, as the crew's presence on Mir continued to foster international cooperation.

June 2 marked another significant milestone with the launch of the U.S. Space Shuttle Endeavour on STS-59. This mission's inaugural Space Radar Laboratory (SRL-1) experiment distinguished it. The experiment aimed to enhance our understanding of Earth's surface through advanced radar imaging techniques. The successful execution of this experiment underscored the shuttle's role in expanding the frontiers of Earth observation.

On July 1, the Russian spacecraft Progress M-23 launched, performing a critical resupply mission to the Mir space station. This mission was vital for sustaining Mir's operations, ensuring that the station remained well-stocked with essential supplies and equipment.

The pace of exploration continued with the September 22 launch of the U.S. Space Shuttle Endeavour on STS-68. This mission followed up on the SRL-1 experiment by deploying the Space Radar Laboratory (SRL-2) experiment, which furthered the study of Earth's surface and atmospheric phenomena.

October 1 brought the launch of Soyuz TM-20, which transported cosmonauts Anatoly Solovyev and Nikolai Budarin to Mir. Their mission was part of the routine rotation of crew members aboard the space station, continuing the collaborative efforts between the United States and Russia.

One of the year's most significant achievements occurred on November 3, when the U.S. spacecraft Galileo released its atmospheric probe. This probe entered Jupiter's atmosphere on December 7, providing invaluable data on the gas giant's atmospheric composition and dynamics, contributing greatly to our understanding of planetary science.

The final notable mission of 1994 was the launch of the U.S. Space Shuttle Endeavour on December 9 for STS-66. This mission featured the Atmospheric Laboratory for Applications and Science-3 (ATLAS-3) experiment, which focused on atmospheric research to understand better the Earth's atmosphere and its interactions with solar radiation.

Several astronauts and cosmonauts made notable contributions to space exploration throughout the year. Viktor Afanasyev and Yuri Usachyev joined the Mir space station crew on January 8, 1994, aboard Soyuz TM-18, continuing their mission until their rotation on July 9, 1994. Valeri Polyakov began his extended stay aboard Mir on Soyuz TM-18, setting a record for the longest single spaceflight in history. Sergei Krikalev's participation in both Soyuz TM-18 and STS-60 marked a significant milestone, as he became the first Russian to fly aboard a U.S. Space Shuttle. Ellen Baker, a U.S. astronaut on STS-60, and Jean-Pierre Haigneré, a French astronaut who participated in the Euromir 94 mission, played crucial roles in advancing scientific research during their respective missions.

The year also saw significant milestones in space science and technology. The NARYAD project, part of the Soviet Union's anti-satellite program, continued its development under Russia's auspices. In 1994, the Russian Ministry of Defense launched the last Rockot booster from Baikonur as part of the NARYAD program. Despite the challenges, the mission placed the Radio-ROSTO satellite into orbit, though the NARYAD component exploded shortly after launch. This event highlighted the evolving nature of Russia's space capabilities and its transition from Soviet-era programs to contemporary space operations.

In 1996, the landscape of global space exploration saw significant expansions through new astronaut selections and international collaborations. On February 9th, Russia made a notable stride by forming the MKS/RKKE–12 Cosmonaut Group. This group welcomed four new cosmonauts: Oleg Kotov and Yuri Shargin from the MKS program, alongside Konstantin Kozeyev and Sergei Revin from the RKKE program. This formation was a testament to Russia's steadfast commitment to manned space missions, reflecting its ongoing adaptation and resilience amidst extensive restructuring and reform.

The momentum continued in March with the addition of Oleg Kononenko to the MKS supplemental cosmonaut group. Kononenko's inclusion bolstered the roster of skilled cosmonauts preparing for future missions, contributing to the growing expertise within Russia's space program.

As the year progressed, May 1st brought the debut of NASA's Group 16, affectionately known as "The Sardines." This diverse group comprised accomplished pilots and mission specialists, including Duane G. Carey, Charles O. Hobaugh, and James M. Kelly. The group also featured mission specialists such as Laurel Clark and Michael J. Massimino. The inclusion of international mission specialists from Spain, Sweden, Italy, Canada, Japan, and other countries underscored NASA's dedication to fostering global collaboration in space exploration. The tragic loss of members such as David McDowell Brown, Laurel Clark, and William Cameron McCool during the final Space Shuttle Columbia mission in 2003 served as a stark reminder of the perils associated with space travel.

June saw Japan making strides in its space program with the selection of Soichi Noguchi by NASDA (National Space Development Agency of Japan). This selection marked Japan's continued advancement and increasing role in international space missions.

By October, China significantly expanded its astronaut program by forming Group 1996. This group included Li Qinglong and Wu Jie, who trained at the Yuri Gagarin Cosmonaut Training Center. This development was a milestone for China's emerging space program, reflecting its growing ambitions and contributions to global space exploration.

In November 1997, Ukraine established its presence in the international aerospace community by welcoming Leonid Kadeniuk and Yaroslav Pustovyi into the Shuttle-97 Group. This expansion was pivotal for Ukraine, highlighting its burgeoning role and commitment to space endeavors.

The year 1997 continued to be marked by noteworthy advancements. Israel introduced its inaugural Shuttle Group in April, comprising Yitzhak Mayo and Ilan Ramon. Ilan Ramon's journey as the first Israeli astronaut to space was historic. However, his mission as a Payload Specialist aboard Space Shuttle Columbia during STS-107 ended in tragedy when the shuttle disintegrated upon re-entry in 2003, underscoring the inherent risks of space exploration.

On July 28, 1997, Russia established the TsPK–12/RKKE-13 Cosmonaut Group, which included Dmitri Kondratyev, Yury Lonchakov, Sergey Volkov from TsPK, and Oleg Skripochka and Fyodor Yurchikhin from RKKE, among others. This new group emphasized Russia's continued dedication to maintaining a robust and capable cosmonaut corps, affirming its active and influential role in international space missions.

The year 1998 was a pivotal one for global space exploration, marked by a series of significant milestones and the strengthening of international collaboration.

In January, China made a notable advancement by announcing the formation of Group 1 of its Astronaut Corps. This group included Yang Liwei, who later achieved historic acclaim as the first Chinese astronaut to journey into space aboard Shenzhou 5 in October 2003. This milestone underscored China's entry into the realm of nations capable of independent manned spaceflight.

On February 24, Mikhail Korniyenko joined Russia's RKKE-14 Cosmonaut Group, further contributing to Russia's storied tradition of manned space missions. Just over a month later, on March 2, Slovakia made a historic leap by establishing the OS "Mir" Stefanik Group. This group included Ivan Bella and Michal Fulier, symbolizing Slovakia's emerging presence in the international space community.

The summer of 1998 brought a series of remarkable advancements. On June 4, NASA introduced Group 17, known as "The Penguins." This diverse group included pilots like Christopher Ferguson and mission specialists like Barbara Morgan, who had been selected as the backup "Teacher-In-Space" for Christa McAuliffe. Their inclusion, alongside international counterparts from France, Italy, Brazil, Germany, and Canada, highlighted NASA's commitment to fostering global collaboration in space exploration.

In October, the European Space Agency (ESA) announced its astronaut group for the year. This cohort included notable figures such as Frank De Winne from Belgium, Léopold Eyharts from France, and Paolo Nespoli from Italy, reinforcing ESA's role in advancing European space capabilities and engaging in multinational missions.

A blend of triumphs and challenges in space missions also marked the year. In February, NASA concluded its pioneering Lunar Prospector mission, launched in January 1998, to determine the presence of water ice on the Moon. The mission's deliberate crash into the Moon's surface ensured valuable data collection on lunar resources, essential for future exploration and potential habitation.

July brought a new challenge with the launch of the Mars Climate Orbiter. This ambitious mission aimed to investigate Martian climate and geology, but it failed due to a navigation error caused by a mix-up in measurement units. The spacecraft's entry into Mars 'atmosphere at a dangerously low altitude resulted in its destruction, underscoring the critical importance of precision in space exploration.

August saw the initiation of the Stardust mission, designed to collect samples from Comet Wild 2's coma and return them to Earth. This mission promised to deepen our understanding of cometary composition and the early solar system.

September marked a historic moment with the launch of the Russian Zarya module, marking the commencement of the International Space Station (ISS) project. Deployed into orbit by a Russian Proton rocket, the Zarya module was the cornerstone for what would evolve into a sprawling space laboratory and international research hub.

October witnessed the crew of the Space Shuttle Discovery undertaking the ISS's first assembly mission. They attached the US-built Unity module to the Russian Zarya module, laying the foundation for the ISS's future expansion and international cooperation.

However, November was marred by the loss of the Mars Climate Orbiter, serving as a stark reminder of the importance of meticulous calculations and effective communication among space agencies.

On December 4, the U.S. contributed to the ISS's construction with the launch of Unity Node 1 during the STS-88 mission. The Space Shuttle Endeavour ferried Unity, the first American-built module for the ISS, into space. Its docking with the already orbiting Zarya module was a significant milestone in the ISS's assembly, a moment captured by the STS-88 crew using a large-format IMAX camera.

In addition to lunar and Martian endeavors, 1998 saw advancements in spacecraft technology. The United States Air Force, NASA, and Boeing collaborated on the X-40A, an 80% scale prototype of the Space Maneuver Vehicle X-37, exploring maneuvering capabilities for future space vehicles. Meanwhile, NASA tested the Scaled Composites X-38, a prototype for a crew return vehicle featuring one of the largest parachutes of its time, a feat later surpassed by the MegaFly parafoil capable of delivering substantial freight.

Several notable missions marked human spaceflight in 1998. On January 22, the Endeavour Space Shuttle launched the STS-89 mission, which involved rotating and resupplying the Mir space station. The crew included Terrence W. Wilcutt, Joe F. Edwards, Bonnie J. Dunbar, Michael P. Anderson, James F. Reilly, and Salizhan Sharipov.

On June 12, Andrew S. Thomas flew aboard the Discovery Space Shuttle for the STS-91 mission, significant for its scientific experiments and as a precursor to the shuttle's final docking with Mir. That year, Léopold Eyharts participated in the Soyuz TM-27 mission on January 29, followed by another mission, Soyuz TM-26, on February 19, focusing on crew rotation aboard Mir.

The STS-90 mission, launched on April 17 aboard the Columbia Space Shuttle, featured a crew including Richard A. Searfoss, Scott D. Altman, Richard M. Linnehan, Dafydd R. Williams, Kathryn P. Hire, Jay C. Buckley, and James A. Pawelzyk. This mission, dedicated to neuroscience research, was the final Spacelab mission. On June 2, the Discovery Space Shuttle embarked on the STS-91 mission, marking the final Shuttle-Mir docking, Mir crew rotation, and resupply, with astronauts Charles J. Precourt, Dominic L. Pudwill Gorie, Wendy B. Lawrence, Franklin R. Chang-Diaz, Janet L. Kavandi, and Valery Ryumin onboard.

In August, Yuri Baturin participated in the Soyuz TM-28 mission, with Gennady Padalka joining the crew in February 1999. The Soyuz TM-27 mission, launched on August 25, 1998, also included Sergei Avdeyev, who later participated in the Soyuz TM-29 mission on August 28, 1999.

The year concluded with the STS-95 mission, which began on October 29 aboard the Discovery Space Shuttle. This mission included Curtis L. Brown, Steven W. Lindsey, Scott E. Parazynski, Stephen K. Robinson, Pedro Duque, Chiaki Mukai, and John H. Glenn, who, at 77, became the oldest person to return to space. This mission was notable for its scientific experiments and for marking Duque as the first Spaniard in space.

On December 4, the Endeavour Space Shuttle launched the STS-88 mission, the first construction mission for the ISS. The crew, including Robert D. Cabana, Frederick W. Sturckow, Nancy J. Sherlock Currie, Jerry L. Ross, James H. Newman, and Sergei Krikalev, delivered the Unity Node Module, a crucial component of the ISS.

Throughout the year, astronauts and cosmonauts made significant contributions. David A. Wolf, an American astronaut, spent 128 days aboard Mir as part of the STS-89 mission, conducting experiments and maintenance. Andrew Thomas, an Australian-born American astronaut, spent 141 days on Mir with the NASA-5 mission, contributing to vital experiments and maintenance. Sergey Avdeyev and Valery Korzun, both Russian cosmonauts, completed extensive missions on Mir, conducting critical experiments and maintenance tasks. Jean-François Clervoy, a French astronaut, contributed to the STS-91 mission, marking the final Space Shuttle mission to Mir. Michael Foale, a British-American astronaut, played a significant role in the NASA-5 mission to Mir and later became the first US astronaut to fly aboard the ISS in 2003.

On May 29, 1999, the Space Shuttle Discovery, on STS-96, became the first shuttle to dock with the ISS. The mission's primary objectives were to deliver supplies and perform the station's first docking operation. Commanded by Kent V. Rominger, with Pilot Rick D. Husband and Mission Specialists Nancy J. Currie, Daniel Tani, and Andreas Mogensen, Discovery fulfilled its goals and undocked on June 3, 1999.

In 1999, the global momentum in space exploration continued to gain strength, characterized by notable astronaut selections and international collaborations. Japan, a key player in this expanding frontier, made significant strides in forming its 1999 astronaut group under NASDA (National Space Development Agency of Japan). In February, Japan welcomed three new astronauts: Satoshi Furukawa, Akihiko Hoshide, and Naoko Sumino. This new cohort was emblematic of Japan's ongoing dedication to advancing its space program and contributing to international space missions. Their selection underscored Japan's growing role in the global space community and its commitment to scientific and exploratory endeavors beyond Earth's atmosphere.

The European Space Agency (ESA) also marked a significant development on November 1, 1999, by expanding its astronaut corps. This expansion incorporated Claudie André-Deshays, Philippe Perrin, and Michel Tognini into ESA's ranks from the French space agency CNES (Centre National d'Études Spatiales). Including these astronauts was a pivotal step in enhancing Europe's space exploration and research capabilities. The move reflected ESA's ongoing efforts to strengthen its position in international space missions and collaborative projects.

As the new millennium approached, NASA announced its Group 18 astronauts in July 2000, a diverse team affectionately known as "The Bugs." This distinguished group included pilots such as Dominic A. Antonelli, Eric A. Boe, Kevin A. Ford, Ronald J. Garan Jr., Douglas G. Hurley, Terry W. Virts Jr., and Barry E. Wilmore, all bringing extensive flight experience and technical expertise. Complementing the pilots were mission specialists like Michael R. Barratt, Robert L. Behnken, Stephen M. Bowen, B. Alvin Drew, Andrew J. Feustel, Michael T. Good, Timothy L. Kopra, K. Megan McArthur, Karen L. Nyberg, and Nicole P. Stott. With its blend of scientific and operational expertise, this team epitomized NASA's rigorous standards and unwavering commitment to advancing human space exploration.

The late 1990s and early 2000s also witnessed notable progress from the Soviet Union, which had by then become Russia. The Soviet space program, originally focused on developing the N-1 rocket for a large space station and lunar landing missions, faced significant setbacks due to repeated failures of N-1 test launches in 1969. This prompted a strategic shift towards developing smaller, more manageable space stations. These stations could be launched using more reliable rockets, reflecting a pragmatic approach to achieving space exploration goals.

Further advancements across the global space community marked the year 2003. Russia's TsPK-13/RKKE-15/IMBP-6 Cosmonaut Group saw the selection of several new members: Anatoli Ivanishin, Aleksandr Samokutyayev, Anton Shkaplerov, Evgeny Tarelkin, and Sergei Zhukov from TsPK; Oleg Artemyev, Andrei Borisenko, and Mark Serov from RKKE; and Sergey Ryazansky from IMBP. This diverse cohort represented Russia's continued investment in its space program, with members skilled in piloting, science, and engineering—essential for future space missions.

Kazakhstan also made its mark by forming Group 1, which included Aydyn Aimbetov and Mukhtar Aymakhanov. Their inclusion highlighted Kazakhstan's growing role in international space exploration efforts and demonstrated a commitment to global collaboration.

In the United States, September 11th, 2003, was a landmark date as it marked the selection of the first commercial astronauts for SpaceShipOne. Brian Binnie and Mike Melvill achieved historic spaceflights aboard SpaceShipOne, underscoring the advances in private-sector aerospace technology and exploration ambitions. Doug Shane and Peter Siebold, also part of the group, contributed to the evolution of commercial spaceflight. Siebold later piloting SpaceShipTwo in subsequent missions, thus continuing the trajectory of private sector involvement in space exploration.

Working and Living in Space

The story about the International Space Station isn't limited to the Expedition crews. It took many missions involving many astronauts and cosmonauts to build and supply the ISS. The included:

The Mir Space Station, a cornerstone of Soviet and later Russian space endeavors, marked a significant chapter in the history of human space exploration. From its inception in 1986 to its deorbit in 2001, Mir hosted a series of pivotal missions that expanded human presence in low Earth orbit and fostered international cooperation, laying the groundwork for future space endeavors.

The journey of Mir began with the launch of Soyuz TM-1 in December 1986, which carried the first crew to the station. This mission was a milestone, heralding the era of long-duration space habitation. The subsequent mission, Soyuz TM-2 in 1987, not only brought additional crew members but also featured the station's first spacewalk, marking a significant advancement in its operational capabilities.

In the same year, Soyuz TM-3 carried Helen Sharman, the first British astronaut to visit Mir. Her mission was a landmark event in the station's history, highlighting the growing international interest in space exploration. Soyuz TM-4, launched in 1988, completed its mission by rotating the crew and introducing new members, while Soyuz TM-5, also in 1988, extended the station's operational period, continuing the tradition of long-duration missions.

The year 1989 saw two noteworthy missions: Soyuz TM-6, which included French astronaut Jean-Loup Chrétien, and Soyuz TM-7, which increased the station's crew size and underscored Mir's role as a collaborative international research platform. In 1990, Soyuz TM-8 and Soyuz TM-9 further rotated the crew and ensured operational continuity, with Soyuz TM-9 featuring German astronaut Ulf Merbold, adding to the station's international representation.

The early 1990s continued to witness significant international collaboration. Soyuz TM-10, launched in 1991, brought a new crew to Mir, while Soyuz TM-11, in the same year, was notable for including Norman Thagard, the first American astronaut to visit Mir. This mission exemplified the expanding international cooperation in space exploration.

In 1992, Soyuz TM-12 hosted international crew members, and Soyuz TM-13 in 1993 introduced additional crew and supported ongoing experiments aboard Mir. Soyuz TM-14 in 1994 brought astronaut Aleksandr Kaleri to the station, and Soyuz TM-15 continued the trend of long-duration missions, further enhancing Mir's research capabilities.

The mid-1990s saw further enhancements in international collaboration with Soyuz TM-16, which included American astronaut William Shepherd, and Soyuz TM-17, continuing collaborative efforts. Soyuz TM-18 and Soyuz TM-19 in 1996 brought new crew members and featured French astronaut Jean-Pierre Haigneré, emphasizing the continued international interest in Mir.

By 1997, Soyuz TM-20 and Soyuz TM-21 solidified Mir's role in long-duration space research, with ongoing operations and crew changes. Soyuz TM-22, launched in 1998, included German astronaut Reinhold Ewald. The final Soyuz mission to Mir, Soyuz TM-23 in 1999, marked the end of an era of continuous human presence on the station, preceding its deorbit.

Parallel to these Soyuz missions, NASA's Space Shuttle program played a crucial role in Mir's history. The first shuttle mission to dock with Mir, STS-71 in June 1995, marked the beginning of a series of joint operations between the United States and Russia. Subsequent missions, including STS-74 and STS-76 in 1995 and 1996, continued to foster this collaboration. STS-79 in 1996 was significant for delivering a new module to Mir, enhancing its research capabilities.

The late 1990s were marked by missions that solidified Mir's operational and collaborative achievements. STS-81 and STS-82 in 1997 brought new crew members, supplies, and conducted essential maintenance work. STS-84 and STS-86 further continued operations with Mir, including crew changes and module deliveries.

The final chapter of the Space Shuttle-Mir partnership was written with STS-91, launched on June 2, 1998. This mission was the eighth and final shuttle flight to Mir, concluding a significant era of international cooperation and laying the groundwork for future space exploration endeavors.

Throughout its operational period, Mir was a symbol of Soviet and Russian space achievements and a testament to the spirit of international collaboration in space. Over approximately 104 Russian cosmonauts visited the station, their missions ranging from scientific research to fostering international cooperation. The station also hosted 28 astronauts from the United States, whose presence was a significant part of the U.S.-Russian collaboration in space.

Seven astronauts from France, Germany, and the United Kingdom participated in European research aboard Mir, adding valuable international perspectives and expertise. Japan's contribution came from Koichi Wakata, who visited the station in 1996 as part of Japan's early efforts in long-term human spaceflight.

With the transition from Mir to the International Space Station (ISS), the international community's involvement in space exploration expanded even further. Since its inception in 1998, the ISS has become a hub for global collaboration, continuing the legacy of Mir and paving the way for future endeavors in space exploration.

First and foremost were the challenges of working in a microgravity environment.

The study of microgravity originates in the early days of space exploration, a period marked by intense curiosity and a drive to understand the effects of gravity—or, more precisely, the absence of it. During the Mercury, Gemini, and Apollo programs, astronauts experienced what was known as weightlessness, a state resulting from the spacecraft's constant freefall toward Earth while traveling at high velocities. This unique condition was due to the spacecraft's orbital trajectory, where it was perpetually falling toward Earth but continuously moving forward at such a speed that it never actually collided with the planet. The intricate balance of gravitational pull and orbital motion created an environment where both the spacecraft and its occupants experienced continuous freefall, effectively counteracting the effects of gravity.

As space exploration progressed, so did our understanding of microgravity. The International Space Station (ISS), orbiting approximately 400 kilometers above Earth, operated in a region where gravity was about 90% of its strength compared to the surface. Despite this significant gravitational force, the ISS and everything within it remain in a constant state of freefall. This phenomenon was due to the station's high orbital velocity, which ensures it continually falls towards Earth but maintains a stable orbit. This delicate orbital dance results in the sensation of weightlessness experienced by astronauts and objects aboard the station.

However, the microgravity environment aboard the ISS was not perfectly uniform. Several factors contributed to slight variations in the sensation of weightlessness. The thin residual atmosphere at this altitude introduced minor drag on the station, causing small perturbations to its otherwise stable orbit. Additionally, the ISS's mechanical systems and crew activities generated vibrations and oscillations that could affect the microgravity environment. For instance, the attitude control moment gyroscopes, essential for maintaining the station's orientation, and thruster firings used to adjust its orbit or attitude could induce slight deviations from a perfect state of weightlessness.

Gravity-gradient effects, or tidal forces, also influenced the variations experienced aboard the ISS. These effects arose from the differences in gravitational pull experienced at different points on the station due to its size and shape. Although these variations were minimal, they could lead to minor deviations in the orbits of unsecured objects, affecting their apparent weightlessness. Despite these perturbations, the ISS's structural integrity and the meticulous design of its systems ensure that it maintains overall rigidity and stability.

The ISS's near-weightlessness has enabled groundbreaking scientific research across various fields. Microgravity conditions allow scientists to investigate how reduced gravity influences biological processes, such as the development and growth of plants and animals. NASA has focused on understanding the impact of microgravity on the formation of three-dimensional human-like tissues and the growth of protein crystals, which often exhibit unique properties in space. These studies are crucial for advancing our knowledge of fundamental biological processes and hold promise for significant improvements in medical treatments and other applications.

Early space missions' exploration of gravity effects laid the foundation for our current understanding of microgravity. The ISS, functioning as a continuous laboratory in orbit, provides a unique environment for studying the complex effects of weightlessness. This ongoing research enhances our scientific knowledge and drives innovations across numerous disciplines, highlighting the ISS's invaluable contribution to space exploration and research.

The International Space Station (ISS) orbits within Earth's magnetosphere, which provides partial protection from the harsh environment of space. Earth's magnetic field acts as a shield, deflecting a significant portion of the solar wind and reducing the impact of space radiation. However, astronauts aboard the ISS are not entirely shielded from solar radiation. For example, in 2005, a significant "proton storm" associated with an X-3 class solar flare prompted the crew of Expedition 10 to take precautionary measures. They sought refuge in a specially shielded section of the Russian Orbital Segment (ROS), designed to minimize radiation exposure during such intense solar events.

Despite the protective influence of the magnetosphere, cosmic rays—primarily high-energy protons—present a continuous radiation risk. These particles travel through space at high velocities and penetrate the ISS without Earth's atmospheric protection. As a result, astronauts on the ISS experience radiation levels about five times higher than those encountered by airline passengers. The average daily exposure aboard the ISS was approximately one millisievert, equivalent to the amount of natural radiation a person would receive in a year on Earth.

This elevated radiation exposure increases the risk of developing health issues such as cancer and could lead to a higher incidence of cataracts among astronauts. To address these risks, the ISS employs protective shielding and medical interventions. Nevertheless, ongoing research into space radiation and its effects remains crucial for safeguarding astronaut health and ensuring the safety of future space missions. This continuous study was vital as we advanced our understanding of space environments and worked to mitigate the potential impacts on human health.

The impact of prolonged spaceflight on human health has been notably underscored by NASA's Astronaut Twin Study, released on April 12, 2019. This groundbreaking study compared astronaut Scott Kelly, who spent an entire year aboard the International Space Station (ISS), with his twin brother, Mark Kelly, who remained on Earth. The study revealed several profound and enduring changes resulting from the space environment, including alterations in DNA and cognitive functions. These findings highlight the significant effects of extended space travel on the human body, illustrating how long-term exposure to the unique conditions of space could induce measurable biological changes.

Further research in November 2019 focused on the cardiovascular challenges faced by astronauts. A six-month study involving eleven healthy astronauts uncovered notable blood flow and clot formation issues. These findings are particularly critical for future long-duration missions, such as those planned for Mars. They underscore the need for enhanced countermeasures to manage these health risks effectively and ensure the safety and well-being of astronauts on extended space journeys.

Psychosocial stressors also represent a significant challenge aboard the ISS, affecting both crew morale and performance. Historical accounts, such as those from cosmonaut Valery Ryumin on the Salyut 6 space station, reveal the intense psychological strain confined living conditions impose. NASA has long acknowledged these stressors, which include isolation from family and the pressures of public visibility.

The psychological stress was particularly pronounced during the initial phases of long-duration missions as astronauts adapted to the novel and extreme space environment. The confined quarters and the necessity of working closely with international crew members from diverse cultural backgrounds further exacerbate psychological strain. Communication barriers and the need for multilingual proficiency add complexity to maintaining effective interpersonal relationships.

Living in microgravity presents additional challenges, such as orientation and spatial awareness. Astronauts often struggle with misjudging distances and making errors during critical operations due to disorientation, highlighting the importance of adaptive strategies and robust psychological support to manage these issues effectively.

The physical effects of prolonged weightlessness on the human body are well-documented. Astronauts experience muscle atrophy, bone density loss (osteopenia), and fluid redistribution, all of which impact cardiovascular health. These physiological changes include reduced red blood cell production, potential balance disorders, facial puffiness, and general body mass loss.

To counteract these effects, the ISS has exercise facilities designed to maintain physical health. These facilities include two treadmills with vibration isolation and stabilization systems (TVIS), the Advanced Resistive Exercise Device (ARED), and a stationary bicycle. Astronauts must exercise for at least two hours daily to combat muscle and bone loss. Bungee cords are used to secure astronauts to the treadmill, simulating weight-bearing activities and mitigating the physical challenges posed by the microgravity environment. Through these measures, the ISS aims to support astronauts in maintaining their health and performance during extended space missions.

The International Space Station (ISS) provides a unique environment for studying fluid physics due to its microgravity conditions. On Earth, gravity significantly influences the behavior of fluids, driving them to settle into predictable patterns. This force creates familiar phenomena such as the separation of liquids based on density and the pronounced effects of surface tension. In stark contrast, the near-absence of gravity on the ISS allows fluids to exhibit behaviors that are not observable on our planet.

In the microgravity environment of the ISS, fluids could demonstrate remarkable properties and engage in complex interactions. Liquids that might separate or resist mixing under Earth's gravitational pull could blend seamlessly in space. This absence of gravitational forces that typically cause density-driven segregation or alter surface tension effects enables scientists to study fluid dynamics with a previously unattainable clarity level. Researchers have observed phenomena such as the formation of uniform mixtures and the behavior of liquids in confined spaces, providing new insights into fluid behavior and interactions.

Furthermore, the ISS offers an ideal platform for examining chemical reactions under reduced gravity and temperature conditions. These slower reaction rates allow scientists to more precisely explore the fundamental principles governing these reactions. The station has been instrumental in investigating material crystallization and the formation of complex compounds, offering valuable insights into mechanisms often obscured by rapid reactions on Earth.

Superconductivity was a notable area of research facilitated by the ISS's microgravity environment. Superconductors, which exhibit zero electrical resistance and the expulsion of magnetic fields when cooled below a critical temperature, could be studied without the interference of gravitational effects. This research holds the potential for significant technological advancements, including more efficient power transmission and developing advanced magnetic systems.

In summary, the ISS's microgravity environment has opened new avenues for research in fluid physics. This research enhances our understanding of fundamental physical principles by enabling scientists to observe and analyze fluid behaviors and chemical reactions in ways impossible on Earth. It paves the way for revolutionary technological advancements.

Materials science was another field significantly advanced by research on the ISS. The station's unique conditions allow scientists to investigate the behaviors of materials that are challenging to study on Earth. This research could potentially improve manufacturing techniques and lead to economic benefits through enhanced processes and new material applications.

One area of focus was the effect of low gravity on combustion processes, including burning efficiency and emission control. Understanding how materials behave under these conditions could advance our energy production knowledge, offering economic and environmental benefits.

The ISS also faces unique challenges related to microbiological environmental hazards, which could affect both the station's structural integrity and crew health. The controlled and confined environment of the ISS could foster the growth of harmful molds and bacteria, which, if left unchecked, could compromise air and water filtration systems, produce corrosive acids, and pose health risks to astronauts.

To address these issues, NASA has developed the LOCAD-PTS (Lab-on-a-Chip Application Development-Portable Test System), a sophisticated device capable of rapidly identifying common bacteria and molds. This system allows for real-time monitoring and prompt intervention, a significant improvement over traditional methods that require sending samples back to Earth.

In 2018, researchers detected five strains of Enterobacter bugandensis on the ISS. While these strains are not pathogenic to humans, their presence highlights the importance of rigorous environmental monitoring. Preventative measures on the ISS include maintaining low humidity, applying mold-resistant coatings, and using antiseptic solutions. Additionally, all materials aboard the ISS undergo stringent testing for resistance to fungal growth.

Since 2016, the European Space Agency (ESA) has sponsored experiments to assess the antibacterial properties of various materials. The Microbial Aerosol Tethering on Innovative Surfaces (MATISS) program involves deploying glass plaques with different coatings on the ISS for six months, which are then analyzed upon their return to Earth. The most recent experiment, launched on June 5, 2023, as part of the SpaceX CRS-28 mission, included plaques with quartz glass for enhanced spectrographic analysis. This research aims to refine our understanding of how different materials could mitigate bacterial growth in space.

In April 2019, NASA completed a comprehensive 14-month study on ISS microorganisms and fungi. This study involved sampling from eight predefined locations within the ISS and utilizing both culturing and molecular methods to catalog microorganisms. The findings contribute to improved health and safety protocols for astronauts and inform similar closed environments on Earth.

Noise aboard the ISS was a significant concern, given that the station was not inherently silent. Astronauts are exposed to noise levels that often exceed acoustic standards set for terrestrial environments. Recognizing this, NASA and its international partners have prioritized noise control to safeguard crew health and performance.

The ISS Multilateral Medical Operations Panel (MMOP) Acoustics Subgroup, comprising acoustical engineers, audiologists, industrial hygienists, and physicians from NASA, Roscosmos, ESA, JAXA, and CSA, has been pivotal in developing and implementing noise control measures. Although noise levels rarely exceed 85 decibels A-weighted (dBA), continuous exposure during six-month missions could impact sleep, communication, and overall crew well-being.

Over nearly two decades, the Acoustics Subgroup has worked to manage and reduce noise. Strategies include setting acoustic limits, selecting quieter payloads, conducting pre-launch acoustic tests, and assessing in-flight noise levels. With the ISS's expansion and the addition of new modules and spacecraft, the subgroup has employed various noise control measures, such as acoustic covers, absorptive materials, noise barriers, and vibration isolators. Aging equipment has been replaced with quieter alternatives to minimize ambient noise.

Damage risk criteria from organizations like the National Institute for Occupational Safety and Health and the World Health Organization guide NASA's noise management approach. The Noise Exposure Estimation Tool (NEET) assesses noise exposure and determines hearing protection needs. The MMOP Acoustics Subgroup monitors noise levels, applies engineering controls, and recommends hearing protection as necessary. Notably, there have been no persistent hearing threshold shifts among ISS crew members, reflecting the effectiveness of these measures. In 2020, the Acoustics Subgroup received the Safe-In-Sound Award for Innovation for its efforts to mitigate noise-related health effects.

Fire and toxic gas hazards present critical safety concerns aboard the ISS. The station's external radiators use ammonia to manage thermal conditions, which could pose severe risks if leaked into the pressurized modules. The ISS employs stringent safety protocols and conducts regular inspections to mitigate these dangers, ensuring the station's continued safety and operational integrity.

The ISS maintains a stable presence in low Earth orbit, traveling along a nearly circular path between 370 and 460 kilometers above Earth's surface. This orbit places the ISS within the thermosphere and inclines at an angle of 51.6 degrees to the equator. This specific inclination facilitates direct access by Russian Soyuz and

Progress spacecraft, minimizing overflights over populated areas and avoiding the drop of spent rocket stages onto land. With an orbital eccentricity of only 0.007, the ISS orbits Earth at a velocity of approximately 28,000 kilometers per hour, completing about 15.5 orbits per day, each in approximately 93 minutes.

The ISS's altitude has been adjusted to meet varying logistical needs. During NASA's Space Shuttle era, the station's altitude was deliberately lowered to facilitate the transfer of heavier cargo. Following the Shuttle's retirement, the orbit was increased to approximately 400 kilometers to accommodate the capabilities of newer supply spacecraft, which do not require altitude adjustments due to their enhanced performance.

The ISS's altitude gradually decreases due to atmospheric drag at an average rate of about 2 kilometers per month. To counteract this decline, orbital boosting maneuvers are conducted using the main engines of the Zvezda service module or Russian or European spacecraft docked to the module. The Automated Transfer Vehicle could also contribute to these adjustments when equipped with a secondary docking port. Boosting the ISS to a higher orbit typically takes about three hours and requires approximately 7.5 tonnes of chemical fuel annually, with an estimated cost of $210 million.

The Russian Orbital Segment's Data Management System (DMS) manages the ISS's guidance, navigation, and control systems, which coordinate the station's position and trajectory using redundant sensors and fault-tolerant computers. Initially, control was provided by the Zarya module, but this role has been taken over by the Zvezda service module, which houses the DMS-R system. This system employs Earth horizon sensors, solar horizon sensors, and sun and star trackers to ensure precise positioning and orbital adjustments. The fault-tolerant computers, each containing three parallel processing units, provide advanced error correction and redundancy.

Orientation and attitude control are maintained through a combination of gyroscopes and thrusters. Gyroscopes manage momentum using electrical power rather than propellant and are supplemented by thrusters when necessary. In February 2005, a malfunction caused an erroneous command that consumed approximately 14 kilograms of propellant before resolution. Although rare, communication failures between the Russian and American segments' attitude control systems could lead to conflicts, necessitating the Russian system to override the American counterpart due to the absence of thrusters on the U.S. segment.

Space debris represents a substantial hazard to the ISS. The station encounters diverse debris, including defunct satellites, spent rocket stages, and fragments from anti-satellite tests. While larger pieces of debris are tracked precisely, smaller particles, numbering in the trillions, pose a significant threat due to their high velocity and kinetic energy. The ISS was equipped with ballistic panels or micrometeorite shields to mitigate collision risks. The U.S. segment uses a composite of aluminum, Kevlar, and stainless steel layers, while the Russian segment employs carbon fiber-reinforced polymers and aluminum honeycomb structures.

When tracked debris poses a risk, the ISS could execute Debris Avoidance Maneuvers (DAMs) to alter its trajectory and avoid potential collisions. These maneuvers involve propellant-driven adjustments to the station's orbit and are carried out when debris was projected to pass within 1.6 kilometers of the ISS. Over the station's operational history, only a few avoidance maneuvers have been necessary, reflecting the effectiveness of tracking and mitigation strategies.

2000 witnessed several significant space flights, marking important milestones in space exploration and international cooperation.

NASA launched the Mars Odyssey spacecraft in April. It aims to study the Martian surface and atmosphere, focusing on water distribution and valuable resources. After entering Mars's orbit in October, it began transmitting crucial data back to Earth.

Another notable event occurred in October with the Deep Space 1 spacecraft launch as part of NASA's New Millennium Program. This mission tested new spacecraft technologies, including an ion engine and a solar concentrator array, pushing the boundaries of spacecraft technology.

August saw the final satellite in the US Air Force's GPS Block II series launched. GPS 2R-11 joined the existing GPS constellation in orbit, enhancing global positioning and navigation capabilities.

During this time, the Russian Federation's space agency, Roscosmos, faced financial challenges and sought new revenue-generating avenues. This led to increased involvement in space tourism and commercial satellite launches. Roscosmos struck a contract with NASA, selling seats on Soyuz spacecraft to NASA for approximately $21 million per person each way. Progress transport flights were also provided for $50 million per Progress vehicle. As part of this agreement, Roscosmos announced an increase in crewed Soyuz flights to four per year and Progress flights to eight per year, commencing in 2008.

In parallel, the Russian anti-satellite program resumed after a ten-year hiatus. Ballistic missiles were converted and housed in well-protected silos, accompanied by maneuverable satellites capable of directing missiles towards enemy satellites. In February, Russia achieved a significant milestone by flying the Soyuz-Fregat, a commercial variant of the Soyuz rocket. The Soyuz-Fregat featured a primary propulsion system capable of multiple restarts, utilizing nitrogen tetroxide and unsymmetrical dimethyl hydrazine as propellants.

In March, Deputy Minister of Defense Vladimir Popovkin announced Russia's retention of the 14F11, an orbital space tug developed by the Russian Federation. The 14F11 was a highly maneuverable rocket stage capable of multiple firings and served as a multiple-missile launch platform. The 14F11, also known as NARYAD-V, was developed at a highly classified KB Tochmash design bureau. The civilian version of 14F11 known as the Briz-K (Breeze) was used for maneuvering and delivering payloads in space.

In November 2000, Expedition 1 began on the International Space Station, marking the first long-duration stay with a three-person crew, one American and two cosmonauts from Roscosmos. This event followed a historical era in space exploration, including launching the first satellite, Sputnik, in 1957 and Yuri Gagarin's successful flight in the Vostok 1 rocket in 1961, making him the first human to reach outer space. These events showcased the competition and subsequent cooperation between the United States and the Soviet Union in space exploration. 1963 President Kennedy proposed a joint US-Soviet lunar expedition during a speech before the United Nations General Assembly.

Before Expedition 1, talks between NASA and the Soviet Academy of Sciences led to agreements on data-sharing in space medicine and satellite meteorological data. The Mir-2, based on the DOS-8 project, became one of the ISS's first modules. The Zvezda Service Module, derived from the Salyut program, formed the core of the early ISS, and the Zarya module derived from Almaz Functional Cargo Block designs.

Leading up to Expedition 1, which commenced shortly before the 2000 U.S. presidential election, there were expectations of a station resembling the Mir space station, largely due to the similarities in their designs and components. However, the ISS was designed to be a more advanced and international collaboration.

At the time of Expedition 1, it was anticipated that the ISS would be fully operational by 2006 and remain continuously inhabited until 2015. This ambitious timeline faced several setbacks, including the tragic Space Shuttle Columbia disaster, which contributed to delays. As a result, the ISS was not completed until 2021 with the arrival of the Nauka laboratory module, marking the end of a prolonged assembly phase and setting the stage for the station's future missions and scientific endeavors.

The ISS Progress 1 cargo spacecraft made its debut on August 6, 2000, marking the beginning of its mission to deliver supplies to the ISS. After a successful docking on August 8, 2000, ISS Progress 1 delivered critical equipment and materials necessary for the station's operation before undocking on October 31, 2000.

The five STS-97 mission astronauts, launched aboard the Space Shuttle Endeavour for a working visit to the International Space Station (ISS). Astronaut Brent W. Jett (front right) and Michael J. Bloomfield (front left) are commander and pilot, respectively. Flanked by those two was astronaut Marc Garneau, mission specialist representing the Canadian Space Agency (CSA). In the rear are astronauts Carlos I. Noriega (left) and Joseph R. Tanner, both mission specialists. Noriega and Tanner are wearing the extravehicular mobility unit (EMU) spacesuits they'll be wearing for spacewalking chores during the flight

The space shuttle STS-105 was a significant mission in the International Space Station's (ISS) history and vital in supporting ISS Expedition 3. Launched on August 10, 2001, aboard the Space Shuttle Discovery, it was a key component in the ISS's ongoing assembly and operational enhancement.

The primary objective of STS-105 was to deliver and install the Express Logistics Carrier-1 (ELC-1) and to rotate the ISS crew. The ELC-1 was a crucial piece of equipment designed to provide additional storage and support for scientific experiments and other station components. Its installation was essential for expanding the ISS's capabilities, allowing it to host a wider range of scientific instruments and facilitating the future growth of the station.

The mission's crew included Commander Kent V. Rominger, Pilot Jeffrey S. Ashby, and Mission Specialists Daniel W. Bursch, Carlos I. Noriega, and Peggy A. Whitson. Among the crew, Peggy Whitson made history as the first female commander of an ISS mission. Her leadership was instrumental in coordinating the various tasks required to achieve the mission's objectives.

During STS-105, the shuttle crew conducted a series of spacewalks to install the ELC-1. These spacewalks were complex operations that required precise maneuvers to attach the carrier to the station's external truss structure. The ELC-1 provided a platform for storing spare parts and scientific experiments, vital for the station's ongoing research and maintenance needs.

In addition to installing ELC-1, STS-105 was responsible for rotating the crew aboard the ISS.

On October 11, 2000, the Space Shuttle Discovery embarked on the STS-92 mission, marking a significant milestone in the International Space Station (ISS) assembly. This mission, which lasted 12 days, 21 hours, 40 minutes, and 25 seconds, was crucial in expanding the ISS's capabilities. It was the 100th shuttle flight and the second dedicated to adding essential hardware to the ISS.

The mission's primary objectives were installing the Zenith Z1 Truss and the third Pressurized Mating Adapter (PMA-3). These components were vital for the ISS's future expansion and functionality. The Z1 Truss would be a central structural component to support future solar arrays and other equipment, while PMA-3 was intended as a docking port for future shuttle missions.

Discovery and her crew docked with the ISS following the launch on the mission's second day. This critical maneuver set the stage for six days of extensive construction and outfitting activities. On the mission's third day, Japanese astronaut Koichi Wakata skillfully operated the shuttle's robotic arm to extract the Zenith Z1 Truss from Discovery's payload bay and position it onto the Unity module of the ISS. Pilot Pam Melroy and astronaut Jeff

Wisoff worked diligently inside the Unity module to secure the truss and establish essential grounding connections.

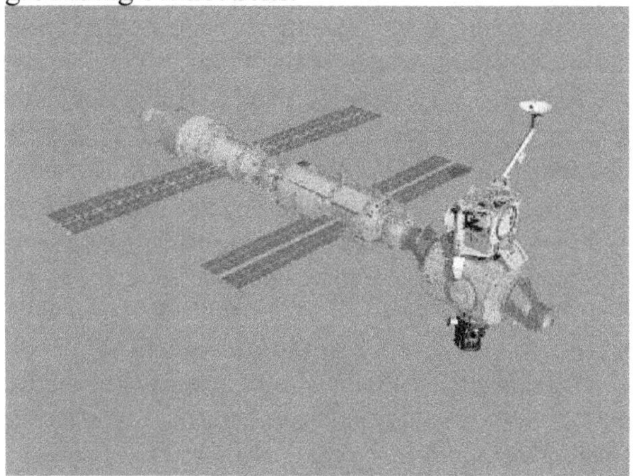

Arriving aboard Space Shuttle Discovery, the STS-92 crew installed the Z1 truss, a third pressurized mating adapter and a Ku-band antenna

The Unity module, also known as Node 1, represented a pivotal component of the International Space Station (ISS) as the first U.S.-built segment of the station. This cylindrical module was a critical link between the Russian and American segments of the ISS, facilitating international collaboration and operations within the station. It was the central hub where astronauts from various space agencies come together during their meal times, fostering a unique environment of unity and shared purpose in space.

Unity's design features a cylindrical structure measuring 4.57 meters (15.0 feet) in diameter and extending 5.47 meters (17.9 feet) in length. This module was constructed from steel and meticulously crafted by Boeing at the Marshall Space Flight Center in Huntsville, Alabama. The module's architecture includes six berthing locations positioned in various orientations— forward, aft, port, starboard, zenith, and nadir— allowing it to connect with other modules and expand the station's capability and functionality.

As the inaugural member of a trio of connecting modules, Unity paved the way for the subsequent installation of Harmony and Tranquility. These modules collectively enhance the ISS's capacity for scientific research, habitation, and international cooperation, reinforcing the station's role as a symbol of global unity and technological achievement in space.

The Zvezda Service Module, also designated as Salyut DOS-8, represents a crucial component of the International Space Station (ISS) and serves as the heart of the Russian Orbital Segment (ROS)."Zvezda," meaning "star" in Russian, aptly reflects its significant role within the ISS.

Launched on July 12, 2000, aboard a Proton rocket, Zvezda was the third module to join the station's assembly. It docked with the Zarya module on July 26, 2000, marking a key milestone in the ISS's development. This module, manufactured by RKK Energia with substantial subcontracting support from GKNPTs Khrunichev, was designed to provide essential life support systems for the station. While some of these systems are augmented by modules from the U.S. Orbital Segment (USOS), Zvezda remains fundamental to the station's operational capabilities.

The Zvezda module also functions as the structural and functional core of the Russian segment, offering vital living quarters for two crew members. The crew assembled here to manage and respond to emergencies, ensuring the smooth operation and safety of the ISS.

The crew conducted four extravehicular activities (EVAs) throughout the mission, each crucial for installing and preparing new components. During EVA #1, which lasted 6 hours and 28 minutes, the astronauts connected electrical umbilicals to power heaters and conduits on the Z1 Truss, relocated communication antennas, and installed a toolbox for future use. EVA #2, extending 7 hours and 7 minutes, involved the attachment of PMA-3 to the ISS and the preparation of the Z1 Truss for the solar arrays scheduled to arrive on a subsequent mission. EVA #3, lasting 6 hours and 48 minutes, saw the installation of two DC-to-DC converter units atop the Z1 Truss to ensure the proper voltage conversion for the station's solar arrays. Finally, EVA #4, lasting 6 hours and 56 minutes, focused on testing the manual berthing mechanism, deploying a power tray for the U.S. Lab, and removing a grapple fixture from the truss. The team also tested two small rescue backpacks designed to assist astronauts in case of accidental drift away from the spacecraft.

On the ninth day of the mission, the crew shifted focus to the ISS's interior, completing connections for the newly installed Z1 Truss and transferring equipment and supplies in preparation for the station's first resident crew, which would arrive in November. The team also tested four control moment gyroscopes, essential for maintaining the ISS's orientation as it orbits Earth.

The configuration of the ISS at the start of Expedition 1. From top to bottom, the three modules are: Unity, Zarya and Zvezda.

The Space Shuttle Atlantis returned to the ISS on October 13, 2000, with mission STS-92. This mission was integral to the ISS's structural development, delivering and installing the Z1 Truss and the Pressurized Mating Adapter (PMA-3). Commanded by Brian Duffy, with Pilot Pamela Melroy, Mission Specialists Koichi Wakata, Leroy Chiao, Peter J. K. (Jack) Arnst, and Payload Specialist David Wolf, the shuttle completed its mission and undocked on October 20, 2000.

On October 31, 2000, at precisely 07:52 UT, the Soyuz TM-31 spacecraft lifted off from the Baikonur Cosmodrome in Kazakhstan, marking a historic milestone in human spaceflight. This mission was particularly significant as it was the first Soyuz flight to dock with the International Space Station (ISS), heralding the start of continuous human occupation aboard the ISS.

The Soyuz TM-31 mission was launched aboard a Soyuz-U rocket and carried the members of Expedition 1, the first crew to embark on a long-duration mission aboard the ISS. On October 31, 2000, at 07:52 UTC, the Soyuz TM-31 spacecraft launched from the Baikonur Cosmodrome in Kazakhstan, embarking on a historic mission as the first Soyuz flight to dock with the International Space Station (ISS).

STS-106 was a pivotal mission in the early history of the International Space Station (ISS), serving as a crucial step in preparing the station for its first long-duration crew. Launched on September 8, 2000, aboard the Space Shuttle Atlantis, STS-106 played a significant role in ISS Expedition 1, the station's inaugural long-duration mission.

The primary objectives of STS-106 were to complete the installation and setup of various components necessary for the ISS's operational readiness. The mission's crew included Commander Terrence W. Wilcutt, Pilot Scott D. Altman, and Mission Specialists Daniel T. Barry, Kristi Yamaguchi, and Edward T. Lu. Their tasks were critical for ensuring the station's infrastructure and systems were fully functional.

Upon arrival at the ISS on September 10, 2000, Atlantis carried essential supplies and equipment, including components for the station's life support systems and additional hardware. The crew's work focused on configuring and activating these systems , vital for sustaining long-term habitation. They also performed necessary maintenance and conducted tests to ensure all systems operated correctly.

During their stay, the Atlantis crew conducted a series of spacewalks to install and configure the station's external components. These tasks included setting up the station's exterior communications antennas and preparing the docking ports for future spacecraft. The successful completion of these tasks was essential for supporting the arrival of Expedition 1's crew and ensuring their safety and comfort.

STS-106 also played a crucial role in delivering and installing supplies that would support the first long-duration crew's mission. This included essential equipment, scientific experiments, and living supplies necessary for the ISS's continuous operation.

The mission concluded with Atlantis's departure from the ISS on September 20, 2000. The successful execution of STS-106's objectives was instrumental in setting the stage for Expedition 1, which began with the arrival of the first long-term ISS crew on November 2, 2000.

Expedition 1

ISS Expedition One Commander William M. (Bill) Shepherd (center) was flanked by Soyuz Commander Yuri P. Gidzenko (right) and Flight Engineer Sergei K. Krikalev (left)

On October 31, 2000, the Soyuz TM-31 spacecraft lifted off from the historic Gagarin's Start launch pad at the Baikonur Cosmodrome in Kazakhstan. This site, named in honor of Yuri Gagarin's groundbreaking flight in 1961, marked the beginning of a new chapter in space exploration.

After a two-day journey, Soyuz TM-31 docked with the Zvezda Service Module on November 2, 2000, at 09:21 UTC. The Zvezda Module, launched in July 2000, provided essential infrastructure for the ISS, making the docking a critical milestone. Ninety minutes after the successful docking, the crew entered the ISS for the first time, marking the official start of continuous human habitation in space.

The arrival of Expedition 1, as the crew was designated, was a landmark moment in the ISS's history. The crew included Commander William Shepherd, Flight Engineer Yuri Gidzenko, and Flight Engineer Sergei Krikalev. Their mission was pivotal, heralding the first long-term crewed presence aboard the International Space Station (ISS). Their mission, which spanned approximately 141 days, was instrumental in transitioning the ISS from a partially constructed structure to a fully operational space habitat. During their tenure, the crew undertook vital tasks, including conducting scientific experiments, activating and calibrating key systems, and performing routine maintenance. They also managed multiple supply deliveries, crucial for sustaining the ISS's operations and supporting ongoing research programs.

The Progress M1-4 supply ship linked up to the orbiting International Space Station (ISS) at 3:48 GMT, November 18, bringing Expedition 1 commander William M. Shepherd, pilot Yuri P. Gidzenko and flight engineer Sergei K. Krikalev two tons of food, clothing, hardware and holiday gifts from their families.

William Shepherd, representing NASA, was the first commander of the ISS and his fourth spaceflight. Despite his previous missions totaling only about two weeks, Shepherd's leadership was critical in integrating American and Russian technologies, a complex task given the ISS's multinational nature. His role was instrumental in guiding the station's initial setup and operational phases.

Yuri Gidzenko, on his second spaceflight, brought valuable experience from his previous 180-day mission aboard the Mir space station. His understanding of long-duration space habitation was vital for adapting to the ISS environment. Sergey Krikalev, on his fifth spaceflight, offered extensive expertise from his previous missions, including a notable tenure aboard Mir. His seasoned insights were crucial for managing ISS operations and maintenance complexities.

Russian Soyuz TM-31 moves to Launch Pad, 29 October 2000. The Soyuz TM-31 launch vehicle, which carried the first resident crew to the International Space Station, moves toward the launch pad at the Baikonur complex in Kazakhstan. The Russian Soyuz launch vehicle was an expendable spacecraft that evolved out of the original Class A (Sputnik).

As the ISS began its new era of continuous human presence, discussions emerged about its designation. Drawing from naval traditions where naming vessels brings good fortune, Commander Shepherd suggested adopting the radio call sign "Alpha" for the station. This choice aimed to honor the station's pioneering spirit with a name that was both concise and symbolic. However, this proposal faced opposition from Yuriy Semenov, then-President of the Russian Space Corporation Energia. Semenov favored alternatives like "Beta" or "Mir 2," believing these names would better reflect the legacy of the Mir space station. This precursor had laid the groundwork for future space stations.

During their early days on the station, the crew focused on vital tasks such as activating life support systems, establishing computer controls, and unpacking supplies left by preceding missions. At this time, the ISS lacked the electrical power to heat all three pressurized modules, leaving the Unity module unheated and unused. Although Unity had been operational for two years, serving as a command and data interface for U.S. flight controllers, it remained inoperative until additional power and systems were fully integrated. As Expedition 1 settled into their mission, logistical support remained crucial.

On November 15, 2000, the Progress M1-2 spacecraft, designated as ISS Progress 2, was launched from the Baikonur Cosmodrome in Kazakhstan. This mission was vital for supporting the newly operational International Space Station (ISS) by delivering essential supplies and scientific equipment.

The spacecraft was launched aboard a Soyuz-U rocket, successfully propelling it into orbit. After reaching orbit, Progress M1-2 followed a carefully planned trajectory to dock with the ISS. The docking occurred on November 17, 2000, at approximately 19:17 GMT, with the spacecraft connecting to the station's Zarya module.

The cargo delivered by Progress M1-2 included vital provisions such as food, water, and oxygen for the crew, as well as equipment necessary for ongoing scientific research and operations aboard the ISS. This resupply mission was crucial in ensuring the continuous operation of the ISS and supporting its crew in conducting various experiments and maintaining the station's systems.

However, the ISS Progress 2 mission was not without its challenges. The spacecraft undocked from the station on December 1, 2000, but faced some difficulties that necessitated a redocking. It re-docked on December 26, 2000, after adjustments were made to ensure proper alignment and secure attachment.

Shortly after, on November 17, 2000, the ISS was further supported by the arrival of the Progress 2 spacecraft. This unmanned cargo vehicle delivered crucial supplies and equipment essential for the station's early operations. The Progress 2's stay was brief but significant, as it undocked on December 1, 2000, to make way for additional cargo and scientific instruments.

On 16 November 2000, the Progress M1-4 spacecraft embarked on its mission to support the International Space Station (ISS). It was launched aboard a Soyuz-U carrier rocket from Site 1/5 at the Baikonur Cosmodrome. The rocket lifted off precisely at 01:32:36 UTC, setting the stage for the spacecraft's crucial supply run.

Progress M1-4 successfully docked with the Nadir port of the Zarya module on 18 November at 03:47:42 UTC. However, this operation was not without its challenges. The primary Kurs docking system, responsible for automated docking, encountered a malfunction. In response, the mission control team employed the manual TORU system to complete the docking procedure, ensuring the spacecraft's secure attachment to the ISS.

Once docked, Progress M1-4 delivered essential supplies to the ISS, including food, water, oxygen, and scientific equipment. This mission marked a significant milestone as it was the first Progress spacecraft to resupply an Expedition crew aboard the ISS. The spacecraft remained attached to the station for two weeks before undocking at 16:22:52 UTC on 1 December 2000.

After the initial docking attempt, the spacecraft's presence was temporarily paused to ensure a clear docking path for Endeavour. On 26 December at 11:03:13 UTC, the spacecraft redocked with the same port on the Zarya module. Although designed for automatic docking, Progress M1-4 encountered a critical malfunction with its docking system. This was due to the irreversible nature of the abort procedure from the initial docking attempt, which involved the retraction of an antenna that could not be redeployed. In response, Commander Yuri Gidzenko took manual control, completing the docking using the TORU system. This manual intervention was not unprecedented, but it carried significant risks. A similar manual docking attempt in 1997 had resulted in a collision with the Mir space station, causing substantial damage and prompting concerns about the reliability of manual docking procedures.

The spacecraft remained docked for 44 days, supporting the ISS before undocking again on 8 February 2001 at 11:26:04 UTC. Progress M1-4 was deorbited at 12:59 UTC the same day following its second undocking. The spacecraft re-entered Earth's atmosphere and burned up over the Pacific Ocean, with any remaining debris landing in the ocean around 13:50 GMT.

Progress M1-4's dual docking with the ISS was a notable achievement, being the first Progress spacecraft to perform this feat. This milestone was not replicated until Progress M-15M in 2012, underscoring the significance of Progress M1-4's mission in the history of spaceflight and the ongoing efforts to sustain and supply the ISS.

On December 2, 2000, the International Space Station (ISS) underwent a significant enhancement with the arrival of the Space Shuttle Endeavour on mission STS-97.

Commanded by Brent Jett, with Michael Bloomfield serving as the Pilot and supported by Mission Specialists Joseph Tanner, Carlos Noriega, and Mark V. Lee, the STS-97 crew was tasked with a key objective: installing the ISS's first set of solar arrays. These arrays were essential for generating power and marked a vital step in the station's development.

Launch of STS-97 Endeavour was transporting the P6 Integrated Truss Structure that comprised Solar Array Wing-3 and the Integrated Electronic Assembly, to provide power to the Space Station. The 11-day mission included two spacewalks to complete the solar array connections.

The STS-97 mission was notable for its technical objectives and the temporary addition of crew members to the Expedition 1 team. The Shuttle Endeavour carried four American astronauts and one Canadian, who joined Expedition 1 for the duration of their mission. This temporary collaboration enriched the ISS's operational capacity and allowed for the integration of crucial equipment, including 17 tons of supplies such as expandable metal girders, batteries, electronics, and cooling systems.

Installing the first pair of U.S.-provided photovoltaic arrays was a highlight of STS-97.

These solar arrays were essential for the station's power generation, providing electricity to support the ISS's ongoing development and future expansions. The crew conducted three spacewalks to complete the installation and other tasks, carefully managing their time to ensure all objectives were met before the hatch between the shuttle and the ISS was opened on December 8, 2000. This cautious approach preserved the distinct atmospheric conditions of each spacecraft until the crews could safely interact.

Upon opening the hatch, the Expedition 1 and STS-97 crews met for the first time. This encounter provided the Expedition 1 crew an opportunity to tour the Space Shuttle Endeavour, an experience that was beneficial for their psychological well-being. The temporary change of environment and interaction with visiting colleagues offered a welcome respite and fostered a sense of connection between the two teams.

After completing its objectives, the Space Shuttle Endeavour undocked from the ISS on December 9, 2000. The mission left behind a significantly enhanced space station equipped with improved energy capabilities and a richer infrastructure, marking a notable advancement in the ISS's development.

As the Expedition 1 crew embarked on their four-month mission, they established the ISS's future by activating and setting up its systems. Their efforts prepared the station for subsequent missions and inhabitants, marking the beginning of a new era for the ISS—a period characterized by continuous human presence, groundbreaking scientific research, and international collaboration in space.

2001 began with further enhancements to the ISS's logistics and functionality. On February 8, 2001, ISS Progress 2 undocked once more, clearing the way for the arrival of a new cargo spacecraft. This was followed by the docking of ISS Progress 3 on February 28, 2001. Launched on February 26, 2001, this mission, like its predecessors, delivered essential supplies and scientific materials to the station, continuing the vital support for ongoing experiments and crew operations.

In parallel with the logistical milestones achieved by Progress M1-4, the Space Shuttle Atlantis mission STS-98, commanded by Kenneth Cockrell, with Pilot Mark Polansky and Mission Specialists Robert Curbeam, Thomas Jones, and Daniel Tani, STS-98 launched into space with a critical mission objective: to deliver and install the Destiny Laboratory.

STS-97 Astronaut Brent W. Jett (front right) and Michael J. Bloomfield (front left) are commander and pilot, respectively. Flanked by those two was astronaut Marc Garneau, mission specialist representing the Canadian Space Agency (CSA). In the rear are astronauts Carlos I. Noriega (left) and Joseph R. Tanner, both mission specialists. Noriega and Tanner are wearing training versions of the extravehicular mobility unit (EMU) spacesuits they'll be wearing for spacewalking chores during the flight.

STS-98 astronauts Kenneth D. Cockrell (right front), mission commander; and Mark L. Polansky (left front), pilot; along with astronauts Marsha S. Ivins, Robert L. Curbeam, Jr., (left rear) and Thomas D. Jones (right rear), all mission specialists.

The mission commenced with Atlantis docking with the ISS on 9 February 2001. The successful installation of the Destiny Laboratory

Module, valued at $1.4 billion, marked a significant advancement in the ISS's infrastructure. Destiny, also known as the U.S. Lab, became a cornerstone of American scientific research aboard the station. This module was designed to significantly enhance the ISS's capacity for scientific experimentation and research.

Destiny's construction began in 1995 at NASA's Michoud Assembly Facility, with final assembly completed at the Marshall Space Flight Center in Huntsville, Alabama. Weighing approximately 14.5 tons (32,000 pounds), it was transported to the Kennedy Space Center in Florida in 1998, where it underwent pre-launch preparations. Its installation was executed with precision using the shuttle's robotic Canadarm, operated by astronaut Marsha Ivins, with support from Thomas Jones and Robert Curbeam during spacewalks. The installation faced initial delays due to concerns over some cables on the shuttle, but these were resolved, and Destiny was successfully integrated into the ISS's structure on 20 February 2001.

Destiny was the primary operating facility for U.S. research payloads on the ISS. It provides essential laboratories for experiments in medicine, engineering, biotechnology, physics, materials science, and Earth science. The data and findings from these experiments contribute significantly to the global scientific community, advancing our understanding and technological capabilities.

The installation process of Destiny was not without its challenges. During one of the spacewalks, an ammonia coolant leak created a temporary contamination scare when Curbeam connected the coolant lines to the module. Despite this, the remaining spacewalks proceeded without further issues. The mission also saw the station's orientation control transition from propellant-based systems to electrically powered gyroscopes, which had been installed in September 2000 but were initially non-operational due to missing navigational electronics.

In addition to their primary tasks, the STS-98 crew also contributed to the IMAX documentary Space Station 3D. This film provided a unique and intimate perspective on life aboard the ISS, capturing significant moments such as the crew's first entry into the Destiny module, their activities in zero gravity, and the docking and crew changeover during STS-102. The documentary offered viewers a rare glimpse into the challenges and successes of space station life and construction.

As STS-98 concluded, the Expedition 1 crew had spent over three months aboard the ISS. Commander William Shepherd expressed his readiness to return home, reflecting the psychological challenges faced by astronauts during extended missions. To support their well-being, NASA implemented strategies such as increased videophone time with family and opportunities to enjoy movies and music. These measures were designed to help the crew navigate the psychological barriers associated with long-duration space missions and maintain their mental health.

ISS Destiny Lab module (NASA)

On December 26, 2000, following Endeavour's departure, Progress M1-4 re-docked with the ISS. Once securely attached, the crew focused on unloading the supplies transported by the Progress spacecraft.

During the holiday season, the Expedition 1 crew enjoyed a brief respite from their rigorous schedule. On Christmas Day, they were given time off to celebrate, opening presents delivered by the Endeavour and Progress supply missions. Each crew member also spoke with their families, providing a personal touch to their remote holiday experience. The crew participated in several video downlinks in the following days, including some with Russian television stations. The New Year was observed quietly, with Shepherd incorporating a Naval tradition into the station's log by providing a poem on behalf of the crew.

The astronauts' workload was intense during the first month. Commander Shepherd remarked on the challenge of managing a demanding schedule, noting that tasks often took longer than planned. For instance, activating a food warmer in Zvezda's galley, initially scheduled for 30 minutes, took a day and a half to complete. To further support the ISS's development and operational needs, the Progress 3 spacecraft was launched on February 26, 2001, and docked with the station on February 28, 2001. This cargo mission provided additional supplies and supported ongoing research activities, ensuring the station's continued functionality and growth.

Endeavour on Launch Pad 39-B before STS-97.

The Tranquility module, or Node 3, was a critical International Space Station (ISS) component. It serves multiple essential functions to support the crew's well-being and mission operations. This module houses the station's environmental control systems, life support systems, a toilet, and exercise equipment to maintain astronaut health. Additionally, Tranquility features an observation cupola, providing a panoramic view of space and Earth.

Thales Alenia Space manufactured Tranquility for the European Space Agency (ESA) and the Italian Space Agency. It was officially transferred to NASA on November 20, 2009. The module was launched into orbit on February 8, 2010, aboard the Space Shuttle Atlantis during the STS-130 mission.

On February 28, 2001, the third Progress spacecraft to visit the International Space Station, Progress M-44, docked with the Zvezda module. This mission was crucial for delivering essential supplies, including air, food, rocket fuel, and various equipment needed for station operations. Progress M-44 remained attached to the ISS until the arrival of Expedition 2. At that point, it was intentionally deorbited and burned up upon reentry into Earth's atmosphere, per standard procedures for decommissioning Progress spacecraft.

STS-102

On March 8, 2001, the Space Shuttle Discovery launched on mission STS-102, carrying a crew of four astronauts: James D. Wetherbee, James M. Kelly, Andrew S. Thomas, and Paul W. Richards. Their mission was pivotal, focusing on crew rotation, outfitting, and resupplying the International Space Station (ISS). This mission marked a key operational milestone and set the stage for a significant historical event.

Discovery arrived at the ISS on March 10, 2001, bringing the new long-duration crew for Expedition 2 and four additional short-term crew members. This was a momentous occasion in the ISS's history, as it brought the total number of astronauts aboard the station to ten, setting a new record for the largest number of people simultaneously present on the ISS. This influx of personnel fostered a strong sense of camaraderie and collaboration among the crew members from both missions.

During their time aboard the ISS, the Expedition 1 crew followed a structured daily schedule to balance their responsibilities and needs. Their routine typically adhered to UTC, with wake-up calls often scheduled around 05:00 UTC, although actual wake-up times usually ranged from 06:00 to 07:00 UTC. The crew's day included:

Due to the microgravity environment, maintaining physical fitness was crucial. The crew had access to a stationary bicycle, a treadmill (TVIS), and a resistance exercise device (IRED). Notably, the bicycle was out of service from mid-December 2000 until March 2001, and the treadmill faced issues towards the end of February but was repaired within a week.

The crew assembled and maintained the ISS, ensuring that systems were operational and equipment was correctly installed.

Conducting scientific experiments was a significant part of their schedule, as the ISS served as a laboratory for various research projects.

Regular communication with ground control was essential for coordinating activities and receiving instructions.

Crew members had designated times for personal activities, meals, and rest.

Entertainment and Psychological Well-being: The team had time for entertainment, such as watching movies, to support mental health and crew bonding. Shepherd noted the unique experience of watching space-themed movies while on a space mission, reflecting on the surreal nature of their situation.

During Expedition 1, the crew's communication capabilities evolved as the mission progressed:

Initially, the crew used Russian VHF communications equipment, known as the "Regul radio link," in the Zvezda and Zarya modules. This system allowed communication with the Russian Mission Control Center (TsUP) in Korolev, outside Moscow. The communications were limited to "comm passes"—brief 10—to 20-minute windows during which the station was within the line of sight of ground stations.

With the arrival of the Unity module on STS-97, the crew gained access to S-band Early Communication gear, which enabled more continuous communication with NASA's Mission Control in Houston. This was made possible through NASA's Tracking and Data Relay Satellites (TDRS) network, which provided broader coverage and more reliable communication.

The crew's schedule was sometimes adjusted to accommodate the arrival and departure of visiting shuttles and resupply vehicles, highlighting the dynamic and multi-faceted nature of living and working in space.

Earlier, during STS-106 in September 2000, equipment for amateur radio, or "ham radio," was delivered to the station. The first ham radio contact occurred on November 13, 2000, with Moscow and shortly after with the Goddard Space Flight Center in Greenbelt, Maryland. The crew reported that the voice quality of the ham radio was superior to other communication links.

As part of the Amateur Radio on the International Space Station (ARISS) project, the crew conducted brief radio contacts with schools and clubs on Earth. The first school contact was with Luther Burbank School in Chicago on December 21, 2000, after a delay due to technical issues. Due to the station's speed, these contacts were limited to 5-10 minutes, typically allowing time for 10 to 20 questions from students.

During Expedition 1, scientific research was limited due to the primary focus on station construction and setup. However, several important experiments and activities were conducted:

PKE-Nefedov (Plasma Crystal Experiment): This was one of the first natural science experiments conducted on the ISS. It was a collaboration between the Max Planck Institute for Extraterrestrial Physics in Germany and the Institute for High Energy Densities (part of the Russian Academy of Sciences). The experiment investigated the behavior of plasma crystals in microgravity.

These communication advancements facilitated better coordination with mission control, enhanced interaction with the public, and supported educational outreach during the mission.

The crew took over 700 photos of Earth, documenting dynamic events such as storms, fires, and volcanic eruptions. For example, on January 23, 2001, they captured a volcanic ash plume from Popocatépetl, Mexico. Photos also documented Mount Cleveland, Alaska, with smoke from its eruption the following month. These images contribute to Earth science by providing unique perspectives on environmental phenomena.

The Protein Crystal Growth Experiment produced better protein crystallizations in microgravity than on Earth, allowing for more accurate models of protein structures. Out of 23 proteins and viruses attempted, only four resulted in successful crystallizations. Notably, the crystallization of Thaumatin, a low-calorie sweetener, was achieved with higher resolution than Earth-grown crystals, improving the accuracy of its protein structure model.

The crew measured their heart rates and the station's carbon dioxide levels to study the effects of exercise in microgravity. This research was crucial for understanding the physiological impacts of long-duration spaceflight.

The successful completion of Expedition 1's mission set the stage for the subsequent crew rotations and the ongoing development of the ISS. On March 18, 2001, Expedition 1 departed, making way for Expedition 2. This transition was supported by three Space Shuttle missions, which provided essential resupply and support, ensuring the ISS's continuous operation. The groundwork laid by Expedition 1 was critical for the future of long-duration missions and international collaboration in space, heralding a new era of continuous human exploration and scientific research aboard the ISS.

The Expedition 1 mission lasted 140 days, 23 hours, and 38 minutes from launch to landing, officially concluded on March 18, 2001, when Space Shuttle Discovery undocked from the International Space Station, marking the end of the crew's four-and-a-half-month stay.

The return trip to Earth concluded with a rare night landing at 2:30 AM local time on March 21, 2001. Just two days after their landing, Mir, the Russian space station in orbit for 15 years, was intentionally deorbited and burned up upon reentry, marking the end of its operational life.

The ISS, taken from Endeavour on 9 December 2000

The Leonardo Multi Purpose Logistics Module rests in Discovery's payload bay in this view taken from the International Space Station by a crew member using a digital still camera. To the left of the MPLM you could see a glimpse of External Stowage Platform-1.

These ten astronauts and cosmonauts represent the base STS-102 crew and the crew members switching out. In the top group are, from the left, astronauts James M. Kelly, pilot; Andrew S.W. Thomas, mission specialist; James D. Wetherbee, mission commander; and Paul W. Richards, mission specialist. The bottom left was the Expedition One crew, from left, cosmonaut Sergei K. Krikalev, flight engineer; astronaut William M. (Bill) Shepherd, commander; and cosmonaut Yuri P. Gidzenko, Soyuz commander. At bottom right was the crew who replaced Shepherd and his collegues aboard the station, from the left, astronaut James S. Voss; cosmonaut Yury V. Usachev, Expedition Two commander; and astronaut Susan J. Helms. Usachev, Krikalev and Gidzenko all represent the Russian Aviation and Space Agency.

Expedition 2

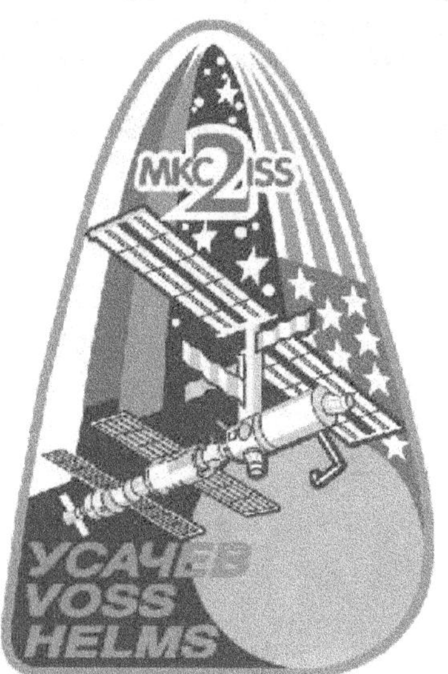

STS-105 Base crew (bottom center): Astronaut Scott J. Horowitz (front right in the bottom grouping) STS-105 crew commander. astronauts Frederick W. (Rick) Sturckow, pilot; and Patrick G. Forrester and Daniel T. Barry, both mission specialists. Expedition Three crew Astronaut Frank L. Culbertson, Jr. (center in the upper right grouping), commander, flanked by cosmonauts Mikhail Tyurin (left) and Vladimir N. Dezhurov representing Rosaviakosmos. Moving from the station over to the Space Shuttle Discovery for their return to Earth after a long stay aboard the ISS: in the upper left gathering was Expedition 2 astronaut James S. Voss, cosmonaut Yury V. Usachev and astronaut Susan J. Helms. Usachev, representing Rosaviakosmos.

Expedition 2, officially designated ISS EO-2, marked the second extended mission aboard the International Space Station (ISS) and began following the departure of the Expedition 1 crew. This transition occurred as Expedition 1 concluded its groundbreaking tenure, with the incoming Expedition 2 team taking over responsibilities for maintaining the station, assisting in its assembly, and conducting essential scientific research.

Expedition 2 of the International Space Station (ISS) unfolded during significant activity and logistical coordination. This expedition, spanning several months in 2001, was marked by a series of key dockings and undockings involving various spacecraft, crucial to the ongoing construction and maintenance of the ISS.

On April 16, 2001, the ISS Progress 3 spacecraft, having completed its mission, undocked from the station, clearing the way for subsequent arrivals and departures. This undocking was followed closely by the arrival of the Soyuz TM-32 on April 30, 2001. The Soyuz TM-32 mission was particularly notable as it carried a new crew of astronauts, including Commander Yuri Gidzenko, Flight Engineer Sergey Krikalev, and NASA astronaut Frank Culbertson. Their arrival was a significant event, as it marked the continuation of Expedition 2 and the rotation of crew members aboard the ISS.

Before the Soyuz TM-32's docking, the Soyuz TM-31 spacecraft had to be relocated on April 18, 2001. This maneuver was necessary to ensure the optimal positioning of the spacecraft for crew transfer and station operations. Soyuz TM-31 had previously docked with the ISS on October 12, 2000, and its relocation was part of the routine logistics of managing spacecraft around the ISS.

The spring of 2001 was particularly busy for the ISS as it hosted several visiting spacecraft. On April 21, 2001, the Space Shuttle Atlantis, on mission STS-100, docked with the ISS. This mission was critical for delivering and installing the Canadian Space Agency's robotic arm, Canadarm2, which was an essential addition to the station's capabilities. The STS-100 mission crew included Commander Chris Hadfield, Pilot Scott Altman, and Mission Specialists Kathleen Thome, and Robert Thirsk, among others. Their work on the ISS was instrumental in expanding the station's operational abilities.

After spending several days at the station, the STS-100 crew undocked from the ISS on April 29, 2001, marking the end of their mission and the successful integration of Canadarm2 into the ISS's infrastructure. The next significant event occurred on May 6, 2001, when the Soyuz TM-31 undocked, completing its mission and paving the way for new arrivals.

Shortly after, the ISS Progress 4 spacecraft arrived on May 20, 2001, bringing necessary supplies and equipment to support the station's operations. This was followed by the docking of the STS-104 mission, which occurred on July 14, 2001. The STS-104 mission, commanded by Steven W. Lindsey and including astronauts Charles D. Gemar, Janet L. Kavandi, and Thomas D. Jones, delivered and installed the Quest Joint Airlock, a critical component for the station's spacewalk operations. The successful docking of STS-104 was followed by its undocking on July 22, 2001, concluding the mission and leaving the station prepared for its next set of activities.

The transition from Expedition 1 to Expedition 2 was finalized by March 14, 2001. During this period, Commander William Shepherd, who had led Expedition 1, retained command of the ISS until the Space Shuttle Discovery undocked. On the morning of the undocking, the crew experienced a memorable wake-up call featuring "Should I Stay or Should I Go" by The Clash, a selection Shepherd's wife chose. The changeover ceremony was marked by a poignant farewell from Shepherd, a former Navy SEAL, who expressed his sentiments with a heartfelt message: "May the good will, spirit, and sense of mission we had enjoyed on board endure. Sail her well."

As Expedition 1 neared its conclusion, the crew prepared to depart on March 19, 2001, at 04:32 UTC. They left the ISS aboard the Space Shuttle Discovery, marking the end of their historic mission and initiating their journey back to Earth. Discovery's commander, Jim Wetherbee, conveyed a message of admiration for the departing crew: "For Captain Shepherd and his crew, we admire you as we prepare to bring you home. This has been an arduous duty for you. This ship was not built in a safe harbor. It was built on the high seas." This exchange highlighted the mutual respect and appreciation between the crews, underscoring the dedication and challenges encountered during their time aboard the ISS.

The day following the departure of Expedition 1, astronauts Jim Voss and Susan Helms conducted a spacewalk that lasted nearly nine hours. This spacewalk set a record for the longest duration, which remained unbroken until August 2010. The extended duration was partially due to complications during the walk, including Voss's accidental release of a small tool, which, despite efforts, could not be retrieved. NASA engineers meticulously tracked the tool's trajectory, and on March 14, 2001, NASA utilized Discovery's thrusters to boost the ISS by four kilometers to avoid potential collision with the drifting debris.

Expedition 2 officially began with the docking of Space Shuttle Discovery on mission STS-102 on March 10, 2001. However, the official start of their mission was marked by the undocking of the previous crew on March 18, 2001. The Expedition 2 crew consisted of Russian Commander Yury Usachev and American flight engineers Susan Helms and James Voss. Their previous experience was considerable: Usachev had completed two long-duration missions aboard the Mir space station (Expedition 15 and Expedition 21). Helms and Voss each brought extensive experience from their previous Space Shuttle missions.

The Space Shuttle Atlantis, on mission STS-101, continued the ISS's development with its docking on May 21, 2000. This mission was dedicated to delivering essential supplies and performing maintenance and upgrades. Under the command of James D. Halsell, with Pilot Susan J. Helms and Mission Specialists John M. Lounge, Jeffrey S. Williams, and Yuri Usachev, Atlantis played a crucial role in ensuring the station's readiness. The shuttle undocked on May 26, 2000, after completing its tasks.

The Expedition 2 crew had first arrived at the ISS during the Space Shuttle mission STS-101 in May 2000, which served as a precursor to their long-duration stay. Their mission was crucial for the ongoing assembly and scientific research on the ISS, significantly contributing to the station's development and operational success. The backup crew for Expedition 2 included Yury Onufrienko as Commander, Daniel W. Bursch as Flight Engineer 1, and Carl E. Walz as Flight Engineer 2, each possessing extensive spaceflight experience.

Expedition 2's members were instrumental in a variety of tasks, including station maintenance, the assembly of additional modules, and the integration of new equipment. They also had the historic opportunity to welcome Dennis Tito, the first space tourist, aboard the ISS.

Throughout their mission, Expedition 2 saw the arrival of additional spacecraft. Two Space Shuttle missions and one Soyuz mission docked with the ISS during this period, underscoring the station's growing international collaboration and its role as a hub for space exploration. The crew's efforts during this time were pivotal in advancing the ISS's operational capabilities and scientific endeavors.

On April 19, 2001, the Space Shuttle Endeavour launched on mission STS-100, marking a pivotal moment in the International Space Station's (ISS) development. The mission, which concluded on May 1, 2001, was notable for its significant contributions to the ISS's assembly and international collaboration in space exploration.

Illustration of the International Space Station during STS-100

Commanded by Kent V. Rominger, the STS-100 crew included Pilot Jeffrey S. Ashby and Mission Specialists John L. Phillips, Scott E. Parazynski, Chris A. Hadfield, Umberto Guidoni, and Yuri Lonchakov. Umberto Guidoni made history as the first European astronaut to visit the ISS, and Canadian astronaut Chris A. Hadfield performed the first Canadian spacewalk, or extravehicular activity (EVA). These milestones underscored the mission's importance in fostering international cooperation in space.

A primary objective of STS-100 was the deployment and installation of the Canadarm2, an advanced robotic arm developed by the Canadian Space Agency. This state-of-the-art arm was designed to enhance the ISS's capabilities for maneuvering and manipulating payloads, representing a significant upgrade from the original Canadarm used on previous Space Shuttle missions. The Canadarm2's installation was essential for future operations, including the docking and undocking of modules, performing spacewalks, and conducting maintenance tasks with improved precision.

The mission was executed closely with the Expedition 2 crew, which had been aboard the ISS since March 2001. Expedition 2 was commanded by Yury Usachev and included Flight Engineers Jim Voss and Susan Helms. This team had been preparing the station for the arrival of new hardware and integrating the Canadarm2 into the station's existing systems.

These international astronauts and cosmonaut trained in a number of venues for the April 2001 visit to the International Space Station (ISS). Seated are astronauts Kent V. Rominger (left) and Jeffrey S. Ashby, commander and pilot, respectively, for the STS-100 mission. Standing, from the left, are cosmonaut Yuri V. Lonchakov, with astronauts Scott F. Parazynski, Umberto Guidoni, Chris A. Hadfield and John L. Phillips, all mission specialists. Guidoni represents the European Space Agency (ESA); Hadfield was with the Canadian Space Agency (CSA) and Lonchakov was affiliated with Rosaviakosmos.

STS-100's crew undertook a series of meticulously planned spacewalks to install and activate the Canadarm2. These extravehicular activities, conducted in the vacuum of space, demonstrated the astronauts' expertise and attention to detail as they attached the arm to the ISS and ensured its proper functionality. In addition to the Canadarm2, the mission delivered a range of supplies and equipment, including scientific experiments and components crucial for the station's ongoing research and operations.

A particularly noteworthy event in April 2001 was the launch of Soyuz TM-32, a mission that marked a significant milestone in space travel. Launched on April 5, 2001, from the Baikonur Cosmodrome in Kazakhstan, Soyuz TM-32 carried three crew members: Russian cosmonaut Yuri Usachev, and American astronauts Susan Helms and James Voss. Their arrival at the International Space Station (ISS) on April 8, 2001, was a crucial development for Expedition 2 and played a significant role in ensuring the continuity and success of the station's operations.

The Soyuz Soyuz TM-32 Taxi Flight crewmembers. Soyuz Commander Talgat Musabayev, Flight Engineer Yury Baturin — both cosmonauts representing Rosaviakosmos — and American Space Flight Participant Dennis Tito blasted off from the Baikonur Cosmodrome in Kazakhstan at 2:37 a.m. CDT (0737 GMT) on April 28, 2001. They docked to the station two days later on April 30, 2001, at 2:58 a.m. CDT (0758 GMT) to begin nearly eight days of docked operations with the Expedition Two crew. The crews transferred gear and equipment from the new Soyuz to the orbital outpost, as well as into the older Soyuz spacecraft in which the visiting crew would return home.

The mission, often referred to as ISS EP-1, ISS-2S, or Soyuz 2 Taxi Flight, was designed not only to transport astronauts to the ISS but also to ensure the ongoing presence of a crew aboard the station. This was a period marked by significant expansion in the ISS's operational capacity and scientific research capabilities. The seamless integration of Soyuz TM-32 with the station's systems facilitated a smooth transition for the incoming crew, who joined the Expedition 2 team already in orbit.

Adding to the historical significance of this mission, Soyuz TM-32 carried Dennis A. Tito, the first private space tourist, alongside cosmonauts Talgat Musabayev and Yuri Baturin. Tito's journey highlighted the increasing involvement of private individuals in space exploration and marked a landmark moment in space history. Upon arriving at the ISS on May 5, 2001, Tito spent a week aboard the station, further emphasizing the evolving nature of space exploration and the expanding role of private participants.

The arrival of Usachev, Helms, and Voss significantly bolstered the ISS's operational team, enabling them to undertake a range of scientific experiments and maintenance tasks that were crucial for the station's growth. Their presence contributed to the station's research capabilities and underscored the collaborative international efforts in space exploration during this era.

During Expedition 2, the crew engaged in numerous scientific experiments, focusing on various fields, including biology, physics, and materials science. Their work contributed to advancing our understanding of the effects of microgravity on biological organisms and physical processes. Additionally, the team maintained and upgraded the station's systems, ensuring its continued functionality and safety.

On July 12, 2001, the Space Shuttle Atlantis embarked on mission STS-104, commanded by Steven W. Lindsey, with pilot Charles O. Hobaugh and mission specialists Michael L. Gernhardt, James F. Reilly, and Janet L. Kavandi.

STS-104 was primarily tasked with delivering and installing the Quest Airlock module, a crucial component for the ISS's continued expansion and functionality. This airlock was pivotal for enabling future spacewalks, facilitating the station's capacity for external maintenance, and supporting the integration of additional modules. The mission also included the installation of the Pressurized Mating Adapter-2 (PMA-2), which was integral to the ISS's growth. The PMA-2 facilitated the docking of future modules and spacecraft, allowing the ISS to accommodate larger equipment and additional crew members.

The mission's crew carried out several intricate spacewalks, essential for the precise placement and attachment of the PMA-2 to the Harmony module. The Expedition 2 crew had previously installed this module, which served as a critical junction for the ISS's expanding infrastructure. The spacewalks conducted during STS-104 were performed in the vacuum of space and demanded meticulous coordination and technical expertise to ensure the successful integration of the new components.

STS-104 astronauts Steven W. Lindsey (right), mission commander; and Charles O. Hobaugh, pilot. Standing, from left, are astronauts Michael L. Gernhardt, Janet L. Kavandi and James F. Reilly, all mission specialists.

The installation of the PMA-2 was not only a technical triumph but also a vital step in the ISS's development. By enabling the future docking of European and Japanese modules, the PMA-2 expanded the station's capability to host a wider array of scientific experiments and support an increased crew. The successful completion of STS-104 on July 24, 2001, represented a significant advancement in the ISS's operational capabilities and set the stage for future growth and scientific discovery.

During Expedition 2, the research focus included developing a better understanding and protective measures against radiation exposure, which poses significant risks to astronauts over long-duration missions. High levels of radiation could damage human cells and increase the risk of cancer and other health issues. The crew's work was instrumental in advancing the station's capability to support ongoing scientific research and crew safety.

Expedition 2 made significant progress in assembling and enhancing the International Space Station (ISS) through a series of critical spacewalks and operations conducted in July 2001.

The first spacewalk of Expedition 2 took place on July 15, 2001, and was dedicated to installing the Quest Joint Airlock. The airlock was a pivotal addition, enabling astronauts to perform spacewalks using both U.S. and Russian spacesuits. Susan Helms was crucial in this operation by utilizing the station's robotic arm to lift the airlock from the Space Shuttle Atlantis' payload bay and maneuver it to the Unity module of the ISS. Jim Reilly supported the installation from a foot platform attached to the shuttle's robotic arm, which Janet Kavandi operated. The spacewalk lasted 5 hours and 59 minutes and involved precise coordination to secure the airlock and prevent thermal damage by connecting essential attachments from inside the station.

The second spacewalk occurred on July 18, 2001, for 6 hours and 29 minutes. This spacewalk focused on installing and connecting oxygen and nitrogen tanks to the new airlock. With the internal hatches between the shuttle and the station closed, Atlantis' cabin pressure was reduced in preparation for the task. Susan Helms again utilized the station's robotic arm to position the tanks from the shuttle's payload bay to the airlock. Once at the airlock, Mike Gernhardt and Jim Reilly secured the tanks and connected the necessary cables and hoses. The team efficiently completed their primary objectives ahead of schedule, allowing them to install an additional oxygen tank, leaving just one tank for the final spacewalk.

The third spacewalk, held on July 21, 2001, marked the first extravehicular activity from the newly installed Quest Joint Airlock. Lasting 4 hours and 2 minutes, this spacewalk's main objective was to install the final nitrogen tank outside the airlock. It also served as a test for a new protocol developed by Mike Gernhardt. It involved exercising while breathing oxygen to purge nitrogen from the astronauts' bodies—a technique designed to reduce the risk of decompression sickness.

As Expedition 2 concluded, Space Shuttle Discovery returned to the station in August 2001 on mission STS-105, carrying the crew members back to Earth. The crew undocked from the ISS on August 20, 2001, marking the end of their mission and the Expedition 2 increment.

During Expedition 2, two crucial science racks, EXPRESS Rack No. 1 and EXPRESS Rack No. 2, were delivered to the International Space Station (ISS) aboard Space Shuttle Endeavour during mission STS-100 in April 2001. These racks played a significant role in advancing the station's research capabilities.

The Quest Joint Airlock module seen attached to the end effector of the Canadarm2 during the module's installation on to the Unity node during STS-104.

Among the experiments housed in EXPRESS Rack No. 2 was the Experiment of Physics of Colloids in Space. This experiment aimed to investigate the behavior and properties of colloidal mixtures in the microgravity environment of space. Colloids are substances where microscopic particles are dispersed throughout another substance, such as gels, emulsions, and suspensions. The unique conditions aboard the ISS allowed for a more detailed study of these mixtures, which behave differently in microgravity than on Earth.

The research focused on understanding how colloidal particles interact and aggregate in space, providing valuable insights into fundamental physical processes. The data collected from these experiments was still being analyzed and continues contributing to the broader understanding of colloid science. This ongoing research has the potential to impact various fields, including material science and pharmaceuticals, by enhancing our knowledge of particle dynamics in space.

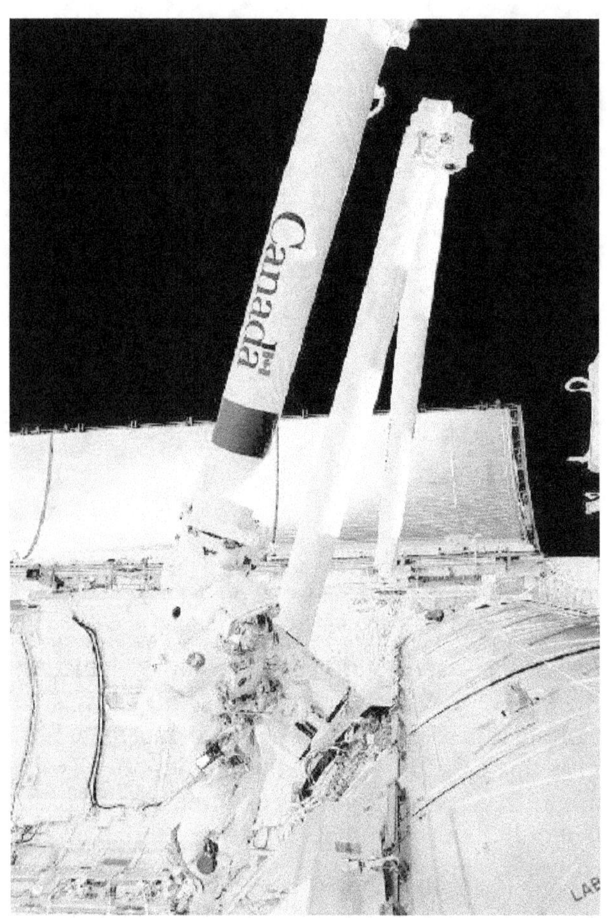

Astronaut Scott E. Parazynski, mission specialist, works with cables associated with the Space Station Remote Manipulator System (SSRMS) or Canadarm2 during one of two days of extravehicular activity (EVA). Parazynski shared both space walks with astronaut Chris A. Hadfield of the Canadian Space Agency (CSA).

The mission concluded on August 20, 2001, with the undocking of Space Shuttle Discovery, which rotated the long-duration crews and brought in the next team of astronauts. This marked the end of Expedition 2, concluding a significant chapter in the ongoing development of the ISS.

Expedition 2, the second long-term International Space Station (ISS) crew, arrived at the station in March 2001 and spent a notable 163 days aboard. They concluded their mission on August 22, 2001, with their return to Earth on Space Shuttle mission STS-105. This period marked an important phase in the ISS's development, as the crew spent 167 days in space, with their time aboard contributing significantly to the station's assembly and research efforts.

View of the International Space Station (ISS) during arrival of STS-127 Space Shuttle Endeavour. Nadir side of the Airlock (A/L) was visible.

James S. Voss, Expedition Two flight engineer, looks over an atlas in the Zvezda Service Module. (NASA)

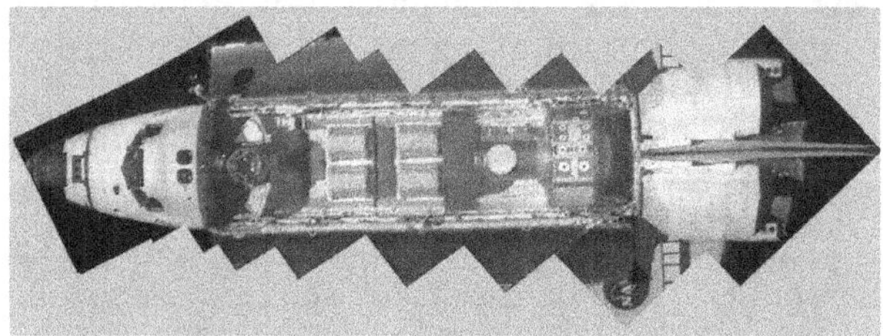

The payload bay of STS-104 imaged by TV camera during its approach to the ISS, no still photography was made of this event

Expedition 3

Astronaut Frank L. Culbertson, Jr. (center),
commander, was flanked by cosmonauts Mikhail
Tyurin (left) and Vladimir Dezhurov, both flight
engineers representing Rosaviakosmos

Expedition 3 to the International Space Station (ISS) began on August 12, 2001, and concluded on December 15, 2001. This mission spanned 124 days, 22 hours, and 47 minutes aboard the ISS, with the entire mission from launch to landing lasting 128 days, 20 hours, 45 minutes, and 58 seconds. During this period, the crew traveled an impressive 85,860,485 kilometers (53,351,232 miles) and completed 2,020 orbits around Earth.

The Expedition 3 crew aboard Space Shuttle Discovery, designated STS-105, arrived at the ISS. This international team was led by American Commander Frank Culbertson, accompanied by Russian crewmates Vladimir Dezhurov, who served as the mission pilot, and Mikhail Tyurin, a flight engineer. They replaced the Expedition 2 crew, taking on the responsibility of continuing the ongoing work and advancing the ISS's mission of scientific research and international cooperation.

Upon arrival, the Expedition 3 crew swiftly integrated into the station's operations, seamlessly continuing the research and maintenance activities initiated by their predecessors. Their mission was marked by a rigorous schedule of scientific experiments and station upkeep, highlighting their commitment to advancing the ISS's objectives.

A notable feature of their tenure was the execution of four spacewalks, which played a crucial role in expanding and enhancing the ISS's capabilities. These extravehicular activities were essential for the station's development and implementation of new technologies and systems.

The mission concluded on December 8, 2001, when the crew's custom Soyuz seat-liners were transferred to Space Shuttle Endeavour as part of the STS-108 mission. This transfer marked the end of Expedition 3 and the beginning of a new chapter for the ISS, which continued to serve as a hub for international collaboration and scientific exploration in space.

The activities began on August 21, 2001, with the launch of ISS Progress 5, a mission that would later play a vital role in supplying the station with necessary cargo. Two days later, on August 23, 2001, ISS Progress 5 docked with the station, delivering equipment and supplies to support the crew aboard.

In the lead-up to Expedition 3, significant preparations and milestones were achieved aboard the International Space Station (ISS). On August 22, 2001, the ISS Progress 4 spacecraft undocked from the station, concluding its mission. This departure was a critical step, clearing the way for the arrival of new cargo and spacecraft essential for the ISS's ongoing operations and expansion.

A major milestone during this period was the arrival of the Pirs Docking Compartment, delivered to the ISS by a Soyuz-U rocket. The Pirs Docking Compartment, also known as DC-1, significantly enhanced the station's capabilities. As a Russian module, Pirs was instrumental in advancing the ISS's operational efficiency.

The Pirs module, with a mass of approximately 8,461 pounds, measures 16 feet long and 8.4 feet in diameter. It serves multiple vital functions: it provides an airlock for cosmonaut spacewalks, enabling astronauts to conduct extravehicular activities (EVAs) outside the station. Additionally, Pirs functions as a docking port for transport and cargo vehicles, facilitating the arrival and departure of spacecraft such as Soyuz and Progress vehicles.

The Pirs docking compartment, with the ISS Progress 77 cargo craft docked to it, attached to Earth-facing port of the International Space Station's Zarya service module.

Moreover, Pirs was critical in fuel transportation between the Zvezda and Zarya modules and docked vehicles.

The station's configuration saw further changes on September 16, 2001, with the delivery and installation of the Pirs Docking Compartment. This new module was a significant addition, providing an additional docking port for spacecraft and enhancing the ISS's capability for conducting spacewalks. The arrival of Pirs was an important milestone in the station's development, expanding its operational flexibility.

The next notable event occurred on September 26, 2001, when the R4 truss segment, an integral part of the ISS's structural framework, undocked. This was part of the station's ongoing assembly and upgrade efforts.

Following the departure of the R4 truss segment, the Soyuz TM-32 spacecraft undocked from the ISS on October 30, 2001. Soyuz TM-32 had previously brought new crew members to the station, and its departure marked the end of its mission. Before this, on October 23, 2001, the Soyuz TM-33 spacecraft docked with the ISS, bringing a new crew onboard. Yuri Gidzenko commanded soyuz TM-33, with Flight Engineers

Sergey Krikalev and Frank Culbertson, continuing the rotation of crew members and ensuring ongoing human presence aboard the station.

A critical maneuver occurred on October 19, 2001, when Soyuz TM-32 was relocated to a different docking port. This relocation was part of the logistical efforts to optimize the station's docking arrangements and prepare for the arrival of new spacecraft.

The Soyuz Taxi crewmembers Flight Engineer Konstantin Kozeev, Commander Victor Afanasyev, and French Flight Engineer Claudie Haignere. Afanasyev and Kozeev represent Rosaviakosmos, and Haignere represents ESA, carrying out a flight program for CNES, the French Space Agency, under a commercial contract with the Russian Aviation and Space Agency.

The final significant event of Expedition 3 occurred on November 22, 2001, when ISS Progress 5 undocked from the station after completing its mission. This was followed by the launch of ISS Progress 6 on November 26, 2001, which carried further supplies and equipment to support the ISS's continued operations. ISS Progress 6 then docked with the station on November 28, 2001, completing a critical phase of resupply and maintenance for the ISS.

One of the most remarkable experiences of Expedition Three was observing the 2001 Leonid meteor storm from orbit. Astronaut Frank Culbertson described the sight with awe: "It looked like we were seeing UFOs approaching the Earth, flying in formation, three or four at a time. There were hundreds per minute going beneath us— really spectacular!" The meteor storm was a highly anticipated celestial event, as Earth was scheduled to pass through a dense trail of debris shed by Comet Tempel-Tuttle. On 18 November 2001, this comet dust entered Earth's atmosphere at speeds of

144,000 miles per hour (64,000 meters per second), creating a stunning meteor shower. While millions on Earth witnessed the display, the view from space was uniquely spectacular as the crew observed the meteors from above, seeing them as they burned up in the lower atmosphere.

The International Space Station (ISS) was backdropped over Miami, Florida,

On 11 September 2001, as the International Space Station (ISS) orbited Earth, it passed directly over New York City. At that moment, astronaut Frank L. Culbertson Jr. captured photographs of the smoke clouds emanating from the World Trade Center site in Downtown Manhattan. This historic imagery from the ISS provided a unique and poignant perspective of the devastating events unfolding below. Culbertson's photographs captured the scale and impact of the attacks on the World Trade Center, offering a rare view of the tragedy from space.

"The world changed today. What I say or do was very minor compared to the significance of what happened to our country today when it was attacked." So said Expedition 3 Commander Frank L. Culbertson upon learning of the September 11, 2001 attack on the World Trade Center. This image was one of a series taken that day of metropolitan New York City by the International Space Station's Expedition 3 crew that shows a plume of smoke rising from the Manhattan skyline. Upon further reflection, Commander Culbertson said, "It's horrible to see smoke pouring from wounds in your own country from such a fantastic vantage point. The dichotomy of being on a spacecraft dedicated to improving life on the earth and watching life being destroyed by such willful, terrible acts was jolting to the psyche, no matter who you are."

During Expedition Three, the International Space Station (ISS) saw a series of pivotal spacewalks, all carried out using the Russian Orlan spacesuits and conducted from the Pirs airlock on the Russian segment of the station. These extravehicular activities were crucial in advancing the ISS's capabilities and maintaining its operational integrity.

ISS picture of New York City on 9/11

On 8 October 2001, the first spacewalk of Expedition Three commenced with Vladimir Dezhurov and Mikhail Tyurin venturing into the vacuum of space. The two astronauts undertook several critical tasks starting at 14:23 UTC and concluding nearly five hours later at 19:21 UTC. Their primary objective was connecting the Pirs airlock to the Zvezda Service Module, facilitating communications between these key station sections. They also installed handrails on the newly added compartment and affixed an exterior ladder to aid future spacewalkers in exiting the Pirs hatch. Additionally, Dezhurov and Tyurin mounted a Strela cargo crane, enhancing the station's cargo handling capabilities.

The second spacewalk, which took place on 15 October 2001, saw Dezhurov and Tyurin once again stepping outside the ISS. This time, their focus shifted to installing Russian commercial experiments on the exterior of Pirs. Among the experiments was MPAC-SEEDS, designed to study how various materials react to the space environment over extended periods. The investigation was housed in three briefcase-sized containers dedicated to different material science inquiries.

On 12 November 2001, a third spacewalk was carried out by Dezhurov, this time joined by American astronaut Frank Culbertson. They began their work at 21:41 UTC and continued into the early hours of 13 November, finishing at 02:45 UTC. This mission involved connecting cables on the exterior of Pirs for the Kurs automated docking system. The duo also checked the Strela cargo crane, simulating cargo handling with one spacewalker stationed at the end of the boom. Their tasks also included inspecting and photographing a small panel on one of the Zvezda Service Module's solar arrays, which had failed to unfold fully.

Vladimir N. Dezhurov, Expedition Three flight engineer in Node 1 on the International Space Station. (Note the speed limit signs on the hatch)

The final spacewalk of Expedition Three occurred on 3 December 2001, with Dezhurov and Tyurin stepping out into space again. Their mission, starting at 13:20 UTC and concluding three hours later at 16:06 UTC, aimed to address an issue with a Progress resupply ship's docking. The astronauts removed an obstruction— specifically a rubber seal from a previous cargo ship—preventing secure docking. They also documented the debris and the docking interface with photographs to assist in future operations.

The Expedition 3 crew conducted extensive scientific research and operations throughout their mission, accumulating 17 hours and 50 minutes of extravehicular activity through their four spacewalks. Their work built on the research initiated by previous crews, advancing our understanding of various scientific disciplines in the unique microgravity environment of space.

The crew's residency aboard the ISS ended on 8 December 2001, when they transferred their custom Soyuz seat-liners to Space Shuttle Endeavour for their return to Earth. This mission exemplified the ongoing commitment to scientific exploration and highlighted the operational challenges of maintaining a continuous human presence in space.

In August 2001, the Space Shuttle Discovery, designated STS-105, played a significant role in facilitating the return of the Expedition 2 crew to Earth, concluding their term aboard the International Space Station (ISS). This mission also marked the arrival of the Expedition 3 crew, who began their stay on August 12, 2001, at 18:41 UTC.

The Expedition 3 crew's tenure on the ISS extended until December 15, 2001, at 17:28 UTC, encompassing a total duration of 124 days, 22 hours, and 47 minutes. Throughout this period, the crew completed 2,020 orbits around Earth, covering a distance of approximately 85,860,485 kilometers (53,351,232 miles). The entire journey, from the launch of the mission to its eventual landing, lasted 128 days, 20 hours, 45 minutes, and 58 seconds.

A setting sun and the thin blue airglow line at Earth's horizon

In 2002, space exploration witnessed a series of pivotal events and advancements, underscoring humanity's ongoing commitment to the cosmos. This period marked notable achievements from various space agencies, including NASA, Russia's Roscosmos, and Japan's Aerospace Exploration Agency (JAXA), each contributing to the expanding frontier of space science and technology.

The year began with NASA's Mars Odyssey mission, which launched on April 1, 2002. This spacecraft, directed by a dedicated team of scientists and engineers, embarked on a journey to Mars. After a successful transit, Mars Odyssey entered orbit around the Red Planet in October, commencing an extensive study of Martian geology and climate. The mission's success provided invaluable data about Mars, paving the way for future planet exploration.

In April 2002, the Soyuz TM-30 mission, also known as Mir EO-28, significantly transitioned from the aging Mir space station to the International Space Station (ISS). This mission, launched by Russia, was notable for being the 39th and final human spaceflight to Mir and the first Soyuz mission to dock with the ISS following Mir's deorbit. The mission marked a significant shift in space exploration and was distinguished by its private funding through MirCorp, a company dedicated to refurbishing and privatizing Mir.

Soyuz TM-30 launched with the dual objectives of rejuvenating Mir and preparing it for its eventual deorbiting. The spacecraft lifted off with a crew of two experienced Russian cosmonauts: Sergey Zalyotin and Aleksandr Kaleri. Their mission was to reactivate and repair Mir and boost its orbit to maintain stability until its planned deorbit in March 2001. The mission also significantly transitioned human spaceflight focus from Mir to the ISS.

The Soyuz TM-30 spacecraft docked with the ISS on April 27, 2002, at 07:55 UTC, connecting to the nadir port of the Zarya module. During their 135-day mission, Zalyotin and Kaleri conducted various activities, including experiments and maintenance tasks, both on Mir and later on the ISS. Their work on Mir was crucial for maintaining the station's functionality until its deorbiting. At the same time, their activities on the ISS contributed to the station's development as a central hub for international collaboration and scientific research.

In August 2002, NASA's Genesis spacecraft was launched, tasked with the unique mission of collecting solar wind samples. The spacecraft's crew precisely operated, gathering particles from the Sun to study their properties and composition. The return of these samples to Earth was a significant milestone in solar science, offering new insights into the formation and behavior of our star.

October saw the launch of NASA's Deep Space 1 mission, a landmark endeavor designed to test innovative technologies for future space exploration. This mission featured an ion propulsion system and autonomous navigation capabilities, both critical for advancing space travel. The crew's work on Deep Space 1 demonstrated cutting-edge technology that would influence future missions.

NASA continued its vital work with the Hubble Space Telescope, which benefited from significant servicing missions in 2002. In March, the space shuttle Atlantis launched on mission STS-109, carrying a team of astronauts including Scott D. Altman, Duane G. Carey, John M. Grunsfeld, Nancy J. Sherlock Currie, James H. Newman, Richard M. Linnehan, and Michael J. Massimino. This mission focused on upgrading and installing new instruments on Hubble, enhancing its ability to capture detailed universe observations.

Japan made a significant contribution to space exploration with the launch of the Hayabusa spacecraft in May. This mission, led by the Japan Aerospace Exploration Agency (JAXA), aimed to explore the asteroid Itokawa. With its sophisticated instruments, the Hayabusa spacecraft gathered valuable data on the asteroid and returned samples to Earth, enhancing our understanding of these distant celestial objects.

Throughout 2002, the ISS saw numerous crew rotations, assembly operations, and supply missions. Notable among these were missions STS-110 and STS-111, carried out by NASA's space shuttles Atlantis and Endeavour. The crews, including Michael J. Bloomfield, Stephen N. Frick, Jerry L. Ross, Steven L. Smith, Ellen L. Ochoa, Lee Morin, Rex J. Walheim, Kenneth D. Cockrell, Paul S. Lockhart, Franklin R. Chang-Diaz, Philippe Perrin, Valeri Korzun, Peggy A. Whitson, Jeffrey S. Ashby, Pamela A. Melroy, David A. Wolf, Piers J. Sellers, Sandra H. Magnus, Fyodor Yurchikhin, Sergei Zalyotin, Yuri Lonchakov, Frank De Winne, and John B. Herrington, played crucial roles in maintaining and expanding the space station.

The year concluded with mission STS-113 in December. The crew, including James D. Wetherbee, Paul S. Lockhart, Michael E. Lopez-Alegria, and John B. Herrington, was instrumental in continuing the assembly of the ISS and rotating the crew, ensuring the station's operations and further development.

Among the notable astronauts and cosmonauts of 2002 were Peggy Whitson, Valery Korzun, Sergei Treshchyov, Pedro Duque, Nikolai Budarin, and Donald Pettit. Peggy Whitson and Valery Korzun, both members of Expedition 5, spent 184 days aboard the ISS, conducting scientific research and maintenance tasks. Sergei Treshchyov, also with Expedition 5, contributed significantly to research and station operations. Pedro Duque, Spain's first astronaut, spent ten days on the ISS as part of the Cervantes mission. Nikolai Budarin and Donald Pettit, serving on Expedition 6, conducted extensive research and captured stunning images of Earth from orbit.

Expedition 4

Expedition 4 of the International Space Station (ISS) began with the launch of Space Shuttle Endeavour on December 5, 2001, at 20:03 UTC, continued through to its conclusion on June 15, 2002, at 14:32 UTC, spanning 195 days, 11 hours, 38 minutes, and 13 seconds. Over this period, the crew orbited Earth 3,068 times, contributing extensively to scientific research and the development of the ISS.

The expedition was notable for its rigorous crew rotations and the arrival of critical modules and supplies. The timeline of Expedition 4 was marked by several key events, starting with the undocking of Soyuz TM-33 on May 4, 2002. Having completed its mission of delivering a crew to the ISS, this spacecraft returned to Earth as part of the routine crew rotation process. Soyuz TM-33's departure was crucial in maintaining the station's continuous human presence.

Earlier, on April 27, 2002, Soyuz TM-34 arrived at the ISS, ushering in a new crew. Commanded by Yuri Gidzenko and accompanied by Flight Engineers Sergey Krikalev and Frank Culbertson, Soyuz TM-34 ensured the uninterrupted staffing of the station. In preparation for this arrival, Soyuz TM-33 underwent a relocation maneuver on April 20, 2002, optimizing

docking arrangements to accommodate the incoming spacecraft.

The Soyuz Taxi crewmembers in the Zvezda Service Module on the International Space Station (ISS). The "taxi" crew arrived at the orbital outpost on April 27, 2002 at 2:56 a.m. (CDT) as the two vehicles flew over Central Asia. Gidzenko represents Rosaviakosmos. From the left are Commander Yuri Gidzenko, who was a member of the first resident crew of the ISS; Flight Engineer Roberto Vittori of the European Space Agency (ESA); and South African space flight participant Mark Shuttleworth. Visible in the background was a portrait of cosmonaut Yuri Gagarin posted in the module.

April 2002 was also marked by the STS-110 mission, during which Space Shuttle Atlantis docked with the ISS on April 10. The primary objective of this mission was the delivery and installation of the S0 truss segment, a vital component of the station's structural framework. Led by Commander Steven W. Lindsey, Pilot Charles O. Hobaugh, and Mission Specialists Ellen Ochoa and Jerry L. Ross, the STS-110 crew played an instrumental role in expanding the ISS's capabilities. Following the successful installation, the crew undocked on April 17, 2002, concluding their mission.

Prior to these events, on March 21, 2002, the ISS Progress 7 spacecraft launched, carrying essential supplies and equipment. It docked with the ISS on March 24, ensuring the station was well-stocked for ongoing research and habitation. This docking followed the undocking of ISS Progress 6 on March 19, which marked the end of its mission and paved the way for new cargo deliveries.

Cosmonaut Yuri I. Onufrienko (standing at center), mission commander, was flanked by astronauts Daniel W. Bursch (left) and Carl E. Walz, both flight engineers.

Space Shuttle Endeavour's mission, STS-108, was also pivotal. Launched in December 2001, Endeavour, commanded by Dominic Gorie and accompanied by Pilot Mark Kelly and Mission Specialists Linda Godwin, Daniel W. Bursch, and David W. Brown, was responsible for delivering and installing the Express Logistics Carrier-2 (ELC-2). This carrier, a significant addition to the ISS's external truss, was designed to enhance the station's scientific research and operational maintenance capabilities by providing additional storage for spare parts and experiments. The mission involved multiple extravehicular activities (EVAs) to ensure the successful integration of the ELC-2 into the station's existing infrastructure.

A key component of Endeavour's mission was the rotation of the ISS crew. The Expedition 4 team, consisting of Commander Yury Onufrienko, Flight Engineer Dan Tani, and NASA astronaut Carl Walz, replaced the Expedition 3 crew, who returned to Earth aboard the same shuttle. This crew change was vital for the ongoing operations and maintenance of the ISS, ensuring that the station continued to function smoothly and effectively.

Space Shuttle Endeavour's mission, STS-108, was pivotal for several reasons. The shuttle, commanded by Dominic Gorie with Pilot Mark Kelly and Mission Specialists Linda Godwin, Daniel W. Bursch, and David W. Brown, carried crucial equipment and facilitated the rotation of the ISS crew. Among the key objectives was the

delivery and installation of the Express Logistics Carrier-2 (ELC-2), a significant addition to the station's external truss. This carrier was designed to bolster the ISS's scientific research and operational maintenance capacity by providing additional storage for spare parts and experiments.

Astronauts Dominic L. Gorie (background), STS-108 commander, and Mark E. Kelly, pilot, are in their respective stations during rendezvous operations with the International Space Station (ISS). Gorie brought Endeavour to a gentle linkup with the ISS at 2:03 p.m. (CST) as the two craft sailed over England. Within minutes, Kelly and astronauts Linda M. Godwin and Daniel M. Tani (out of frame), mission specialists, began to conduct post-docking checks of the mechanical interface between Endeavour and the station's Destiny Laboratory prior to the opening of the hatches on the two vehicles

On December 10, 2001, astronauts Linda Godwin and Daniel Tani of the Space Shuttle Endeavour conducted a spacewalk lasting four hours and twelve minutes. This extravehicular activity (EVA) was pivotal for the International Space Station (ISS) as it involved the installation of insulation on the mechanisms responsible for rotating the station's main solar arrays. During the spacewalk, Godwin and Tani also retrieved a cover from a stowage bin that had been removed from a station antenna on a previous flight. This cover was intended for reuse after its return to Earth. Additionally, they completed a preparatory task by positioning two switches on the station's exterior, which were slated for installation on a future shuttle mission, STS-110.

The EVA contributed to a record-setting year in spacewalks, with eighteen conducted in 2001—twelve from the shuttle and six from the station. Endeavour's mission further included the meticulous installation of the Express Logistics Carrier-2 (ELC-2). The spacewalks required careful planning and execution to ensure the carrier's successful attachment and integration into the ISS's existing infrastructure.

In addition to the EVA activities, Endeavour's crew, including Pilot Mark Kelly and Mission Specialist Linda Godwin, utilized the shuttle's robotic arm to lift and attach the Raffaello Multi-Purpose Logistics Module (MPLM) to a berth on the station's Unity node. The module, retrieved from the shuttle's payload bay, was subsequently unloaded on the same day, facilitating the transfer of essential supplies to the ISS.

The astronauts and cosmonauts completed the transfer of more than 5,000 pounds of supplies and material from Endeavour's mid-deck and the Raffaello Multi-Purpose Logistics Module to the station. The transferred items included more than 850 pounds of food, 1,000 pounds of clothing and other crew provisions, 300 pounds of experiments and associated equipment, 800 pounds of spacewalking gear, and 600 pounds of medical equipment. In turn, the crew packed up the Raffaello module with items bound for a return trip to Earth.

Endeavour's mission also played a crucial role in crew rotation. The Expedition 4 team, comprising Commander Yury Onufrienko, Flight Engineer Dan Tani, and NASA astronaut Carl Walz, replaced the Expedition 3 crew. This crew change was vital for ensuring continuous operations and effective management aboard the ISS, with the returning Expedition 3 crew returning to Earth aboard the same shuttle.

On December 15, 2001, the departure of Space Shuttle Endeavour from the International Space Station (ISS) was carefully orchestrated to address an impending collision risk with space debris. Flight controllers implemented slight modifications to the shuttle's departure plan, allowing time for a small jet firing to adjust the station's trajectory and avoid a piece of space debris—a spent Russian rocket upper stage from the 1970s—that was projected to come within three miles of the station. This precautionary maneuver, known as a reboost, effectively increased the distance between the ISS and the debris to over 40 miles, ensuring the safety of the station.

Due to the additional propellant required for the reboost, Endeavour did not perform its typical full-circle flyaround of the station. Instead, after undocking, the shuttle executed a quarter-circle flyaround, positioning itself approximately 400 feet directly above the station. At 12:20 a.m. EST, Endeavour's engines were fired in a final separation burn, marking the shuttle's departure from the orbiting outpost.

Endeavour's mission also included returning scientific experiments conducted during Expedition 3. The shuttle's middeck carried results from several significant studies, including the Advanced Protein Crystallization Facility and the Dynamically Controlled Protein Crystal Growth experiment. Additionally, cells from the Cellular Biotechnology Operations Support System (CBOSS) were transported back to Earth. The CBOSS equipment continued to operate during Expedition 4, growing ovarian and colon cancer cells, as well as kidney cells, in the unique microgravity environment of the ISS.

In the shuttle's payload bay, a range of experiments was returned for analysis by researchers worldwide. The Multiple Application Customized Hitchhiker-1 (MACH-1) contained diverse experiments, such as the Prototype Synchrotron Radiation Detector, the Collisions into Dust Experiment-2, the Capillary Pump Loop, and the Space Experiment Module (SEM). The SEM included experiments from Argentina, Portugal, Morocco, Australia, and U.S. schoolchildren, reflecting the international scope of the ISS research program.

As part of its mission, Endeavour's crew deployed STARSHINE 2, a small satellite designed to engage students in scientific research. The satellite, launched from a canister in the payload bay, was tracked by over 30,000 students from 660 schools across 26 countries. The students, who had assisted in polishing STARSHINE's 845 mirrors, used the data collected from its orbit to calculate the density of the Earth's upper atmosphere, furthering their involvement in space science education.

During their stay, Expedition 4's crew undertook a series of notable spacewalks. The first occurred on January 14, 2002, when Yury Onufrienko and Carl Walz conducted a six-hour and three-minute EVA. Their tasks included relocating the cargo boom for the Russian Strela crane from Pressurized Mating Adapter 1 to the Pits Docking Compartment's exterior, which enhanced the crane's functionality. Additionally, they installed an amateur radio antenna on the

Zvezda Service Module, utilizing Russian Orlan spacesuits from the Pirs Airlock.

Onufrienko and Daniel Bursch performed the second spacewalk on January 25, 2002. Lasting nearly six hours, this EVA focused on installing deflector shields on the Zvezda Service Module's jet thrusters, mounting another amateur radio antenna, attaching four scientific experiments, and replacing a device designed to measure material ejected by the thrusters. This spacewalk, also conducted from the Pirs Airlock, further emphasized integrating Russian technology and expertise into ISS operations.

The final spacewalk of Expedition 4 took place on February 20, 2002, involving Carl Walz and Daniel Bursch. This EVA, conducted from the Quest Airlock, was significant as it was the first such activity not supported by a space shuttle. The spacewalk aimed to test the Quest Airlock's functionality and prepare for the upcoming STS-110 mission to install the S0 Truss onto the ISS. Walz and Bursch's efforts were crucial in laying the groundwork for future expansions of the station.

The Quest Airlock, an integral component of the International Space Station (ISS), consists of two distinct segments: the "Equipment Lock" and the "Crew Lock." The Equipment Lock was designed to house spacesuits and other essential equipment, providing a space for preparation before spacewalks. The Crew Lock was the segment from which astronauts exit into the vacuum of space.

Derived from the Space Shuttle's airlock, the Quest Airlock underwent significant modifications to improve its efficiency and minimize atmospheric gas loss. These enhancements were crucial for optimizing the airlock's performance during operations.

STS-108 crew, with Expeditions 3 and 4. Standing at rear (from the left) are the four STS-108 crew members Daniel M. Tani and Linda M. Godwin, both mission specialists; Dominic L. Gorie and Mark E. Kelly, commander and pilot, respectively. In front, from the left, are Daniel W. Bursch, Yuri Onufrienko, Carl E. Walz, Mikhail Tyurin, Frank L. Culbertson and Vladimir N. Dezhurov. :Culbertson, ISS Expedition Three commander, as well as flight engineers Tyurin and Dezhurov, used the Space Shuttle Discovery on STS-105 to reach the station for a lengthy 128 day stay and then returned to Earth aboard Endeavour on STS-108. They were replaced aboard the orbital outpost by Onufrienko, Expedition Four commander, along with Bursch and Walz, both flight engineers. The Expedition Four crew accompanied the STS-108 crew into Earth orbit. Dezhurov, Tyurin and Onufrienko represent Rosaviakosmos.

The Quest Airlock was equipped with mountings for four high-pressure gas tanks—two containing oxygen and two containing nitrogen. These tanks play a vital role in replenishing the atmospheric conditions on the American segment of the ISS. Specifically, they address the loss of gas that occurs when the airlock's hatch was opened for spacewalks, ensuring that the station's atmosphere remains stable and suitable for both astronauts and onboard equipment.

The Quest Airlock offers astronauts a specialized environment to prepare for spacewalks, particularly by allowing them to "camp out" before venturing into space. This preparation involves spending time in a reduced-nitrogen atmosphere to purge nitrogen from the bloodstream, which helps prevent decompression sickness when transitioning into their spacesuits' low-pressure, pure-oxygen environment.

Quest Joint Airlock Module. Crew lock with EVA hatch on right, and equipment lock with three attached HP gas tanks on left

Astronaut Linda M. Godwin, STS-108 mission specialist, near the end of the Space Shuttle Endeavour's remote manipulator system (RMS) arm during the four-hour session of extravehicular activity (EVA). Astronaut Daniel M. Tani (out of frame), mission specialist, joined Godwin on the space walk.

The mission concluded with the return of the Expedition 4 crew on June 15, 2002, aboard Space Shuttle Endeavour during mission STS-111. The successful completion of Expedition 4 highlighted the collaborative efforts of international space agencies and the ongoing commitment to advancing human knowledge and capabilities in space.

The Soyuz TM-30 mission concluded with undocking from the ISS on November 9, 2002, at 20:44 UTC. This mission marked a significant milestone in space exploration, symbolizing the end of an era for Mir and the beginning of a new chapter with the ISS. The successful integration of Soyuz TM-30 into the ISS's operations underscored the evolving nature of crewed space missions and highlighted the ISS's role as the focal point for future space exploration and international cooperation.

Expedition 5

Expedition 5 of the International Space Station (ISS), conducted during the latter half of 2002, was a period of significant activity and progress for the station. Crucial spacecraft maneuvers and crew rotations marked this expedition, all contributing to the ISS's ongoing development and operational efficiency.

The expedition began with the arrival of the ISS Progress 8 spacecraft on June 29, 2002. This supply mission was vital for sustaining the station's daily operations and research activities, delivering essential cargo including scientific equipment and personal supplies for the crew. After completing its mission, ISS Progress 8 undocked from the station on September 24, 2002, clearing the way for the arrival of its successor.

On September 25, 2002, the ISS Progress 9 spacecraft launched and successfully docked with the ISS just four days later, on September 29. This cargo spacecraft was instrumental in providing additional supplies and resources necessary for the station's continued functioning, reinforcing its operational capability.

As autumn approached, the ISS witnessed a notable increase in activity with the arrival of the Space Shuttle Atlantis. On October 9, 2002, Atlantis, on mission STS-112, docked with the ISS. This mission was pivotal for the station's expansion, as it involved delivering and installing the S1 truss segment, a crucial component of the station's structural framework. The STS-112 crew, commanded by Jeffrey Ashby and accompanied by Pilot Pamela Melroy and Mission Specialists David Wolf, Piers Sellers, and Fyodor Yurchikhin, played an essential role in enhancing the station's infrastructure. Their work significantly advanced the ISS's structural capabilities, vital for the ongoing construction and optimization of the station. On October 8, 2002, the second day of the STS-112 mission, the crew aboard Space Shuttle Atlantis commenced their preparations for the crucial rendezvous and docking with the International Space Station (ISS). The day began early, with the crew waking up at 4:46 a.m. CDT to start their first full day in orbit.

Pilot Pamela Melroy, alongside Mission Specialists David Wolf and Piers Sellers, conducted a thorough checkout of their spacewalk suits and equipment. Meanwhile, Commander Jeffrey Ashby and Mission Specialist Sandra Magnus worked together to ensure the shuttle's robotic arm was fully operational. They powered up the arm and conducted a video survey of Atlantis 'payload bay. The team also completed essential tasks, including setting up the orbiter docking system's centerline camera, extending the orbiter's spring-loaded ring designed for initial contact, and verifying the rendezvous tools.

Throughout the day, the STS-112 crew performed three Orbital Maneuvering System (OMS) burns to adjust the shuttle's trajectory and fine-tune its approach to the ISS. Additionally, astronaut David Wolf checked out the Spatial Heterodyne Imager for Mesospheric Radicals (SHIMMER) experiment, sponsored by the Naval Research Laboratory. The SHIMMER experiment, designed to use ultraviolet sensing to observe the Earth's atmosphere and detect possible ozone loss, encountered some issues. However, with assistance from Mission Control, the crew managed to prepare the equipment for its observations during the mission.

On October 9, 2002, the third flight day, Atlantis reached a significant milestone as it docked with the ISS at 15:17 GMT. Commander Jeffrey Ashby expertly guided the shuttle's docking system to engage with the Pressurized Mating Adapter-2, located at the forward port of the Destiny Laboratory. As the two spacecraft glided 245 miles above central Asia at a speed of five miles per second, the docking was executed flawlessly.

The arrival of Atlantis marked the beginning of a week of joint operations between the STS-112 and Expedition 5 crews. This docking was particularly notable as the Atlantis crew became the first visitors to Expedition 5 team, which had arrived at the station earlier in June 2002.

Following the docking, the pressure checks were completed successfully, and a lighthearted exchange occurred when Station Science Officer Peggy Whitson inquired about a request she had made earlier. Upon confirming that Commander Ashby had brought the salsa she had requested, Whitson humorously gave the green light to open the hatches.

The hatches between Atlantis and the ISS were opened at 16:51 GMT, and Commander Ashby floated into the Destiny Module to a warm welcome from Whitson, who greeted him with an embrace. Mission Specialist Sandra Magnus followed, and the rest of the Atlantis crew soon joined them. The three-member ISS crew welcomed their visitors as the new arrivals integrated with the station team.

After a safety briefing from Station Commander Valery Korzun, the combined crews began their busy schedule. Astronaut Pamela Melroy, Commander Korzun, and Mission Specialists Dave Wolf, Piers Sellers, and Fyodor Yurchikhin prepared spacesuits for their upcoming extravehicular activity (EVA) 1. Meanwhile, Magnus and Whitson reviewed the procedures for operating the robotic arm to position the new truss segment, a critical component for the station's expansion.

On October 10, 2002, Flight Day 4 of the STS-112 mission, the day's activities began at 3 a.m. CDT with a wake-up call from Mission Control in Houston. The crew of Space Shuttle Atlantis was greeted with a musical reminder that the day's tasks were about to commence.

The focus of the day was on the first spacewalk of the mission, known as EVA 1, which marked the 44th spacewalk dedicated to the assembly and maintenance of the International Space Station (ISS). Before the spacewalk began, astronauts Peggy Whitson and Sandra Magnus utilized the station's Canadarm2 robotic arm to grapple the S1 truss segment from Atlantis' payload bay. They maneuvered the truss to its designated position at the starboard end of the S0 truss segment. At 8:36 a.m. CDT, four remotely operated motorized bolts successfully connected the two truss segments, marking a significant step in the station's expansion.

As these operations proceeded, astronauts Dave Wolf and Piers Sellers prepared for their spacewalk. EVA 1 commenced with the two astronauts exiting the Quest Airlock at approximately 11:21 a.m. EDT. Wolf, identifiable by his spacesuit's solid red stripes, and Sellers, clad in an all-white spacesuit, set to work on their tasks. Wolf focused on connecting power, data, and fluid lines between the S0 and S1 trusses, while Sellers, on his inaugural spacewalk, worked to release the locks on three folded-up radiators mounted on the S1 truss. This adjustment allowed the radiators to be oriented for optimal cooling of the new equipment.

The spacewalk, which lasted seven hours and one minute, extended beyond the planned duration by 31 minutes due to an unexpected glitch with the Canadarm2. The malfunction required Wolf to install a television camera system on the far end of the truss without the aid of the robotic arm. Additionally, near the end of the EVA, Wolf reported a minor issue with his helmet earphones losing power.

Throughout the spacewalk, Pilot Pamela Melroy provided guidance and advice to the spacewalkers while keeping them on schedule. Shuttle Commander Jeff Ashby operated the robotic arm from Atlantis, offering camera views

to document the proceedings. Following the spacewalk, Wolf and Sellers conducted a tool inventory check and performed cleanup activities before re-entering the Quest Airlock. The airlock was re-pressurized at 5:22 p.m. CDT, marking the conclusion of EVA 1.

On October 11, 2002, Flight Day 5, the combined STS-112 and Expedition 5 crews took some well-deserved off-duty time before resuming their work. They began transferring items between the shuttle and the ISS, preparing for the second spacewalk of the mission. Among the transferred items were scientific experiments and equipment, including liver cell tissue samples that had been studied in microgravity and were now ready for return to Earth. Additionally, experiments from the Marshall Space Flight Center, including protein crystal growth thermal enclosures, were moved to and from the station.

The crew also facilitated the transfer of seven water containers to the ISS and initiated a nitrogen transfer process, moving approximately 15 pounds of the gas from Atlantis to the station by the end of the day.

With assistance from Pilot Pamela Melroy, STS-112 spacewalkers David Wolf and Piers Sellers prepared the extravehicular activity (EVA) equipment for the upcoming spacewalk. They recharged water on their extravehicular mobility units (EMU), configured their tools, and prepared the Quest airlock for use.

In addition to their technical tasks, the astronauts participated in several live media interviews. Magnus, Wolf, and Sellers discussed their experiences with CBS Radio Network and Cable News Network (CNN), reflecting on the challenges and accomplishments of EVA 1. Wolf noted the physical intensity of the manual work performed during the spacewalk, which saw their heart rates soar to over 170 beats per minute. Russian cosmonauts Valery Korzun, Sergei Treshchev, and Fyodor Yurchikhin also engaged with the Russian press. As the day concluded, the crew reviewed procedures for the forthcoming EVA 2, ensuring they were prepared for the next phase of their mission.

After completing its mission, Atlantis undocked from the ISS on October 16, 2002.

The transition into November saw further changes in crew and spacecraft. Soyuz TM-34, which had delivered a previous crew to the ISS, undocked on November 9, 2002, marking the end of its mission. Its place was taken by Soyuz TMA-1, which had docked with the ISS on November 1, 2002. Soyuz TMA-1 was the first of the TMA series, designed to enhance crew safety and comfort. This spacecraft brought a new crew to the ISS, including Commander Yuri Gidzenko, Flight Engineer Sergey Krikalev, and NASA astronaut Frank Culbertson. Their arrival was significant, ensuring a continuous human presence aboard the station.

The culmination of Expedition 5's achievements came with the Soyuz TMA-1 mission in early 2003, marking the inaugural flight of the TMA model. This mission represented a significant advancement in spacecraft design, focusing on improving crew safety and comfort. Shortly after that, Soyuz TMA-2 continued the momentum, facilitating crew exchanges and supporting the ISS's operational flow.

STS-112 Astronauts Jeffrey S. Ashby and Pamela A. Melroy, commander and pilot, respectively, are in the center of the photo. The mission specialists are, from left to right, astronauts Sandra H. Magnus, David A. Wolf and Piers J. Sellers and cosmonaut Fyodor Yurchikhin, who represents Rosaviakosmos.

Expedition 5 represented the fifth long-duration mission on the ISS, continuing the uninterrupted human presence in space that began with Expedition 1 in 2000. This mission was a collaborative effort involving astronauts from NASA, the Russian space agency Roscosmos, and the European Space Agency (ESA).

The backup crew comprised of Commander Aleksandr Kaleri (Russia, Roscosmos), a veteran of four spaceflights, Flight Engineer 1 Scott J. Kelly (United States, NASA) on his second spaceflight, and Flight Engineer 2 Dmitri Kondratyev (Russia, Roscosmos) on his first spaceflight.

The crew for Expedition 5 was composed of three astronauts who arrived aboard the Space Shuttle Endeavour during mission STS-111, which launched on June 5, 2002. Endeavour docked with the ISS on June 7, 2002. The primary objective of STS-111 was the delivery and installation of the Mobile Base System (MBS), a key component of the ISS's External Truss Structure. The MBS, designed to enhance the station's capabilities for future construction and maintenance tasks, serves as a platform for the Canadarm2, an advanced robotic arm. This system allowed for increased flexibility in assembling and maintaining the station's exterior modules, improving its ability to conduct large-scale operations in space. The mission also included the Integrated Cargo Carrier (ICC), which delivered various supplies and equipment necessary for the station's continued operation and expansion.

Upon arriving at the ISS aboard the Space Shuttle Endeavour during the STS-111 mission, the Expedition 5 crew took over from their predecessors, the Expedition 4 crew, consisting of one Russian and two American astronauts, for the Expedition 5 team, comprising two Russian and one American.

During their stay aboard the ISS, the STS-111 crew worked closely with the Expedition 5 astronauts to facilitate the installation of the MBS and to ensure a smooth transition of responsibilities. On June 10, 2002, a formal handover ceremony took place, symbolized by the ceremonial ringing of the station's brass bell, marking the official transition of command from the previous crew. Throughout their mission, the crew conducted approximately 25 new scientific investigations and continued ongoing research, contributing significantly to the scientific goals of the ISS.

A notable achievement during this period was the progress made in developing the station's scientific capabilities. The Expedition 5 crew was crucial in setting up the European Space Agency's "European Drawer Rack," a key component for experiments in fluid physics, materials science, and other fields. Their efforts significantly contributed to the ISS's ability to support cutting-edge research.

Expedition 5 also included a variety of international scientific experiments aimed at expanding human knowledge in multiple disciplines. The crew conducted research in areas such as biology, physics, and astronomy, which provided valuable data on the effects of long-duration spaceflight on the human body, among other scientific inquiries.

Expedition 5 was also notable for its spacewalks, during which the crew conducted two notable extravehicular activities (EVAs) from the Pirs Docking Compartment, wearing Russian Orlan space suits. The first spacewalk occurred on August 16, 2002, with Valery Korzun and Peggy Whitson working together for over four hours. They focused on installing six debris panels on the Zvezda Service Module, designed to protect the module from potential impacts by space debris. The panels had been temporarily stored on the Pressurized Mating Adapter 1 before being securely attached to Zvezda. Ultimately, 23 such shields were planned to safeguard the Service Module.

ISS Expedition 5 crew: Commander Valery Korzun (Russia, Roscosmos): A spaceflight veteran on his second and final mission, Flight Engineer 1 Peggy Whitson (United States, NASA): On her inaugural spaceflight and honored as NASA's first ISS Science Officer, and Flight Engineer 2 Sergei Treshchev (Russia, Roscosmos): Embarking on his only spaceflight.

The second spacewalk, conducted on August 26, 2002, saw Korzun and Sergei Treshchev work for over five hours. Their tasks included installing

a frame on the Zarya Module to support future assembly operations. They also mounted new material samples for Japanese Space Agency experiments on the Zvezda Module and added two amateur radio antennas to enhance communication capabilities. Additionally, they installed the Kromka hardware to measure residue emissions from Zvezda's jet thrusters, a task initially scheduled for their first spacewalk.

The Expedition 5 crew spent 184 days in space, 178 aboard the ISS, and returned to Earth on December 7, 2002, aboard the Space Shuttle Endeavour during mission STS-113.

During this mission, the Expedition 5 crew—Commander Valery Korzun, Flight Engineer Peggy Whitson, and European Space Agency (ESA) Flight Engineer Andre Kuipers—completed their six-month tenure aboard the ISS and were replaced by the new Expedition 6 crew.

These four astronauts comprise the prime crew for NASA's STS-111 mission. Astronaut Kenneth D. Cockrell (front right) was mission commander, and astronaut Paul S. Lockhart (front left) was pilot. Astronauts Philippe Perrin (rear left), representing the French Space Agency, and Franklin R. Chang-Diaz are mission specialists, assigned to extravehicular activity (EVA) work on the International Space Station (ISS).

Peggy A. Whitson, Expedition Five flight engineer, wears a Russian Orlan spacesuit as she prepares for an EVA. (NASA)

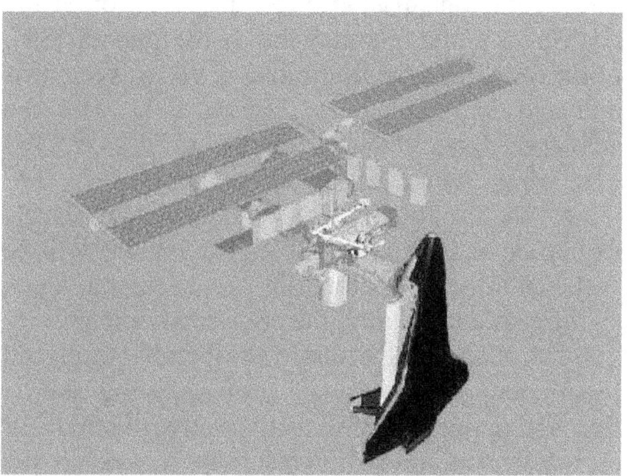

STS-111/UF2. Space Shuttle Endeavour; with a Multi-Purpose Logistics Module (MPLM); delivered more payload and experiment racks to the Destiny laboratory.

Atlantis during rendezvous and docking operations. The steel truss segment and radiators are in the payload bay. Notice the Canadarm near the top of the cargo bay.

The Mobile Remote Servicer Base System (MBS) was moved by the Canadarm2 for installation on the International Space Station (ISS).

In 2003, human space exploration was marked by a profound and tragic turning point that deeply impacted our efforts beyond Earth. The year began with a heart-wrenching disaster on February 1, when the Space Shuttle Columbia disintegrated upon re-entry into Earth's atmosphere. This catastrophic event resulted in the loss of all seven astronauts aboard, casting a somber shadow over the history of space exploration.

Commander Rick D. Husband led the ill-fated mission, with William C. McCool serving as the pilot. The crew also included mission specialists David M. Brown, Kalpana Chawla, Michael P. Anderson, Laurel B. Clark, and Ilan Ramon, the first Israeli astronaut to fly in space. The mission was designed to advance scientific research and contribute to our understanding of space, but it became a poignant reminder of the inherent risks of spaceflight.

The disaster was caused by damage to Columbia's thermal protection system, a crucial component that shields the spacecraft from the intense heat experienced during re-entry. During launch, a piece of foam insulation from the shuttle's external tank detached and struck the left wing of Columbia. This impact compromised the wing's protective coating, but the damage went undetected until it was too late. As the shuttle re-entered the atmosphere, the integrity of the wing was fatally compromised, leading to the vehicle's disintegration.

In response to the Columbia tragedy, NASA took decisive and immediate action. The space agency suspended the Space Shuttle program for two years to conduct a thorough investigation and implement comprehensive safety reforms. This period was a pause and a critical phase of introspection and restructuring to improve future missions' safety. The Columbia disaster underscored the significant perils and complexities of human space exploration, serving as a somber reminder of the high stakes in pushing the boundaries of our reach into space.

In early 2003, the International Space Station (ISS) saw a series of significant events that marked a period of transition and activity.

On February 1, 2003, the crew of the Space Shuttle Columbia faced a tragic end as the shuttle disintegrated during its re-entry into Earth's atmosphere. This somber event profoundly impacted subsequent space missions, including those involving the ISS.

Expedition 6

Soyuz 5 crew Belgian Flight Engineer Frank DeWinne left, of the European Space Agency ESA, Cosmonaut Soyuz Commander Sergei Zalyotin center and Cosmonaut Yuri Lonchakov, Flight Engineer in the Service Module SM/Zvezda during Expedition Five on the International Space Station ISS.

One of the notable events during Expedition 6 was the successful docking of the ISS Progress

M1-10 spacecraft on February 4, 2003. Launched on February 2, 2003, Progress M1-10 was a vital resupply mission, delivering essential provisions, scientific equipment, and research materials to support the crew's ongoing work. This docking ensured the station remained well-stocked with the necessary resources for its experiments and daily operations.

Before Progress 10's arrival, the Progress 9 spacecraft was undocked from the ISS on February 1, 2003. After completing its mission, Progress 9 made way for the new cargo delivery, exemplifying the efficient logistics crucial for maintaining the ISS's functionality.

In November 2002, the Space Shuttle Endeavour, on mission STS-113, transported the sixth crew to the International Space Station (ISS). Initially planned for four months, this mission was intended to conclude in March 2003 with the arrival of Space Shuttle Atlantis on mission STS-114, which would deliver the Expedition 7 crew. However, the tragic disaster of Space Shuttle Columbia on February 1, 2003, led to the indefinite suspension of Shuttle flights, necessitating a significant alteration in the crew rotation plan.

The Columbia disaster profoundly impacted ISS operations, effectively grounding the Shuttle fleet and requiring a shift to alternative methods for crew transport and station resupply. As a result, the Expedition 6 crew, originally scheduled to return in March, remained aboard the ISS until May 2003. Their return was facilitated by the Soyuz TMA-1 spacecraft, marking a historic event as it was the first time U.S. astronauts returned to Earth aboard a Russian Soyuz spacecraft. However, Dennis Tito, an American space tourist, had previously done so in 2001.

The Soyuz TMA-1 mission, officially designated Soyuz TM-35, played a crucial role during this transition period. Launched from the Baikonur Cosmodrome in Kazakhstan on October 30, 2002, the mission was notable for several reasons. It was the fifth Russian Soyuz spacecraft to visit the ISS and marked the debut of the TMA-class Soyuz vehicles, succeeding the Soyuz-TM series with Soyuz TM-34 as the last of its kind.

The mission's importance was underscored by the need to maintain a continuous human presence on the ISS despite the Shuttle program's suspension. With the Shuttle grounded for an extended period, Soyuz and Progress spacecraft became the sole means for logistical support and crew rotation. This situation highlighted Soyuz TMA-1's critical role in ensuring the ISS's operational continuity and demonstrated the resilience and adaptability of international spaceflight operations.

Soyuz TMA-1 docked with the ISS on November 1, 2002, and remained attached until May 3, 2003, when it undocked, completing its mission. The crew's extended stay aboard the ISS was a testament to their flexibility and the collaborative efforts of the international space community during a challenging time.

The Soyuz TMA-1 mission's crew comprised six astronauts from five different nations, each bringing a unique set of experiences and expertise to the mission. Commander Sergei Zalyotin from Russia led the team, representing the Russian Federal Space Agency (RKA). Zalyotin was on his second and final spaceflight, bringing seasoned leadership to the mission. Joining him was Flight Engineer Frank De Winne from Belgium, representing the European Space Agency (ESA), who was embarking on his first spaceflight. Russian cosmonaut Nikolai Budarin, also from RKA, served as Flight Engineer 2 and was on his third and final spaceflight. From the United States, Kenneth Bowersox, representing NASA, was on his fifth and final spaceflight, having previously served as both Expedition 6 ISS Commander and Soyuz Flight Engineer. Yury Lonchakov, another Russian cosmonaut from RKA, was on his second spaceflight. Rounding out the crew was Donald Pettit from the United States, also representing NASA, who was on his first spaceflight and served as an Expedition 6 Flight Engineer.

Expedition 6 crew: Donald Pettit, Ken Bowersox, and Nikolai Budarin

The Soyuz TMA-1 spacecraft docked with the ISS on November 1, 2002, at 05:01 UTC, connecting to the Pirs docking module. This successful docking facilitated the seamless integration of the Expedition 6 crew with the ISS operations, enhancing the station's capability to continue its scientific and operational objectives. The spacecraft remained docked until May 3, 2003, when it undocked from the ISS at 22:43 UTC, marking the end of its operational tenure.

Originally, Soyuz TMA-1 was intended as a taxi mission to transport the crew to the ISS and return them to Earth. However, the mission faced an unexpected alteration when American musician Lance Bass expressed interest in joining the mission as a space tourist for a one-week stay aboard Soyuz TMA-1. Despite initial enthusiasm, Bass's participation was ultimately canceled by September 2002 due to contractual and training complications. Consequently, Russian cosmonaut Yury Lonchakov was assigned to replace Bass, ensuring that the mission retained its intended crew composition.

During Expedition 6, the International Space Station (ISS) crew conducted two significant spacewalks, both originating from the Quest Airlock. These extravehicular activities (EVAs) were performed using U.S. spacesuits called Extravehicular Mobility Units (EMUs). Originally, the mission was slated for just one spacewalk, but a second was added to address additional tasks and prepare for future assembly missions.

The two spacewalks conducted by the Expedition 6 crew increased the total number of spacewalks performed in support of ISS assembly and maintenance to 51. Of these, 26 were based out of the station itself, with 17 originating from the Quest Airlock. Commander Kenneth Bowersox and Flight Engineer Donald Pettit accumulated 13 hours and 17 minutes of spacewalking time throughout their mission.On January 15, 2003, the first spacewalk of the Soyuz TMA-1 mission commenced at 12:50 UTC and concluded at 19:41 UTC, spanning 6 hours and 51 minutes. This extravehicular activity (EVA) was marked by significant efforts to enhance and activate the International Space Station's (ISS) latest addition, the P1 (P-One) Truss.

Commander Kenneth Bowersox and Flight Engineer Donald Pettit dedicated their efforts to deploying the radiator assembly of the P1 Truss. Their work involved releasing the final launch locks that secured the radiator, allowing it to be fully extended and operational. In addition to this, the astronauts addressed several maintenance and installation tasks. They meticulously removed debris from a sealing ring on the Unity Module's Earth-facing docking port and conducted a test on an ammonia reservoir attached to the P6 Truss.

Although the crew intended to install a light fixture on one of the station's Crew and Equipment Translation Aid (CETA) carts, this task was deferred to a subsequent spacewalk. The operation was briefly delayed due to a thermal cover strap obstructing the rotation of the Quest Airlock's hatch, promptly resolved.

The second spacewalk lasted 6 hours and 26 minutes on April 8, 2003, beginning at 12:40 UTC and ending at 19:06 UTC. This EVA focused on reconfiguring cables on the S0 (S-Zero), S1, and P1 Trusses, a preparatory step for future installations of Integrated Truss Structure components.

During this spacewalk, Bowersox and Pettit replaced a Power Control Module on the Mobile Transporter and provided the Control Moment Gyro No. 2 with a redundant power channel through cable rerouting. They also installed Spool Positioning Devices on the Destiny Laboratory's heat exchangers and reinstalled a thermal cover on the S1 Radiator Beam Valve Module.

During this pivotal period for the International Space Station (ISS), the Soyuz spacecraft played a crucial role, encompassing three distinct modules: the Orbital Module, the Descent Module, and the Instrumentation/Propulsion Module. The crew was housed in the central Descent Module throughout the mission, while the Orbital and Instrumentation/Propulsion Modules were separated and jettisoned prior to re-entry. These discarded modules burned up upon re-entering Earth's atmosphere, leaving only the Descent Module to complete its journey back to Earth.

On May 4, 2003, Soyuz TMA-1 commenced its return to Earth, marking the inaugural re-entry and descent for this class of Soyuz spacecraft. The deorbit burn, which lasted 258.1 seconds, reduced the spacecraft's mass by two-thirds. As the spacecraft approached the Entry Interface—approximately 400,000 feet (121.9 kilometers) above Earth—it encountered increasing atmospheric density and friction. With just 23 minutes remaining before landing, the focus shifted to managing the descent rate to ensure a safe landing.

Eight minutes into the descent, Soyuz TMA-1 was traveling at a speed of 755 feet (230 meters) per second. To decelerate, the spacecraft deployed its parachute system. Initially, two pilot parachutes were released, followed by a drogue chute with a surface area of 24 square meters (258 square feet), which reduced the descent rate to 262 feet (80 meters) per second. The final parachute, the main chute with a surface area of 10,764 square feet (1,000 square meters), adjusted the spacecraft's orientation to a 30-degree angle relative to the ground to manage heat dissipation and then transitioned to a vertical descent position. Despite these measures, the descent rate remained at 24 feet (7.3 meters) per second, still too fast for a smooth landing. To cushion the impact, two sets of small engines fired one second before touchdown, further reducing the descent speed.

However, Soyuz TMA-1 faced significant challenges during its descent. A technical malfunction forced the spacecraft into a steeper, less controlled ballistic trajectory, leading to a landing approximately 300 miles (483 kilometers) from the intended site. This deviation subjected the crew to substantial gravitational forces. Communication was lost during descent due to damage and malfunction of the spacecraft's antennas, though contact was later re-established through an emergency transmitter after landing. Following this incident, subsequent missions were equipped with satellite phones to ensure reliable communication with recovery teams.

The return of Expedition 6 aboard Soyuz TMA-1 was historically significant, as it marked the first instance of U.S. astronauts returning to Earth in a Russian Soyuz spacecraft. While U.S. space tourist Dennis Tito had previously completed a similar return in 2001, this was the first time professional astronauts had done so. Soyuz TMA-1 landed approximately 276 miles (444 kilometers) from its intended landing zone. Russian Mission Control confirmed that support helicopters reached the crew around 2:45 a.m. on May 4, and all three astronauts were reported to be in good health despite the ordeal.

The Expedition 7 crew, including Commander Yuri Malenchenko and Flight Engineer Edward T. Lu, was prepared to continue the work of their predecessors and advance the scientific research aboard the ISS. Their spacecraft docked with the ISS on April 28, 2003, following a launch on April 25, ensuring a smooth transition and uninterrupted station operations.

The successful return of Expedition 6 on May 3, 2003, marked the conclusion of their mission. Their departure signified the end of their time aboard the ISS, where they had made significant contributions to the ongoing scientific research and station operations. The transition to Expedition 7 was executed seamlessly, allowing the ISS to maintain its operational and research activities without disruption.

Expedition 7

ISS Expedition 7 began on April 28, 2003, at 05:56:20 UTC, setting the stage for a mission that would span 182 days, 16 hours, 20 minutes, and 49 seconds aboard the International Space Station (ISS). The total duration from launch to landing was 184 days, 22 hours, 46 minutes, and 28 seconds. During this time, the crew traveled approximately 123,133,253 kilometers (76,511,456 miles) and completed 2,895 orbits of Earth.

Expedition Seven Commander Yuri I. Malenchenko (left), and NASA ISS Science Officer and Flight Engineer Edward T. Lu pose representing Rosaviakosmos, the Russian Aviation and Space Agency.

Expedition 7 to the International Space Station (ISS) was a notable chapter in the history of space exploration, led by a team of highly experienced astronauts amidst significant challenges. Commander Yuri Malenchenko, representing the Russian space agency Roscosmos, spearheaded this mission. This was Malenchenko's third journey into space, reflecting his extensive experience and deep understanding of the complexities of spaceflight. His prior missions had provided him with invaluable insights and skills crucial for managing the intricacies of space operations aboard the ISS.

NASA flight Engineer Ed Lu was another key member of the team. This mission marked Lu's third and final spaceflight, where he applied the vast knowledge he had accumulated from his previous space endeavors. His contributions were instrumental in ensuring the success of the mission.

Originally, Expedition 7 was planned to feature three astronauts from Russia and the United States. However, the tragic loss of the Space Shuttle Columbia in early 2003 led to a major shift in crew configuration. The subsequent grounding of the Shuttle fleet necessitated a reduction in the ISS crew size. Consequently, the mission was adjusted to include only two members from each country. This change significantly altered the scope and execution of the mission.

Before the Columbia disaster, Expedition 7's planned crew included Malenchenko as Commander, Ed Lu as Flight Engineer 1, and Aleksandr Kaleri as Flight Engineer 2. The tragic event and its aftermath required adjustments to the mission's objectives and operations, leading to the revised crew composition.

The Soyuz TMA-2 mission, launched on April 25, 2003, at 05:56:20 UTC from the Baikonur Cosmodrome in Kazakhstan, marked the seventh crewed mission to the ISS. The Soyuz spacecraft, designated Союз TMA-2 in Russian, was the sixth in the Soyuz series to visit the ISS and the second flight of the TMA modification. It docked with the ISS on April 28, 2003, initiating Expedition 7. The successful docking allowed the crew to begin their work, exchange responsibilities, and support ongoing scientific experiments and station operations.

The Soyuz TMA-2 spacecraft remained docked with the ISS until its undocking on October 28, 2003. During this period, the crew significantly contributed to advancing the station's research and operational capabilities.

In addition to the primary crew, the backup team for the mission included seasoned astronauts Aleksandr Kaleri and Michael Foale. Kaleri, a veteran of four spaceflights, and Foale, with six impressive space missions to his name, were prepared to step in if needed, ensuring continuity and readiness for the mission's success.

Initially, the backup crew was planned to consist of Commander Gennady Padalka from Roscosmos, Flight Engineer Pedro Duque from Spain's European Space Agency (ESA), and Flight Engineer Oleg Kotov from Russia. However, these arrangements were altered following the Columbia disaster to accommodate the revised mission requirements.

The Soyuz TMA-2 spacecraft, docked to the functional cargo block (FGB) nadir port on the International Space Station (ISS)

Soyuz TMA-2 was a critical component of the ISS's operational and research activities during its time. The mission facilitated the exchange of international crew members. It supported the continuous operation and scientific research aboard the ISS, contributing to its status as a hub of human space exploration and international collaboration.

The grounding of the Space Shuttle fleet following the Columbia disaster had significant implications for Expedition 7. The tragedy necessitated a reduction in crew size, with the mission ultimately accommodating only two members per space agency, as opposed to the three that the Shuttle could have transported. This adjustment reflected space agencies' ongoing challenges in maintaining station operations under constrained circumstances.

Expedition 7's crew, including European Space Agency (ESA) astronaut Pedro Duque, concluded their mission with a return to Earth on October 27, 2003. They undocked from the ISS at 23:17 UTC and landed in Kazakhstan at 02:41:20 UTC. The mission's duration underscored the resilience and adaptability required by the crew and mission control in the wake of unexpected developments.

The reduced crew size directly impacted the scope of scientific activities aboard the ISS. With only two primary crew members, the mission's scientific endeavors were scaled down, completing 15 distinct experiments. This reduction also led to a focus on essential maintenance tasks and spacewalk training despite no actual spacewalks being planned during the mission.

The Soyuz TMA-2 spacecraft successfully docked with the aft port of the Zvezda module on August 31, 2003, at 03:40:45 UTC. This docking marked the commencement of Expedition 7, which would see the spacecraft remain attached to the International Space Station (ISS) for 150 days. During this time, the crew engaged in various scientific experiments and station operations.

On January 28, 2004, at 08:35:56 UTC, Soyuz TMA-2 undocked from the ISS to make way for Progress M1-11's arrival. This changeover was key to maintaining the ISS's operational and logistical support.

Throughout this period, the ISS received vital resupply missions. Progress M1-10, which arrived in June 2003, and Progress M-48, which arrived in August, played crucial roles in ensuring the station's sustainability. Progress M-48, in particular, delivered essential supplies, including food, water, and oxygen for the crew and equipment necessary for ongoing scientific research.

Following its mission, Progress M-48 was deorbited on August 28, 2003, at 13:11 UTC. The spacecraft re-entered Earth's atmosphere and burned up upon re-entry, with any remaining debris landing in the Pacific Ocean.

On October 15, 2003, ISS Spacecraft Communicator Mike Fossum shared the exciting news with Expedition 7 Commander Yuri Malenchenko and Science Officer Ed Lu: the successful launch of the Long March rocket carrying the Shenzhou 5 spacecraft, with Chinese astronaut Yang Liwei aboard. This event marked China's entry into the community of spacefaring nations.

In response, Lu congratulated, "The more people that go into space, the better off we all are. This was a great achievement and good for everyone in the long run." He further extended a welcome to space in Chinese, wishing the new arrivals a safe journey.

Malenchenko also conveyed his enthusiasm, highlighting the significance of having an additional international presence in space. "I also know this was great for thousands and thousands of people from China. I congratulate all of them," he remarked.

Both Malenchenko and Lu had previously collaborated on the STS-106 Shuttle mission, during which they conducted a spacewalk together, further showcasing their extensive experience and cooperative spirit in space exploration.

The Soyuz TMA-2 spacecraft carrying cosmonaut Yuri I. Malenchenko, Expedition 7 mission commander; astronaut Edward T. Lu, NASA ISS science officer and flight engineer; and European Space Agency (ESA) astronaut Pedro Duque of Spain after landing in Kazakhstan on

Initially, Soyuz missions to the International Space Station (ISS) were designed as taxi flights, primarily intended to deliver a new Soyuz spacecraft to serve as the station's lifeboat every six months. These missions included a visiting crew but were not initially planned for regular crew exchanges. However, the tragic Space Shuttle Columbia disaster in February 2003 necessitated a significant shift in mission protocols and objectives.

Originally, Soyuz TMA-2 was scheduled to adhere to this taxi mission model. The plan was to send a visiting crew consisting of Commander Gennady Padalka and ESA astronaut Pedro Duque to the ISS for approximately one week. Following their brief stay, they were to return to Earth aboard the preceding Soyuz TMA-1 spacecraft. The third seat on Soyuz TMA-2 had been considered for Klaus von Storch, a cosmonaut from the Chilean space agency Agencia Chilena del Espacio. However, due to the uncertainties and disruptions caused by the Columbia disaster, von Storch's flight was never realized. Instead, the third seat was allocated to Russian cosmonaut Oleg Kotov, or it may have been used to transport additional cargo to the ISS.

During the Soyuz TMA-2 mission, Commander Yuri Ivanovich Malenchenko achieved a historic milestone by becoming the first person to marry in space. His marriage was conducted via long-distance communication with his bride in Texas, where such marriages are legally recognized.

The Soyuz TMA-2 spacecraft returned to Earth on October 28, 2003, marking the end of its mission. Onboard were the "Expedition 7" crew, including Yuri Malenchenko and Edward Tsang Lu, along with Pedro Duque, who had joined the ISS on Soyuz TMA-3 for a brief one-week mission.

NASA Astronaut Lu and Cosmonaut Yuri I. Malenchenko, commander, launched onboard a Soyuz rocket from Baikonur, Kazahstan.

On June 2, 2003, the European Space Agency embarked on an ambitious mission to Mars with the launch of Mars Express. This mission sought to provide a comprehensive exploration of the Red Planet. However, the mission encountered a setback when the Beagle 2 lander, an integral component of the mission, failed to re-establish contact with Earth after separating from the orbiter, leaving the mission's objectives unfulfilled.

In Martian exploration, NASA's Spirit and Opportunity rovers continued their pioneering work on the Martian surface. Spirit faced a notable setback on May 1, 2009, when it became ensnared in the sand and eventually lost contact with Earth on March 22, 2010. Despite this challenge, Opportunity far exceeded expectations, continuing its mission until June 10, 2018, well beyond its original 90-day duration.

Several notable astronauts and cosmonauts made significant contributions in 2003. Yang Liwei, China's pioneering astronaut, made history with his 21-hour orbital flight aboard Shenzhou 5. Michael Foale, a British-American astronaut, was crucial in Expedition 8 aboard the International Space Station (ISS), where he conducted experiments and performed spacewalks during his 194-day stay. Aleksandr Kaleri, a Russian cosmonaut, was also part of Expedition 8, contributing to experiments, system maintenance, and spacewalks. Charles "Charlie" Camarda, an American astronaut, was instrumental in evaluating safety measures for the Space Shuttle Discovery during the STS-114 mission, the first space shuttle flight following the Columbia disaster. Tragically, Ilan Ramon and Kalpana Chawla lost their lives during the STS-107 mission aboard the Space Shuttle Columbia.

In addition to these landmark missions, 2003 saw significant advancements in aeronautical research. NASA, in collaboration with Lockheed Martin, developed the X-56A Multi-Utility Technology Testbed, a remotely piloted aircraft designed to test active aeroelastic control technologies for flutter suppression and gust-load alleviation on flexible wing structures. The X-56A also supported research on lightweight structures and advanced control technologies for future efficient and environmentally friendly transport aircraft.

Furthermore, the year was marked by several noteworthy projects pursued by DARPA (Defense Advanced Research Projects Agency) and the US Navy. DARPA showcased the Boeing X-50 Dragonfly, a Canard Rotor/Wing, emphasizing its commitment to experimental aircraft development. The Boeing X-46, an Uncrewed Combat Air Vehicle (UCAV), was initiated but ultimately canceled without taking flight. NASA's involvement in the classified X-41 spaceplane project, part of the FALCON (Force Application and Launch from Continental US) program sponsored by DARPA and NASA, also highlighted the year. DARPA and the US Navy also succeeded

with the Northrop Grumman X-47A Pegasus, a UCAV designed for naval applications.

Expedition 8

Astronaut C. Michael Foale (right), Expedition 8 mission commander, and cosmonaut Alexander Y. Kaleri, flight engineer,

Expedition 8 marked the eighth mission to the International Space Station, spanning 192 days, 13 hours, 36 minutes, and 11 seconds aboard the station. From launch to landing, the mission extended over 194 days, 18 hours, 33 minutes, and 12 seconds. During this period, the crew traveled approximately 129,123,519 kilometers (80,233,635 miles) and completed around 3,036 orbits of Earth.

Soyuz TMA-3 was crucial in Expedition 8, serving transport vehicle and a lifeboat for the crew aboard the International Space Station (ISS). Launched on 18 October 2003, Soyuz TMA-3 carried Expedition 8 Commander Michael Foale, Flight Engineer Alexander Kaleri, and ESA Astronaut Pedro Duque from the Baikonur Cosmodrome in Kazakhstan to the ISS.

On 20 October 2003, at precisely 07:15:58 UTC, the Soyuz TMA-3 spacecraft, carrying

Expedition 8 Commander Michael Foale, Flight Engineer Alexander Kaleri, and ESA Astronaut Pedro Duque, docked with the International Space Station. Both spacecraft were orbiting the Earth above Russia at the moment of docking, marking the start of a significant mission in the ISS's ongoing development and operation.

This docking was instrumental in facilitating the crew's transition to the station. Soyuz TMA-3 provided essential support for the mission, including the means for crew transport to and from the ISS, and served as a backup vehicle in case of emergencies.

Commander Michael Foale of NASA, representing the United Kingdom and the United States, led the crew for Expedition 8. Foale, who had previously flown on five space missions, undertook his sixth and final spaceflight during this expedition. His extensive experience and leadership were pivotal in ensuring the mission's success.

Joining Foale was Flight Engineer Aleksandr Kaleri from the Russian space agency RSA. This mission was Kaleri's fourth spaceflight, and his expertise in space operations was crucial for the mission's objectives. The backup crew initially planned for this expedition included William McArthur, also from NASA, who was to serve as Commander and Flight Engineer Valeri Tokarev from Russia. However, the tragic Columbia disaster led to a reconfiguration of the crew.

During Expedition 8, the Soyuz TMA-3 spacecraft ensured that the crew had a reliable means of return to Earth.

Once aboard the station, the crew began their work immediately after the Expedition 7 crew undocked, setting into motion a more than six-month mission centered on station operations and maintenance.

Astronaut C. Michael Foale (right), Expedition 8 mission commander and NASA ISS science officer, and cosmonaut Alexander Y. Kaleri, flight engineer, conduct a teleconference with the Moscow Support Group for the Russian New Year celebration, via Ku- and S-band, with audio and video relayed to the Mission Control Center (MCC) at Johnson Space Center (JSC). Kaleri represents Rosaviakosmos.

An overall view of the station flight control room (BFCR) in the Mission Control Center (MCC) at Johnson Space Center (JSC), Houston, Texas during the Expedition 8 mission.

A notable highlight of Expedition 8 was the execution of the first two-person spacewalk at the ISS. Following the Space Shuttle Columbia accident, station spacewalks continued. Still, only from the Russian segment with the added complication that with the resident crew size was reduced to two, the pair of spacewalking crew members left no one inside to monitor its systems. Although this posed a slightly increased risk should something go wrong, these "two-person" spacewalks proved essential during the shuttle hiatus. Expedition 8 crew members Aleksandr Y. Kaleri and Mike Foale conducted the first of these EVAs on Feb. 26, 2004. Foale had prior experience with the Orlan suit, as he had completed an EVA during his long-duration stay aboard Mir in 1997. The crew had to cut the spacewalk short due to Kaleri's suit overheating and water droplets forming inside his helmet. The crew later identified the problem as a kink in the water line in his liquid cooling garment. The incident previewed a more serious problem, which would occur in an EMU during an EVA more than nine years later.

The spacewalk, which lasted 3 hours and 55 minutes, marked a notable achievement in ISS operations as it was conducted without a crew member inside the station, a first for such an undertaking.

During this extravehicular activity (EVA), Foale and Kaleri were initially focused on several key tasks despite encountering a setback. The spacewalk was cut short due to a malfunction in Kaleri's spacesuit's cooling system, which affected their ability to complete all planned activities. Nevertheless, the astronauts accomplished several important objectives.

One of their primary tasks was replacing cassette containers that held samples for an experiment designed to study the effects of long-duration exposure to microgravity. These samples were crucial for understanding how extended stays in space impact various materials and biological specimens.

Foale also attended to two additional cassette replacements on the exterior of the Zvezda Service Module. These cassettes were part of the Matryoshka experiment, which aimed to provide valuable data on radiation exposure during spaceflight by measuring radiation levels on the station's hull.

In addition to these tasks, the spacewalkers removed one of the suitcase-sized devices associated with the Japanese Aerospace Exploration Agency's MPAC-SEEDS experiment. This experiment, set up to study micro-meteor impacts and material exposure in space, had been installed by Expedition 3 spacewalkers on 15 October 2001. Foale and Kaleri also relocated a second device from the same experiment, furthering the research into space debris and its effects on spacecraft materials.

However, the astronauts could not complete all their planned activities due to the cooling system issue. Specifically, they did not finish removing laser light retroreflector devices from the aft end of Zvezda , being tested as potential navigation aids for the European Space Agency's Automated Transfer Vehicle. This vehicle would later make its first trip to the ISS in 2008. Additionally, the crew did not work on the Kromka experiment, designed to measure the residue emitted from Zvezda's jet thruster firings.

Despite the interruptions, the spacewalk contributed significantly to the ongoing maintenance and scientific research aboard the ISS, showcasing the resilience and adaptability of the Expedition 8 crew in addressing unexpected challenges.

On 29 April 2004, Soyuz TMA-3 undocked from the ISS, completing its mission with a departure time of 20:52:09 UTC. The spacecraft then safely transported the departing crew—Foale, Kaleri, and ESA Astronaut André Kuipers, who had joined the ISS with Expedition 9—back to Earth. The mission lasted 192 days, 13 hours, 36 minutes, and 11 seconds docked with the station.

Elsewhere in the space race, NASA's Mars Exploration Rover mission continued to make significant strides as the Spirit and Opportunity rovers roamed the Martian surface. These rovers, equipped with a suite of scientific instruments, conducted detailed investigations of the planet's geology and atmospheric conditions. Their findings contributed to a deeper understanding of Mars' history and its potential to support life. Concurrently, the European Space Agency's Mars Express mission made valuable contributions by studying the Martian atmosphere and surface, further expanding our knowledge of the Red Planet and its complex geological features.

In NASA's Space Shuttle program, the tragic loss of the Space Shuttle Columbia in February 2003 led to a suspension of shuttle flights as NASA undertook a comprehensive review of its safety protocols. Now, the space agency was focused on implementing critical safety improvements and preparing for the return of shuttle missions, which would resume following this rigorous reform period.

Astronaut C. Michael Foale (left), Expedition 8 commander and NASA ISS science officer; cosmonaut Alexander Y. Kaleri (center), Soyuz flight engineer representing Russia's Federal Space Agency; and European Space Agency (ESA) astronaut Andre Kuipers (right) of the Netherlands, successfully landed in north central Kazakhstan on April 30, 2004, in their Soyuz TMA-3 capsule. Foale and Kaleri completed 195 days in space aboard the International Space Station (ISS), while Kuipers returned after an 11-day research mission as part of a commercial agreement between ESA and Russia's Federal Space Agency. Photo Credit: NASA/Bill Ingalls

A landmark achievement occurred in 2004 with the successful suborbital flights of SpaceShipOne, a privately funded spacecraft developed by Scaled Composites. SpaceShipOne's flights represented a significant milestone in commercial spaceflight, demonstrating the viability of private sector involvement in space exploration. This achievement paved the way for future commercial ventures and highlighted the increasing role of private entities in the space industry.

In spacecraft technology, Russia made strides with the development and testing of Kosmos-2410, a spacecraft designed for military payloads. Developed by TsSKB Progress and OAO MZ Arsenal, Kosmos-2410 showcased Russia's ongoing commitment to advancing its space capabilities and expanding its portfolio of space technologies.

NASA's aeronautical advancements were exemplified by the X-43A, which achieved a historic first by operating as the first scramjet-powered aircraft. This milestone marked a significant advancement in hypersonic flight technology, offering promising prospects for future high-speed and efficient transportation.

Expedition 9

Astronaut Edward M. (Mike) Fincke (left), Expedition 9 NASA ISS science officer and flight engineer, and cosmonaut Gennady I. Padalka, commander representing the Federal Space Agency.

Expedition 9, spanning from April 21, 2004, to October 23, 2004, marked the ninth mission of human operations aboard the International Space Station (ISS). This expedition was distinguished by a series of significant events, including critical crew changes, essential maintenance tasks, and a series of spacewalks aimed at enhancing the station's capabilities and preparing it for future missions.

The journey began with the arrival of the Soyuz TMA-4 spacecraft at the ISS on April 21, 2004. This spacecraft, launched on April 19, 2004, from the Baikonur Cosmodrome aboard the Soyuz FG launch vehicle, carried a distinguished crew composed of Russian Commander Gennady Padalka, American Flight Engineer Michael Fincke, and European Space Agency (ESA) Astronaut André Kuipers. The mission was the first spaceflight for both Fincke and Kuipers, while Padalka brought prior space experience to the team. Upon arrival, the Expedition 9 crew undertook a week-long handover with the outgoing Expedition 8 crew, allowing them to acclimate to the ISS's systems and procedures.

Soyuz TMA-4, a Russian passenger spacecraft launched by a Soyuz-FG rocket from Baikonur at 03:19 UT on April 19, 2004. It carried three astronauts (a Russian, an American and a Dutchman) to the International Space Station (ISS). It docked with the Zarya module of the ISS automatically on April 21 at 05:01 UT.

On April 29, 2004, Padalka and Fincke formally assumed command of the ISS from Expedition 8 Commander Michael Foale and Flight Engineer Alexander Kaleri, who departed the station on the Soyuz TMA-3 spacecraft. The Expedition 9 crew's tenure lasted 185 days, 15 hours, and 7 minutes, during which they traveled approximately 121,802,083 kilometers (75,684,306 miles) and completed 2,940 orbits of Earth. The total mission duration from launch to landing was 187 days, 21 hours, and 16 minutes.

The Soyuz TMA-4 spacecraft approaches the International Space Station (ISS). Onboard the spacecraft are cosmonaut Gennady I. Padalka, Expedition 9 commander representing Russia's Federal Space Agency; astronaut Edward M. (Mike) Fincke, NASA ISS science officer and flight engineer; and European Space Agency (ESA) astronaut Andre Kuipers of the Netherlands. The Soyuz linked to the nadir docking port of the Zarya Control Module at 12:01 a.m. (CDT) on April 21, 2004 as the two spacecraft flew 230 miles above central Asia. The docking followed Monday's launch from the Baikonur Cosmodrome in Kazakhstan.

The mission was marked by several pivotal activities, including the arrival of the ISS Progress 14 spacecraft on May 25, 2004. This cargo spacecraft, which carried vital scientific equipment and supplies, docked automatically at the rear port of the Zvezda module three hours and twenty minutes after launch. Progress 14 delivered 2,528 kilograms of cargo, including 1,358 kilograms of dry cargo. Its arrival followed the undocking of the ISS Progress 13 spacecraft on May 24, 2004, which was part of routine logistical operations.

Astronaut Edward M. (Mike) Fincke (left), NASA International Space Station (ISS) science officer and flight engineer; cosmonaut Gennady I. Padalka (center), Russia's Federal Space Agency Expedition 9 mission commander; and European Space Agency (ESA) astronaut Andre Kuipers (right) of the Netherlands

In July 2004, the ISS Progress 14 spacecraft undocked on July 30, clearing the docking port for new arrivals. Shortly after that, on August 14, 2004, the ISS Progress 15 spacecraft docked with the station, continuing the essential flow of supplies necessary for the ISS's operations and research.

One of the key aspects of Expedition 9 was the series of four spacewalks (extravehicular activities or EVAs) undertaken by the crew. These spacewalks, conducted using Russian Orlan spacesuits and launched from the Pirs Docking Compartment, were crucial for the maintenance and assembly of the ISS. Unfortunately, the first spacewalk, on June 24, 2004, was cut short due to a pressure issue with Fincke's primary oxygen tank. The spacewalk was rescheduled for June 29, 2004, when Padalka and Fincke completed the second EVA, lasting 5 hours and 40 minutes. During this spacewalk, they replaced a failed Remote Power Controller Module (RPCM), which had caused the temporary loss of Control Moment

Gyroscope No. 2, a critical component for the ISS's orientation control.

The third spacewalk, on August 3, 2004, lasted 4 hours and 30 minutes. This EVA focused on preparing the ISS for the arrival of the European Space Agency's Automated Transfer Vehicle (ATV). Tasks included removing and updating laser retroreflectors, installing new antennas, and replacing materials science experiments exposed to the space environment.

The final spacewalk of Expedition 9 took place on September 3, 2004, and extended for 5 hours and 20 minutes. This EVA involved several preparatory tasks for future assembly operations and support for the ATV. The crew replaced the Zarya Control Module's flow control panel, installed safety tether fairleads on Zarya's handrails, and prepared various components for the ATV's arrival, including equipment for air-to-air radio link antennas.

Expedition 9 concluded with the handover to the Expedition 10 crew on October 16, 2004. On this date, Leroy Chiao from the United States assumed the role of Commander, and Salizhan Sharipov from Russia joined as Flight Engineer. This transition marked the end of Expedition 9, led by Russian Commander Gennady Padalka and American Flight Engineer Michael Fincke.

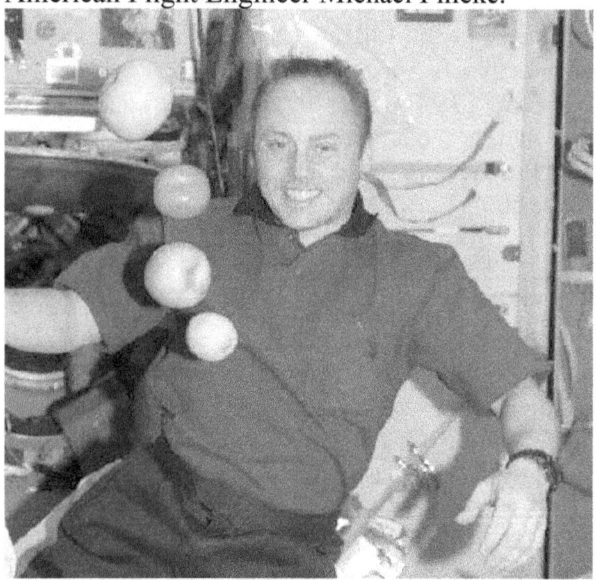

Astronaut Edward M. (Mike) Fincke, Expedition 9 NASA ISS science officer and flight engineer, near fresh fruit floating freely in the Zvezda Service Module of the International Space Station (ISS).

Expedition 10

Astronaut Leroy Chiao, left, was commander and NASA ISS science officer. Cosmonaut Salizhan S. Sharipov, representing Russia's Federal Space Agency, was flight engineer.

Expedition 10, which commenced on October 16, 2004, was a significant chapter in the International Space Station (ISS) history, marked by notable scientific progress and key crew transitions. The mission was led by Commander Leroy Chiao and Flight Engineers Salizhan Sharipov and Robert Thirsk, who were tasked with advancing the ISS's research programs and maintaining the station's operations.

As Expedition 10 unfolded, the ISS witnessed a series of critical logistical maneuvers. On December 22, 2004, the ISS Progress 15 spacecraft undocked after completing its resupply mission.

This action cleared the docking port for the arrival of new cargo. The following day, December 23, 2004, the ISS Progress 16 launched from Baikonur Cosmodrome, carrying essential supplies, scientific experiments, and equipment. It docked with the ISS on December 25, ensuring the crew had the necessary resources to continue their research and daily tasks.

A key operational activity during this period involved the relocation of the Soyuz TMA-5 spacecraft on November 29, 2004. This maneuver was part of routine operations to optimize docking configurations and ensure efficient use of docking ports.

Soyuz TMA-5 crew. Chiao and Sharipov will launch with Shargin for a six-month mission on the International Space Station, while Shargin will spend eight days on the Station, returning with the Expedition 9 crew.

The crew of Soyuz TMA-5 included Tokarev, McArthur, and Vittori, who spent a significant portion of their mission integrated with the Expedition 10 team. Their presence bolstered the ISS's human resources and brought new expertise to the station. The Expedition 10 crew, led by Commander Leroy Chiao and Flight Engineer Salizhan Sharipov, benefited from this additional support, which was crucial for maintaining the ISS's objectives.

Commander Leroy Chiao, from NASA, was on his fourth and final spaceflight and overseeing the mission's objectives. Salizhan Sharipov, representing the Russian Space Agency (RSA), was on his second and final spaceflight, playing a key role as both Soyuz commander and flight engineer. Upon their arrival on October 16, 2004,

aboard the Soyuz TMA-5, they relieved the previous crew members, Mike Fincke and Gennady Padalka, who had been part of Expedition 9.

During their tenure, the crew undertook several important tasks. They replaced critical hardware in the Quest Joint Airlock, essential for conducting spacewalks, and repaired U.S. spacesuits to ensure their functionality for future missions. They also contributed to the Advanced Diagnostic Ultrasound in Microgravity Project by submitting a paper on applying ultrasound technology in space.

A notable historical milestone during Expedition 10 was Leroy Chiao becoming the first astronaut to vote in a U.S. Presidential election from space, underscoring the mission's blend of scientific achievement and personal milestones.

The mission faced an unexpected challenge during preflight testing when an explosive bolt on the Soyuz TMA-5 was accidentally activated. This caused damage that required repairs before launch. Despite this setback, the mission launched successfully from Baikonur Cosmodrome on October 14, 2004.

During Expedition 10, the International Space Station (ISS) benefitted from a series of crucial resupply missions provided by the Progress spacecraft. These missions were integral to sustaining the station's operations and supporting the crew's daily needs.

The first of these vital missions was conducted by Progress 15. This spacecraft, a Russian cargo vessel designed to deliver supplies, docked with the ISS on December 25, 2004. The docking occurred at the Pirs Docking Compartment, a key docking port on the station. Progress 15 carried a comprehensive load of essential materials, including scientific experiments, spare parts, and other necessary provisions to support the station's ongoing research and daily functions. The arrival of Progress 15 ensured that the ISS had an ample supply of consumables, such as food and water, as well as critical equipment for maintaining the station's systems and conducting experiments.

Following Progress 15, the ISS welcomed another important cargo vessel, Progress 17, which arrived on February 28, 2005. Progress 17 docked with the ISS at the same Pirs Docking Compartment. This mission was equally vital, as it delivered a fresh batch of supplies and scientific materials necessary for the continued success of the station's experiments and operations. The cargo included scientific research materials, technological equipment, and additional provisions that the crew relied upon to conduct their extensive research program and maintain the station's systems.

During Expedition 10, the crew undertook two notable spacewalks that significantly contributed to the advancement and preparation of the International Space Station (ISS).

The first spacewalk took place on January 26, 2005, and was a pivotal event for the mission. During this extravehicular activity (EVA), astronauts Leroy Chiao and Salizhan Sharipov ventured outside the ISS to perform essential tasks. Their primary objectives included installing scientific experiments and preparing the station's exterior for future missions. The work carried out during this EVA was crucial for maintaining and upgrading the station's systems, which are essential for its ongoing research and operational capabilities.

The second spacewalk, conducted on March 28, 2005, built upon the progress made in the earlier EVA. In this mission, Chiao and Sharipov continued their work to enhance the ISS's readiness for future upgrades. This spacewalk was particularly focused on preparing the station for the European Automated Transfer Vehicle (ATV), a spacecraft designed to deliver cargo and provide additional support to the ISS. Although the ATV was initially scheduled to arrive in 2006, it was ultimately postponed until 2008. The preparatory work done during this spacewalk ensured the station was well-positioned to integrate the ATV's capabilities when it arrived.

As Expedition 10 concluded, preparations were underway for a smooth transition to Expedition 11. On April 14, 2005, the new crew for Expedition 11, comprising Commander Sergey Krikalev, Flight Engineer John Phillips, and Flight Engineer Roberto Vittori, launched from the Baikonur Cosmodrome in Kazakhstan. Their mission was to continue the essential work and scientific research aboard the International Space Station (ISS). The crew successfully docked with the ISS on April 16, 2005, marking the beginning of their tenure and the official handover of responsibilities from Expedition 10.

This transition was a critical moment for the ISS, ensuring the continuity of its operations and research endeavors. Expedition 11's arrival not only marked the end of a significant chapter in the ISS's history but also set the stage for future progress and discoveries.

Expedition 10 spanned 190 days, 11 hours, and 23 minutes, during which the crew accomplished many tasks. Over this period, they traveled approximately 122 million kilometers (76 million miles) and completed around 2,975 orbits of Earth. These extensive metrics underscore the demanding nature of their mission, which was vital for the ISS's ongoing scientific and operational success.

The Soyuz TMA-5 spacecraft, integral to Expedition 10, served as both a transport vehicle and an emergency evacuation craft, ensuring the crew's safety and supporting their mission objectives. After spending 192 days, 19 hours, and 2 minutes in space, the Soyuz TMA-5 spacecraft re-entered Earth's atmosphere on April 24, 2005. The re-entry was executed flawlessly, and the spacecraft landed approximately 53 miles northeast of Arkalyk, Kazakhstan.

Salizhan Sharipov (left) and Leroy Chiao (right) work with their Russian Orlan spacesuits in the Pirs Docking Compartment of the International Space Station (ISS). (NASA)

The Soyuz TMA-5 spacecraft blasts off from the Baikonur Cosmodrome in Kazakhstan October 14, 2004, carrying astronaut Leroy Chiao, Expedition 10 commander and NASA International Space Station (ISS) science officer, cosmonaut Salizhan S. Sharipov, Russia's Federal Space Agency flight engineer and Soyuz commander, and Russian Space Forces cosmonaut Yuri Shargin to the ISS. The crew docked to the Station on October 16, and Chiao and Sharipov replaced the current Station crewmembers, cosmonaut Gennady I. Padalka, Expedition 9 commander, and astronaut Edward M. (Mike) Fincke, NASA ISS science officer and flight engineer, who returned to Earth October 24 with Shargin.

In 2005, space exploration witnessed a series of pivotal activities and milestones, underscoring a period of dynamic advancement and expanding international collaboration. This year saw significant developments from space agencies and private entities, reflecting the escalating global interest in space exploration and scientific discovery.

NASA made significant strides in planetary exploration with the launch of the Mars Reconnaissance Orbiter. This spacecraft, which would reach Mars the following year, was tasked with studying the Martian surface and atmosphere. Its observations provided valuable insights into Mars' geological and atmospheric conditions, furthering our understanding of the Red Planet.

NASA's Cassini spacecraft, orbiting Saturn since 2004, continued its groundbreaking work in 2005. Cassini's observations provided a wealth of information about Saturn and its intricate system of moons, enhancing our knowledge of the planet's complex dynamics and the various processes occurring in its atmosphere and rings.

In October 2005, China achieved a significant milestone with the launch of the Shenzhou 6 mission. This mission marked China's second manned spaceflight and underscored the country's growing capabilities in human space exploration. The successful execution of Shenzhou 6 demonstrated China's advancing expertise and commitment to expanding its space program.

The year 2005 also witnessed a landmark event in private spaceflight. Space Adventures, a private spaceflight company, facilitated the first privately funded human spaceflight in September. Businessman Greg Olsen became the third space tourist to visit the ISS, traveling aboard a Russian Soyuz spacecraft. This mission highlighted the increasing involvement of private entities in space exploration and underscored the burgeoning opportunities for commercial participation in space travel.

On the scientific research front, NASA's Dryden Flight Research Center conducted extensive tests with the X-56A, a remotely piloted aircraft developed by Lockheed Martin. These tests aimed to advance research on lightweight structures and cutting-edge control technologies for future aircraft, emphasizing the drive towards more efficient and environmentally friendly transport solutions. Furthermore, the X-37 program, originally under NASA's management, was transferred to the Defense Advanced Research Projects Agency (DARPA) in 2005. This transition marked the program's evolution into a classified project, reflecting its significance in national security and ongoing development.

Expedition 11

ISS Expedition 11 Commander Sergei Krikalev and
Flight Engineer John Phillips

Expedition 11, launched on April 17, 2005, was a critical mission in the International Space Station (ISS) program, representing the eleventh crewed operation of the ISS and lasting until October 10, 2005. This mission was conducted aboard the Soyuz TMA-6 spacecraft, which, while primarily intended for emergency evacuation, played a crucial role throughout its stay at the ISS.

Commander Sergei K. Krikalev, an esteemed Russian cosmonaut, led the Expedition 11 crew. Known for his remarkable career, Krikalev was undertaking his sixth and final spaceflight during this mission. His extensive experience and leadership were vital in navigating the complexities of this mission. Flight Engineer 1 John L. Phillips from the United States joined him, representing NASA. Phillips was on his second spaceflight and significantly contributed to the mission's scientific research and operational objectives.

European Space Agency astronaut Roberto Vittori also launched with Expedition 11 aboard Soyuz TMA-6. However, his tenure was brief, as he returned to Earth on April 24, 2005, aboard Soyuz TMA-5 with the previous Expedition 10 crew.

Two spacewalks were initially scheduled during Expedition 11, but only one was completed. On August 18, 2005, the crew conducted a four-hour and fifty-eight-minute extravehicular activity, beginning at 19:02 UTC (3:02 p.m. EDT). This spacewalk involved several critical tasks aimed at advancing the scientific research and operational efficiency of the International Space Station (ISS).

The astronauts first removed and brought a Russian Biorisk experiment container inside the station. This container housed bacteria samples that had been collected from the exterior of the Pirs docking compartment, providing valuable data for understanding the impact of space conditions on microbial life. Additionally, they addressed the micrometeoroid and orbital debris collector (MPAC) and the materials exposure array (SEED) panel, both of which had been mounted on the Zvezda Service Module. These instruments were essential for studying the effects of micrometeoroids and orbital debris and the behavior of various materials exposed to the harsh environment of space.

The crew also attended to the Matroska experiment, which involved radiation dosimeters designed to measure radiation exposure in materials that simulate human tissue. Furthermore, they installed a new television camera on the Zvezda module, which enhanced the station's visual monitoring capabilities. During the spacewalk, they also checked a Korma contamination-exposure experiment tablet and replaced a materials exposure experiment container.

Despite the initial plan for two spacewalks, this was the only one successfully executed during Expedition 11.

One of the pivotal moments of Expedition 11 was the arrival of Space Shuttle Discovery (STS-114) on July 28, 2005, at 11:18 UTC. This marked the shuttle's return to spaceflight following the tragic Columbia disaster. Discovery's mission included delivering critical supplies to the ISS, such as a replacement Control Moment Gyroscope and an External Stowage Platform 2, as part of approximately 4,100 kilograms of cargo in its payload bay and the Multi-Purpose Logistics

Module (MPLM) Raffaello. The shuttle's successful docking and subsequent undocking on August 6, 2005, with the MPLM were significant for maintaining the station's operational capabilities.

During Expedition 11, Commander Sergei Krikalev set a new record for cumulative time spent in space. On August 16, 2005, surpassing the previous 747.593 days held by Sergei Avdeyev, Krikalev achieved 803 days, 9 hours, and 39 minutes in space by the end of the mission. This achievement highlighted his exceptional contributions to space exploration and set a new benchmark in human spaceflight.

The mission also included various logistical operations to support the ISS. On September 7, 2005, the unpiloted Progress Spacecraft 53 (P18) was deorbited, clearing the way for Progress 54 (P19), which arrived later that month. Progress 54 delivered approximately 2,300 kilograms of essential cargo, including fuel, water, dry supplies, and oxygen generators necessary for sustaining the crew and maintaining station operations.

Soyuz TMA-7 launched atop a Soyuz-FG rocket from Baikonur Cosmodrome, docked with the ISS on October 3, 2005, at 05:27 UTC, initially connecting with the Pirs module. On November 18, 2005, it undocked from Pirs at 08:46 UTC and then re-docked at 09:05 UTC to the nadir port of the Zarya module. After another undocking on March 20, 2006, from Zarya's nadir port, Soyuz TMA-7 re-docked to the aft port of the Zvezda module. Finally, it undocked from the ISS on April 8, 2006, at 20:28 UTC.

Upon its ascent, Soyuz TMA-7 carried a trio of distinguished individuals. The primary crew included ISS Commander William McArthur and Flight Engineer Valery Tokarev, who were tasked with joining and eventually replacing the outgoing Expedition 11 team members. Their journey marked an essential transition in the ISS's operational phases. In addition to McArthur and Tokarev, the mission also featured Spaceflight Participant Gregory Olsen, an American scientist and entrepreneur. Olsen's involvement in the mission was notable as he brought a fresh perspective and conducted various scientific experiments during his eight-day stay aboard the ISS.

STS-114 crew portrait. In front are astronauts Eileen M. Collins (right), commander; Wendy B. Lawrence, mission specialist; and James M. Kelly, pilot. In back are astronauts Stephen K. Robinson (left), Andrew S. W. Thomas, Charles J. Camarda, and Soichi Noguchi, all mission specialists. Noguchi represents Japan Aerospace Exploration Agency (JAXA). Image courtesy NASA.

Space Shuttle Discovery at night before the launch of STS-114 mission.

On October 1, 2005, Soyuz TMA-7 was launched from Baikonur Cosmodrome aboard a Soyuz-FG rocket, marking a significant moment in the ongoing collaboration at the International Space Station (ISS). This mission was tasked with transporting vital personnel for ISS Expedition 12. The crewed spacecraft delivered Commander William McArthur and Flight Engineer Valery Tokarev to the station, relieving the Expedition 11 team. In addition, the mission included Spaceflight Participant Gregory Olsen, who joined the TMA-7 crew for the journey to the ISS.

Soyuz TMA-7 arrived at the ISS on October 3, 2005, docking with the Pirs docking compartment at 05:27 UTC. The docking marked the start of a new phase in the Expedition 12 mission. After approximately eight days aboard, Gregory Olsen conducted a series of experiments before departing with the outgoing Expedition 11 crew on Soyuz TMA-6. McArthur and Tokarev, now part of Expedition 12, remained at the station, continuing their duties in orbit.

Their time at the ISS saw a notable addition to their team when Marcos Pontes, a Brazilian astronaut, joined them. Pontes had launched aboard Soyuz TMA-8 and spent about seven days onboard, working on experiments for the Brazilian Space Agency. His mission provided valuable contributions to the station's research efforts.

Several significant docking and undocking events marked the Soyuz TMA-7 spacecraft's journey. After initially docking to the Pirs module, Soyuz TMA-7 undocked on November 18, 2005, at 08:46 UTC, only to dock again later the same day at 09:05 UTC, this time to the nadir port of the Zarya module. Following further missions, the spacecraft undocked from Zarya on March 20, 2006, at 06:49 UTC, and subsequently docked to the aft port of Zvezda at 07:11 UTC. The mission concluded with its undocking from the ISS on April 8, 2006, at 20:28 UTC, representing the 28th crewed flight to the station.

Soyuz TMA-7 was also the final mission covered under the 1996 "balance" agreement, which mandated Russia's provision of eleven Soyuz spacecraft to ferry U.S.-Russian crews to and from the ISS. Subsequent missions required new negotiations between NASA and the Russian space agency and modifications to the Iran Nonproliferation Act of 2000.

The mission faced a critical challenge during re-entry when the main parachute took an unusually long time to deploy after the pilot chute was released at 10 km altitude. This delay raised concerns about the crew's safety and underscored the importance of the parachute system's reliability in ensuring a safe landing.

Soyuz TMA-7's mission was a significant chapter in the ISS's history and a milestone in the broader context of U.S.-Russian cooperation in space. This flight was the 28th crewed mission to the ISS, and it fulfilled the last obligation of the 1996 "balance" agreement, which mandated the provision of 11 Soyuz spacecraft for ferrying joint U.S.-Russian crews. This agreement marked a

critical era in international space collaboration, and subsequent missions required new negotiations and adjustments to international regulations.

Upon its return to Earth, Soyuz TMA-7 faced a challenging re-entry. During descent, the deployment of the pilot parachute at 10 kilometers altitude was followed by a delayed opening of the main parachute. This delay raised concerns among the crew, as it could result in a catastrophic failure if the main parachute had not deployed in time. Fortunately, the spacecraft landed safely, concluding a mission that had played a vital role in maintaining the ISS's continuous human presence and operational success.

The crew of spacecraft Soyuz TMA-7: Expedition 12 Commander William S. McArthur, Jr. (right), Flight Engineer and Soyuz Commander Valery I. Tokarev (center) and U.S. Spaceflight Participant Gregory Olsen

This mission marked the 28th crewed flight to the ISS and was the last to be covered by the 1996 "balance" agreement, which mandated Russia to provide 11 Soyuz spacecraft for ferrying U.S.-Russian crews.

Expedition 11, which lasted 176 days, 19 hours, and 30 minutes aboard the ISS, completed 2,817 orbits of Earth. This mission was notable for its duration and contribution to the ISS program, including the groundbreaking Rendezvous Pitch Maneuver conducted by Space Shuttle Discovery. This maneuver was a historic first in spaceflight, underscoring human space exploration's ongoing evolution and milestones.

ISS Expedition 12 crew, Soyuz TMA-7, launched a few days earlier from the Baikinour Cosmodrome, approaches the International Space Station. The Aral Sea was clearly visible in the background. Backdropped by the blackness of space and Earth's horizon, the Soyuz TMA-7 spacecraft approaches the international space station. Onboard the spacecraft are astronaut William S. McArthur, Jr., Expedition 12 commander and NASA science officer; cosmonaut Valery I. Tokarev, Expedition 12 flight engineer and Soyuz commander; and U. S. Spaceflight Participant Gregory Olsen. The Soyuz linked up to the Pirs Docking Compartment at 12:27 a.m. (CDT) on Oct. 3, 2005 as the two spacecraft flew over eastern Asia. The docking followed Friday's launch from Baikonur Cosmodrome in Kazakhstan.

Expedition 12

Expedition 12 marked a significant chapter in the ongoing saga of the International Space Station (ISS) as it completed its twelfth rotation of astronauts. The mission officially began with the launch of the Russian Soyuz TMA-7 spacecraft on September 30, 2005, from the Baikonur Cosmodrome in Kazakhstan. Over 189 days, 19 hours, and 53 minutes, the crew of Expedition 12 spent a total of 187 days, 14 hours, and 1 minute aboard the ISS, orbiting Earth nearly 3,000 times.

Throughout 2005, the ISS continued its vital role in hosting both crew and cargo missions, with Soyuz TMA-6 and Soyuz TMA-7 serving as key components of its operational framework. Soyuz TMA-6, which launched on April 9, 2005, carried a distinguished crew including Russian cosmonaut Sergei Krikalev, making his sixth and final spaceflight, American astronaut John Phillips on his second flight, and European Space Agency (ESA) astronaut Roberto Vittori. Krikalev's extensive experience, coupled with Phillips' and Vittori's backgrounds, formed a highly capable team dedicated to advancing the ISS's research and maintaining its operations. This mission, the 26th crewed flight to the ISS, was pivotal, serving both as a transport vehicle and an emergency evacuation craft.

ISS Expedition 12 Commander William McArthur and flight engineer Valery Tokarev

Soyuz TMA-6 was launched aboard a Soyuz FG rocket and docked with the ISS on April 17, 2005, connecting with the Pirs docking compartment. On April 24, 2005, Vittori returned to Earth on Soyuz TMA-5 with the previous Expedition 10 crew, exemplifying the seamless transitions typical of ISS operations. The mission also witnessed the crucial replacement of the Expedition 10 crew—Commander Leroy Chiao and Flight Engineer Salizhan Sharipov—with the Expedition 11 crew.

Expedition 12 commenced with the arrival of Soyuz TMA-7, which brought aboard William S. McArthur as the mission commander, Valeri I. Tokarev as Flight Engineer 1, and Gregory Olsen, an American entrepreneur and space tourist. McArthur, in his fourth and final spaceflight, was joined by Tokarev, who was on his second and final spaceflight. Olsen's participation marked his role as the third space tourist to visit the ISS. During the mission, Soyuz TMA-6 undocked from the Pirs module on July 19, 2005, and re-docked to the nadir port of the Zarya module later the same day. The spacecraft finally undocked from the Zarya port on October 10, 2005.

The crew for spacecraft Soyuz TMA-6. From left : John Phillips (USA), Sergei Krikalev (Russia), Roberto Vittori (Italy).

A notable aspect of Soyuz TMA-6's mission was its return flight, which experienced a cabin pressure leak in the descent module, prompting further investigation. The mission also facilitated the transition of the Expedition 10 crew with the Expedition 11 crew, ensuring the continuity of the ISS's operations.

Expedition 12's mission on the International Space Station (ISS) was marked by routine maintenance and significant scientific research in the unique microgravity environment. Among the key activities were two notable spacewalks, or extravehicular activities (EVAs), which played crucial roles in the ongoing development and upkeep of the station.

The first spacewalk occurred on November 7, 2005, and lasted 5 hours and 22 minutes. During this EVA, Commander William S. McArthur and Flight Engineer Valeri Tokarev undertook essential tasks to enhance the ISS's infrastructure. Their primary objective was to install a new camera on the P1 Truss, an important component of the station's external structure. This camera was intended to facilitate future installations of additional truss segments, which are crucial for the station's expanding research capabilities and for supporting new modules. Additionally, the crew addressed a technical issue by jettisoning the Floating Potential Probe. This instrument, designed to measure the station's electrical potential relative to the surrounding plasma, had failed to perform its intended function and was removed to prevent potential interference with other equipment.

The second spacewalk took place on February 3, 2006, and lasted 5 hours and 43 minutes. This EVA was notable for its focus on several critical tasks, including the disposal of SuitSat-1. SuitSat-1 was an old Russian Orlan spacesuit repurposed to serve as a floating educational tool equipped with a radio that broadcasted educational messages to students worldwide. Its mission had concluded, and it was safely ejected from the station to make way for new scientific endeavors. During this spacewalk, McArthur and Tokarev also retrieved the Biorisk experiment, which was part of a study assessing the impact of space conditions on biological systems. Additionally, they documented a sensor for a micrometeoroid experiment, which was used to study the effects of micrometeoroids on the station's surface. Lastly, they secured the remaining umbilical of the mobile transporter, ensuring that the station's moving parts were properly maintained and operational.

A remarkable highlight of Expedition 12 was witnessing the total solar eclipse on March 29, 2006. The crew captured breathtaking images of the Moon's shadow cast upon Earth, providing a unique perspective from space.

On November 3, 2005, Expedition 12 experienced an extraordinary event as Paul McCartney performed a live concert link-up from the Arrowhead Pond in Anaheim, California. This unique musical touch included McCartney's performances of "Good Day Sunshine" and "English Tea," broadcast live on NASA TV, adding a special moment to the astronauts 'daily routine.

The mission concluded with a successful landing in Kazakhstan on April 8, 2006. This landing was historically significant because it included Brazil's first astronaut, Marcos Pontes, symbolizing the growing global engagement in space exploration.

Elsewhere in the space race, the international space community saw significant advancements in the broader landscape of space exploration in 2005. Virgin Galactic's Astronaut Pilots Group, consisting of Steve Johnson, Alistair Hoy, David MacKay, and Alex Tai, signaled the burgeoning era of commercial spaceflight. Malaysia's Angkasawan Group selection on September 4th, which included Sheikh Muszaphar Shukor, Faiz Khaleed, Siva Vanajah, and Mohammed Faiz Kamaludin, marked Malaysia's entry into space exploration. Sheikh Muszaphar Shukor later achieved historical significance as Malaysia's first astronaut aboard Soyuz TMA-11.

Russia's TsPK-14/RKKE-16 Cosmonaut Group, announced on October 11th, included Aleksandr Misurkin, Oleg Novitskiy, Aleksey Ovchinin, Maksim Ponomaryov, Sergey Ryzhikov, Yelena Serova, and Nikolai Tikhonov, representing a new generation of skilled cosmonauts essential for future missions. South Korea's Astronaut Program Group, introduced in December with Yi So-yeon and Ko San, exemplified the country's rapid strides in space exploration. Yi So-yeon would become South Korea's first astronaut, marking another milestone in the global space exploration narrative.

In the space race, on December 14, 2006, the United States launched a Delta II rocket from Vandenberg Air Force Base to place the A-193 American reconnaissance satellite into orbit. Unfortunately, the satellite failed to function properly and began to decay from orbit a month after its launch. Due to the presence of approximately 1000 pounds of toxic hydrazine fuel onboard, the US Navy executed a RIM-161 Standard Missile 3 anti-ballistic missile test on February 20, 2007. The missile destroyed the malfunctioning satellite, creating an explosion consistent with the destruction of the hydrazine fuel tank, thus addressing the potential environmental hazard.

STS-115, launched by Space Shuttle Atlantis, marked a pivotal moment in the ongoing assembly of the International Space Station (ISS). As the first assembly mission following the tragic Columbia disaster, it was a significant milestone in NASA's Return to Flight program, which had already seen success with missions STS-114 and STS-121. The mission commenced on September 9, 2006, at 11:14:55 EDT (15:14:55 UTC), with Atlantis lifting off from Launch Complex 39B at the Kennedy Space Center.

Also known as ISS-12A within the ISS program, STS-115 delivered and installed the second port-side truss segment, designated ITS P3/P4. This critical segment, weighing over 17.5 short tons (approximately 16 metric tons), included a pair of solar arrays and batteries. The payload was so substantial that the crew complement was reduced from the usual seven to six astronauts to accommodate the additional weight and balance the spacecraft.

The mission involved a series of complex operations, including three spacewalks, which were instrumental in connecting the systems on the newly installed trusses and preparing them for deployment. During these extravehicular activities, the crew undertook essential maintenance work on the station, ensuring the new components were fully integrated into the ISS's growing structure.

The original launch date for STS-115 was set for April 2003. However, the devastating Columbia accident in February 2003 necessitated a postponement, pushing the launch to August 27, 2006. This date was further delayed due to various factors, including a threat from Tropical Storm Ernesto and an unprecedented lightning strike on the shuttle launchpad.

The crew for STS-115 consisted of six astronauts with diverse roles and expertise. Commander Brent W. Jett Jr. led the mission, marking his fourth and final spaceflight. Pilot Christopher Ferguson, making his inaugural spaceflight, worked alongside Mission Specialist 1 Steven MacLean from the Canadian Space Agency, who was on his second and final spaceflight. Mission Specialist 2 and Flight Engineer Daniel C. Burbank, with his second spaceflight, joined Mission Specialist 3 Joseph R. Tanner, who was on his fourth and final mission. Lastly, Mission Specialist 4 Heidemarie M. Stefanyshyn-Piper made her debut spaceflight.

Space Shuttle mission STS-115 Astronauts Brent W. Jett, Jr. (right) and Christopher J. Ferguson, commander and pilot, respectively, flank the mission insignia. The mission specialists are, from left to right, astronaut Heidemarie M. Stefanyshyn-Piper, Joseph R. (Joe) Tanner, Daniel C. Burbank, and Steven G. MacLean, who represents the Canadian Space Agency.

A noteworthy achievement of STS-115 was MacLean's role in operating the Canadarm2 and its Mobile Base, making him the first Canadian astronaut to do so. Ferguson and Burbank, who controlled the original Canadarm, handed him a new set of solar arrays and performed a spacewalk, making him only the second Canadian astronaut to undertake such an activity, following Chris Hadfield.

The primary objectives of STS-115 were to deliver and install the P3/P4 truss segment, deploy two new solar arrays (4A and 2A), and prepare the ISS for future assembly missions, particularly the upcoming STS-116. This mission represented a crucial step in the ISS's expansion, enhancing its capabilities and setting the stage for continued progress in space station construction.

Atlantis approaching the ISS with the P3/P4 truss segment and solar arrays in the payload bay.

The mission's second day was marked by an important pre-docking task: a thorough inspection of the Space Shuttle Atlantis. The crew utilized the Orbiter Boom Sensor System, a 15-meter (50-foot) extension of the shuttle's robotic arm, to meticulously examine the reinforced carbon-carbon panels along the leading edges of Atlantis' wings and nose cap. Pilot Chris Ferguson and Mission Specialists Dan Burbank and Steve MacLean conducted this critical inspection, ensuring that the shuttle was in prime condition for its operations.

The crew's activities for the day extended beyond the inspection as they prepared the shuttle for its upcoming docking with the International Space Station (ISS) and the mission's planned spacewalks. Mission Specialists Joe Tanner and Heide Stefanyshyn-Piper reviewed the spacesuits and tools used during the spacewalks scheduled for Days 4, 5, and 7. These spacewalks were set to include the installation of the P3/P4 truss, deploying new solar arrays, and preparing for operational integration.

Meanwhile, aboard the ISS, Expedition 13 Flight Engineer Jeffrey Williams was preparing for Atlantis' arrival on Day 3. He set up digital cameras to capture high-resolution images of the shuttle's heat shield. With the assistance of Commander Pavel Vinogradov, Williams pressurized the Pressurized Mating Adapter 2, located at the end of the Destiny Laboratory Module, where Atlantis would later dock. Vinogradov also prepacked equipment for return to Earth.

Atlantis docked to the ISS as seen during EVA 1 on Day 4.

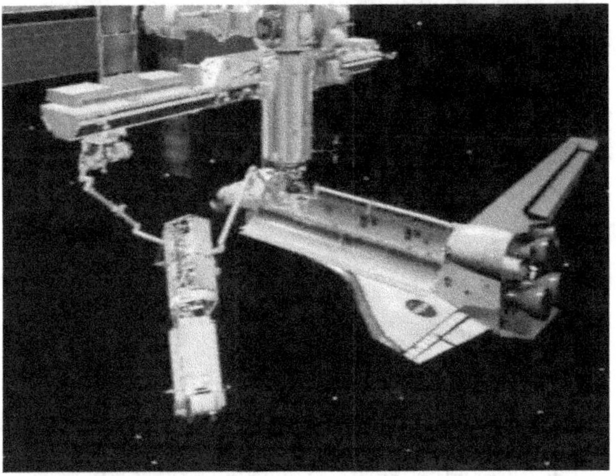

Atlantis's Canadarm hands the P3/P4 Truss segment to the station's Canadarm2 on Day 3

September 11 (Flight Day 3), the third day began with a spectacular maneuver: Atlantis executed an orbital backflip approximately 180 meters (600 feet) below the ISS. This maneuver allowed the Expedition 13 crew to capture high-resolution photographs of the orbiter's heat shield. At 10:46 UTC, Atlantis successfully docked with the ISS, and nearly two hours later, the hatch between the two spacecraft was opened, allowing the crew to board the station at 12:35 UTC.

Following the docking, Ferguson and Burbank utilized the shuttle's robotic Canadarm to lift the 17.5-ton P3/P4 truss segment from the payload bay and maneuver it for handover to the station's Canadarm2. This handoff was carried out by MacLean, the first Canadian to operate the Canadarm2 in space, and Expedition 13 Flight Engineer Jeff Williams.

Mission Specialists Tanner and Stefanyshyn-Piper then began preparations for their spacewalk on Day 4 by entering the Quest Airlock for their "camp out" period. This new procedure involved spending the night in the airlock at reduced pressure and wearing oxygen masks to acclimate their bodies to the low pressures they would encounter during their upcoming extravehicular activities (EVAs).

Joseph Tanner installing the P3/P4 Truss on Day 4's extra-vehicular activity

On Day 4, Tanner and Stefanyshyn-Piper conducted the mission's first spacewalk. Their tasks included installing the P3/P4 Truss and its associated power and data cables and removing launch restraints. The spacewalk commenced at 09:17 UTC and concluded successfully at 15:43 UTC, with the astronauts completing several tasks planned for later EVAs. However, during their activities, a bolt, spring, and washer assembly from a launch lock were lost into space.

Preparations for Day 5's spacewalk began after EVA. Burbank and MacLean entered the Quest Airlock for their own "camp out," gearing up for their scheduled spacewalk at 09:15 UTC.

On Day 5, first-time spacewalkers Burbank and MacLean performed the second spacewalk. Their focus was on activating the Solar Alpha Rotary Joint (SARJ), an essential component that allows the station's solar arrays to orient towards the sun. The spacewalkers encountered several challenges, including a malfunctioning helmet camera, a broken tool, and a stubborn bolt that required the combined effort of both astronauts to remove. Despite these issues, Burbank and MacLean completed their tasks and spent 7 hours and 11 minutes outside the station.

Engineers encountered a glitch during the SARJ activation, which temporarily delayed the deployment of the new solar arrays. The issue, which was attributed to software, was addressed overnight, allowing the array deployment to proceed as scheduled.

The new P3/P4 truss during EVA 2. Also, the new undeployed solar arrays could be seen on the bottom right

Day 6 involved the continuation of the solar array deployment. Despite the previous day's software issues, the panels were unfurled in stages throughout the morning to avoid sticking. While some panels did exhibit minor sticking, this did not impact the overall deployment. The solar arrays were not expected to provide power until the next shuttle mission, STS-116, scheduled for December 2006, including a major electrical system rewiring.

Additional activities for the day included a "double walk off" of the station's Canadarm2 from the Mobile Base System to the Destiny Laboratory Module and preparations for the third spacewalk. Later, astronauts Jett and MacLean participated in interviews with Canadian Prime Minister Stephen Harper and students.

The final spacewalk of mission STS-115 took place on Day 7. Initially delayed by a circuit-breaker issue affecting the airlock's depressurization pump, the spacewalk began at 10:00 UTC after a 45-minute delay. Tanner and Stefanyshyn-Piper conducted a 6-hour and 42-minute spacewalk, performing maintenance and repair tasks, including removing hardware from the P3/P4 radiator and retrieving a materials exposure experiment. They also completed "get-ahead" tasks and conducted a test to evaluate using infrared video to detect debris damage on Atlantis' wing.

Following the spacewalk, the station's mobile transporter was repositioned to inspect portions of the P3 truss.

Day 8 was primarily dedicated to preparing for the upcoming undocking. The crew used the day to rest and prepare for their departure by transferring ISS equipment and science experiments to Atlantis. A traditional joint-crew news conference was held, where Mission Commander Brent Jett praised the mission's success and discussed the future challenges of upcoming assembly missions.

On the final day of STS-115, Atlantis undocked from the ISS at 12:50 UTC. After closing and locking the hatch at 10:27 UTC, Atlantis performed a 360-degree flyaround of the ISS to document its new configuration, capturing the results of the successful mission and the enhanced capabilities of the expanded station.

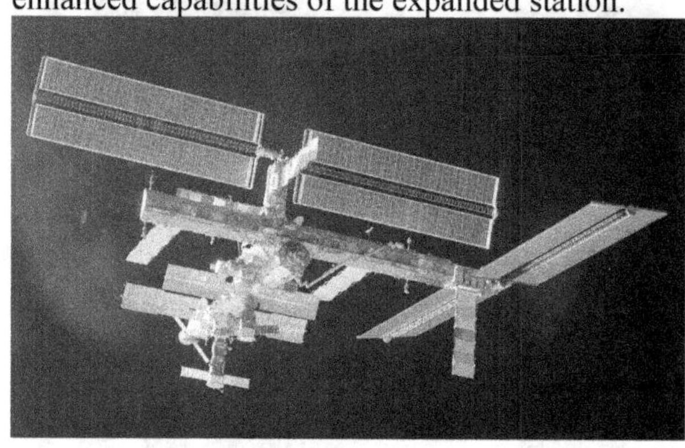

This view of the International Space Station, backdropped against the blackness of space, was taken shortly after the Space Shuttle Atlantis undocked from the orbital outpost at 7:50 a.m. CDT. The unlinking completed six days, two hours and two minutes of joint operations with the station crew. Atlantis left the station with a new, second pair of 240-foot solar wings, attached to a new 17.5-ton section of truss with batteries, electronics and a giant rotating joint. The new solar arrays eventually doubled the station's onboard power when their electrical systems were brought online during the shuttle flight launched in December.

Additionally, the ISS Progress 21 spacecraft, responsible for resupplying the station, undocked on September 18, 2006. Before this, Progress 22 had arrived on June 24, 2006, and docked with the ISS on June 26, 2006. The Progress 22 mission was critical in delivering necessary supplies and equipment for the station's operations.

Before the arrival of Expedition 13, the International Space Station (ISS) marked the end of Expedition 12, which had fulfilled its objectives. The departure of Expedition 12 was a significant event, concluding their successful mission and preparing the station for the incoming crew.

Expedition 12's departure began with the meticulous process of readying the station and the crew for undocking. The crew members, including Commander William "Bill" McArthur and Flight Engineers Valery Tokarev and Robert Curbeam, conducted final checks and ensured that all necessary preparations were in place for their departure. This involved securing equipment, transferring scientific experiments, and performing final safety inspections.

On the scheduled day, the crew of Expedition 12 bid farewell to their successors, Expedition 13, who had arrived and were now acclimating to the station. The farewell was marked by a traditional ceremony, with heartfelt exchanges between the two crews. As the time for undocking approached, the hatch separating the departing spacecraft, the Soyuz TMA-8 for Expedition 12, from the station was closed and securely locked.

At precisely the planned time, the Soyuz TMA-8 spacecraft, which had been Expedition 12's lifeline, undocked from the ISS. The spacecraft maneuvered away from the station, leaving behind a well-functioning outpost in orbit. As the Soyuz TMA-8 performed a series of controlled burns, it began its deorbit trajectory, setting course for re-entry into Earth's atmosphere.

During its departure, the spacecraft executed a series of precise maneuvers to ensure a safe distance from the ISS. This included a 360-degree flyaround of the station, capturing images of the ISS from various angles and documenting its current configuration. This visual documentation provided valuable data and showcased the progress made during Expedition 12.

The successful undocking of the Soyuz TMA-8 marked the official end of Expedition 12's mission, and the spacecraft's descent through Earth's atmosphere concluded with a controlled landing in the designated recovery area. The crew members of Expedition 12 were safely recovered and returned to their respective home bases, concluding their historic mission aboard the International Space Station.

With Expedition 12's departure, the ISS prepared for the new crew of Expedition 13, ready to continue its role as a hub for scientific research and international collaboration in space.

Expedition 13

Thomas Reiter, German ESA astronaut and flight engineer, Cosmonaut Pavel V. Vinogradov, mission commander representing Russia's Federal Space Agency and astronaut Jeffrey N. Williams, NASA space station science officer and flight engineer, crew of ISS Expedition 13

Expedition 13, the thirteenth mission to the International Space Station (ISS), commenced with the launch of the Soyuz TMA-8 spacecraft at 02:30 UTC on March 30, 2006. This mission marked a significant milestone, particularly with the inclusion of astronaut Marcos Pontes, who became the first Brazilian to journey into space. Pontes's mission was a brief nine-day sojourn; he returned to Earth with Expedition 12 aboard the Soyuz TMA-7.

In July 2006, Thomas Reiter, representing the European Space Agency (ESA), joined Expedition 13, enhancing the international collaboration on the ISS. Reiter's journey was part of the second "Return to Flight" mission, conducted by the Space Shuttle Discovery (STS-121), which launched on July 4, 2006. His arrival at the ISS on July 6, 2006, at 19:13 UTC, marked a historic moment as he became the first European astronaut to serve as a long-duration crew member on the ISS. This addition restored the station's crew size to three members for the first time since May 2003, following the reduction to two crew members due to the suspension of shuttle flights after the Space Shuttle Columbia accident on February 1, 2003.

The Expedition 13 crew consisted of: Commander: Pavel Vinogradov from Russia, embarking on his second spaceflight, Flight Engineer 1: Jeffrey Williams from the United States, also on his second spaceflight, and Flight Engineer 2: Thomas Reiter from Germany, marking his second and final spaceflight.

Cosmonaut Dmitri Kondratyev was originally slated for the mission but was replaced by Reiter months before the launch. The backup crew included Fyodor Yurchikhin (Commander and Soyuz Commander from Russia), Michael Fincke (Flight Engineer and NASA Science Officer from the United States), and Léopold Eyharts (Flight Engineer from France).

Expedition 13 was characterized by two significant spacewalks (Extravehicular Activities, or EVAs) that showcased the crew's dedication to maintaining and enhancing the ISS.

On June 1, 2006, the first involved Pavel Vinogradov and Jeffrey Williams. This EVA, scheduled initially from 23:48 UTC on June 1 to 06:19 UTC on June 2, was extended to 6 hours and 31 minutes due to unforeseen delays. Key objectives included replacing a camera, repairing a vent associated with the station's oxygen-producing Elektron unit, and retrieving experiment packages.

Jeffrey Williams and Thomas Reiter participated in the second spacewalk on August 3, 2006. This EVA lasted 5 hours and 54 minutes, beginning at 14:04 UTC and concluding at 19:58 UTC. Tasks included installing the Floating Potential Measurement Unit (FPMU), essential for measuring electrical potential in space, and two Materials on Materials International Space Station Experiment (MISSE) containers to study material degradation. They also installed a controller for a thermal radiator rotary joint, replaced a computer, and performed various other critical maintenance.

Expedition 14

Astronaut Michael E. Lopez-Alegria (center), Expedition 14 commander and NASA space station science officer; cosmonaut Mikhail Tyurin (right), flight engineer representing Russia's Federal Space Agency; and European Space Agency (ESA) astronaut Thomas Reiter, flight engineer,

In December 2006, a significant transition occurred within Expedition 14 of the International Space Station (ISS). The STS-116 mission, conducted by Space Shuttle Discovery, was pivotal in this transition. The mission brought astronaut Sunita Williams to the ISS, replacing Thomas Reiter as the third member of Expedition 14. Reiter's departure marked the conclusion of his second and final spaceflight as he returned to Earth.

The Expedition 14 crew, during the first part of the mission from September to December 2006, consisted of Commander Michael López-Alegría from NASA, who was undertaking his fourth and final spaceflight. Flight Engineer Mikhail Tyurin from the Russian Space Agency joined him, on his second spaceflight, and Thomas Reiter from the European Space Agency (ESA), completing his second and last spaceflight.

STS-116 crew for the Space Shuttle Discovery are, front row (from the left), astronauts William A. Oefelein, pilot; Joan E. Higginbotham, mission specialist; and Mark L. Polansky, commander. On the back row (from the left) are astronauts Robert L. Curbeam, Nicholas J.M. Patrick, Sunita L. Williams and the European Space Agency's Christer Fuglesang, all mission specialists. Williams joined Expedition 14 in progress to serve as a flight engineer aboard the International Space Station. The crewmembers are attired in training versions of their shuttle launch and entry suits.

With Reiter's departure in December 2006, Sunita Williams joined the station, marking her first spaceflight and taking over as a Flight Engineer. The latter half of Expedition 14, from December 2006 to April 2007, featured Williams alongside Commander López-Alegría and Flight Engineer Tyurin. This new configuration continued the crucial tasks of maintaining and advancing the ISS's scientific objectives.

The backup crew for Expedition 14 included Peggy Whitson as the Commander, Yuri Malenchenko and Leopold Eyharts as Flight Engineers, with Clayton Anderson designated as the backup Flight Engineer for Williams. This comprehensive crew arrangement ensured the continued success and progression of the ISS's mission objectives during this important period in its history.

Astronaut Sunita L. Williams (left), Expedition 14 flight engineer; cosmonaut Mikhail Tyurin, flight engineer representing Russia's Federal Space Agency; and astronaut Michael E. Lopez-Alegria, commander and NASA space station science officer. Lopez-Alegria and Tyurin launched to the International Space Station in a Soyuz spacecraft in mid-September and Williams joined Expedition 14 in progress and served as a flight engineer after traveling to the station on space shuttle mission STS-116 in December.

A significant milestone during this period was the STS-116 mission, conducted by the Space Shuttle Atlantis. Atlantis docked with the ISS on December 11, 2006, in a mission crucial for the station's expansion. The primary objectives of STS-116 were to deliver and attach the P5 truss segment, which was vital for augmenting the station's structural framework and power generation capabilities. The successful docking was a key step in enhancing the ISS's functionality to support its growing research and operational demands.

Before Atlantis undocked on December 19, 2006, the crew undertook a series of critical tasks. Installing the P5 truss segment expanded the station's structural framework, facilitating the addition of future modules and improving the ISS's overall capabilities. This component was the third steel truss segment added to the station, complementing the previous P3/P4 solar arrays installed during the earlier STS-115 mission.

In addition to the truss installation, STS-116 involved extensive ISS power system rewiring. This upgrade was essential for integrating the new solar arrays, ensuring the station's power system could meet its increasing demands. The mission also saw the deployment of the third SPHERES testbed, a spherical device designed for experiments in microgravity, and several small satellites launched from the Integrated Cargo Carrier. Among these were the ANDE technology demonstrator, developed by the Naval Research Laboratory, and various CubeSats used for diverse research and technology demonstrations.

The STS-116 mission launched aboard Space Shuttle Discovery, began with a night launch on December 9, 2006, the first of its kind since STS-113 in November 2002. Discovery's mission, designated ISS-12A.1, was notable not only for its technical achievements but also for its historical significance. This mission marked the first spaceflight of Swedish astronaut Christer Fuglesang, a milestone for Scandinavian space exploration.

Discovery's crew comprised a mix of seasoned and rookie astronauts. Commander Mark L. Polansky and Pilot William Oefelein led the team, which included Mission Specialist Robert Curbeam. Curbed became the first astronaut in his third spaceflight to perform four spacewalks during a single mission. Mission Specialists Nicholas Patrick, Joan E. Higginbotham, and Sunita Williams joined him. Williams, who later set a record for the longest single spaceflight by a female astronaut, was a key member of Expedition 14. The crew also included Thomas Reiter of the European Space Agency, who transitioned from Expedition 14 crew to return to Earth with the STS-116 mission.

STS-116 was also notable for being the last Space Shuttle mission scheduled for launch from Kennedy Space Center's Pad 39B, as NASA prepared the pad for future Ares I missions. The successful completion of the mission led to the temporary reconfiguration of Discovery for maintenance before its next flight, STS-120, which was scheduled to commence on October 23, 2007.

Following STS-116, Discovery modified its Assembly Power Converter Units (APCUs), replacing them with Station-Shuttle Power Transfer System (SSPTS) components. This upgrade allowed for more efficient use of onboard power. It extended the duration of docked missions by reducing the consumption of hydrogen and oxygen used by the orbiter's fuel cells.

December 11, Flight Day 3 commenced for the astronauts aboard Space Shuttle Discovery at 15:18 UTC, setting the stage for a pivotal moment in their mission. As the shuttle executed its rendezvous pitch maneuver, the docking with the International Space Station (ISS) occurred precisely at 22:12 UTC. This maneuver involved aligning the shuttle with the ISS and carefully guiding it into place. The docking was a significant milestone, marking a crucial step in the mission's objectives.

While flying east of New Zealand, Robert L. Curbeam Jr. and Christer Fuglesang participate in the mission's first spacewalk.

The hatch connecting Discovery to the ISS was opened at 23:54 UTC, allowing the shuttle and the station crews to meet and begin their collaborative tasks. The first order of business was a detailed inspection of the orbiter, followed by the careful unloading of the P5 truss segment from Discovery's payload bay. The truss was successfully transferred from the shuttle's robotic arm to the station's arm, a complex procedure requiring precise coordination.

In preparation for the next day's extravehicular activity (EVA), astronauts Robert Curbeam and Christer Fuglesang entered the airlock for a "campout" sleep session. This pre-breathe practice involves spending time in a lower-pressure environment to purge nitrogen from the astronauts' bodies, mitigating the risk of decompression sickness.

Flight Day 4 began for the crew at 15:47 UTC. The first EVA of the mission marked the day, a critical task designed to advance the ISS's construction. The EVA commenced at 20:31 UTC, with astronauts Robert Curbeam and Christer Fuglesang venturing outside the station. Their primary task was to integrate the P5 truss segment into its designated position on the ISS.

The spacewalk involved removing launch restraints from the P5 truss and directing Mission Specialist Joan Higginbotham, who operated the station's robotic arm, Canadarm2, to maneuver the truss into alignment with the P4 truss. Curbeam and Fuglesang provided visual guidance to ensure the truss was correctly positioned and securely attached.

As seen through windows on the aft flight deck of Space Shuttle Discovery, the payload bay was featured in this image

Following the successful attachment of the P5 truss, the astronauts connected power, data, and heater cables and replaced a malfunctioning video camera on the S1 truss. Their efficiency allowed them to complete additional tasks beyond the original schedule. The EVA concluded at 03:07 UTC on December 13, after a productive 6 hours and 36 minutes.

During the spacewalk, mission managers reviewed imagery and confirmed that the shuttle's heat shield was intact and would support a safe return. Consequently, they decided against performing a more detailed inspection planned for later in the mission.

The crew also initiated power reconfiguration on the ISS to utilize the solar arrays installed during STS-115. This involved retracting the port solar array on the P6 truss to activate and rotate the Solar Alpha Rotary Joint (SARJ) on the P4 truss, which allows the solar arrays to track the Sun.

Flight Day 5 started at 15:21 UTC, with the day's focus on reconfiguring the ISS's solar arrays. The most challenging task was retracting the P6 port-side solar array, which began at 18:28 UTC. However, problems arose as the array experienced kinks and billows during retraction. The mission controllers made numerous attempts to resolve the issue, issuing over 40 commands to adjust the array.

By 00:50 UTC, the retraction efforts were halted for the day. Despite retracting 14 of the 31 bays, the port-side arrays remained partially extended but in a safe position. This allowed the activation of the SARJ at 01:00 UTC, enabling the solar arrays on the P3/P4 truss to begin tracking the Sun.

Flight Day 6 began at 15:19 UTC, with the day's highlight being the second EVA of the mission. This EVA focused on reconfiguring the ISS's electrical power systems. The spacewalk commenced at 19:41 UTC, with astronauts Bob Curbeam and Christer Fuglesang exiting the Quest airlock. Their task was to activate channels 2 and 3 of the station's four-channel electrical system.

The EVA proceeded smoothly, with the astronauts successfully energizing the system from the P4 solar arrays. The spacewalk was completed in exactly 5 hours, concluding at 00:41 UTC.

Flight Day 7 was relatively light, allowing the crews to recover from the intense activities of previous days. Spacewalkers Bob Curbeam and Christer Fuglesang enjoyed a day of rest, while rest of the crew focused on cleanup and preparations for the next EVA. The day featured a joint photo session and news conference, during which Crown Princess Victoria interviewed Swedish astronaut Christer Fuglesang. Fuglesang also set a 20-second Frisbee world record in space, broadcast live on Swedish TV4.

Thomas Reiter attempted to free a stuck solar panel by using exercise equipment known to cause oscillations in the arrays. However, this effort was unsuccessful, and mission controllers continued to explore other solutions for the solar panel issue.

Flight Day 8 began at 14:48 UTC, with astronauts Bob Curbeam and Sunita Williams embarking on the third EVA of the mission at 17:12 UTC. The primary goal was to complete the rewiring of the ISS and address the ongoing solar array retraction issues. During the EVA, Curbeam and Williams successfully retracted six more sections of the P6 solar array, leaving 11 bays remaining.

The EVA also saw the unfortunate loss of Sunita Williams' digital camera, which fell due to a snagged tether. Despite this, mission controllers determined that the lost images were not critical and did not require replacement. The astronauts concluded their EVA and re-entered the ISS through the Quest airlock.

Flight Day 9 was devoted to preparations for the upcoming fourth EVA. The crew focused on preparing their space suits, adjusting sizes, and replacing lithium hydroxide canisters. They reviewed new procedures for addressing the solar array retraction issues and took precautions to prevent damage to the arrays during the EVA.

Space Shuttle Discovery's Canadarm-1 robotic arm hands off the P5 truss section to the International Space Station's Canadarm-2 during shuttle mission STS-116 in December 2006.

Leonardo Permanent Multipurpose Module

Flight Day 10 commenced at 14:17 UTC, with Bob Curbeam and Christer Fuglesang undertaking a planned EVA at 17:12 UTC. The objective was to retract the remaining bays of the problematic P6-port Solar Array Wing. The EVA lasted 6 hours and 38 minutes and was successful, with the astronauts retracting the final eleven bays. At the end of the spacewalk, Curbeam achieved a notable milestone, ranking fifth in total EVA time for U.S. astronauts.

Flight Day 11 began at 14:47 UTC, with the crews of Discovery and the ISS preparing for undocking. The hatches between the ISS and Discovery were closed, and undocking was completed at 22:10 UTC. Due to the extended duration of EVA #4, the shuttle did not complete a full orbit around the ISS but flew slightly more than a quarter of the way before performing its departure burn.

In early 2007, the ISS Progress 22 spacecraft, which had been docked since June 26, 2006, undocked on January 16, 2007. This undocking made way for the arrival of ISS Progress 24, which docked with the ISS on January 19, 2007. The Progress spacecraft was crucial in resupplying the station with necessary materials and equipment.

Expedition 14 of the International Space Station (ISS) began with a notable event in human spaceflight history. On September 18, 2006, at 04:10 UTC, the Soyuz TMA-9 spacecraft launched from the Baikonur Cosmodrome in Kazakhstan. This mission, propelled by a Soyuz FG rocket, was pivotal for the ISS, as it delivered a new crew and a spaceflight participant, Anousheh Ansari, marking the start of Expedition 14.

The Soyuz TMA-9 crew included Russian cosmonaut Mikhail Tyurin, who served as the commander on his inaugural spaceflight; American astronaut Michael Lopez-Alegria, who was on his third spaceflight and served as the flight engineer; and European Space Agency (ESA) astronaut Thomas Reiter, who was making his first spaceflight. The crew's arrival at the ISS was marked by a successful docking with the station's Pirs docking compartment on September 20, 2006, at 02:54 UTC. The hatch to the ISS opened at 05:21 UTC, welcoming the new crew members aboard. They were greeted by the remaining Expedition 13 crew, including Pavel Vinogradov and Jeff Williams, as well as Thomas Reiter, who had transitioned from Expedition 13 to Expedition 14 following a change in his Soyuz seat-liner.

The STS-116 and Expedition 14 crewmembers gather for a group portrait during a joint crew press conference in the Destiny laboratory of the International Space Station while Space Shuttle Discovery was docked with the station. From the left (front row) are European Space Agency (ESA) astronaut Thomas Reiter, Nicholas J. M. Patrick, Joan E. Higginbotham, all STS-116 mission specialists; and William A. (Bill) Oefelein, STS-116 pilot. From the left (center row) are astronaut Robert L. Curbeam, Jr., European Space Agency (ESA) astronaut Christer Fuglesang, STS-116 mission specialists; and astronaut Mark L. Polansky, STS-116 commander. From the left (back row) are astronaut Michael E. Lopez-Alegria, Expedition 14 commander and NASA space station science officer; cosmonaut Mikhail Tyurin, Expedition 14 flight engineer representing Russia's Federal Space Agency; and astronaut Sunita L. Williams, Expedition 14 flight engineer. Shortly after the two spacecraft docked on Dec. 11, Williams became a member of the station crew. At the same time, Reiter became a Discovery crewmember for his ride home, completing about six months in space.

Soyuz TMA-9 played a crucial role beyond transporting the new crew. It served as an emergency evacuation vehicle throughout Expedition 14, providing a vital safety measure for the ISS crew. This capability underscored the importance of the Soyuz spacecraft in maintaining the station's operational continuity and crew safety.

During their seven-month mission, the Expedition 14 crew contributed significantly to the ISS's scientific research and operational activities. Their time aboard the station involved conducting various experiments, advancing the station's research capabilities, and supporting international collaboration in space exploration. Throughout their mission, the crew orbited Earth 3,401 times, highlighting their stay's extensive duration and impact.

Expedition 14 concluded its mission with Soyuz TMA-9 undocking from the ISS on April 21, 2007. The spacecraft safely returned to Earth, marking the end of the expedition and ensuring a smooth transition to Expedition 15, which had launched on April 7, 2007, and docked with the ISS on April 9, 2007. Expedition 15, commanded by Fyodor Yurchikhin, continued the station's scientific research and assembly, building on the foundation laid by Expedition 14.

A photograph of the International Space Station after STS-116 with the new P5 truss segment. Backdropped by the blackness of space and Earth's horizon, the International Space Station moves away from Space Shuttle Discovery. Earlier the STS-116 and Expedition 14 crews concluded eight days of cooperative work onboard the shuttle and station. Undocking of the two spacecraft occurred at 4:10 p.m. (CST) on Dec. 19, 2006. Astronaut William A. (Bill) Oefelein, STS-116 pilot, was at the controls for the fly-around, which gave Discovery's crew a look at its handywork, a new P5 spacer truss segment and a fully retracted P6 solar array wing. During their stay on orbital outpost, the combined crew installed the newest piece of the station's backbone and completely rewired the power grid over the course of four spacewalks.

The day following their arrival, the Expedition 14 crew had the unique opportunity to observe Space Shuttle Atlantis re-entering Earth's atmosphere at the end of the STS-115 mission. Positioned a few hundred miles ahead of Atlantis, the station's inhabitants witnessed the shuttle create a luminous contrail as it descended. Commander López-Alegría and Flight Engineer Williams provided live commentary of the re-entry to Mission Control in Houston, enhancing the mission's coordination and outreach.

Expedition 14's mission objectives were multifaceted and crucial for the continued development and operation of the ISS. Key tasks included advancing the space station assembly through three spacewalks and coordinating with the Space Shuttle mission STS-116. A notable logistical maneuver involved relocating Soyuz TMA-9 from the Zvezda module's aft port to the Zarya module's nadir port. The expedition also managed three Progress spacecraft visits, which delivered essential supplies such as food, fuel, and water, supplementing the provisions brought by the Space Shuttles.

Additionally, the crew reconfigured the station's power systems from its solar arrays and cooling systems and removed and discarded the protective shrouds from the truss system, a necessary step for the station's ongoing construction and maintenance.

On November 22, 2006, Expedition 14 conducted its first spacewalk, beginning at 19:17 EST (23:17 UTC). This extravehicular activity (EVA) was briefly delayed from its planned start time due to a cooling issue with Tyurin's suit. After resolving the issue, the spacewalk commenced with Tyurin performing an unusual and memorable experiment: hitting a golf ball from the Pirs airlock. Sponsored by the Toronto-based company Element 21, which specializes in scandium golf clubs, this stunt involved a lightweight 3-gram ball, in contrast to the standard 48-gram ball. The experiment was intended to showcase the company's innovative technology, and although only one shot was taken, it was captured for potential use in a commercial. E21's "Track the Ball in Space" website monitored the ball's progress, which provided an estimated distance based on a constant speed rather than real tracking.

Following this unique event, the crew focused on several critical tasks. They first addressed an issue with a Kurs antenna on the Progress 23 unpiloted cargo carrier, which had docked at the aft end of the station's Zvezda Service Module on October 26, 2006. The final latching of the spacecraft was delayed due to uncertainty about whether the antenna was fully retracted. Tyurin and López-Alegría photographed the antenna, still extended. Despite their efforts and those of

Russian flight controllers, the antenna could not be retracted, and the task was ultimately abandoned.

Subsequently, the crew relocated a WAL antenna to facilitate the docking of the Automated Transfer Vehicle (ATV), Jules Verne, which later arrived at the ISS on April 3, 2008. The relocation was necessary to prevent interference with a cover for a Zvezda booster engine. They also installed a BTN neutron experiment on the Zvezda module, designed to characterize charged and neutral particles in low Earth orbit. The data collected, particularly during solar bursts, continued to provide valuable scientific insights.

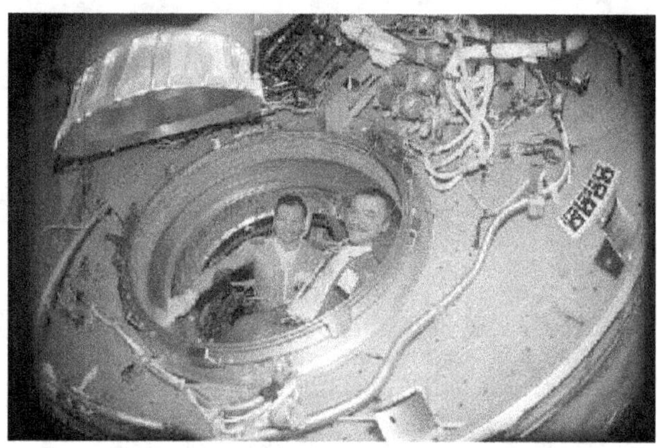

Commander Michael E. López-Alegría (left) and cosmonaut Mikhail Tyurin conduct pre-spacewalk operations in the Pirs airlock (February 2007)

The spacewalk concluded with the jettisoning of two thermal covers from the BTN experiment, and the EVA ended at 00:55 EST (04:55 UTC) on November 23, lasting 5 hours and 38 minutes. A final scheduled task, the inspection of bolts on one of the Strela hand-operated cranes on the docking compartment, was postponed to a future spacewalk.

Expedition 14 marked a significant chapter in the history of the International Space Station (ISS) as it became the longest mission to date. Additionally, the Soyuz capsule used for this mission was the oldest model ever employed, and Commander Michael López-Alegría not only continued to hold the record for the most spacewalks by a U.S. astronaut but also set a new record for the longest spaceflight by a NASA astronaut.

During Expedition 14, five spacewalks were conducted, accumulating 33 hours and 42 minutes. Commander López-Alegría took part in all five spacewalks, setting a record for the highest number of EVAs (extravehicular activities) within a single ISS mission, later matched by Peggy Whitson during Expedition 16. Flight Engineer Sunita Williams participated in three of these spacewalks, while Flight Engineer Mikhail Tyurin joined in two.

The first spacewalk of the expedition, conducted on November 22, 2006, began at 23:17 UTC and concluded at 04:55 UTC the following day. This EVA, notable for the "orbiting golf shot" event sponsored by a Canadian golf company, involved López-Alegría and Tyurin performing a unique experiment. Tyurin hit a specially designed lightweight golf ball while López-Alegría managed the setup. The spacewalk also included inspecting and photographing a Kurs antenna on the Progress 23 cargo carrier, relocating an Automated Transfer Vehicle (ATV) WAL antenna, installing a BTN neutron experiment, and jettisoning thermal covers from the BTN.

The second spacewalk occurred on January 31, 2007, from 15:14 UTC to 23:09 UTC. During this EVA, López-Alegría and Williams reconfigured one of the two cooling loops serving the Destiny laboratory from a temporary to a permanent system. They also connected a cable for the Station-to-Shuttle Power Transfer System (SSPTS), installed several components to secure the P6 Truss radiator, and removed one fluid line from the Early Ammonia Servicer (EAS), which would be jettisoned in a later spacewalk.

The third spacewalk occurred on February 4, 2007, from 13:38 UTC to 20:49 UTC. This EVA focused on reconfiguring the second cooling loop, working on the Early Ammonia Servicer (EAS) on the P6 Truss, and preparing the P6 solar wing for retraction during STS-117. The crew also removed a sunshade from a data relay device and continued work on the SSPTS.

On February 8, 2007, the fourth spacewalk, spanning from 13:26 UTC to 20:06 UTC, involved removing thermal shrouds from Rotary Joint Motor Controllers (RJMC) on the P3 Truss and deploying an Unpressurized Cargo Carrier Assembly Attachment System (UCCAS). The team also removed launch locks from the P5 Truss and connected cables of the SSPTS to the Pressurized Mating Adapter (PMA-2) on Destiny.

Expedition 14's final spacewalk occurred on February 22, 2007, from 10:27 UTC to 16:45 UTC.

Tyurin and López-Alegría retracted the Progress cargo carrier's antenna at the Zvezda service module's aft port, inspected and photographed various antennas, replaced a Russian materials experiment, and carried out other detailed inspections and photographic documentation of hardware connectors and experiments.

The crew's journey included the collaborative efforts of Space Shuttle Discovery on STS-116, which was crucial in the mission's success. Following the shuttle's departure, the crew continued their important work aboard the station until April 21, 2007, when the expedition ended at 12:31 UTC. The team departed aboard Soyuz TMA-9; Commander López-Alegría and Flight Engineer Tyurin concluded their mission and re-entering Earth's atmosphere aboard Soyuz TMA-9. Their spacecraft touched down at 12:31:30 UTC, bringing a successful conclusion to their time aboard the ISS.

In 2007, space exploration witnessed a series of significant events that underscored the dynamic advancements and milestones achieved across various nations. This year marked notable achievements in satellite technology, planetary exploration, space station operations, and international collaboration.

The year commenced with a landmark event as China conducted its Anti-Satellite (ASAT) test, demonstrating its capacity to destroy satellites in orbit. This test raised substantial international concerns regarding the militarization of space, highlighting the potential risks and the need for regulatory measures to ensure space remains a domain for peaceful exploration and cooperation.

In planetary exploration, the United States embarked on two ambitious missions that expanded our understanding of the solar system. The Phoenix mission was launched to explore Mars' surface, studying its geological features and searching for signs of water and life. Concurrently, the Dawn mission was dispatched to investigate the asteroid belt, targeting two of the largest asteroids, Vesta and Ceres, to provide insights into the early solar system's formation and evolution.

Japan made substantial contributions to lunar research with the launch of the Kaguya Lunar orbiter. This mission was designed to survey the Moon's surface and geological characteristics, offering valuable data that advanced our comprehension of Earth's natural satellite.

China also made significant strides in lunar exploration with successfully deploying the Chang'e 1 Lunar probe. This mission, China's first lunar exploration endeavor, was tasked with gathering data about the Moon's surface, marking a pivotal step in China's space exploration ambitions.

India achieved a milestone in its space program with the SRE-1 spacecraft's attempt to recover a satellite after re-entry. This ambitious project showcased India's growing technological prowess and its commitment to advancing its space capabilities.

In aeronautical research, Boeing Phantom Works, in collaboration with NASA and the USAF Research Laboratory, made strides in aircraft design, developing the X-48B and X-48C Hybrid/Blended Wing Body prototypes represented significant progress in improving aerodynamic efficiency and fuel consumption, reflecting ongoing innovations in aviation technology.

The year also witnessed several successful maiden flights of newly modernized carrier rockets, including the PSLV-CA, Long March 3B/E, Shavit-2, Zenit-2M, and Proton-M Enhanced. These launches underscored the continuous efforts to enhance the reliability and performance of launch vehicles.

In military technology, Russia showcased its advanced ballistic missile capabilities with the inaugural launch of the RS-24 missile. Additionally, the Atlas V rocket, known for its versatility and power, made its first flight in the 421 configuration, representing a significant development in launch vehicle technology.

A pivotal development in 2007 was the launch of the final Defense Support Program (DSP) satellite, Defense Support Program Satellite 23. This satellite was crucial for providing early warnings of ballistic missile threats, bolstering national security efforts. It was launched from Cape Canaveral Air Force Station aboard a Titan-IV Heavy booster.

China achieved a significant milestone in September 2007 with the launch of its first space laboratory, Tiangong-1. This accomplishment marked a critical step in China's efforts to establish a sustained presence in space and advance its space program.

At the International Space Station, the year began with the undocking of the ISS Progress 24 cargo spacecraft on August 1, 2007. This maneuver marked the end of its mission delivering essential supplies and equipment to the ISS. A few weeks later, on August 5, the ISS welcomed the arrival of the ISS Progress 26 spacecraft. This new cargo vehicle brought further supplies and continued the flow of resources necessary for the station's operations and research activities.

Expedition 15

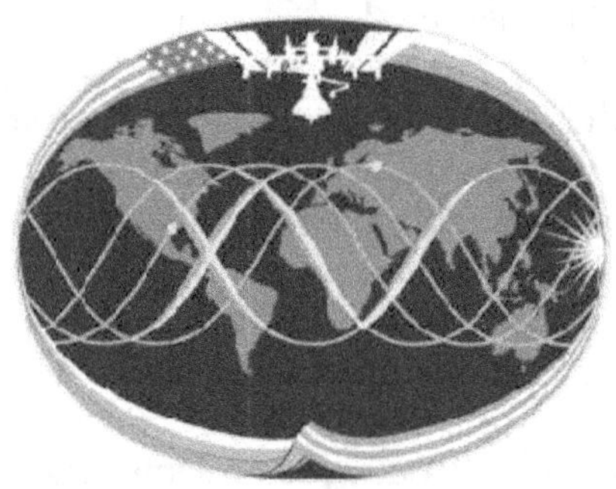

Expedition 15 crewmembersfFrom the left are astronaut Clayton C. Anderson, flight engineer; along with cosmonauts Oleg Kotov, flight engineer and Soyuz commander; and Fyodor Yurchikhin, commander. Kotov and Yurchikhin represent Russia's Federal Space Agency.

Expedition, 15 of the International Space Station (ISS), embarked on its significant journey on April 7, 2007, marking a critical chapter in the annals of human spaceflight. This mission, extending over six months and fourteen days, was a period of profound development and activity aboard the ISS, ultimately spanning 196 days, 17 hours, and 17 minutes from launch to landing.

A rotating crew of four astronauts characterized the expedition, though only three were stationed on the ISS at any given time. The crew for this mission included Russian cosmonauts Fyodor Yurchikhin and Oleg Kotov, and American astronauts Sunita Williams and Clayton Anderson. Fyodor Yurchikhin, serving as the commander, brought extensive spaceflight experience to the mission. His leadership was essential in navigating the complex tasks required for the station's operations and ensuring its smooth functioning.

Oleg Kotov, a Russian flight engineer, was embarking on his first spaceflight during Expedition 15. Despite being new to space travel, Kotov's contributions were pivotal. He worked closely with Yurchikhin to manage the station's operations and adapt to the challenges of his inaugural mission.

Sunita Williams, who had joined the ISS as part of Expedition 14, was the first member of Expedition 15 to arrive. She reached the ISS on December 11, 2006, aboard Space Shuttle Discovery (STS-116). Her early arrival allowed her to acclimate to the ISS environment and prepare for the ongoing mission objectives. Clayton Anderson, another American astronaut, joined the crew later on June 10, 2007, arriving aboard Space Shuttle Atlantis during mission STS-117. Anderson's inclusion resulted from a shift in the mission schedule; Williams's return was rescheduled from STS-118 to STS-117, allowing Anderson to join Expedition 15.

Expedition 15's mission was particularly notable for its significant contributions to expanding the ISS's Integrated Truss Structure. This crucial component of the station's architecture was vital for the station's growth and operational capabilities. During their tenure, the crew conducted two spacewalks, totaling 11 hours and 2 minutes, essential for maintaining and enhancing the ISS's infrastructure.

A view photographed from the International Space Station shows the Space Shuttle Atlantis backdropped over terrain as the two spacecraft were nearing their much-anticipated link-up in Earth orbit.

Throughout Expedition 15, the ISS saw significant infrastructure improvements with the expansion of its Integrated Truss Structure. The STS-117 mission, launched aboard the Space Shuttle Atlantis, was pivotal in expanding the International Space Station (ISS). The mission, which began with the shuttle's liftoff from Launch Pad 39A at the Kennedy Space Center on June 8, 2007, was a landmark achievement in human spaceflight. It marked the 250th orbital human spaceflight and was noted for being the heaviest shuttle flight ever undertaken. The launch was originally scheduled for March 15, 2007, but was delayed due to damage caused by a hailstorm, which required additional preparations and adjustments.

The Atlantis docked with the ISS on June 10, 2007, successfully delivering the truss segment and other crucial elements.

STS-117's primary objective was to deliver the S3/S4 truss segment, the second starboard truss segment of the ISS's truss structure. Manufactured by Boeing at the Michoud Assembly Facility, the S3/S4 truss segment was a significant addition to the ISS's infrastructure. Measuring approximately 44 feet 9.6 inches (13.655 meters) in length, 16 feet 3.4 inches (4.963 meters) in width, and 15 feet 2.3 inches (4.630 meters) in height, and weighing 35,678 pounds (16,183 kilograms), it was the heaviest payload ever transported to the ISS by the shuttle. Constructed from stainless steel, the truss was engineered to endure the harsh conditions of space while performing essential functions.

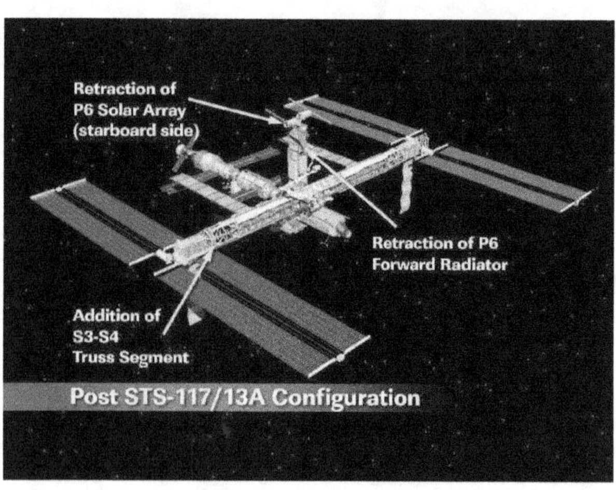

Post STS-117 station configuration with the newly installed S3/S4 truss segment.

The S3/S4 truss segment's design included vital components that expanded the ISS's capabilities. It was equipped to enhance the station's electrical power and data interfaces, converting sunlight into electricity and providing active thermal protection for the station's electronic components. Additionally, it offered attachment points for platforms that store spare parts, further contributing to the ISS's operational efficiency.

The ISS seen from Atlantis

During the mission, on the fourth day of flight, the S3/S4 truss segment was extracted from Atlantis's payload bay using the shuttle's robotic arm. It was then transferred to the station's Canadarm2, which carefully maneuvered and attached it to the outboard end of the S1 truss. This process marked a significant milestone in the ISS's assembly.

The S3 segment, a hexagonal-shaped structure composed of stainless steel, includes four bulkheads and six longerons—beams connecting the bulkheads. It also features a range of secondary structures, such as brackets, fittings, and attach platforms. Among its major subsystems are the Solar Array Rotary Joint (SARJ), which allows the Solar Array Wings (SAWs) to remain oriented toward the sun, and the Segment-to-Segment Attach System (SSAS), which provides a passive attachment point to the S1 segment. The Payload Attach System (PAS) facilitates the attachment of additional platforms for science payloads or spare Orbital Replacement Units (ORUs). The S3 segment was delivered to the Space Station Processing Facility at Kennedy Space Center on December 7, 2000.

The S4 segment houses several critical subsystems, including the Photovoltaic Module (PVM) and the Photovoltaic Radiator (PVR). The PVMs collect, convert, store, and distribute electrical power within the segment and to other station modules. Each Solar Array Wing (SAW) on the S4, measuring 115 by 38 feet (35 by 12 meters), was equipped with two solar blankets and mounted to a common mast made from shape memory alloy. The Beta Gimbal Assembly (BGA) adjusts the pitch of the wings to maximize solar exposure. The PVR, capable of dissipating up to 14 kW of heat, measures 44 by 12 by 7 feet (2.1 meters) when deployed. The Alpha Joint Interface Structure (AJIS) provides the necessary structural transition between the S3 and S4 segments, and the Modified Rocketdyne Truss Attachment System (MRTAS) includes the passive attachment mechanism for the S5 truss. The S4 segment was delivered to the Space Station Processing Facility on January 15, 2001.

STS-117 crewmembers take a moment to pose for a photo while working various tasks in the Destiny laboratory of the International Space Station during flight day five activities while Space Shuttle Atlantis was docked with the station. Pictured at left are astronauts Patrick Forrester (foreground), Jim Reilly, both mission specialists; and Rick Sturckow, commander. On the right are astronauts Steven Swanson (foreground) and John "Danny" Olivas, both mission specialists.

In addition to the primary truss components, Atlantis also carried a Hydrogen Vent Valve on its mid-deck. This valve was part of the Oxygen Generating System (OGS), which produces oxygen to replace that lost during experiments, airlock depressurization, and venting.

Other items on the mid-deck included a contingency water container for transferring water to the ISS, a "modified torque multiplier" tool for removing launch restraints from the S3 segment, and audio interface hardware to address communication issues between the shuttle and station experienced during previous missions. The total mid-deck payload for STS-117 was under 1,000 pounds.

Among the cargo was a historically significant artifact: a nearly 400-year-old metal cargo tag etched with "Yames Towne," the historic name for Jamestown, and four commemorative coins honoring early American explorers. The tag, prepared by a 17th-century metalworker, was unearthed in 2006 at Jamestown, the site of the first permanent English settlement in the Americas established in 1607. This artifact highlighted the mission's connection to both historical exploration and the advancement of space exploration.

The mission was extended by two days, and a decision was made after the launch due to uncertainties about the initial timeline and the potential for additional extravehicular activities (EVAs). The extension included a fourth EVA, critical for repairing a gap in the Orbital Maneuvering System (OMS) thermal blanket.

The crew conducted several extra-vehicular activities (EVAs) to support the station's expansion throughout their mission. The EVAs included three notable spacewalks: on May 30, 2007, Yurchikhin and Kotov conducted a 5-hour and 25-minute spacewalk; on June 6, 2007, they performed another spacewalk lasting 5 hours and 37 minutes; and on July 23, 2007, Yurchikhin and Anderson undertook a 7-hour and 41-minute spacewalk.

The mission's successful completion was marked by the shuttle's undocking from the station on June 19, 2007, and its return to Earth. Following the departure of STS-117, the ISS Progress 25 spacecraft arrived on May 15, 2007. This resupply mission was crucial in maintaining the station's operational capabilities and supporting the crew onboard.

In the summer of 2007, the International Space Station (ISS) continued its dynamic evolution with the rotation of spacecraft and crews. A pivotal moment occurred on August 10, when the Space Shuttle Endeavour arrived at the ISS, marking the beginning of the STS-118 mission. This mission was particularly notable for the delivery of the S5 truss segment and for including Sunita Williams, the first American female astronaut of Indian descent. This mission was a significant milestone in the ISS's ongoing expansion.

Initially, the STS-118 mission was slated to be conducted by the Space Shuttle Columbia, which would have marked its 29th mission and its first and only visit to the ISS. However, due to Columbia's heavier weight, Endeavour was chosen for this mission instead. The primary objectives of STS-118 were to deliver critical components to the ISS and provide essential supplies for the crew. This mission played a crucial role in constructing and maintaining the ISS, a platform dedicated to microgravity research and international collaboration in low Earth orbit.

The shuttle and space station crew poses for a press photo.

During and after the mission, a notable concern emerged regarding a small puncture in Endeavour's heat shield caused by a piece of insulation foam detached from the external tank during liftoff. This issue drew considerable media attention, although it was distinct from the more critical foam impact that led to the tragic failure of Columbia during STS-107. Despite these concerns, KSC Launch Director Michael D. Leinbach reassured the public during a post-flight news conference that, upon initial inspection, Endeavour was in remarkably good condition. He described it as the "cleanest" post-flight orbiter since the Return to Flight program. On August 31, NASA confirmed that the damaged tiles had been removed and inspected in the Orbiter Processing Facility, with no evidence of heat-related damage found on the orbiter itself.

The STS-118 crew was a blend of seasoned and novice astronauts, each bringing unique skills to the mission. The flight was commanded by Scott Kelly, marking his second spaceflight. The pilot, Charles O. Hobaugh, was also on his second spaceflight and had previously served as the entry team CAPCOM for STS-107. Tracy Caldwell, making her first spaceflight, served as Mission Specialist 1. Mission Specialist 2 and Flight Engineer Richard Mastracchio, on his second spaceflight, and Mission Specialist 3 Dafydd Williams from the Canadian Space Agency, who was making his second and final spaceflight, were integral to the mission's success. Barbara Morgan, Mission Specialist 4, undertook her only spaceflight on this mission, while Alvin Drew, on his first spaceflight, replaced astronaut Clayton Anderson, who had been reassigned to STS-117.

The mission's primary tasks involved delivering and assembling critical ISS components, including the starboard S5 truss segment, the External Stowage Platform 3 (ESP-3), and a replacement Control Moment Gyroscope (CMG). Additionally, STS-118 marked the final flight of the Spacehab Logistics Single Module, a pressurized aluminum habitat with a capacity of 6,000 pounds (2,700 kilograms) designed for transporting cargo and research projects. This module also returned valuable items, including the MISSE PEC 3 and 4 payloads, which had been installed on the ISS to study the effects of prolonged space exposure on over 850 material specimens.

The mission's payload configuration was meticulously planned to ensure efficiency. The Orbiter Docking System, located in Bay 1–2, weighed approximately 1,800 kilograms (4,000 pounds). The Tunnel Adapter in Bay 3 had a mass of 112 kilograms (247 pounds). The Spacehab Logistics Module, positioned in Bays 5–7, weighed 5,480 kilograms (12,080 pounds), while the Truss Segment S5 in Bay 8–10 was 1,584 kilograms (3,492 pounds). ESP-3, stored in Bays 11–12, weighed 3,400 kilograms (7,500 pounds), bringing the total cargo mass to 14,036 kilograms (30,944 pounds).

On August 10, Endeavour successfully docked with the ISS at 18:02 UTC (14:02 EDT) on the third day of the mission. Prior to docking, the shuttle performed a Rendezvous Pitch Maneuver (RPM), rotating at one degree per second. This maneuver allowed ISS crew members to capture digital images of Endeavour's heat shield, which were analyzed by NASA's Image Analysis Team and Mission Management Team to assess its integrity. Following a series of leak checks, the hatches were opened at 20:04 UTC (16:04 EDT), and the STS-118 crew was welcomed aboard by the Expedition 15 team.

The following day, August 11, saw the commencement of the first Extravehicular Activity (EVA), during which astronauts Rick Mastracchio and Dave Williams installed the S5 truss onto the ISS. This installation increased the ISS's total mass to 232,693 kilograms (513,000 pounds). The EVA, which lasted 6 hours and 17 minutes, achieved all of its objectives successfully.

On August 19, the final day of the mission, Endeavour undocked from the ISS at 11:56 UTC.

To provide the shuttle crew with some well-deserved rest, NASA managers decided to forgo the traditional station fly-around that typically follows each mission, ensuring that the crew had adequate time to rest before the landing process began.

(19 Aug. 2007) --- Backdropped by a blue and white Earth, the International Space Station moves away from Space Shuttle Endeavour. Earlier the STS-118 and Expedition 15 crews concluded nearly nine days of cooperative work onboard the shuttle and station. Undocking of the two spacecraft occurred at 6:56 a.m. (CDT) on Aug. 19, 2007.

As the ISS Progress 25 spacecraft departed on September 18, 2007, the ISS continued its regular operations and preparations for the upcoming crew change. On September 27, 2007, the Soyuz TMA-10 spacecraft was relocated to an alternative docking port, making way for the arrival of new crew members.

Throughout Expedition 15, the crew engaged in several significant extravehicular activities (EVAs), essential for the station's assembly and maintenance. These spacewalks were critical in advancing the station's construction and infrastructure. The first EVA, conducted by Yurchikhin and Kotov on May 30, 2007, lasted 5 hours and 25 minutes and was pivotal in progressing the station's construction. A subsequent EVA on June 6, 2007, involving Yurchikhin and Kotov, lasted 5 hours and 37 minutes, focusing on installing necessary equipment. The third EVA, held on July 23, 2007, saw Yurchikhin and Anderson working together for 7 hours and 41 minutes, addressing and resolving complex technical challenges.

In addition to these key spacewalks, the crew conducted several other EVAs throughout the mission, further contributing to the ISS's development. On August 13, 2007, Richard Mastracchio and David Williams spent 6 hours and 28 minutes in space, followed by another spacewalk on August 11, 2007, with the same astronauts working for 6 hours and 17 minutes. On August 15, 2007, Mastracchio and Anderson engaged in a spacewalk lasting 5 hours and 28 minutes, while David Williams and Clayton Anderson performed a spacewalk on August 18, 2007, lasting 5 hours and 2 minutes. These activities were crucial for installing and repairing key components and enhancing the station's operational capabilities.

The official conclusion of Expedition 15 occurred on October 19, 2007, with the arrival of Expedition 16 Commander Peggy Whitson aboard Soyuz TMA-11. A formal change of command ceremony was held to mark this transition, symbolizing the end of Expedition 15's tenure on the ISS. The formal handover was completed on October 21, 2007, when the outgoing Expedition 15 crew departed the station aboard Soyuz TMA-10.

The transition between expeditions was marked by Expedition 16's arrival on October 10, 2007, with the docking of the Soyuz TMA-11 spacecraft.

Astronaut Patrick Forrester participate in the mission's second planned session of extravehicular activity, as construction resumes on the International Space Station

TMA-11 Astronaut Peggy Whitson (right), Expedition 16 commander; Russia's Federal Space Agency cosmonaut Yuri Malenchenko, flight engineer and Soyuz commander; and Malaysian spaceflight participant Sheikh Muszaphar Shukor

Astronauts Rick Mastracchio and Canadian Space Agency's Dave Williams (out of frame), both STS-118 mission specialists, participate in the mission's first planned session of extravehicular activity (EVA), as construction continues on the International Space Station. During the 6-hour, 17-minute spacewalk Mastracchio and Williams attached the Starboard 5 (S5) segment of the station's truss, retracted the forward heat-rejecting radiator from the station's Port 6 (P6) truss, and performed several get-ahead tasks.1

The year 2008 marked a period of significant developments and milestones in space exploration, showcasing the growing capabilities and ambitions of various spacefaring nations.

In China, 2008 saw the country's ascent in space technology with the successful Shenzhou-7 mission. This mission was a landmark event for China, not only because it demonstrated the nation's increasing prowess in space exploration but also because it marked China's first extravehicular activity (EVA). The crew of Shenzhou-7, consisting of taikonauts Zhai Zhigang, Liu Boming, and Jing Haipeng, conducted a spacewalk that underscored China's growing capabilities in space. The mission also involved the release of the BX-1 micro-satellite, which, though it did not collide with the International Space Station (ISS), flew perilously close to it. The BX-1's proximity to the ISS highlighted China's advancements in developing space technology with anti-satellite potential, signaling a new phase in the country's ambitious space program, known as Project 921.

Expedition 16

Expedition 16, the sixteenth mission to the International Space Station (ISS), commenced on October 10, 2007, with the launch of Soyuz TMA-11. On October 10, 2007, the Soyuz TMA-11 mission commenced with a launch from the Baikonur Cosmodrome at 13:22 UTC. This mission utilized the Soyuz FG launch vehicle to transport its crew to the International Space Station (ISS). The spacecraft's primary objectives were to ferry two ISS Expedition 16 crew members and carry Sheikh Muszaphar Shukor, the first Malaysian to journey into space. Soyuz TMA-11 was not only a transport vehicle but also served as an escape craft during its tenure at the ISS. The spacecraft's docking with the ISS was successfully executed on October 12, 2007, at 14:50 UTC.

International Space Station during Expedition 16. Astronaut Peggy Whitson (front row, right), station commander; and Russia's Federal Space Agency cosmonaut Yuri Malenchenko (front row, left), flight engineer and Soyuz commander, NASA astronaut Clay Anderson (back row, left), flight engineer, astronaut Dan Tani (back row, second from left), flight engineer, Leopold Eyharts of the European Space Agency (back row, third from left), astronaut Garrett Reisman (back row, far right), flight engineer.

Soyuz TMA-11's crew comprised Commander Yuri Malenchenko from Russia, Flight Engineer Peggy Whitson from the United States, and Spaceflight Participant Sheikh Muszaphar Shukor from Malaysia. The backup crew included Commander Salizhan Sharipov, Flight Engineer Michael Fincke, and Spaceflight Participant Faiz Khaleed from Malaysia. Notably, South Korean astronaut Yi So-Yeon was also a spaceflight participant, marking her first venture into space.

Sheikh Muszaphar's mission was part of a cooperative program between Malaysia and Russia. As part of a multi-billion-dollar deal involving the purchase of Russian fighter jets, Malaysia secured the training of two Malaysians for space travel, with Sheikh Muszaphar being the first to fly to the ISS. Although the term "Spaceflight Participant" used in official documents and press briefings may have suggested a space tourist, Sheikh Muszaphar was regarded by many, including Russian ambassador Alexander Karchava and retired NASA astronaut Robert "Hoot" Gibson, as a fully qualified astronaut.

During the early phase of Expedition 16, the International Space Station (ISS) underwent a notable transition in its crew composition. Flight Engineer Clayton Anderson, who had served on

Expedition 15, continued his mission aboard the ISS for several weeks into Expedition 16. His tenure concluded with the arrival of Space Shuttle Discovery on mission STS-120. Launched on October 23, 2007, Discovery docked with the ISS on October 25, 2007, bringing new crew member Daniel Tani. Tani replaced Anderson, who then joined the STS-120 crew, marking a pivotal shift in the station's crew dynamics.

STS-120 crew from the left are astronauts Scott E. Parazynski, Douglas H. Wheelock, Stephanie D. Wilson, all mission specialists; George D. Zamka, pilot; Pamela A. Melroy, commander; Daniel M. Tani, Expedition 15 flight engineer; and Paolo A. Nespoli, mission specialist representing the European Space Agency (ESA). Tani joined Expedition 15 as flight engineer after launching to the International Space Station on mission STS-120 and returned home on mission STS-122.

Expedition 16's crew rotation continued with further changes. Léopold Eyharts arrived on February 9, 2008, to replace Tani. Eyharts' assignment was brief, as he was succeeded by Garrett Reisman, who arrived aboard the Space Shuttle Endeavour with mission STS-123 on March 11, 2008. Reisman's involvement extended into Expedition 17, reflecting the fluid nature of crew assignments aboard the ISS.

A significant milestone for Expedition 16 occurred on October 26, 2007, when STS-120 delivered and attached the Harmony module, also known as Node 2, to the ISS. The Harmony module was a major addition to the station's infrastructure, increasing its living volume by over 2,500 cubic feet (71 cubic meters). This expansion was a quantitative enhancement and a qualitative one, as Harmony served as a crucial "utility hub" for the

ISS. It provided essential connections between the laboratory modules of the United States, Europe, and Japan, and facilitated the distribution of electrical power and electronic data throughout the station. Harmony also included sleeping quarters for four crew members, thus contributing significantly to the station's capacity for long-duration missions.

(25 Oct. 2007) --- Backdropped by a blue and white Earth, Space Shuttle Discovery approaches the International Space Station during STS-120 rendezvous and docking operations. Docking occurred at 7:40 a.m. (CDT) on Oct. 25, 2007. The Harmony node was visible in Discovery's cargo bay.

The module's journey began with its launch on October 23, 2007, aboard Space Shuttle STS-120. Upon arrival, it was temporarily attached to the port side of the Unity module. However, on November 14, 2007, Harmony was relocated to its permanent position on the forward end of the Destiny laboratory. This strategic relocation integrated Harmony into the ISS's infrastructure, enabling it to connect the various international laboratory modules efficiently.

The installation of Harmony increased the ISS's habitable volume from 424.8 cubic meters (15,000 cubic feet) to 500.2 cubic meters (17,666 cubic feet), marking an almost 20 percent increase in available space. For NASA, this expansion was a significant milestone, symbolizing the completion of the "U.S. Core" of the ISS. The successful installation of Harmony was a critical step in establishing a fully functional American segment of the station's infrastructure and enhancing the overall capabilities of the ISS.

Among the notable activities were the EVAs conducted by astronauts Peggy Whitson and Yuri Malenchenko. On November 9, 2007, they performed the first EVA of the mission, preparing docking ports for future relocations. Additional EVAs were carried out to relocate and outfit the Pressurized Mating Adapter (PMA-2) and Harmony. A significant milestone occurred on December 18, 2007, when Peggy Whitson set a new record for the most cumulative EVA time by a female astronaut in NASA history, surpassing Sunita Williams. This achievement highlighted Whitson's extensive contributions to the station's assembly and maintenance.

Astronauts Pam Melroy (left), STS-120 commander; and Peggy Whitson, Expedition 16 commander while Space Shuttle Discovery was docked with the station.

The Space Shuttle Discovery completed its mission and undocked from the ISS, returning to Earth on November 5, 2007. As the year progressed, crucial cargo and supply missions continued. The ISS Progress 26 spacecraft undocked on December 21, 2007, clearing the way for the ISS Progress 27 spacecraft, which launched on December 23, 2007, and docked with the ISS on December 26, 2007. This resupply mission was vital for sustaining the station's ongoing operations and scientific experiments.

Entering 2008, Expedition 16 marked several key milestones. The ISS Progress 27 spacecraft undocked on February 4, 2008, and was succeeded by the ISS Progress 28 spacecraft, which launched on February 5, 2008, and docked with the ISS on February 7, 2008. This resupply mission further supported the station's needs.

April 3, 2008, witnessed the arrival of the European Automated Transfer Vehicle (ATV-1) Jules Verne, the first of its kind to dock with the ISS. The ATV-1's mission was a significant achievement in European spaceflight technology, marking a major advancement in the station's supply chain. The mission underscored the increasing international collaboration in space exploration.

Additionally, the arrival of Soyuz TMA-12 brought Yi So-yeon, the first Korean astronaut to visit the ISS, adding a new dimension to the station's diverse international crew. The mission included five EVAs, each contributing to the station's growth and operational readiness. Notably, the fifth EVA on January 30, 2008, involved the replacement of a Bearing Motor Roll Ring Module in the S4 starboard Beta Gimbal Assembly. This critical task addressed ongoing maintenance needs, ensuring the optimal functionality of the station's solar arrays.

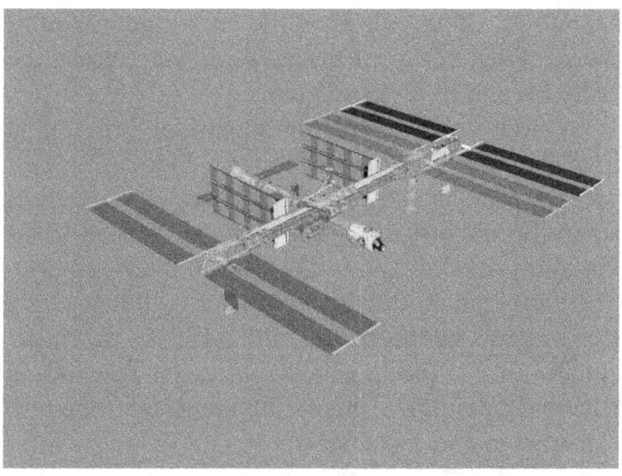

Illustration of the ISS after STS-120, highlighting addition of the Harmony node.

On February 7, 2008, Space Shuttle Atlantis embarked on its STS-122 mission, marking a pivotal moment in the International Space Station's (ISS) ongoing expansion. This mission was dedicated to delivering and installing the Columbus laboratory module, a significant addition to the ISS that would support a decade of scientific research in orbit.

The Columbus module, constructed by the European Space Agency (ESA) at an investment of approximately €1.4 billion (about US$1.6 billion), represented a major commitment to advancing space research and international collaboration. This substantial expenditure covered the construction and integration of the module and the development of ground control infrastructure essential for its operation. Columbus was designed to enhance the ISS's capabilities with advanced facilities for conducting experiments across various scientific disciplines, including biology, physics, and materials science.

Upon its arrival in space, the Columbus module was placed under the management of the Columbus Control Centre, operated by the German Aerospace Center in Oberpfaffenhofen, near Munich. This center was responsible for overseeing the module's operations and ensuring its successful integration into the ISS.

The mission began with the precise launch of Atlantis, carrying a crew of seven astronauts: Commander Stephen Frick, Pilot Alan Poindexter, and Mission Specialists Hans Schlegel, Leopold Eyharts, Leland Melvin, Sandra Magnus, and Timothy Kopra. Each crew member brought unique expertise to the mission, ensuring the successful completion of their complex objectives.

The STS-122 mission reached a notable milestone on December 18, 2007, when astronaut Peggy Whitson, during her fourth spacewalk, set a record for the most cumulative Extravehicular Activity (EVA) time by a female astronaut in NASA history. By the conclusion of her fifth EVA, Whitson had accumulated a total of 32 hours and 36 minutes, placing her 20th on the overall EVA time list.

The Atlantis crew's primary task involved docking with the ISS and installing the Columbus module. This intricate process required meticulous coordination and advanced technology. The docking was executed with remarkable precision, followed by a series of critical operations. The Columbus module was carefully maneuvered from Atlantis's payload bay into its designated position on the ISS using the shuttle's robotic arm, operated by astronauts from inside the shuttle. Simultaneously, two astronauts conducted extravehicular activities (EVAs) outside the station to secure the module and connect its systems.

The successful installation of Columbus represented more than just a technical achievement; it marked a significant enhancement to the ISS's research capabilities and underscored the importance of international cooperation in space exploration. With the new laboratory in place, the ISS could now facilitate a broader range of experiments, contributing to scientific advancement across multiple fields.

STS-122 crew from the left (front row) astronauts Stephen N. Frick, commander; European Space Agency's (ESA) Leopold Eyharts; and Alan G. Poindexter, pilot. From the left (back row) astronauts Leland D. Melvin, Rex J. Walheim, Stanley G. Love and European Space Agency's (ESA) Hans Schlegel, all mission specialists. Eyharts who joined Expedition 16 in progress to serve as a flight engineer aboard the International Space Station.

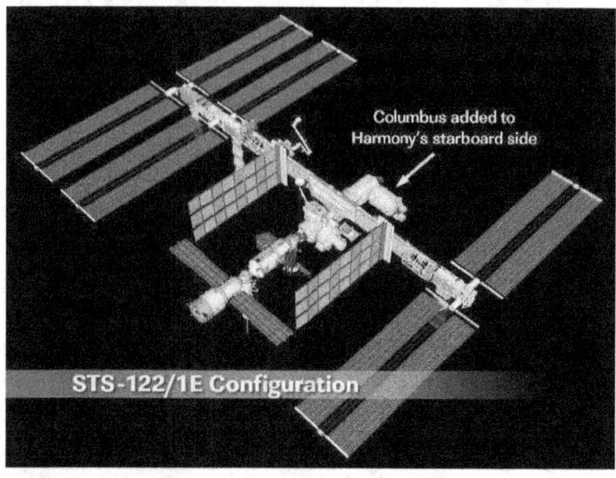

Illustration of the ISS after STS-122, highlighting the addition of the Columbus laboratory module.

The STS-123 mission, launched on March 11, 2008, aboard the Space Shuttle Endeavour, represented a key milestone in the ongoing development of the International Space Station (ISS). This mission was pivotal in advancing the ISS's construction and expanding its scientific research capabilities.

The primary objective of STS-123 was to deliver and install the Japanese Experiment Module-Pressurized Section (JEM-PS), also known as Kibo, and the Express Logistics Carrier-1 (ELC-1), a critical component for mounting additional scientific experiments. The mission also featured a significant international collaboration, as it included contributions from the Japanese Aerospace Exploration Agency (JAXA) and the Canadian Space Agency (CSA).

Endeavour's crew, led by Commander Dominic Gorie, included Pilot Gregory Johnson and Mission Specialists Richard Linnehan, Michael Foreman, Robert Behnken, and Koichi Wakata. During this mission, Wakata, a JAXA astronaut, became the first Japanese astronaut to live and work on the ISS, underscoring the mission's international dimension.

The mission began with a successful launch, followed by a series of meticulous docking maneuvers. Endeavour docked with the ISS on March 13, 2008, allowing the crew to commence the complex series of tasks required to install the JEM-PS and ELC-1 successfully.

The installation of the Japanese Experiment Module was a highlight of STS-123. This pressurized laboratory, designed for various scientific experiments, was mounted to the ISS's Harmony module using the shuttle's robotic arm. The Japanese module, later known as Kibo, was a significant addition to the ISS's scientific capabilities, offering new facilities for research in various disciplines, including materials science, astronomy, and life sciences.

STS-123 crew from the right (front row) are astronauts Dominic L. Gorie, commander; and Gregory H. Johnson, pilot. From the left (back row) are astronauts Richard M. Linnehan, Robert L. Behnken, Garrett E. Reisman, Michael J. Foreman and Japan Aerospace Exploration Agency's (JAXA) Takao Doi, all mission specialists.

The Kibo Japanese Pressurized Module and Kibo Japanese logistics module of the International Space Station while Space Shuttle Discovery was docked with the station. The blackness of space and Earth's horizon provide the backdrop for the scene.

In addition to the JEM-PS installation, the crew conducted three spacewalks to perform necessary tasks outside the ISS. These extravehicular activities were essential for attaching the ELC-1, which would serve as a platform for future experiments and equipment. The spacewalks required precise coordination and demonstrated the astronauts' skill and expertise in challenging space environments.

One of STS-123's key achievements was the successful deployment of the Japanese Experiment Logistics Module-Pressurized Section, which marked the beginning of a series of planned Japanese contributions to the ISS. The mission also laid the groundwork for subsequent missions to expand and enhance the station's research capabilities.

In June, the Discovery shuttle launched the STS-124 mission featuring astronaut Garrett E. Reisman. This mission, part of the ongoing ISS assembly and crew rotation efforts, delivered significant components to the station and ensured its continued smooth operation. The Discovery shuttle's STS-124 mission in May, with its crew comprising Mark Kelly, Kenneth T. Ham, Karen L. Nyberg, Ronald J. Garan, Michael E. Fossum, and Akihiko Hoshide, was crucial in delivering the Kibo Japanese Experiment Module, enhancing the ISS's research capabilities.

STS-124 space shuttle crew; From the left astronauts Gregory E. Chamitoff, Michael E. Fossum, both STS-124 mission specialists; Kenneth T. Ham, pilot; Mark E. Kelly, commander; Karen L. Nyberg, Ronald J. Garan and Japan Aerospace Exploration Agency's (JAXA) Akihiko Hoshide, all mission specialists. Chamitoff joined Expedition 17 as flight engineer after launching to the International Space Station on mission STS-124.

At their crew quarters in Baikonur, Kazakhstan, Expedition 17 Commander Sergei Volkov (center), Flight Engineer Oleg Kononenko (right) and South Korean spaceflight participant So-yeon Yi on April 7, 2008, the eve of their launch to the International Space Station. Volkov, Kononenko and Yi launched to the station on the Soyuz TMA-12 spacecraft from the Baikonur Cosmodrome on April 8 and arrived at the ISS on April 10 to begin what will be six months in space for Volkov and Kononenko. Yi remained in space nine days on the complex, returning to Earth with two of the Expedition 16 crewmembers currently on the station.

April 2008 saw the Soyuz TMA-12 spacecraft transporting Yi So-yeon, the first South Korean in space, to the ISS. The Soyuz TMA-12 mission, launched on April 8, 2008, was a crucial component of the International Space Station (ISS) program, marking a significant event in the ongoing assembly and operation of the ISS. This mission exemplified the collaborative efforts of international space agencies and underscored the importance of crew rotations in maintaining the station's operations.

The Soyuz TMA-12 spacecraft was a Russian-built vehicle designed to transport astronauts to and from the ISS. This mission's primary goal was to deliver a new crew to the station and ensure the continuity of its operations, including scientific research and maintenance tasks.

The crew for Soyuz TMA-12 consisted of three astronauts: Commander Gennady Padalka of the Russian space agency Roscosmos, Flight Engineer Yuri Lonchakov, also from Roscosmos, and NASA astronaut and Mission Specialist Garrett Reisman. Their arrival at the ISS was highly anticipated, as it would replace the departing crew and bring new skills and expertise to the station.

The spacecraft launched from the Baikonur Cosmodrome in Kazakhstan, embarking on a journey that would last approximately six hours. During this time, Soyuz TMA-12 executed a series of orbital maneuvers to align itself with the ISS. This precise docking process required meticulous planning and coordination to ensure a safe and successful arrival.

On April 9, 2008, Soyuz TMA-12 successfully docked with the ISS, marking a smooth and well-executed insertion into the station's crew rotation schedule. Padalka, Lonchakov, and Reisman's arrival was a significant moment for the ISS, as it ensured that the station could continue its vital research and operational tasks without interruption.

Once docked, the new crew members transferred to the ISS and were greeted by the outgoing crew, including Commander Peggy Whitson, Flight Engineer Yuri Malenchenko, and NASA astronaut Daniel Tani. The crew exchange process involved thorough handovers of ongoing research and operational responsibilities, ensuring a seamless transition.

During their stay aboard the ISS, Padalka, Lonchakov, and Reisman contributed to various scientific experiments, station maintenance, and upgrades. Their presence was essential for continuing the station's research programs, which spanned diverse fields from biology to physics.

In addition to these shuttle missions, Expedition 16 was marked by the arrival of the Automated Transfer Vehicle (ATV) Jules Verne on March 9, 2008. Named after the renowned science fiction author Jules Verne, it was the first of its kind to dock with the ISS. This cargo spacecraft significantly contributed to the station's resupply operations, bringing essential supplies and experiments to support ongoing research.

The Columbus laboratory, a key International Space Station (ISS) component, was the European Space Agency's (ESA) largest single contribution to the orbiting laboratory. This advanced science module, integral to the ISS's international collaboration, embodies Europe's commitment to space research and exploration.

Constructed by Thales Alenia Space in Turin, Italy, Columbus represents a marvel of modern engineering and space technology. EADS in Bremen, Germany, spearheaded the design and development of its functional equipment and software. After assembly, Columbus was integrated at the EADS facility in Bremen before its transatlantic journey. The module was transported to the Kennedy Space Center in Florida aboard an Airbus Beluga jet, a specially designed aircraft for carrying oversized cargo.

Between November 2007 and January 2008, the International Space Station (ISS) saw a series of spacewalks that was crucial in its ongoing assembly and maintenance. Astronauts performed these extravehicular activities (EVAs) Peggy Whitson, Yuri Malenchenko, and Daniel M. Tani, each contributing to advancing the ISS's infrastructure and systems.

The first EVA, conducted on November 9, 2007, began at 09:54 UTC and concluded at 16:49 UTC, lasting 6 hours and 55 minutes. During this spacewalk, Whitson and Malenchenko focused on the disconnection and stowage of the SSPTS (Space Shuttle Power Transfer System) cables and the PMA-2 (Pressurized Mating Adapter 2) umbilicals. They also addressed the temporary stowage of the Node 2 avionics umbilical, essential tasks for preparing the station for subsequent installations and operations.

Harmony shown connected to Columbus, Kibo, and Destiny. PMA-2 faces. The nadir and zenith locations are open.

On November 20, 2007, Whitson and Tani embarked on their second EVA, which began at 10:10 UTC and ended at 17:26 UTC, extending for 7 hours and 16 minutes. This spacewalk was pivotal for the external configuration of PMA-2 and the Harmony module. The astronauts meticulously connected various fluid, electrical, and data lines, ensuring avionics and heater cables hook up properly. Additionally, they relocated a fluid tray, crucial for integrating these modules into the ISS's existing framework.

Expedition commander Peggy Whitson during the increment's third EVA. Behind Whitson, was the Destiny Laboratory Module, and Harmony.

The third EVA took place on November 24, 2007, starting at 09:50 UTC and concluding at 16:54 UTC, lasting 7 hours and 4 minutes. During this session, Whitson and Tani completed the installation of fluid, electrical, and data lines for PMA-2 and Harmony. They connected the Loop B Fluid Tray to the port side of the Destiny module. They conducted a photographic analysis of the starboard Solar Alpha Rotary Joint (SARJ) to assist with ground-based troubleshooting. The spacewalk also involved the re-installation of the CETA (Crew Equipment Translation Aid) cart from a temporary stowage location.

The fourth EVA occurred on December 18, 2007, from 09:50 UTC to 16:46 UTC, lasting 6 hours and 56 minutes. Whitson and Tani inspected the S4 starboard Solar Alpha Rotary Joint (SARJ) and the Beta Gimbal Assembly (BGA). Notably, this EVA marked the 100th in support of the ISS's construction, underscoring the significance of their work in the station's ongoing development.

Finally, on January 30, 2008, Whitson and Tani conducted their fifth EVA, starting at 09:56 UTC and ending at 17:06 UTC, lasting 7 hours and 10 minutes. This spacewalk was dedicated to the replacement of a Bearing Motor Roll Ring Module (BMRRM) in the S4 starboard Beta Gimbal Assembly (BGA) and further inspection of the Solar Alpha Rotary Joint (SARJ). Their efforts were vital for maintaining the operational integrity of the station's solar array mechanisms, which are essential for its power generation and overall functionality.

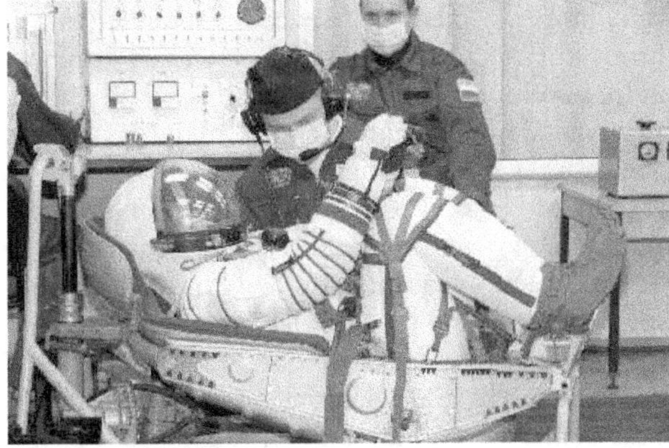

Peggy Whitson suited up in a pressurized Sokol space suit getting ready for Expedition 16.

STS-123 concluded with the undocking of Endeavour from the ISS on March 26, 2008, and its safe return to Earth on March 29, 2008.

Expedition 16 concluded with the crew's undocking and return on April 19, 2008. This marked the end of their mission, which was characterized by significant contributions to the station's scientific research and infrastructure.

On October 12, 2008, Soyuz TMA-12 undocked from the ISS, carrying the outgoing crew back to Earth. The spacecraft safely landed in Kazakhstan, concluding a successful mission that reinforced the operational continuity and scientific progress of the ISS.

The mission concluded with the return of Soyuz TMA-11 to Earth on April 19, 2008. The spacecraft experienced a challenging, ballistic reentry, which was steeper and more abrupt than standard reentries. This occurred due to a malfunction that prevented the service module from separating correctly, resulting in the spacecraft landing 475 kilometers from its intended location. The reentry, characterized by its intense heat, caused burn damage to the spacecraft's hatch and antenna. Yi So-Yeon, the South Korean astronaut, sustained minor injuries during the rough return and was hospitalized upon her arrival in South Korea, suffering from neck muscle strain and spinal bruising.

In the aftermath of the reentry, Russian space agency head Anatoly Perminov speculated that the unusual event might be linked to a Russian superstition concerning the predominance of women in space missions. This flight marked the first instance of two women being onboard a Soyuz spacecraft together and the first time women outnumbered men in space since Valentina Tereshkova's solo mission in 1963. Perminov addressed concerns by suggesting that future missions should avoid having a majority female crew to prevent similar incidents.

On May 24, 2008, an investigation into the malfunction revealed that the primary cause of the ballistic reentry was the failure of one of five pyro-bolts responsible for the service module separation. The investigation also suggested that long-term exposure to the ISS's electrical environment might have contributed to the failure, reminiscent of the Soyuz 5 reentry anomaly in 1969.

Elsewhere in international space endeavors, in July 2008, Virgin Galactic expanded its roster by adding three new pilots to its Astronaut Pilots Group from the UK. Robert Bendall, Rich Dancaster, and Brad Lambert joined the ranks of commercial astronauts selected to pioneer the next phase of private space exploration. Their expertise and dedication represented Virgin Galactic's ongoing commitment to pushing the boundaries of space tourism and advancing the accessibility of space travel beyond traditional governmental programs. These selections reinforced Virgin Galactic's position at the forefront of commercial spaceflight initiatives, marking a pivotal moment in the evolution of private-sector involvement in human space exploration.

Expedition 17

NASA astronauts Greg Chamitoff (left), Garrett Reisman, both Expedition 17 flight engineers; Russian Federal Space Agency cosmonauts Sergei Volkov, commander; and Oleg Kononenko, flight engineer. Reismanl launched to the International Space Station on the STS-123 mission of Endeavour in March 2008, joining Expedition 16 in progress and will provide Expedition 17 with an experienced flight engineer for the first part of its increment. Volkov and Kononenko launched to the complex in the Soyuz TMA-12 spacecraft from the Baikonur Cosmodrome in Kazakhstan in April for a six-month mission. Chamitoff launched to the station on the STS-124 mission of Discovery in June, joining Expedition 17 in progress.

Expedition 17, which commenced in April 2008, marked a pivotal period of transition and progress for the International Space Station (ISS). This expedition was distinguished by significant events and the arrival of new crew members, setting the stage for an era of enhanced scientific exploration and international collaboration.

The expedition began with Sergey Volkov and Oleg Kononenko's launch aboard the Soyuz TMA-12 spacecraft on April 8, 2008. Their arrival at the ISS heralded the start of their mission, where they joined Garrett Reisman, who had transitioned from Expedition 16. Reisman's tenure was crucial as he played a key role in the station's operations and research during this period.

One of Expedition 17's first major milestones was the docking of the ISS Progress 29 spacecraft on May 16, 2008. This cargo vehicle delivered essential supplies and equipment necessary for the station's continued operation and the advancement of its scientific experiments.

The Space Shuttle Discovery's mission, STS-124, arrived at the ISS on June 2, 2008. This mission was particularly notable for delivering and installing the Japanese Pressurized Module (JPM), also known as Kibō, and the associated Kibo laboratory. These additions were critical in expanding the ISS's research capabilities. Following the successful installation, STS-124 undocked from the ISS on June 11, 2008, concluding its mission.

In September 2008, the ISS continued to receive logistical support by docking the ISS Progress 30 spacecraft on September 17. This mission ensured the ongoing flow of supplies necessary for maintaining the station's daily operations. Earlier in the month, the ATV-1 (Automated Transfer Vehicle) completed its mission and undocked from the ISS on September 5, 2008, alongside the 29P spacecraft, which undocked on September 1.

NASA astronaut Michael Fincke (right), Expedition 18 commander; Russian Federal Space Agency cosmonaut Yury Lonchakov (center), flight engineer; and American spaceflight participant Richard Garriott launches to the International Space Station in a the TMA-13 Soyuz spacecraft in October.

As Expedition 17 drew to a close, it was marked by the arrival of Expedition 18. The Soyuz TMA-13 spacecraft, which carried a new crew, launched on October 12, 2008, and docked with the ISS on October 14, 2008. This mission was significant as it included the transportation of Richard A. Garriott, contributing to the rotation of the station's crew. Another crew rotation occurred in April 2009 with the Soyuz TMA-13, involving Yuri Lonchakov and E. Michael Fincke, marking the beginning of Expedition 18. The new crew members settled into their roles, continuing the ISS's ongoing research and operational activities.

The conclusion of Expedition 17 was marked by the departure of its crew on October 23, 2008. Sergey Volkov and Oleg Kononenko returned to Earth, while Gregory Chamitoff remained aboard the ISS as part of Expedition 18 until November 2008. Volkov, at 35 years old, became the youngest individual to command the ISS, leading a team that demonstrated dynamic international cooperation. Notably, Garrett Reisman was the first Jewish astronaut to reside on the ISS and was involved in critical transitions between the STS-123 and STS-124 missions. Gregory Chamitoff's contributions included a unique personal touch, as he brought three bags of sesame seed bagels—an unprecedented addition for ISS crew members.

The backup crew for Expedition 17 played a crucial role in ensuring mission continuity. Maksim Surayev was designated as Sergey Volkov's backup commander, Oleg Skripochka prepared to substitute for Oleg Kononenko, and Timothy Kopra was the backup flight engineer for Gregory Chamitoff. Their readiness was a testament to the meticulous planning and support integral to the success of each space mission.

Among Expedition 17's key advancements was the addition of the Japanese Experiment Module (JEM), or Kibō, which represents Japan's significant contribution to the ISS's scientific infrastructure. Developed by the Japan Aerospace Exploration Agency (JAXA), Kibō was the largest single module on the ISS, highlighting its importance in expanding the station's research capabilities and enhancing its role as a hub for international scientific endeavors.

Kibō was attached to the Harmony module, which facilitates its integration into the broader framework of the ISS. Its construction and deployment were accomplished in stages, with the initial pieces of the module being launched on Space Shuttle missions STS-123 and STS-124. These missions delivered the first two segments of Kibō to the ISS, setting the stage for its full operational capacity.

Kibō Exposed Facility on the right

The final component of Kibō was transported to the ISS on the STS-127 mission. This mission completed the assembly of the Japanese Experiment Module, allowing it to function as a comprehensive research facility. Kib'ōs installation and subsequent operations reflect Japan's commitment to advancing space science and contributing to the collaborative research efforts aboard the ISS.

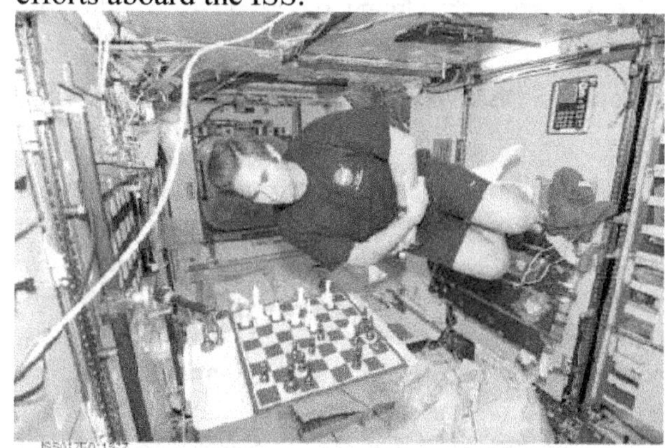

NASA astronaut Greg Chamitoff, Expedition 17 flight engineer, ponders his next move as he plays a game of chess in the Harmony (Node 2) module of the International Space Station.

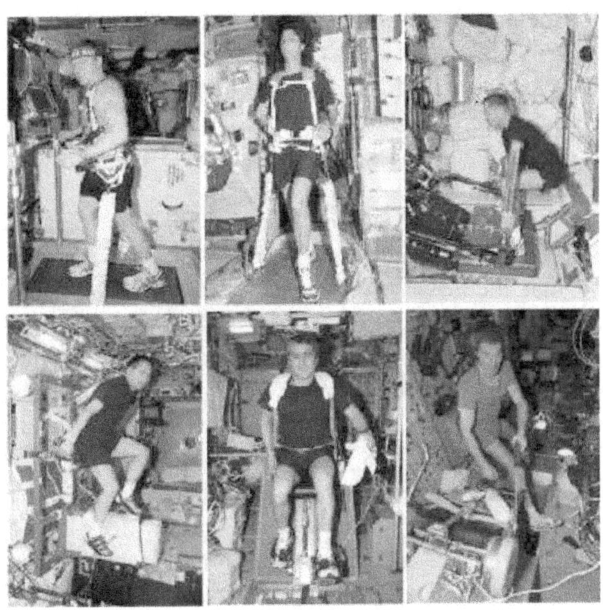

(11 Oct. 2007) --- Cosmonaut Oleg D. Kononenko, Expedition 17 flight engineer representing Russia's Federal Space Agency, participates in an Extravehicular Mobility Unit (EMU) spacesuit fit check in the Space Station Airlock Test Article (SSATA) in the Crew Systems Laboratory at the Johnson Space Center. Cosmonaut Sergei A. Volkov (left), commander representing Russia's Federal Space Agency, and United Space Alliance (USA) suit technician James Lemmon assisted Kononenko.

Exercise devices in the ISS (top left to bottom right): T2/COLBERT, TVIS, ARED, CEVIS, VELO Ergometer, VELO Ergometer with cable (NASA)

Expedition 18

Expedition 18 NASA astronaut Michael Fincke and Russian Federal Space Agency cosmonaut Yury Lonchakov, commander and flight engineer, respectively. Japan Aerospace Exploration Agency astronaut Koichi Wakata, NASA astronauts Sandra Magnus and Greg Chamitoff, all flight engineers. Chamitoff launched to the International Space Station (ISS) on the STS-124 mission, joining Expedition 17 in progress and provided Expedition 18 with an experienced flight engineer for the first part of its increment. Fincke and Lonchakov launched to the station in the Soyuz TMA-13 spacecraft from the Baikonur Cosmodrome in Kazakhstan in October for a six-month mission. Magnus joined Expedition 18, replacing Chamitoff, as flight engineer after launching to the ISS on mission STS-126. Wakata joined Expedition 18, replacing Magnus, as flight engineer after launching to the orbital complex on mission STS-119.

Expedition 18, the 18th long-term crew aboard the International Space Station (ISS), commenced its mission on October 12, 2008, with the arrival of Soyuz TMA-13, a spacecraft designed to transport astronauts to and from the station. This crewed mission marked a pivotal chapter in the ISS's continuous development and operational history.

The initial members of Expedition 18 included Commander Michael Fincke from NASA and Flight Engineer Yuri Lonchakov from the Russian Space Agency (RSA). Their journey to the ISS set the stage for a series of critical operations and research endeavors. Sandra Magnus, another NASA astronaut, joined the crew shortly after, arriving aboard the Space Shuttle Discovery on mission STS-124. Magnus played a crucial role in the expedition, contributing significantly to scientific experiments and station operations. Her tenure on the ISS extended until March 25, 2009, when she returned to Earth aboard Space Shuttle Endeavour during mission STS-119.

On March 17, 2009, JAXA astronaut Koichi Wakata joined Expedition 18, arriving with STS-119. Wakata, an experienced space traveler, continued to contribute to the mission until its conclusion. Meanwhile, Gregory Chamitoff, also from NASA, replaced Magnus and became part of Expedition 18 following the end of Expedition 17. Chamitoff's involvement extended through the arrival and departure of the STS-126 crew, who played a crucial role in the station's ongoing operations before leaving with STS-127.

Throughout Expedition 18, the crew conducted a range of significant scientific experiments and technological advancements to improve life in space. Their research focused on various aspects crucial to long-duration spaceflight, including studying the effects of medications and vitamins on crew health, exploring novel fire suppression techniques, and monitoring air quality aboard the station. Additionally, they tested innovative solar cell technologies to enhance the ISS's power systems, contributing to the station's sustainability and efficiency.

Expedition 18 on the International Space Station (ISS) was marked by several significant events in early 2009. These events reflected a period of active crew rotations and spacecraft operations that highlighted the dynamic nature of the ISS program.

The expedition began its transition on February 10, 2009, with the launch of ISS Progress 32, a Russian cargo spacecraft designed to resupply the station with essential goods and equipment. This mission was pivotal for maintaining the station's operational capabilities and ensuring the crew's ongoing access to supplies.

The docking of ISS Progress 32 occurred on February 13, 2009. This successful docking was crucial, allowing the crew to receive the much-needed cargo and ensuring the station's continued scientific and living activities. Shortly before this, on February 5, 2009, ISS Progress 31 undocked from the ISS, making way for its successor and completing its resupply mission.

November 14, 2008, saw the undocking of ISS Progress 30, which had been delivering cargo to the station. This event was part of the routine rotation of cargo spacecraft, ensuring that the station's supply chain remained uninterrupted.

The end of November brought another important event with the docking of the Space Shuttle Endeavour, designated STS-126, on November 16, 2008. STS-126, the one hundred and twenty-fourth mission of NASA's Space Shuttle program, marked the twenty-second orbital flight of Space Shuttle Endeavour (OV-105) to the International Space Station (ISS). Launched on November 15, 2008, at 00:55:39 UTC from Launch Pad 39A at Kennedy Space Center, the mission, also known as ULF2, was pivotal in the ongoing construction and maintenance of the ISS. Endeavour's launch proceeded without delays or issues, setting the stage for a crucial operation in space.

The primary objectives of STS-126 were to deliver essential equipment and supplies to the ISS, service the Solar Alpha Rotary Joints (SARJs), and address a malfunction in the starboard SARJ that had limited its functionality since the previous STS-120 mission. After a flawless launch, Endeavour successfully docked with the ISS on November 16, 2008. The crew spent 15 days, 20 hours, 30 minutes, and 30 seconds attached to the station, performing four spacewalks and transferring a significant amount of cargo. The mission concluded with Endeavour undocking on November 28, 2008, and landing at Edwards Air Force Base on November 30, 2008, due to unfavorable weather conditions at Kennedy Space Center.

The crew of STS-126 was composed of experienced astronauts and one first-timer. Commander Christopher J. Ferguson, in his second spaceflight, led the team. Pilot Eric A. Boe, making his inaugural spaceflight, supported Ferguson. Mission Specialists included Donald R. Pettit, in his second flight; Stephen G. Bowen, on his first flight; Heidemarie M. Stefanyshyn-Piper, on her second and final spaceflight; Robert S. Kimbrough, also on his first spaceflight; Sandra H. Magnus, in her second spaceflight; and Gregory E. Chamitoff, who was serving as an ISS Flight Engineer for Expedition 18 during his first spaceflight.

STS-126 Astronaut Christopher J. Ferguson, commander, was at center; and astronaut Eric A. Boe, pilot, was third from the right. Remaining crewmembers, pictured from left to right, are astronauts Sandra H. Magnus, Stephen G. Bowen, Donald R. Pettit, Robert S. (Shane) Kimbrough and Heidemarie M. Stefanyshyn-Piper, all mission specialists. Magnus joined Expedition 18 as flight engineer after launching to the International Space Station on mission STS-126.

The mission's objectives encompassed extensive service and repair tasks, notably the maintenance of the Solar Alpha Rotary Joints. These joints, crucial for the station's solar arrays, had exhibited anomalous behavior since August 2007. The crew lubricated both SARJs and replaced all trundle bearings in the starboard SARJ. This repair was vital for ensuring the continued operation of the station's solar power system.

STS-126 also carried the Leonardo Multi-Purpose Logistics Module (MPLM), which was making its fifth spaceflight. Leonardo contained over 6,400 kilograms (14,100 pounds) of supplies and equipment, including new crew quarters racks, a second galley for the Destiny laboratory, an additional Waste and Hygiene Compartment, and the advanced Resistive Exercise Device (aRED). The module's cargo included two water reclamation racks, spare hardware, new

experiments, and the General Laboratory Active Cryogenic ISS Experiment Refrigerator (GLACIER), a double locker cryogenic freezer designed to transport and preserve scientific experiments. In addition, the shuttle transported a special Thanksgiving meal for the station crew and an Official Flight Kit containing mementos for those who contributed to the mission.

Another significant payload was the Lightweight MPESS Carrier (LMC), which included a Flex Hose Rotary Coupler and returned a Nitrogen Assembly Tank for refurbishment. The cargo in the shuttle's bays totaled 17,370 kilograms (38,290 pounds), including the Orbiter Docking System, the Shuttle Power Distribution Unit, and other components essential for the mission.

Educational outreach was a key component of STS-126, with over 500,000 student signatures flown as part of the 2008 Student Signatures in Space program, a collaboration between NASA and Lockheed Martin. This initiative, celebrating Space Day in May 2008, involved students from over 500 schools whose signatures were scanned and carried aboard the shuttle.

The mission also featured the Agricultural Camera (AgCam), developed by students and faculty at the University of North Dakota. Installed in the Destiny module, AgCam was designed to assist farmers and provide educational opportunities by capturing visible and infrared images of crops and landscapes. The data collected would benefit agricultural producers and support environmental protection efforts, including disaster management.

STS-126 also carried the first bovine embryos and porcine embryonic stem cells on an American spacecraft. This experiment, a collaboration between ZeroGravity Inc., the University of Florida, and USDA ARS, aimed to study the effects of the space environment on embryonic development, contributing to our understanding of biological processes in space.

On November 23, 2008, the International Space Station (ISS) mission proceeded into its tenth flight day. The day's activities were marked by Mission Specialist Stephen Bowen's involvement in the mission's final spacewalk, the third Extravehicular Activity (EVA), which commenced at 18:01 UTC. This spacewalk was dedicated to essential maintenance tasks, specifically focusing on the starboard Solar Alpha Rotary Joint (SARJ). The tasks included cleaning, lubricating, and replacing the trundle bearing assemblies. While the installation of the final trundle bearing assembly was deferred to the subsequent EVA, all other scheduled tasks were completed without incident. By the end of the spacewalk, Stefanyshyn-Piper's cumulative EVA time totaled thirty-three hours and forty-two minutes, positioning her twenty-fifth in the overall EVA time rankings.

The station crews continued their routine transfer operations and focused on the water reclamation system. During the Mission Status briefing, Lead ISS Flight Director Ginger Kerrick provided an update on the Water Processor Assembly. A sample, which was a mix of 10% urine and 90% condensate, was taken for return to Earth with the shuttle. Kerrick noted that if the system activation proceeded as planned, a sample from the potable water dispenser would be collected on flight day eleven, November 24, 2008. Concurrently, ground engineers were troubleshooting the Urine Processor Assembly, investigating a possible sensor interference with the system's centrifuge, which could be causing intermittent slowdowns.

On November 24, the eleventh flight day, the two crews prepared for the mission's final spacewalk. This EVA commenced at 18:24 UTC and was executed by Bowen and Kimbrough. Just before this spacewalk began, mission managers informed the crew of extending Endeavour's mission by one docked day to sixteen days. This additional time was granted to address ongoing issues with the Urine Processing Assembly. The spacewalk was completed after 6 hours and 7 minutes, bringing the total EVA time for the mission to 26 hours and 41 minutes.

The following day, November 25, saw continued efforts in transferring supplies between the ISS and Endeavour. A notable achievement was the starboard Solar Alpha Rotary Joint successfully tracking the Sun for the first time in over a year during a three-hour test. The Urine Processor Assembly also completed its second full five-hour operational run. During the Mission Status briefing, International Space Station Program Manager Mike Suffredini reported that the water recycling system was functioning correctly following crew modifications. Additional

water samples were collected, to be analyzed on Earth before the crew could use the water. Initial results from the EVA maintenance tasks on the SARJ were promising, with the potential for more routine operations than initially anticipated, though a comprehensive assessment would take weeks.

On November 26, the thirteenth flight day, Commander Chris Ferguson continued the tradition of placing the STS-126 mission patch in the Zvezda module of the ISS. Pettit and Kimbrough used the Canadarm2 to relocate the Leonardo module from the Harmony module to Endeavour's cargo bay. Stefanyshyn-Piper packed equipment and supplies used during the spacewalks for return, while Magnus continued work on the new regenerative life support system, including draining a condensate collection tank and gathering additional water samples for testing.

November 27 marked the final day of joint operations. The crews enjoyed some off-duty time, participated in media interviews, and shared a Thanksgiving meal. They finalized last-minute transfers and gathered in the Harmony node to bid farewell. The hatches between the two spacecraft were closed at 23:31 UTC, with Endeavour's crew preparing for undocking scheduled for the next day.

On November 28, Endeavour undocked from the ISS at 14:47 UTC. The shuttle's docked time had totaled 11 days, 16 hours, and 46 minutes, making it the second-longest docked shuttle mission to the station. Pilot Eric Boe maneuvered the shuttle for a flyaround inspection of the station, although the final separation burn was delayed to avoid debris from a Russian Cosmos satellite. The burn was eventually completed at 23:23 UTC. Following separation, the Endeavour crew conducted a heat shield inspection using the robotic arm, Canadarm, ensuring the shuttle's readiness for re-entry.

Following its successful mission, STS-126 undocked from the ISS on November 28, 2008, returning to Earth and marking the conclusion of its cargo and assembly tasks.

The new year continued to see busy activity with the undocking of STS-119 on March 25, 2009. This shuttle mission, which had previously docked on February 17, 2009, brought additional solar arrays to the station, further enhancing its power generation capabilities.

In March 2009, the Space Shuttle Discovery embarked on STS-119, a pivotal mission in the assembly of the International Space Station (ISS). This mission, designated as ISS assembly flight 15A, was instrumental in delivering and installing crucial components to the growing orbital outpost.

Launched on March 15, 2009, at 19:43 EDT, STS-119 marked a significant milestone in the ISS's construction. The primary objective of this mission was to deliver the fourth starboard Integrated Truss Segment (S6), along with a new set of solar arrays and batteries. These additions were vital for enhancing the station's power generation capabilities. Discovery concluded its mission with a successful landing on March 28, 2009, at 15:13 EDT.

Endeavour lands back at Kennedy Space Center atop the 747 Shuttle Carrier Aircraft

The crew of STS-119 comprised a highly skilled team of astronauts, each bringing unique expertise to the mission. Commander Lee Archambault led the flight, marking his second spaceflight. Pilot Dominic A. Antonelli, making his debut spaceflight, worked alongside Mission Specialists Joseph M. Acaba, Steven Swanson, Richard R. Arnold, John L. Phillips, and Koichi Wakata from JAXA, each contributing their first or final spaceflight experience to the mission. Additionally, Sandra Magnus, who had previously served on Expedition 18, participated in the mission as an ISS Flight Engineer.

STS-119 was not solely focused on assembly tasks; it also carried a range of scientific experiments. These included the Shuttle Ionospheric Modification with Pulsed Local Exhaust (SIMPLEX), Shuttle Exhaust Ion Turbulence Experiments (SEITE), and Maui Analysis of Upper Atmospheric Injections (MAUI). A noteworthy experiment, the "Boundary Layer Transition Detailed Test Objective," involved raising a tile of the shuttle's thermal protection system by 0.25 inches (6.4 mm) to initiate a boundary layer transition during reentry, a test later repeated with an increased height in STS-128.

Attired in training versions of their shuttle launch and entry suits, STS-119 crew from the right (front row) are NASA astronauts Lee Archambault, commander, and Tony Antonelli, pilot. From the left (back row) are NASA astronauts Joseph Acaba, John Phillips, Steve Swanson, Richard Arnold and Japan Aerospace Exploration Agency astronaut Koichi Wakata, all mission specialists. Wakata joined Expedition 18 as flight engineer after launching to the International Space Station on STS-119.

Upon reaching orbit, the crew's initial tasks included preparing for docking with the ISS. Discovery performed the rendezvous pitch maneuver (RPM), allowing the Expedition 18 crew to photograph the shuttle's underside. The successful docking at 21:20 UTC was followed by hatch leak checks and opening at 23:09 UTC, leading to a warm exchange between the two crews and a mandatory safety briefing.

During the mission, Discovery's cargo included essential items such as the Orbiter Docking System, the Shuttle Power Distribution Unit (SPDU), and the large S6 Truss, which weighed 14,088 kilograms (31,059 pounds). The installation of the S6 truss completed the assembly of the ISS's Integrated Truss Structure, which was crucial for supporting the station's solar arrays.

On March 18, the crew began preparing the S6 truss segment for installation. Due to clearance constraints, a complex series of handoffs were required to move the truss into position between the station's robotic arm and the shuttle's robotic arm. This setup was necessary for the following day's spacewalk.

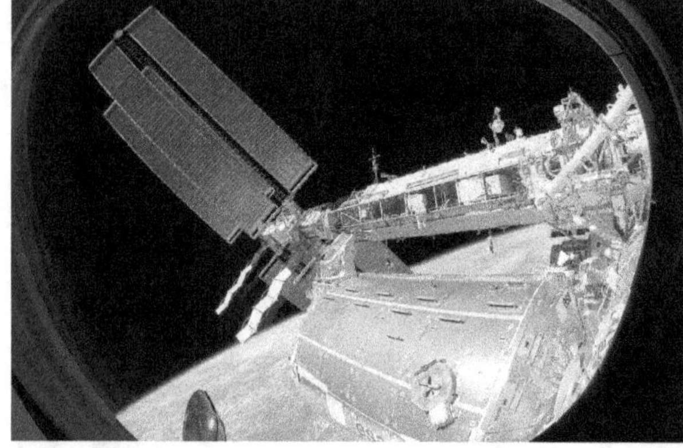

Astronaut Joseph Acaba, STS-119 mission specialist, participates in the mission's third scheduled session of extravehicular activity (EVA) as construction and maintenance continue on the International Space Station. During the six-hour, 27-minute spacewalk, Acaba and Richard Arnold (out of frame), mission specialist, helped robotic arm operators relocate the Crew Equipment Translation Aid (CETA) cart from the Port 1 to Starboard 1 truss segment, installed a new coupler on the CETA cart, lubricated snares on the "B" end of the space station's robotic arm and performed a few "get

The first spacewalk of STS-119 commenced on March 19, with Swanson and Arnold exiting the Quest airlock to begin installing the S6 truss segment. Their tasks included securing the truss, connecting power and data cables, and deploying the solar array blanket boxes. This spacewalk lasted 6 hours and 7 minutes and marked a significant achievement in the mission.

The deployment of the solar arrays, initially scheduled for March 21, was advanced to March 20, allowing for earlier activation of the new power systems. The arrays were carefully unfurled, with the station maneuvered into position to maximize sunlight exposure and minimize friction issues. The successful extension of the arrays increased the station's power output to 120 kilowatts and doubled its scientific power to 30 kilowatts.

The second spacewalk on March 21 involved Swanson and Acaba performing maintenance tasks to facilitate future work on the Port 6 truss batteries. They also installed a second Global Positioning Satellite antenna on the Japanese Kibo laboratory and addressed various technical issues, including deploying the Unpressurized Cargo Carrier Attachment System (UCCAS).

The final significant event of Expedition 18 occurred on April 7, 2009, when the expedition's crew undocked, marking the end of their mission. The landing on April 8, 2009, concluded their time aboard the ISS, allowing the next crew to continue the station's ongoing research and maintenance.

The expedition also featured a series of spacewalks integral to the station's maintenance and enhancement. On November 18, 2008, Heidemarie Stefanyshyn-Piper and Stephen Bowen conducted a spacewalk lasting 6 hours and 52 minutes, followed by another EVA on November 20, 2008, where they spent 6 hours and 45 minutes outside the station. This was part of a series of spacewalks to install and test new hardware. The pace continued with further spacewalks on November 22, 2008, and December 23, 2008, where the astronauts undertook tasks such as replacing equipment and conducting routine maintenance.

March 2009 saw an equally significant series of spacewalks. On March 10, 2009, Yury Lonchakov and Michael Fincke completed a 4-hour and 49-minute EVA. This was followed by a series of spacewalks involving Steve Swanson, Joseph Acaba, Richard Arnold, and others, focusing on various technical tasks and system checks. These activities culminated by a spacewalk on March 23, 2009, where Joseph Acaba and Richard Arnold spent 6 hours and 27 minutes working outside the ISS.

The crew's mission included two significant extravehicular activities (EVAs). The first, on 22 December 2008, extended into the early hours of 23 December, during which Lonchakov and Fincke conducted a variety of tasks outside the station. Their work included installing an electromagnetic energy measuring device, a Langmuir probe on the Pirs module, removing the Russian Biorisk experiment, and installing the Expose-R experiment package. Unfortunately, the Expose-R package malfunctioned and had to be removed.

The second EVA occurred on 10 March 2009. Lonchakov and Fincke embarked on this spacewalk at 16:22 UTC and concluded it at 21:11 UTC, spending 4 hours and 49 minutes outside the station. Their activities included installing the EXPOSE-R experiment on the universal science platform of the Zvezda module, clearing tape straps from the Pirs airlock, and conducting an exterior inspection and photography of the Russian segment of the ISS.

A significant event during Expedition 18 was the debris incident on 12 March 2009. A fragment from a Delta II rocket launched in 1993 approached the ISS. The debris was not detected in time to execute a collision avoidance maneuver, prompting the crew to prepare for an emergency evacuation. Fortunately, the debris passed the station at 16:38 UTC without collision, and the crew could resume their activities shortly after that.

The mission concluded on 8 April 2009, with the crew returning to Earth aboard Soyuz TMA-13.

Elsewhere in international space activities, the global landscape of space exploration saw significant advancements and selections across multiple astronaut programs, underscoring diverse international collaboration and advancements in human spaceflight capabilities.

Japan's JAXA Group, announced on February 25th, introduced Takuya Onishi and Kimiya Yui. This selection highlighted Japan's continued commitment to expanding its presence in space exploration through skilled astronauts prepared for missions aboard the International Space Station (ISS). On May 13th, Canada's CSA Group welcomed Jeremy Hansen and David Saint-Jacques, demonstrating Canada's ongoing participation in international space missions and contributions to scientific research aboard the ISS.

ESA's Group, aptly named "The Shenanigans," announced on May 20th, featured Samantha Cristoforetti (Italy), Alexander Gerst (Germany), Andreas Mogensen (Denmark), Luca Parmitano (Italy), Timothy Peake (United Kingdom), and Thomas Pesquet (France). This diverse cohort, selected from over 8,413 European applicants, underscored the ESA's role in fostering a unified approach to space exploration among member states.

NASA's Group 20, known informally as "Chumps," was unveiled on June 29th, marking a significant milestone as the first cohort of astronauts chosen for the post-Space Shuttle era. Comprising mission specialists Serena M. Auñón, Jeanette J. Epps, Jack D. Fischer, Michael S. Hopkins, Kjell N. Lindgren, Kathleen (Kate) Rubins, Scott D. Tingle, Mark T. Vande Hei, and Gregory R. (Reid) Wiseman, this group was selected from a pool of over 3,500 applicants.

Notably, Fischer, Tingle, and Wiseman were initially designated as pilots, reflecting NASA's transition toward a unified mission specialist role in the absence of the Shuttle program.

By 2009, Roscosmos had provided space tourism services for six fare-paying passengers to the ISS through the Space Adventures company for an estimated fee of at least $20 million each. The agency continued to operate the Mir space station past its planned lifespan, contribute to the ISS, and fly Progress and Soyuz missions.

Expedition 19

Cosmonaut Gennady Padalka (center), Expedition 19 commander; NASA astronaut Michael Barratt (left) and Japan Aerospace Exploration Agency (JAXA) astronaut Koichi Wakata, both flight engineers,

Expedition 19, the nineteenth long-duration mission to the International Space Station (ISS), commenced on March 26, 2009, with the launch of the Soyuz TMA-14 spacecraft. This mission marked a transitional phase in the ISS's history, as it was the final expedition to operate with a crew of three before the ISS's crew complement expanded to six with the arrival of Expedition 20.

On March 26, 2009, the Soyuz TMA-14 spacecraft embarked on a significant journey to the International Space Station (ISS), marking a pivotal moment in space exploration history. This mission was notable not only for its operational role but also for its distinguished crew. Soyuz TMA-14 was the 101st crewed Soyuz flight overall, and the 100th to successfully launch and land with a crew on board, following a previous uncrewed flight, Soyuz 34, which had replaced Soyuz 32.

The crew of Soyuz TMA-14 included Commander Gennady Padalka from Russia, Flight Engineer Michael Barratt from the United States, and spaceflight participant Charles Simonyi, a Hungarian-American making his second self-funded trip to the space station. The mission also featured Canadian space tourist Guy Laliberté, participating in his only spaceflight. The backup crew comprised Commander Maksim Surayev of Russia, Flight Engineer Shannon Walker from the United States, and spaceflight participants Esther Dyson and Barbara Barrett, both from the United States.

Upon its arrival at the ISS, Soyuz TMA-14 docked to the station and played a critical role throughout the Expedition 20 increment, primarily serving as an emergency escape vehicle. On July 2, 2009, the spacecraft conducted a docking port swap, moving from the Zvezda Service Module's aft port to the Pirs Docking Compartment's nadir port. This maneuver was essential to facilitate the docking of Progress 34P at the SM aft port on July 29, 2009.

During the early phase of Expedition 19, which began with the arrival of Soyuz TMA-14, the crew encountered significant operational challenges that impacted their daily routines and overall morale. One of the primary issues emerged regarding allocating and using resources and facilities on the International Space Station (ISS), a situation that Commander Gennady Padalka brought to the forefront on March 31, 2009.

Commander Padalka voiced concerns about the American exercise equipment available on the ISS. Initially, this equipment was permitted for all crew members, but restrictions were later imposed. These restrictions were part of broader operational guidelines that mandated the separation of Russian and American crew members' activities, including using facilities and supplies. Such measures were intended to streamline resource management and maintain organizational efficiency. However, they inadvertently created challenges for the crew members who had previously enjoyed more collaborative and flexible arrangements.

The separation of facilities and supplies extended beyond exercise equipment. Crew members from different space agencies had to use designated areas and supplies specific to their national contingents. This separation complicated the daily operational logistics and affected the international crew's interpersonal dynamics. The directive to avoid sharing resources and facilities led to division and frustration among the astronauts and cosmonauts.

The implications of these restrictions were evident in the decline in crew morale. The psychological and social impact of the enforced separation became a notable issue during this period. The camaraderie and teamwork that had been a hallmark of previous expeditions were strained as the crew adapted to a more segmented approach to station operations.

Throughout their mission, Expedition 19's crew engaged in a range of scientific experiments and maintenance tasks. Their research focused on understanding the effects of microgravity on human health, including studies on bone density, muscle atrophy, and immune system function. These investigations were crucial for developing countermeasures to mitigate the adverse effects of long-duration spaceflight. The crew also examined space nutrition, assessing how dietary intake impacts health and performance in the unique environment of space.

A significant achievement during Expedition 19 was the launch of Japan's Kuonotori cargo craft, also known as the H-II Transfer Vehicle (HTV), which marked a new era in international cargo resupply capabilities. The HTV was designed to deliver essential supplies and experiments to the ISS, enhancing the station's ability to conduct long-term research and support its crew.

The logistical operations of Expedition 19 included managing the arrival and departure of cargo spacecraft. On May 6, 2009, the ISS Progress 32 spacecraft undocked from the station after fulfilling its role in delivering vital supplies and scientific equipment. This was followed by the arrival of the ISS Progress 33 on May 12, 2009. The docking of Progress 33 was a crucial event for Expedition 19, as it brought additional food, equipment, and experiments necessary for sustaining the station's operations and research.

The mission concluded with Soyuz TMA-14 successfully undocking and landing on October 11, 2009. Notably, space tourist Guy Laliberté, who had traveled to the ISS aboard Soyuz TMA-16 with Expedition 21, was onboard for the return flight. Laliberté was the first Canadian to experience space tourism, marking a significant milestone in the broader context of space exploration and private space travel.

The Soyuz TMA-15 spacecraft approaches the International Space Station, carrying cosmonaut Roman Romanenko, European Space Agency astronaut Frank De Winne and Canadian Space Agency astronaut Robert Thirsk, all Expedition 20 flight engineers. The crew launched from the Baikonur Cosmodrome in Kazakhstan at 5:34 a.m. (CDT), May 27, 2009, and arrived at the station at 7:34 a.m. (CDT) on May 29. Later that day, hatches opened between the two spacecraft, beginning Expedition 20 and six-person crew operations.

U.S. spaceflight participant Charles Simonyi (left), cosmonaut Gennady Padalka (center), Expedition 19 commander, and astronaut Michael Barratt (right), NASA Expedition 19 flight engineer, shake hands after an inspection of their Soyuz TMA-14 spacecraft March 12, 2009 in its integration facility at the Baikonur Cosmodrome in Kazakhstan. The trio were launched March 26 on a two-day trip to the International Space Station.

Expedition 20

Expedition 20 crew members from the right (front row) are cosmonaut Gennady Padalka, Expedition 19/20 commander; and European Space Agency astronaut Frank De Winne, Expedition 20 flight engineer and Expedition 21 commander. From the left (back row) are NASA astronaut Michael Barratt, Expedition 19/20 flight engineer; Canadian Space Agency astronaut Robert Thirsk, Expedition 20/21 flight engineer; Japan Aerospace Exploration Agency (JAXA) astronaut Koichi Wakata, 18/19/20 flight engineer; and cosmonaut Roman Romanenko, Expedition 20/21 flight engineer.

Expedition 20, which began in late May 2009, was a dynamic period in the International Space Station's history, marked by significant docking and undocking operations that facilitated the advancement of space research and station logistics.

On May 27, 2009, the Expedition 20/21 crew launched aboard Soyuz TMA-14, marking the start of their mission. This crew rotation was crucial for maintaining the operational capability of the ISS. The crew comprised Russian Commander Gennady Padalka, along with American Flight Engineers Michael Barratt and Timothy Creamer. Their arrival was pivotal for the ongoing research and maintenance of the ISS.

Following their launch, the Expedition 20 crew experienced a flurry of docking and undocking events. On May 29, 2009, Soyuz TMA-14 docked with the ISS, joining the station's existing modules and crew.

The operational tempo continued with a series of cargo and shuttle missions. On July 2, 2009, the crew facilitated the relocation of Soyuz TMA-14 to a new docking port, optimizing the station's configuration for upcoming missions and arrivals. This move was part of routine station operations to accommodate new spacecraft and maintain docking port availability.

Later in July, the station hosted the arrival of the STS-127 mission. Space Shuttle Endeavour, carrying astronauts Christopher Cassidy, David Wolf, and others, docked with the ISS on July 17, 2009. This mission was part of the assembly of the station's Japanese Experiment Module (JEM), enhancing the ISS's research capabilities. The shuttle undocked on July 28, 2009, concluding its mission and leaving behind critical new components and research modules.

The following month, the ISS welcomed a new Progress spacecraft. Progress 34 docked with the station on July 29, 2009, bringing supplies and equipment essential for the ongoing scientific experiments and daily operations on the ISS.

August brought further activity with the docking of the STS-128 mission. Space Shuttle Discovery, with a crew of six including Astronauts John Olivas and Robert Behnken, arrived at the station on August 30, 2009. This mission delivered and installed the Permanent Multipurpose Module (PMM) to the ISS. After a successful mission, STS-128 undocked from the station on September 8, 2009.

As part of Expedition 20's resupply operations, the HTV-1 Kounotori, Japan's first cargo spacecraft, launched on September 10, 2009. The HTV-1 arrived at the ISS and docked on September 17, 2009, delivering vital supplies and experiments.

The final significant event of Expedition 20 was the undocking of Soyuz TMA-14 on September 21, 2009. This departure marked the end of the Expedition 20 crew's stay aboard the ISS, concluding their vital contribution to the station's ongoing research and operations.

Expedition 20, the twentieth long-duration mission to the International Space Station (ISS), marked a significant milestone as it introduced a six-member crew for the first time. This mission required two separate Soyuz-TMA spacecraft launches due to the limited capacity of each spacecraft. The first launch occurred on March 26, 2009, with Soyuz TMA-14, followed by the launch of Soyuz TMA-15 on May 27, 2009. Soyuz TMA-15, launched from the Baikonur Cosmodrome at 10:34 UTC, docked with the ISS on May 29, 2009. This docking officially transitioned the crew of Soyuz TMA-14 from Expedition 19 to Expedition

20, signifying the beginning of a new chapter in the station's crew operations.

On March 26, 2009, the Soyuz TMA-14 spacecraft launched from the Baikonur Cosmodrome in Kazakhstan, embarking on its mission to the International Space Station (ISS). This flight marked a significant milestone as it was the 101st crewed mission of a Soyuz spacecraft, though it was the 100th successful crewed launch and landing. The distinction lies in the fact that Soyuz 34, a previous flight, had launched uncrewed to replace Soyuz 32, which had landed without a crew.

The Soyuz TMA-14 mission carried a diverse crew. Russian Commander Gennady Padalka was at the helm, embarking on his third spaceflight. NASA Flight Engineer Michael Barratt joined him, making his debut in space. The mission also included Charles Simonyi, a Hungarian-American spaceflight participant and self-funded space tourist returning to the ISS for his second and final spaceflight. Additionally, the mission was notable for carrying Canadian space tourist Guy Laliberté, marking him as the first Canadian to visit space.

Soyuz TMA-14's primary role on the ISS was to serve as an emergency escape vehicle, ensuring the safety of the crew during Expedition 20. On July 2, 2009, the spacecraft relocated from its initial docking port at the Zvezda Service Module's aft port to the Pirs Docking Compartment's nadir port. This maneuver was essential to free up the Zvezda port for the docking of Progress 34P, which arrived on July 29, 2009.

The mission concluded with Soyuz TMA-14 undocking from the ISS and safely landing back on Earth on October 11, 2009. The return journey included the space tourist Guy Laliberté, who had previously launched with Expedition 21 aboard Soyuz TMA-16. His flight marked a historic moment as he was the first Canadian to journey into space.

Soyuz TMA-15 was a notable mission in the Soyuz program, marking a significant advancement in the history of crewed spaceflight. Launched on May 27, 2009, from the Baikonur Cosmodrome in Kazakhstan, Soyuz TMA-15 was the 102nd crewed flight of a Soyuz spacecraft since the program's inception with Soyuz 1 in 1967. The mission was pivotal in transitioning to a six-person crew operation aboard the International Space Station (ISS), a milestone that expanded the scope of research and daily operations on the station.

The spacecraft was launched atop a Soyuz-FG carrier rocket at 10:34 UTC and successfully docked with the ISS at 12:34 UTC on May 29, 2009. The crew of Soyuz TMA-15 consisted of three members who joined the Expedition 20 team. Roman Romanenko of Russia, the mission commander, was making his first spaceflight and became the third second-generation space traveler, following in the footsteps of his father, who had previously flown on Soyuz 26. Romanenko chose the mission callsign "Taymyr" in homage to his father's mission, although "Parus," meaning "Sail" in Russian, was ultimately used for communications.

The other crew members included Frank De Winne from Belgium, representing the European Space Agency (ESA), and Robert Thirsk from Canada, representing the Canadian Space Agency (CSA). De Winne, in his second and final spaceflight, achieved a significant milestone by becoming the first European to command the ISS. Thirsk, also on his second and final flight, became the first Canadian to fly aboard a Soyuz spacecraft, a notable departure from his fellow Canadians who had previously only flown aboard Space Shuttles.

Soyuz TMA-15 was docked to the ISS throughout the Expedition 20 and Expedition 21 increments, serving as an emergency escape vehicle for the station's crew. The mission concluded with a safe return to Earth on December 1, 2009, returning the crew after a successful and historically significant mission. The successful operation of Soyuz TMA-15 underscored the ongoing evolution of international collaboration in space exploration and marked a new era of expanded crew capabilities aboard the ISS.

At the Baikonur Cosmodrome in Kazakhstan, Expedition 20 Flight Engineer Bob Thirsk of the Canadian Space Agency, Russian Cosmonaut Roman Romanenko and Flight Engineer Frank De Winne of the European Space Agency prepared for a check of the Soyuz TMA-15 spacecraft they will launch in on 27 May on a trip to the International Space Station. The trio will join three other residents on the station to form a six-person crew for the first time.

Colonel Gennady Padalka of Russia, who had commanded Expedition 19, continued his role as the commander for Expedition 20, becoming the first to lead two consecutive expeditions. Nicole Stott, the final astronaut to be launched on the Space Shuttle before its retirement, concluded her shuttle mission and joined the ISS crew, marking a significant shift in crew dynamics.

During Expedition 20, Japanese astronaut Koichi Wakata conducted a notable experiment involving wearing a specially designed pair of underpants for an entire month without changing or washing them. This test aimed to evaluate the effectiveness of the garment in preventing body odor, demonstrating that the special underwear maintained its performance under prolonged use.

The ISS did not maintain a constant six-member crew throughout Expedition 20. Following the departure of Roman Romanenko, Frank De Winne, and Robert Thirsk in November 2009, the station temporarily housed only two crew members, Jeff Williams and Maksim Surayev, for about two weeks. The crew number increased to five in early December with the arrival of Oleg Kotov, Timothy Creamer, and Soichi Noguchi aboard Soyuz TMA-17. By March 2010, the crew size was again reduced to three as Williams and Surayev departed. The crew complement finally returned to six with the arrival of Soyuz TMA-18 in April 2010, bringing Aleksandr Skvortsov, Mikhail Korniyenko, and Tracy Caldwell Dyson to the station.

The rollout of Soyuz TMA-15 on 25 May, with an American flag painted on the capsule

During Expedition 20, a series of important crew and cargo missions took place, marking significant events in the International Space Station's operations and logistics.

On May 27, 2009, the Expedition 20/21 crew launched from Earth, embarking on their journey to the ISS. Their arrival on May 29, 2009, marked the beginning of their mission, which involved crew operations and various tasks to enhance the station's capabilities.

STS-127, also known as ISS assembly flight 2J/A, was a NASA Space Shuttle mission dedicated to expanding the International Space Station (ISS) and furthering scientific research. Launched on July 15, 2009, this mission marked the twenty-third flight of the Space Shuttle Endeavour, aiming to deliver and install the final components of the Japanese Experiment Module (JEM). The mission's primary goals included the deployment of the Exposed Facility (JEM EF) and the Exposed Section of the Experiment Logistics Module (ELM-ES), both integral parts of the Japanese contribution to the ISS.

The Endeavour crew, under the command of Mark L. Polansky, successfully docked with the ISS on July 17, 2009. This historic docking set a record for the highest number of humans present in space simultaneously, with thirteen individuals aboard the station. This event tied the previous record set in 1995 and underscored the collaborative spirit of the ISS program, which includes contributions from all its international partners.

STS-127 crew astronauts Mark Polansky (right), commander, and Doug Hurley, pilot. Remaining crewmembers, pictured from left to right, are astronauts Dave Wolf, Christopher Cassidy, Canadian Space Agency's Julie Payette, Tom Marshburn and Tim Kopra, all mission specialists. Kopra joined Expedition 19 as flight engineer after launching to the International Space Station with the STS-127 crew.

The mission was initially fraught with challenges, as Endeavour's launch on July 15 was the sixth attempt, after several delays. During ascent, pieces of foam were observed falling off the External Tank—an issue reminiscent of the 2003 Columbia disaster. Fortunately, Endeavour sustained only minor scuffs to its heat shield, which did not pose a significant risk to reentry. The shuttle landed safely at Kennedy Space Center 16 days later on July 31, 2009.

The crew of STS-127 comprised a diverse team of astronauts. Commanded by Mark L. Polansky, the crew included Pilot Douglas G. Hurley, Mission Specialists Christopher J. Cassidy, Julie Payette, Thomas H. Marshburn, David Wolf, and Timothy Kopra, along with Japanese astronaut Koichi Wakata. This mission was notable for marking the first time two Canadian astronauts, Julie Payette and Robert Thirsk, were in space simultaneously. Christopher Cassidy became the 500th person to fly in space during this mission.

Endeavour's payload was substantial and varied, reflecting the mission's multifaceted objectives. The largest components included the

Kibō Japanese Experiment Module Exposed Facility (JEM EF) and the Kibō Japanese Experiment Logistics Module – Exposed Section (ELM-ES). The JEM EF was designed to facilitate scientific experiments in the vacuum of space, while the ELM-ES served as a non-pressurized component similar to the logistics module on Kibō.

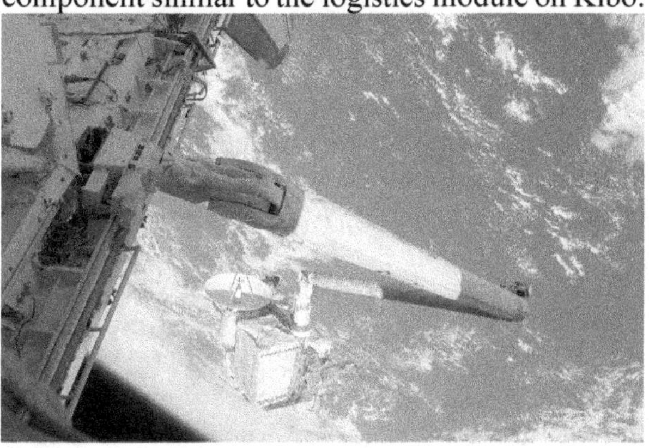

Backgrounded by a blue and white Earth, the remote manipulator system (RMS) arm of the Space Shuttle Endeavour, was about to hand off the Integrated Cargo Carrier (ICC) to the International Space Station (out of frame). The ICC was an unpressurized flat bed pallet and keel yoke assembly that was carried into space in the shuttle's payload bay.

In addition to these components, Endeavour carried an Integrated Cargo Carrier-Vertical Light Deployable (ICC-VLD), which contained six new batteries for the P6 truss, a spare space-to-ground antenna, and other critical equipment. The mission also included two satellites for deployment upon mission completion: the Dual Autonomous Global Positioning System On-Orbit Navigator Satellite (DRAGONSAT) and the Atmospheric Neutral Density Experiment (ANDE-2). DRAGONSAT consisted of two picosatellites, AggieSat2 and PARADIGM (BEVO-1), which gathered GPS data to improve spacecraft docking capabilities. ANDE-2, a Department of Defense project managed by the Naval Research Laboratory, aimed to measure the density and composition of the low Earth orbit atmosphere to aid in predicting the movement and decay of orbital objects.

STS-127 also carried a set of experiments, including the Dosimetry for Biological Experiments in Space, Validation of Procedures for Monitoring Crew Member Immune Function, Image Reversal in Space, Nutritional Status Assessment, NASA Biological Specimen Repository, and Tomatosphere-II.

The mission's official flight kit included unique mementos such as water samples from the five Great Lakes, a resin statue for the One Drop Foundation, and a copy of Beethoven's Fifth Symphony for the Montreal Symphony Orchestra.

Upon docking, Endeavour's crew conducted a series of operations, including a Rendezvous Pitch Maneuver (RPM), allowing ISS crew members to capture high-resolution images of the shuttle's thermal protection system. The docking occurred at the Pressurized Mating Adapter 2 (PMA-2) on the Harmony module. Following this, a crew swap occurred, with Koichi Wakata replaced by Tim Kopra.

On July 18, 2009, astronauts Dave Wolf and Tim Kopra initiated their first extravehicular activity (EVA) of the mission at 16:19 UTC, marking a significant milestone for the Space Shuttle Endeavour's mission. The spacewalk began with both astronauts switching their spacesuit power to internal battery, a standard procedure to ensure mobility and power supply outside the spacecraft.

The primary task of EVA 1 was the installation of the Japanese Exposed Facility (JEF) on the Japanese Experiment Module (JEM). This operation required a meticulous coordination between the robotic arms of both the station and the shuttle. The JEF, a substantial component weighing 4.1 tons, was first unberthed by the station's robotic arm from the shuttle's payload bay. Subsequently, the shuttle's robotic arm took over, maneuvering the JEF to its designated location on Node-2 (Harmony). The facility was then successfully latched onto the Experiment Module.

During this EVA, the astronauts also managed to deploy the port Unpressurized Cargo Carrier Attach System (UCCAS), a task that had previously failed during the STS-119 mission due to a stuck detent pin. Engineers had developed a custom tool to resolve this issue, which was successfully used to deploy the mechanism.

In other operational news, shuttle managers determined that no focused inspection of the shuttle's heat shield was necessary. Although there was one notable gouge, most of the impact sites on the shuttle were found to be minor, primarily involving a loss of coating. The planned installation of a starboard-side cargo carrier was postponed due to time constraints. A pre-launch issue with a fuel cell was analyzed, but it continued to function as expected without affecting the mission.

The following day, the focus shifted to installing the Integrated Cargo Carrier-Vertical Light Deployable (ICC-VLD) on the station's port side. Utilizing both the shuttle and station robotic arms, the cargo pallet, which contained spares and fresh batteries, was successfully transferred. The shuttle's robotic arm lifted the pallet from its bay and handed it off to the Canadarm2, the station's robotic arm, which positioned it for future use.

Meanwhile, a malfunction in a new toilet in the Destiny laboratory required the crew to use the facilities in the Russian segment while repairs were underway. Despite this issue, the shuttle was cleared for reentry.

On July 20, Flight Day 6 EVA 2 began at 15:27 UTC with astronauts Wolf and Marshburn exiting the Quest airlock. This spacewalk focused on transferring spare components from the ICC-VLD to the External Stowage Platform-3 (ESP-3). Wolf utilized the station's robotic arm to transport the spares, including a Ku-Band Space-to-Ground antenna, a coolant pump module, and a drive unit for the station's robotic arm, to the P3 truss stowage platform. Marshburn assisted in securing these items for long-term storage.

Due to time constraints, the installation of a camera on the Japanese Experiment Facility was deferred. The malfunctioning toilet was also repaired with new internal components and restored to normal operation.

July 21, Flight Day 7 saw a more relaxed schedule as the Japanese logistics carrier was attached to the Japanese Exposed Facility. The cargo pallet, which included various experiments such as an X-ray astronomy payload and a space environment monitor, was unberthed from the shuttle and temporarily secured to the facility by the station's robotic arm. This cargo would be returned to Earth after the experiments were conducted.

July 22, Flight Day 8: EVA 3, astronauts Wolf and Cassidy began EVA 3 at 14:32 UTC. Cassidy worked on removing thermal covers from the experiment carrier on the Japanese external science deck. At the same time, Wolf cleared obstructions from the Harmony node to facilitate the docking of a Japanese automated resupply ship.

The planned replacement of four batteries in the P6 truss did not proceed as intended. During the EVA, elevated CO2 levels in Cassidy's suit, though never exceeding safety limits, led to an early termination of the spacewalk. Only two batteries were replaced, and the remaining tasks were rescheduled for future EVAs.

July 23, Flight Day 9, the Kibō robotic arm was employed to install three experiments on the Japanese Exposed Facility, including the Monitor of All-sky X-ray Image and the Inter-orbit Communication System. This operation was part of a revised plan for EVA 4, which included replacing the remaining four batteries on P6 and completing the deferred camera installation.

July 24, Flight Day 10: EVA 4conducted by Cassidy and Marshburn, focused on replacing the final four batteries in the P6 truss. After securing the old batteries in the ICC-VLD, the cargo pallet was returned to Endeavour's payload bay. Elevated CO2 levels experienced by Cassidy during EVA 3 were attributed to the rapid pace of work.

On July 25, Flight Day 11, the crew enjoyed a day off. There were only minor issues, such as a temporary shutdown of the station's American CO2 removal system, which did not significantly affect operations.

On July 26, Flight Day 12, the Japanese Exposed Section cargo carrier was berthed by the shuttle's robotic arm in Endeavor's payload bay. The day also featured a joint news conference with the shuttle and station crews. Meanwhile, the malfunctioning CO2 removal system had been switched to manual mode to ensure continued operation.

July 27, Flight Day 13: EVA 5 began at 11:33 UTC with Cassidy and Marshburn switching their suits to internal battery power. This spacewalk involved reconfiguring power channels in the Zenith 1 patch panel to ensure the proper operation of the control moment gyroscopes, which are crucial for station orientation. Cassidy also completed the installation of video cameras on the Japanese Exposed Facility, aiding in the docking of Japanese cargo crafts and overall operation. Due to limitations with Cassidy's METOX, the deployment of the Portable Airlock System (PAS) was deferred, and alternative tasks, such as installing handrails, were completed.

July 28, Flight Day 14: ISS Undocking

The mission concluded on July 28, with Endeavour undocking from the ISS at 17:26 UTC. The hatch closure and undocking occurred on the same day due to launch delays and the arrival of the Progress 34 cargo craft. Following undocking, the shuttle conducted a flyaround of the station, allowing the crew to capture photographs of the ISS's current configuration. A final separation burn was completed at 15:09 EDT, marking the end of a successful mission.

Endeavour shortly after the shuttle and station post-undocking separation

In addition to the shuttle operations, on July 29, 2009, the ISS Progress 34 spacecraft docked with the station. Progress spacecraft deliver supplies, including fuel, equipment, and scientific experiments. The docking of Progress 34 was essential for maintaining the station's resources and supporting ongoing missions.

STS-128, officially known as ISS Assembly Flight 17A, was a significant mission conducted by NASA to enhance the International Space Station (ISS). Launched on August 28, 2009, aboard Space Shuttle Discovery, this mission was pivotal in advancing the ISS's capabilities and crew support infrastructure.

The core of the STS-128 mission was the Multi-Purpose Logistics Module (MPLM) Leonardo, designed to bolster the station's crew capacity to six members. Leonardo, a versatile cargo module, carried an array of vital equipment and scientific experiments essential for studying microgravity's effects on physics and chemistry. This included three racks for life support systems, a new treadmill named COLBERT, which was initially placed in Node 2 before being relocated to Node 3, and an Air Revitalization System (ARS) intended for future use in Node 3 after an initial deployment in the Kibo module.

Leonardo was an imposing structure measuring 21 feet in length and 15 feet in diameter. Its launch mass was 27,510 pounds, and its return mass was slightly less, at 16,268 pounds. When empty, Leonardo weighed 9,810 pounds. The module was crucial in transporting essential supplies and scientific equipment to the ISS, facilitating the station's operational expansion.

In addition to Leonardo, the shuttle's payload included a Lightweight Multi-Purpose Experiment Support Structure Carrier (LMC), which housed an Ammonia Tank Assembly (ATA) to replace an existing empty tank. The mission also saw the deployment of the TriDAR, a cutting-edge 3D dual-sensing laser camera designed for potential autonomous rendezvous and docking operations. The TriDAR successfully tracked the ISS's position and orientation during docking maneuvers.

STS-128 was notable for its array of scientific experiments. Among them was the Fluids Integrated Rack (FIR), which allowed detailed examination of fluid behavior in microgravity—critical for understanding many chemical processes. The Materials Science Research Rack (MSRR-1) and the Minus Eighty Laboratory Freezer for ISS (MELFI-2) were also installed. The FIR investigated colloids' behavior without gravity interference, while the MSRR-1 focused on materials science research, and MELFI-2 provided advanced cryogenic storage for scientific samples.

The mission team, commanded by Frederick W. Sturckow and piloted by Kevin A. Ford, comprised a diverse crew including Patrick G. Forrester, José M. Hernández, John D. Olivas, Christer Fuglesang, Nicole Stott, and Timothy Kopra. Notably, STS-128 marked the first instance of two Hispanic Americans—John "Danny" Olivas and José M. Hernández—serving on the same crew. Olivas was making his second spaceflight, and Hernández, on his first, represented a significant milestone in NASA's history.

Throughout the mission, Discovery's crew conducted three spacewalks. The first spacewalk, on September 1, involved removing an empty ammonia tank assembly and installing new experiments. During the second spacewalk, on September 3, Olivas and Fuglesang replaced the old ammonia tank and performed additional tasks

to enhance the station's systems. The mission also included various transfer operations, activating new equipment, and preparing for future scientific endeavors.

STS-128 crew NASA astronauts Rick Sturckow (right), commander; and Kevin Ford, pilot. From the left (standing) are astronauts Jose Hernandez, John "Danny" Olivas, Nicole Stott, European Space Agency's Christer Fuglesang and Patrick Forrester, all mission specialists. Stott joined Expedition 20 as flight engineer after launching to the International Space Station on STS-128.

STS-128 was the final Space Shuttle flight dedicated to ISS crew rotation, with Nicole Stott replacing Tim Kopra. This mission contributed to the ISS's operational capabilities and marked a milestone in the diversity and expertise of NASA's astronaut corps. The successful completion of STS-128 demonstrated the ongoing commitment to expanding the ISS's capacity and scientific potential, paving the way for future missions and international collaborations.

The HTV-1 Kounotori, a Japanese cargo spacecraft, debuted in September. The HTV-1 launched on September 10, 2009, marking Japan's entry into cargo resupply missions for the ISS. On September 17, 2009, it berthed with the station, delivering critical supplies and equipment. The HTV-1 undocked from the station on September 21, 2009, completing its mission.

Lastly, on June 30, 2009, the ISS Progress 33 undocked from the station, having completed its supply mission. This event was part of the regular cargo resupply and removal cycle to ensure the station's operational efficiency.

During Expedition 20, which took place from March 26, 2009, to May 29, 2009, the International Space Station (ISS) crew undertook two critical extravehicular activities (EVAs) to prepare the station for forthcoming modules and upgrades.

The first EVA occurred on June 5, 2009, and was conducted by astronauts Gennady Padalka and Michael R. Barratt. Beginning at 07:52 UTC and concluding at 12:46 UTC, this spacewalk lasted 4 hours and 54 minutes. The primary focus of this mission was to ready the Zvezda service module's transfer compartment for the anticipated arrival of the Poisk module. The astronauts installed a docking antenna on the Zvezda module to facilitate this. This antenna enabled the Poisk module to dock with the ISS smoothly. During the EVA, Padalka and Barratt also conducted a detailed photographic survey of the newly installed antenna, sent back to mission control for evaluation. They also captured images of the Strela-2 crane, a piece of equipment used to move large payloads in space. This spacewalk was essential in ensuring that the docking infrastructure was correctly set up and functioning, thus facilitating the seamless integration of new modules into the ISS.

The second EVA of Expedition 20 took place on June 10, 2009. This brief spacewalk, lasting just 12 minutes, was also performed by Padalka and Barratt. It was conducted inside the depressurized Zvezda transfer compartment, where the astronauts carried out an internal procedure to replace one of the Zvezda hatches with a docking cone. This modification was a preparatory step for the Poisk module, slated to dock with the ISS later in the year. The Poisk module, which ultimately docked to the zenith port of Zvezda on November 12, 2009, was designed to provide an additional docking port for Russian spacecraft. This enhancement significantly increased the station's capacity to accommodate multiple vehicles and improved its operational flexibility.

Several notable spacewalks occurred during this period. On July 22, 2009, astronauts David Wolf and Chris Cassidy spent 5 hours and 59 minutes outside the station, performing critical tasks. Another EVA followed this on July 24, 2009, where Cassidy and Thomas Marshburn conducted a 7-hour and 12-minute spacewalk. On July 27, 2009, Chris Cassidy and Thomas Marshburn engaged in a 4-hour and 54-minute EVA. Earlier in the month, on July 18, 2009, David Wolf and Tim Kopra completed a spacewalk lasting 5 hours and 32 minutes. The month of July was also marked by spacewalks on July 20, 2009, with David Wolf and Thomas Marshburn spending 6 hours and 53 minutes outside, and on July 1, 2009, with John Olivas and Nicole Stott spending 6 hours and 35 minutes. The crew also conducted significant spacewalks on September 1, 2009, September 3, 2009, and September 5, 2009, with durations ranging from 6 hours and 39 minutes to 7 hours and 1 minute.

On July 3, 2009, the crew undocked Soyuz TMA-14 from the aft port of the Zvezda service module and relocated it to the Pirs docking compartment. This maneuver was essential to clear space for the arrival of a Progress supply spacecraft, illustrating the ongoing logistical coordination required for maintaining the ISS.

Surrounded by medical personnel, left to right, Guy Laliberte, Expedition 20 Commander Gennady Padalka, and Expedition 20 Flight Engineer Michael Barratt.They landed aboard the Soyuz capsule near the town of Arkalyk, Kazakhstan.

Expedition 21

Expedition 21 crew front row: European Space Agency astronaut Frank De Winne (center), commander; NASA astronaut Nicole Stott and Russian cosmonaut Roman Romanenko, both flight engineers. Back left: Russian cosmonaut Maxim Suraev, NASA astronaut Jeffrey Williams and Canadian Space Agency astronaut Robert Thirsk.

The Soyuz TMA-16 mission, a pivotal chapter in the annals of space exploration, unfolded as a landmark event for the International Space Station (ISS) program. Launched from the Baikonur Cosmodrome in Kazakhstan on September 30, 2009, this mission carried with it two key members of the Expedition 21 crew and a distinguished Canadian space tourist.

As the 103rd flight of the venerable Soyuz spacecraft—a series that first ventured into space in 1967—Soyuz TMA-16 achieved a notable milestone. It marked the first time since 1969 that three Soyuz vehicles orbited the Earth simultaneously. Onboard were Commander Maksim Surayev from Russia and Flight Engineer Jeffrey Williams from the United States, who joined the ISS to contribute to their respective expeditions. The mission also featured Guy Laliberté, the founder and CEO of Cirque du Soleil, as a spaceflight participant. Laliberté's journey to space was a significant milestone in the burgeoning field of space tourism, for which he had paid approximately $35 million through Space Adventures, an American company. His presence on Soyuz TMA-16 highlighted the expanding frontier of private space travel.

Upon reaching the ISS, Soyuz TMA-16 initially docked at the aft port of the Zvezda module. On January 21, 2010, Surayev and Williams performed a crucial maneuver to relocate the spacecraft to the zenith-facing port of the Poisk module. This intricate procedure began with the Soyuz TMA-16 undocking from the Zvezda service module at 5:03 a.m. EST, while the station was approximately 213 miles above the southwest coast of Africa. After a brief separation, the spacecraft backed away to about 100 feet before re-docking at 5:24 a.m. EST. Surayev skillfully guided Soyuz TMA-16 around the station, aligning it with the Poisk module using the spacecraft's maneuvering thrusters.

The ISS had been a hive of activity in the preceding months. On October 14, 2009, the Progress 35 cargo spacecraft was launched to deliver vital supplies and equipment to the station, docking successfully on October 17, 2009. This operation ensured the crew had the necessary resources for their extended stay. In October, the HTV-1 (Kounotori) cargo spacecraft also departed the ISS on October 30, 2009, after delivering critical supplies and marking Japan's first successful cargo resupply mission.

Significant events continued into November, including the undocking of the Space Shuttle Atlantis (STS-129) on November 25, 2009. The shuttle had arrived on November 18, 2009, delivering additional equipment and components for the ISS, including parts for the station's exterior and an external truss structure.

The transition from Expedition 20 to Expedition 21 was marked by the landing of Expedition 20 on December 1, 2009. The Expedition 20 crew, including Commander Gennady Padalka of Roscosmos, Flight Engineer Michael Barratt of NASA, and European Space Agency (ESA) Astronaut Robert Thirsk, returned to Earth, completing their mission and handing over to the new Expedition 21 crew.

Expedition 21, which commenced on October 11, 2009, was notable for several reasons, including the historic appointment of Belgian astronaut Frank De Winne as the first ESA astronaut to command the ISS. This mission was also significant for the simultaneous docking of three Soyuz spacecraft at the ISS—a first in the station's history. Soyuz TMA-16, which arrived in October, delivered the final members of Expedition 21 along with Laliberté, who returned to Earth aboard Soyuz TMA-14 on October 11, 2009, alongside two members of Expedition 20.

Nicole P. Stott, the last ISS crew member to travel aboard the Space Shuttle, concluded her mission in November 2009 with her return on STS-129. This marked the end of an era as the Shuttle program transitioned out of its operational phase.

The initial crew of Expedition 21 included Frank De Winne as Commander, Roman Romanenko as Flight Engineer 1, Robert Thirsk as Flight Engineer 2, Jeffrey Williams as Flight Engineer 3, and Nicole Stott as Flight Engineer 5. Thirsk, Romanenko, and Stott each achieved significant career milestones during this mission.

Following a successful 12-day mission, Guy Laliberté returned to Earth aboard Soyuz TMA-14, which had been left as an emergency vehicle during a previous flight. Surayev and Williams concluded their mission with a landing in Kazakhstan on March 18, 2010, aboard Soyuz TMA-16.

The Soyuz TMA-16 mission also marked a pivotal moment for space tourism, as the imminent retirement of the Space Shuttle and the ISS's expansion to accommodate a six-member crew signaled a shift in the focus of Soyuz flights. Future Soyuz missions would be dedicated exclusively to expedition crews, with space tourism to the ISS on hiatus until the development of new crewed spacecraft, such as SpaceX's Dragon 2 or Boeing's Starliner, would resume such missions.

Elsewhere in the international space race, In 2010, the global landscape of space exploration continued to evolve with significant selections and advancements across various astronaut programs.

China's Group 2, announced in March, introduced Cai Xuzhe, Chen Dong, Liu Yang, Tang Hongbo, Wang Yaping, Ye Guangfu, and Zhang Lu. This cohort represented China's commitment to expanding its human spaceflight capabilities and preparing astronauts for future missions aboard Chinese spacecraft and space stations.

The Association of Spaceflight Professionals unveiled its Group 1 on April 12th, comprising Jim Crowell, Bruce Davis, Kristine Ferrone, Amnon Govrin, Chad Healy, Ryan Kobrick, Joseph Palaia, Luís Saraiva, Brian Shiro, Laura Stiles, and Veronica Ann Zabala-Aliberto. This group highlighted the growing influence of private sector involvement in space exploration, focusing on scientific research, technology development, and commercial spaceflight initiatives.

On June 7, 2009, the Association of Spaceflight Professionals introduced its Group 2 of Commercial Astronauts, including Ben Corbin, José Miguel Hurtado Jr., Jason Reimuller, Todd Romberger, Erik Seedhouse, and Alli Taylor, highlighting the increasing role of commercial entities in space travel.

Additionally, Russia's TsPK-15/RKKE-17 Cosmonaut Group, announced on October 12, 2009, included Aleksey Khomenchuk, Denis Matveev, Sergey Prokopyev from TsPK, and Andrei Babkin, Ivan Vagner, Sergey Kud-Sverchkov, and Svyatoslav Morozov from RKKE. This group underscored Russia's commitment to excellence in space exploration and international collaboration.

Expedition 21 Soyuz-FG carrying Soyuz TMA-16 at Baikonur Cosmodrome, Kazakhstan to the ISS.

Expedition 22

Expedition 22 crew members. From the left (front row) are NASA astronaut Jeffrey Williams, commander; and Russian cosmonaut Oleg Kotov, flight engineer. From the left (back row) are NASA astronaut T.J. Creamer, Russian cosmonaut Maxim Suraev and Japan Aerospace Exploration Agency (JAXA) astronaut Soichi Noguchi, all flight engineers.

Expedition 22 embarked on a notable chapter in the history of the International Space Station (ISS), marking significant milestones that shaped its operations and research endeavors. The mission commenced on December 1, 2009, following the departure of the Expedition 21 crew. For the initial three weeks, the ISS functioned with a reduced crew of just two members: Commander Jeff Williams from NASA and Flight Engineer Maksim Surayev from the Russian Space Agency (RSA). This period of operation with a minimal crew was an unusual scenario since the days following the Space Shuttle mission STS-114.

On December 20, 2009, Expedition 22 officially began its journey with the launch of the crewed spacecraft, heralding a new phase for the ISS. The mission brought new personnel and supplies vital for sustaining the station's operations. The crew of Expedition 22 included Russian cosmonauts Oleg Kotov and Sergey Ryschikov, alongside NASA astronaut Timothy Creamer. The full complement of Expedition 22 followed their arrival on December 22, 2009, which included Flight Engineers Soichi Noguchi from the Japan Aerospace Exploration Agency (JAXA) and Timothy Creamer from NASA. This integration brought the crew to five members, enhancing the station's operational capabilities and research potential.

Throughout their time aboard the ISS, the crew engaged in a range of scientific experiments and maintenance tasks essential for the station's functionality. The crew conducted one significant spacewalk, or Extravehicular Activity (EVA), on January 14, 2010. Spacewalkers Oleg Kotov and Maksim Surayev, equipped with Orlan spacesuits, spent 5 hours and 44 minutes preparing the Poisk module for future dockings, a crucial step in integrating this module into the ISS's infrastructure.

Expedition 22's mission was characterized by its diverse international participation, encompassing 167 days aboard the ISS and 169 days from launch to landing. The mission's completion was marked by the return of the Expedition 21/22 crew on March 18, 2010, ending their term and transitioning to the next phase of the ISS's journey with Expedition 23.

Among the significant developments during Expedition 22 was the arrival of the Poisk module, which launched on November 10, 2009, aboard a modified Progress spacecraft known as Progress M-MIM2. The module, whose name means "Search" in Russian, was sent into orbit from Launch Pad 1 at the Baikonur Cosmodrome in Kazakhstan. Poisk, a critical component of the Russian Orbital Segment (ROS), functions as an airlock and a storage facility for Orlan spacesuits, supporting various spacewalks. Its design features inward-opening EVA hatches, addressing issues experienced with outward-opening hatches on the Mir space station, thus enhancing safety during spacewalks.

Poisk also provides a docking port for Soyuz and Progress spacecraft, facilitating the automatic transfer of propellants and maintaining the station's systems. Following the Pirs module's departure on July 26, 2021, Poisk became the sole airlock for the ROS, underscoring its essential role in the station's operations and scientific activities.

In addition to Poisk, the Nauka module, translating to "Science" in Russian, significantly advanced the ISS's capabilities. Officially known as the Multipurpose Laboratory Module-Upgrade (MLM-U), Nauka was launched on July 21, 2021, and docked to the nadir port of the Zvezda module on July 29, 2021. Originally intended to replace the Docking and Stowage Module (DSM), Nauka instead replaced the Pirs module, marking a major upgrade for the ISS. Equipped with a temporary docking adapter, Nauka facilitated both crewed and uncrewed missions, preparing for the future arrival of the Prichal module.

Expedition 22's conclusion came with the undocking of Soyuz TMA-16 from the ISS on March 18, 2010. This transition smoothly led to Expedition 23, marking the end of one chapter and the beginning of another in the ISS's ongoing journey. Through its diverse crew, significant modules, and essential maintenance, Expedition 22 contributed notably to the ISS's mission of advancing space research and international collaboration.

The Expedition 22 crew landed on Thursday, 18 March 2010.

The Bigelow Expandable Activity Module (BEAM) represents a pioneering approach to space station design. Bigelow Aerospace developed it under a NASA contract. This experimental module was designed to test expandable habitat technology and was temporarily incorporated into the International Space Station (ISS) from 2016 to at least 2020.

BEAM arrived at the ISS on April 10, 2016, and was berthed to the station at Tranquility Node 3 on April 16, 2016. The module underwent its expansion and pressurization processes on May 28, 2016, marking a key milestone in its deployment.

This expandable design was intended to demonstrate the feasibility and benefits of expandable habitats in space, providing valuable data on their performance and utility.

In December 2021, following Bigelow Aerospace's cessation of activity, the ownership of BEAM was transferred to NASA. This transition marked the end of Bigelow's direct involvement with the module, but it underscored its continued relevance and contribution to space station research and technology testing.

The Soyuz TMA-17 rocket lifts off headed for the ISS on Expedition 22.

In the year 2010, space exploration continued to advance, characterized by a series of significant events and milestones. The global landscape of space activities saw both technological innovation and strategic developments, with numerous countries making notable contributions to space exploration and defense.

January 2010 marked a pivotal moment in space defense when China tested its SC-19 missile by targeting and destroying a moving CSS-X-11 medium-range ballistic missile. This action suggested a potential shift in China's space strategy, hinting at the SC-19's potential role as an anti-ballistic missile (ABM). Concurrently, Russia's announcement of its development of inspection and strike spacecraft underscored the increasing capabilities of nations to address space-based threats. These advancements were part of a broader context where nations like China, India, Russia, and the United States demonstrated their anti-satellite (ASAT) capabilities by neutralizing decommissioned satellites. While ASAT systems were designed for defense and strategic advantages, they also raised concerns about space debris and its potential to cause a cascading effect of collisions, known as the Kessler syndrome.

Throughout 2010, the international space community grappled with the implications of these technological advancements. The increasing capability of remotely controlled flying robots to launch missiles at ground targets highlighted the growing complexity of space and defense technologies. In response to the evolving threats from potentially adversarial regimes, NATO adopted a new anti-ballistic missile program during the Lisbon Summit. While the program did not specify any particular country as a target, the United States identified Iran as a potential threat and moved to implement an anti-ballistic missile system. Russia, however, perceived the NATO system as a threat and proposed that any such system should be universal, covering all of Europe while maintaining nuclear parity. The United States sought NATO's support in establishing this system, but concerns arose among NATO officials about its potential impact on their primary responsibility for collective defense.

By 2010, two major space stations were operational: the International Space Station (ISS) and, although not yet launched that year, the Chinese Tiangong-1 station. The ISS, a product of collaboration among NASA, Roscosmos, JAXA, ESA, and CSA, continued to be a hub of international scientific and technological activity. During the year, astronauts and cosmonauts aboard the ISS engaged in a variety of tasks, including scientific experiments and maintenance work. In January, a spacewalk was conducted to replace a malfunctioning ammonia pump in the station's cooling system, showcasing the crew's problem-solving capabilities in space. February saw the arrival of a new supply module from Japan's JAXA agency, further supporting the station's operations. By August, the crew welcomed a Russian Soyuz spacecraft, transporting three new members to the ISS while another Soyuz craft returned three crew members to Earth.

Expedition 23

In 2010, the International Space Station (ISS) experienced a series of pivotal events that marked both a transition in crew and significant advancements in its scientific and logistical operations.

The year began with the conclusion of Expedition 22 and the initiation of Expedition 23. On March 18, 2010, Expedition 22 ended with the undocking of Soyuz TMA-16, making way for the arrival of Expedition 23. This transition was marked by the launch of Soyuz TMA-18 from the Baikonur Cosmodrome in Kazakhstan on April 2, 2010, at 00:04 EST. The spacecraft, commanded by Oleg Kotov and carrying flight engineers Timothy Creamer and Soichi Noguchi, was pivotal in ensuring the continuity of scientific research aboard the ISS. The new crew arrived at the station on April 4, 2010, and immediately began their mission of maintaining station operations and advancing ongoing research.

Expedition 23 crew members. From the left are Russian cosmonaut Mikhail Kornienko, NASA astronaut Tracy Caldwell Dyson, Russian cosmonaut Alexander Skvortsov, all flight engineers; Russian cosmonaut Oleg Kotov, commander; NASA astronaut T.J. Creamer and Japan Aerospace Exploration Agency (JAXA) astronaut Soichi Noguchi, both flight engineers.

The three astronauts of STS-131 and Tracy Caldwell (bottom left) of ISS Expedition 23, the first time four women being at the same time in space.

Expedition 23 was notable for its diverse international composition. The crew included Commander Oleg Kotov of Russia, embarking on his second spaceflight; Flight Engineer Soichi Noguchi of Japan, also on his second mission; and Flight Engineer Timothy Creamer of the United States, on his inaugural spaceflight. Joining them were Russian cosmonauts Aleksandr Skvortsov and Mikhail Korniyenko, both on their first missions, and American astronaut Tracy Caldwell Dyson, making her second spaceflight. This crew configuration was significant as it marked the first instance of three Russian cosmonauts aboard the ISS simultaneously, reflecting the strengthened collaboration among international partners in the ISS program.

Following Expedition 23's arrival, several crucial logistical events unfolded. On April 7, 2010, the Space Shuttle Discovery (STS-131) docked with the ISS, bringing vital supplies and scientific experiments. On April 5, 2010, Space Shuttle Discovery embarked on its historic mission, STS-131, also known as ISS assembly flight 19A. The launch occurred at 6:21 AM EDT from Launch Complex 39A at Kennedy Space Center. After an exhilarating 8.5-minute ascent, Discovery and its seven-member crew began their journey to the International Space Station (ISS). The mission concluded with a landing at 9:08 AM EDT on April 20, 2010, on Runway 33 at the Kennedy Space Center's Shuttle Landing Facility. STS-131 stands out as the longest flight in the history of Space Shuttle Discovery.

The primary objective of STS-131 was to deliver critical supplies and equipment to the ISS. The payload included the Multi-Purpose Logistics Module (MPLM) Leonardo, loaded with essential items such as food, scientific equipment, and the third and final Minus Eighty Degree Laboratory Freezer for ISS (MELFI). Also onboard were the Window Orbital Research Facility (WORF), a Crew Quarters Rack, the Muscle Atrophy Resistive Exercise (MARES) rack, Resupply Stowage Racks (RSRs), and Resupply Stowage Platforms (RSPs). Additionally, the mission involved the exchange of an ammonia tank assembly on the S1 truss of the ISS and included a Lightweight Multi-Purpose Equipment Support Structure Carrier (LMC), which transported a refurbished Ammonia Tank Assembly (ATA) to be swapped out.

STS-131 was notable for several reasons. It was the final Space Shuttle mission to feature a crew of seven, as well as the last to include "rookie" astronauts, with all subsequent missions consisting of veteran crews. The launch was also the final night launch in the Space Shuttle program. This mission was the third and final flight in the program to feature three female astronauts, following STS-40 and STS-96. Importantly, STS-131 marked the first occasion when two Japanese astronauts, Naoko Yamazaki and Soichi Noguchi, were simultaneously in space, with Yamazaki on the shuttle and Noguchi on the ISS. Moreover, Expedition 23 Flight Engineer Tracy Caldwell

Dyson was on the ISS, making this the first time four women were in space concurrently. Notably, Naoko Yamazaki was the last Japanese astronaut to fly aboard the Space Shuttle.

Upon Discovery's successful docking with the ISS at 7:44 AM UTC on April 7, 2010, the spacecraft and station were positioned 220 miles above the Caribbean. Commander Alan Poindexter and Pilot Jim Dutton, supported by Mission Specialist Dorothy Metcalf-Lindenburger, performed engine burns to adjust their trajectory. The docking was preceded by a Rendezvous Pitch Maneuver (RPM), where Station Commander Oleg Kotov and Flight Engineer T.J. Creamer captured over 350 images of Discovery's heat shield. Following the docking, crews conducted leak checks and opened the hatches at 9:11 AM UTC, thirty minutes ahead of schedule. The shuttle crew transferred essential items, including the two Extravehicular Mobility Units (EMUs) needed for the upcoming spacewalks. The Orbiter Boom Sensor System (OBSS) was also grappled by the Space Station Remote Manipulator System (SSRMS) or Canadarm2.

On April 8, Stephanie Wilson and Naoko Yamazaki grappled and berthed the MPLM Leonardo at 4:24 AM UTC. Soichi Noguchi and Clayton Anderson opened the hatches at 11:58 AM UTC, marking the beginning of the cargo transfer. Commander Poindexter participated in several in-flight interviews during the day. Mastracchio and Anderson prepared for the mission's first spacewalk by undergoing a pre-breathe protocol to mitigate the risk of decompression sickness.

On April 9, flight day 5 of the STS-131 mission, astronauts Rick Mastracchio and Clayton Anderson undertook the first of three planned spacewalks. This extravehicular activity (EVA) began early in the mission timeline and was marked by a series of critical tasks essential to maintaining the functionality of the International Space Station (ISS). During their seven-hour and twenty-six-minute spacewalk, Mastracchio and Anderson successfully released a new ammonia tank assembly from the Space Shuttle's payload bay, preparing it for installation during a subsequent EVA. They also removed an experiment from the Kibo Exposed Facility, replaced a Rate Gyro Assembly (RGA), and accomplished several other tasks to streamline future operations.

The spacewalk was aided by the Space Station Remote Manipulator System (SSRMS), operated by pilot Jim Dutton and mission specialist Stephanie Wilson. While Mastracchio and Anderson were outside, Naoko Yamazaki, supported by commander Alan Poindexter and the rest of Expedition 23, moved several large science racks from the Multi-Purpose Logistics Module (MPLM) Leonardo to their designated locations within the ISS.

On the following day, April 10, flight day 6, the focus shifted to transferring supplies from the MPLM Leonardo and the Space Shuttle mid-deck. Key transfers included the relocation of the Windows Observational Research Facility (WORF) to the Destiny laboratory and the final placement of Express Rack 7 (ER7) by mission specialist Naoko Yamazaki and flight engineer Soichi Noguchi. During the morning, a false smoke alarm in the Russian segment prompted a brief emergency response, but normal operations resumed swiftly.

The crew also engaged in public affairs activities, with mission specialists Clay Anderson, Rick Mastracchio, and Stephanie Wilson conducting interviews with various media outlets. Commander Alan Poindexter, pilot Jim Dutton, and mission specialist Dorothy Metcalf-Lindenburger participated in a live discussion with students from the Naval Postgraduate School. The day concluded with a detailed review of procedures for the upcoming second spacewalk. For this preparation, spacewalkers Clay Anderson and Rick Mastracchio entered the Quest airlock, closed the hatch, and conducted a pre-spacewalk campout by lowering the air pressure and breathing pure oxygen.

On April 11, flight day 7, Mastracchio and Anderson saw the execution of the second spacewalk. They commenced their EVA at 05:30 UTC, an advance of 45 minutes from the planned time, and spent 7 hours and 26 minutes outside the ISS. Their primary objectives included the removal of the old Ammonia Tank Assembly (ATA) from the S1 truss and installing a new ATA. Despite a minor challenge with one of the attachment bolts, the team successfully secured the tank after adjusting the other three bolts. They also installed two radiator grapple fixture stowage beams on the P1 truss and temporarily stowed the old ATA using the SSRMS. Due to time constraints, the connection of fluid lines was deferred to the final spacewalk.

Meanwhile, the rest of the crew continued transferring items from Shuttle Discovery's mid-deck and the MPLM Leonardo, completing approximately half of the scheduled transfers.

Flight day 8, April 12, was largely devoted to off-duty activities and public outreach. The joint STS-131 and Expedition 23 crews had the morning off, following which they resumed transfer activities, now over seventy percent complete. The day featured several public affairs events, including VIP meetings with Roscosmos, Russian President Dmitry Medvedev, RSC Energia, and the Japanese Aerospace Exploration Agency (JAXA), among others. Later, crew members participated in interviews with American media outlets and engaged with students from Eastern Guilford High School and local elementary schools.

In preparation for the final spacewalk, Mastracchio and Anderson reviewed their procedures, then entered the Quest airlock to undergo a pressure reduction and oxygen breathing regimen.

On April 13, flight day 9, the third and final spacewalk of the STS-131 mission took place. Mastracchio and Anderson's tasks included connecting ammonia and nitrogen lines to the newly installed Ammonia Tank Assembly, installing the old ATA in Discovery's payload bay, retrieving Micro-Meteoroid Orbital Debris (MMOD) shields, and preparing cables for future missions. Despite encountering difficulties securing a bolt during the installation of the old ATA, the spacewalk was completed in 6 hours and 24 minutes, bringing the total EVA time to 20 hours and 19 minutes. During the spacewalk, other crew members continued with the transfer operations, which were now more than seventy-five percent complete.

Flight day 10, April 14, was focused on completing the remaining transfer activities. By mid-day, the crew had transferred almost all items to the MPLM Leonardo, with only a few remaining to be moved to the Shuttle mid-deck. The day included a shared meal with the Expedition 23 crew and a joint crew photo followed by a news conference with international media. The crew also interacted with students and spent the majority of the afternoon off-duty.

April 15 (Flight Day 11 – MPLM Unberthing)

On April 15, flight day 11, the MPLM Leonardo was unberthed from the Harmony node. After closing the MPLM's hatches at 07:38 UTC, it was carefully maneuvered to a low hover above Discovery's payload bay at 20:24 UTC due to a delay caused by stuck bolts on the Common Berthing Mechanism (CBM). The crew completed approximately ninety-four percent of the transfer operations by this stage.

April 16 (Flight Day 12 – Late Inspection)

Flight day 12, April 16, involved securing the MPLM Leonardo in Discovery's payload bay for its return to Earth. Mission specialist Dorothy Metcalf-Lindenburger activated the latches at 07:15 UTC. Following this, Metcalf-Lindenburger and pilot Jim Dutton, joined by commander Alan Poindexter and mission specialist Naoko Yamazaki, conducted a comprehensive inspection of Discovery's heat shield, including the Reinforced Carbon-Carbon (RCC) panels and heat-resistant tiles. This inspection, completed three hours ahead of schedule, was performed while the shuttle remained docked to the ISS due to the loss of the shuttle's Ku-Band antenna.

On the final flight day, April 17, Space Shuttle Discovery undocked from the ISS at 12:52 UTC. The shuttle had been docked to the ISS for ten days, five hours, and eight minutes. Following undocking, pilot Jim Dutton performed a fly-around of the space station. The undocking was preceded by a farewell ceremony where shuttle commander Alan Poindexter and station commander Oleg Kotov bid farewell to their respective crews. After undocking, the shuttle crew stowed the Orbiter Boom Sensor System (OBSS) and the Shuttle Remote Manipulator System (SRMS) as they prepared for the mission's concluding phases.

In the latter part of April, the ISS Progress 35 spacecraft undocked on April 22, 2010, clearing the way for the arrival of ISS Progress 37. This cargo spacecraft launched on April 28, 2010, and successfully docked with the ISS on May 1, 2010, delivering essential supplies to support the station's operations.

Further adjustments in crew and vehicle logistics were evident in May. On May 10, 2010, ISS Progress 36 undocked, and shortly after that, on May 12, 2010, Soyuz TMA-17, which had been part of the Expedition 22 crew, was relocated within the station to optimize docking ports for upcoming missions.

A significant event occurred on May 14, 2010, when the Space Shuttle Atlantis embarked on the STS-132 mission from Kennedy Space Center, marking a significant moment in the history of the International Space Station (ISS) assembly. The primary objective of this mission was to deliver and install the Russian Mini-Research Module-1 (MRM-1) Rassvet, a critical component for the ISS, along with an Integrated Cargo Carrier-Vertical Light Deployable (ICC-VLD). The shuttle successfully docked with the ISS on May 16, 2010, and landed back at Kennedy Space Center on May 26, 2010.

This view of the space shuttle Discovery was provided by an Expedition 23 crew member during a survey of the departing vehicle following undocking from the International Space Station on April 17. The Leonardo Multi-Purpose Logistics Module was visible in Discovery's cargo bay.

STS-132 was initially planned as Atlantis's final mission. However, in February 2011, NASA decided to extend the shuttle program, leading to the announcement that STS-135 would be the final mission of Atlantis and the entire Space Shuttle program. Despite the uncertain funding situation, STS-135 was confirmed and scheduled.

The crew of STS-132 was led by Commander Kenneth Ham, who was on his second and final spaceflight. The pilot, Dominic A. "Tony" Antonelli, mission specialists Garrett Reisman, Michael T. Good, and Stephen G. Bowen were also on their final spaceflights, while Piers Sellers made his third and last spaceflight. Their mission was to manage and install diverse equipment and modules.

Among the cargo carried was the Rassvet Mini-Research Module-1, also known as "Rassvet," which translates to "dawn" in Russian. Built by the Russian aerospace company Energia, Rassvet arrived at Kennedy Space Center on December 17, 2009, aboard an Antonov 124 cargo plane. Once unloaded and processed, it was prepared for its space journey. This module was equipped with various components, including an airlock and a radiation heat exchanger for the Nauka Module, a spare elbow part of the European Robotic Arm (ERA), and a portable work platform. It had a total cargo and science volume of 5 cubic meters and was outfitted with standard grapple fixtures for easy handling by the ISS's robotic arm.

STS-132 crew NASA astronauts Ken Ham (bottom), commander; Garrett Reisman and Michael Good, both mission specialists; Tony Antonelli, pilot; Piers Sellers and Steve Bowen both mission specialists.

In addition to the Rassvet module, the shuttle carried the Integrated Cargo Carrier-Vertical Light Deployable (ICC-VLD2) pallet. This pallet included essential items such as a Ku-band Space-to-Ground Antenna (SGANT), an Enhanced Orbital Replacement Unit (ORU) Temporary Platform (EOTP) for the Canadian Dextre robotic arm, and six new battery ORUs. These new batteries were used to replace older ones on the ISS's P6 truss, with the old batteries returning to Earth aboard the ICC-VLD2.

The ICC-VLD2, constructed from aluminum, was about 8 feet long, 13 feet wide, and 10 inches thick, weighing approximately 2,645 pounds empty. The pallet, along with its cargo, totaled around 8,330 pounds. The mission also included the transport of two Light Weight Tool Stowage

Assemblies, which were modified to carry memorabilia. A compact disk containing digital copies of entries from NASA's Space Shuttle Program Commemorative Patch Contest, won by Blake Dumesnil, was flown aboard Atlantis. Additionally, seventeen handcrafted beads from various North American artists were included as part of the Beads in Space project, and a small wood sample from Sir Isaac Newton's apple tree was carried into orbit, paying homage to Newton's contributions to science.

On May 16, the crew of STS-132 began their day early, preparing for the critical docking operation with the ISS. Commander Ham executed a series of precise orbital burns to align Atlantis's orbit with the ISS's. The final docking burn, the terminal initiation (TI) burn, was performed with the left OMS engine. By 13:26 UTC, Atlantis was in position for a 360-degree flip maneuver to allow ISS crew members to photograph the shuttle's underside to inspect its thermal protection system.

At 14:28 UTC, Atlantis docked with the ISS Pressurized Mating Adapter-2 while the two spacecraft were orbiting 220 miles above the South Pacific Ocean. Following docking, Atlantis was reoriented by its thrusters to minimize the risk of micrometeoroid orbital debris impacts. After thorough leak checks and a brief welcoming ceremony, the shuttle crew commenced the transfer of equipment and supplies, starting their critical work aboard the ISS.

On May 22, during the STS-132 mission, the shuttle Atlantis and its crew engaged in a series of critical activities. Among these was reinstalling the Intermodule Ventilation System's Ventilation and Lighting Device (ICC-VLD) back into Atlantis's payload bay. This task marked the completion of its mission objectives. The Canadarm2, operated by Mission Specialists Piers Sellers, Garrett Reisman, and Space Station Flight Engineer Tracy Caldwell Dyson, played a crucial role in this operation. The ICC-VLD re-installation began promptly at 4:30 a.m. EDT and was successfully completed by 5:50 a.m. EDT.

During this period, the shuttle crew and Caldwell Dyson engaged in a unique educational outreach. They responded to questions from elementary and middle school students across the United States who had submitted their inquiries via video. This interaction was part of NASA's efforts to inspire young minds through space exploration.

Later in the day, the combined crew of Atlantis and the International Space Station (ISS) shared a meal, fostering camaraderie before the shuttle crew enjoyed two and a half hours of well-deserved off-duty time starting at 11:05 a.m. EDT.

On May 23, the tenth flight day, the STS-132 and Expedition 23 crews began their final day of the docked mission. The astronauts completed the last time-sensitive transfers, including scientific research samples that required refrigeration. Following these essential tasks, the crew held a joint news conference and took a commemorative crew photograph. As the mission neared its conclusion, the crews participated in a mutual farewell ceremony, marking the end of their collaborative efforts.

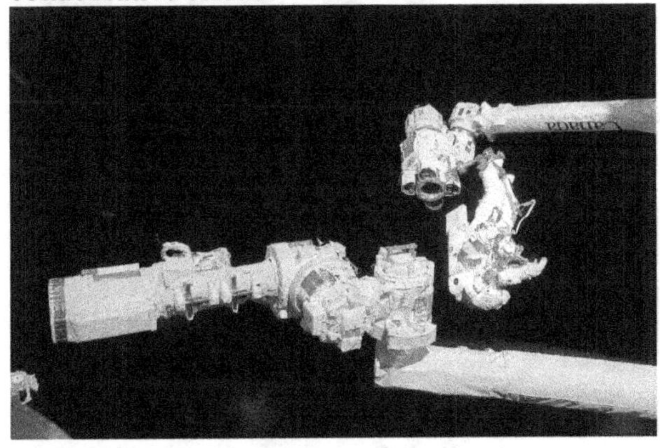

Anchored to a Canadarm2 mobile foot restraint, NASA astronaut Garrett Reisman, STS-132 mission specialist, participates in the mission's first session of extravehicular activity (EVA) as construction and maintenance continue on the International Space Station. During the seven-hour, 25-minute spacewalk, Reisman and NASA astronaut Steve Bowen (out of frame), mission specialist, loosened bolts holding six replacement batteries, installed a second antenna for high-speed Ku-band transmissions and adding a spare parts platform to Dextre, a two-armed extension for the station's robotic arm.

Subsequently, the hatches between Atlantis and the ISS were closed, and a thorough leak check was conducted to ensure that all seals were secure. At 15:22 UTC, just over two hours after closing the hatches, Atlantis undocked from the ISS. The spacecraft was orbiting approximately 220 miles (350 kilometers) above the Southern Ocean, southwest of Perth, Australia. Under the guidance of Pilot Tony Antonelli, Atlantis maneuvered away from the ISS, maintaining a distance of about 400 feet (120 meters). Antonelli then conducted a flyaround of the space station, allowing the ISS and Atlantis crews to capture photographs of their respective spacecraft. Following the fly-around, the shuttle crew executed two separation burns to further distance Atlantis from the ISS, concluding the mission's docking phase.

The culmination of these activities was marked by the departure of the Expedition 22/23 crew on June 1, 2010. Their departure marked the end of their mission and the continuation of the ISS's operations under the stewardship of the new crew.

The space shuttle Atlantis's cabin and forward cargo bay and part of the International Space Station while the two spacecraft remain docked, during STS-132's flight day four extravehicular activity of astronauts Garrett Reisman and Steve Bowen (both out of frame). Though three sessions of extravehicular activity (EVA) will involve only three astronauts (two on each occasion) who actually leave the shirt-sleeve environments of the two docked spacecraft, all twelve astronauts and cosmonauts on the two combined crews have roles in supporting the EVA work.

Expedition 24

Expedition 24, which spanned from June to November 2010, was a critical period for the International Space Station (ISS), marked by significant crew rotations and scientific advancements. During this expedition, the Soyuz missions TMA-18 and TMA-19 played essential

roles in maintaining the ISS's operational and research capabilities.

Soyuz TMA-18, launched on April 2, 2010, was a key element in the crew exchange process. This mission carried three astronauts: Russian cosmonauts Alexander Skvortsov and Mikhail Kornienko, and NASA astronaut Tracy Caldwell Dyson. Upon docking with the ISS on April 4, Soyuz TMA-18 facilitated the replacement of the outgoing Expedition 23 crew. The spacecraft's docking at the Poisk module marked the beginning of Expedition 24 and was a testament to the ongoing international cooperation aboard the ISS.

Expedition 24 crew members clockwise are Russian cosmonaut Alexander Skvortsov (bottom), commander; NASA astronauts Tracy Caldwell Dyson and Doug Wheelock; Russian cosmonauts Mikhail Kornienko and Fyodor Yurchikhin; along with NASA astronaut Shannon Walker, all flight engineers.

The docking process for Soyuz TMA-18 was a complex and critical operation. Despite initial communication issues shortly after launch, the spacecraft successfully achieved orbit and aligned itself for docking. At 05:25 UTC on April 4, Soyuz TMA-18 docked three minutes ahead of schedule. The automatic closure of the docking port and subsequent system transitions demonstrated the spacecraft's reliability. After conducting necessary leak checks and preparations, the crew entered the ISS at 07:21 UTC. Family members and dignitaries from the Russian mission control center in Korolyov celebrated their arrival. The Soyuz TMA-18 remained docked throughout Expedition 24, serving as a crucial emergency escape vehicle.

As Expedition 24 progressed, the crew conducted a variety of scientific research and maintenance tasks, including experiments in material science, biology, and human physiology. These activities were essential for understanding space's impact on living organisms and materials, furthering the ISS's mission of advancing space science.

Soyuz TMA-19, launched on September 30, 2010, continued the important role of crew rotation. This mission brought three additional astronauts to the ISS: Russian cosmonaut Dmitry Kondratyev, European Space Agency (ESA) astronaut Paolo Nespoli, and NASA astronaut Catherine Coleman. Docking with the ISS on October 2, Soyuz TMA-19 facilitated the exchange of crew members, allowing the Expedition 24 team to expand and ensure the station's smooth operation.

Soyuz TMA-19's arrival was a significant milestone for Expedition 24, as it ensured the continuity of scientific research and station operations. The crew included seasoned astronauts and newcomers, contributing to the diverse and collaborative environment aboard the ISS. Dmitry Kondratyev and Paolo Nespoli brought new perspectives and expertise, while Catherine Coleman joined the ongoing efforts to advance the ISS's research objectives.

Throughout Expedition 24, the crew undertook a series of spacewalks and addressed technical challenges while ensuring the ISS's smooth operation. Notably, on June 28, 2010, the crew successfully relocated Soyuz TMA-19 from one docking port to another, a routine maneuver to optimize port usage and prepare for future arrivals.

Critical supply missions during the expedition met the ISS's logistical needs. On August 31, 2010, the Progress 38 spacecraft undocked after completing its cargo delivery mission, making way for Progress 39. This spacecraft, launched from Baikonur on September 10, 2010, arrived and docked with the ISS on September 12, 2010. Progress 39 delivered essential supplies, including food, equipment, and scientific experiments, vital for the station's continued operations and research.

The crew faced several technical incidents during the mission. The most significant occurred on July 31, 2010, when an ammonia pump module malfunction triggered an alarm that interrupted the crew's rest. The malfunction led to a remote power controller tripping and cutting power to certain sections of the ISS. Astronauts Tracy Caldwell Dyson and Doug Wheelock worked tirelessly to assist ground controllers in reactivating critical systems, including two main power buses and a Control Moment Gyroscope. Despite their efforts, another alarm disrupted their rest as ground teams attempted to restart the faulty pump module.

In addition to the pump module problem, the crew encountered a delay with the docking ring on the Mini-Research Module 2 (MRM2) Poisk. This technical glitch led to a postponement of the Soyuz TMA-18 spacecraft's undocking, originally scheduled for September 24, 2010, to September 25, 2010. The delay was caused by a faulty micro-switch and a damaged drive gear in the docking mechanism, necessitating an adjustment to the spacecraft's departure schedule.

The Mini-Research Module 1 (MRM-1), known as Rassvet, played a crucial role in Expedition 24. Rassvet, which translates to "dawn" in Russian, was a cargo storage facility and a docking port for visiting spacecraft. Its design resembles the Mir Docking Module, launched in 1995 aboard Space Shuttle mission STS-74, underscoring its versatility. Rassvet was transported to the ISS aboard Space Shuttle Atlantis on the STS-132 mission, launching on May 14, 2010.

Expedition 24 included three planned spacewalks, originally scheduled for Russian Orlan suits and U.S. Extravehicular Mobility Units (EMUs). However, the mission expanded to include additional spacewalks to address the malfunctioning ammonia pump module.

The first spacewalk occurred on July 27, 2010, involving Russian astronauts Mikhail Korniyenko and Fyodor Yurchikhin. This extravehicular activity (EVA) lasted 6 hours and 42 minutes and focused on routing data cables between the Rassvet and Zvezda modules and relocating a camera on Rassvet.

On August 7, 2010, American astronauts Douglas Wheelock and Tracy Caldwell Dyson conducted their second EVA, which lasted 8 hours and 3 minutes. Their tasks included disconnecting electrical and fluid connectors, though they faced a problem with a stuck quick disconnect that limited the completion of their activities. After their work, they underwent thorough decontamination to ensure no ammonia residue remained on their suits before re-entering the station.

On August 11, 2010, the third spacewalk saw Wheelock and Caldwell Dyson working for 7 hours and 26 minutes to address the failed pump module. Wheelock removed the quick disconnect valve and detached the final fluid line, while Caldwell Dyson handled the removal of electrical cables and bolts from the malfunctioning pump. The old pump was prepared for future analysis, and the spare pump was readied for installation.

The final spacewalk, conducted on August 16, 2010, involved Wheelock and Caldwell Dyson working for 7 hours and 20 minutes. Wheelock removed the spare pump module from its stowage platform and installed it on the ISS's S1 Truss, securing it with bolts and connecting electrical components.

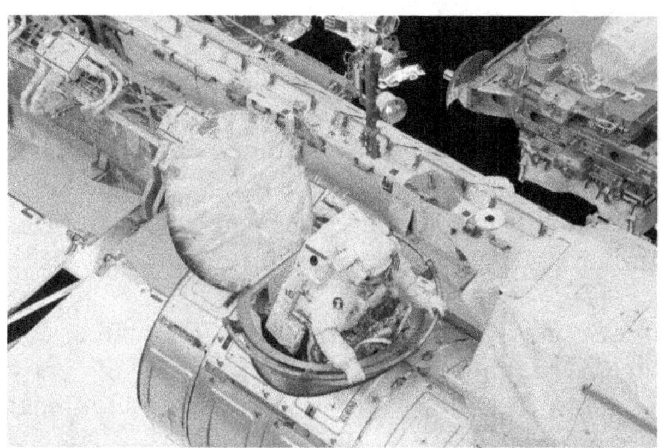

Wheelock egresses the Quest Airlock hatch on the ISS during the EVA 3 on 11 August 2010. Notice what looks like scorch marks on the hatch thermal cover, the effect of vacuum and atomic oxygen on the threads and thread sealant used on the thermal cover. The 'smell' of space follows suit, He described it like burnt cake or cookies, or like the smell of an extinguished match."

The Soyuz TMA-19 spacecraft (foreground), docked to the Rassvet Mini-Research Module 1 (MRM1), and Progress 37 resupply vehicle, docked to the Pirs Docking Compartment,

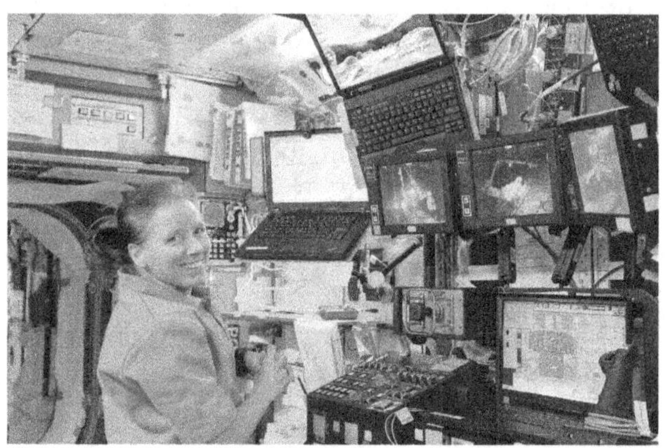

NASA astronaut Shannon Walker, Expedition 24 flight engineer, near a robotic workstation in the Destiny laboratory of the International Space Station.

Soyuz TMA-19 crewmembers -- NASA astronaut Doug Wheelock, (second left in blue), Expedition 25 commander, along with Russian cosmonaut Fyodor Yurchikhin, Soyuz commander, and NASA astronaut Shannon Walker, flight engineer -- are seen after being removed from the Soyuz TMA-19 capsule near the town of Arkalyk, Kazakhstan Nov. 25, 2010 (USA time or Nov. 26 in local Kazakhstan time). The three are returning from over five months onboard the International Space Station where they served as members of the Expedition 24 and 25 crews

Expedition 25

Expedition 25 - Russian cosmonauts Oleg Skripochka and Alexander Kaleri; NASA astronauts Scott Kelly, Shannon Walker; Russian cosmonaut Fyodor Yurchikhin, all flight engineers, and NASA astronaut Doug Wheelock, commander

Expedition 25, launched in the fall of 2010, represents a significant chapter in the ongoing story of the International Space Station (ISS) and its role in international space collaboration. The mission commenced with the launch of the Soyuz TMA-19 spacecraft on October 7, 2010, from the Baikonur Cosmodrome in Kazakhstan. This event signaled the beginning of a new phase in the ISS's operational history, as the spacecraft carried a new crew to the station, ready to advance the scientific and exploratory objectives aboard.

The Soyuz TMA-19 docked with the ISS on October 9, 2010, bringing onboard Commander Scott Kelly from NASA and Flight Engineers Alexander Kaleri and Oleg Skripochka, both from Roscosmos. Their arrival marked the official start of Expedition 25, which was set to focus on a range of scientific experiments and the continued maintenance of the station.

In the days leading up to the arrival of the Soyuz TMA-19, another pivotal event occurred on October 27, 2010, with the launch of the ISS Progress 40 spacecraft. This resupply mission was crucial for delivering essential cargo and scientific equipment necessary for ongoing experiments and the day-to-day operations of the ISS. The Progress 40 successfully docked with the ISS on October 30, 2010, ensuring that the station remained well-stocked and fully operational.

Expedition 25 followed the conclusion of Expedition 24, which ended with the departure of its crew on November 25, 2010. This changeover allowed for the full operational commencement of the new expedition. A noteworthy milestone during this period was commemorating the 10th anniversary of human presence aboard the ISS. On November 2, 2000, the first crew, Expedition 1 Commander William Shepherd, along with Flight Engineers Sergei Krikalev and Yuri Gidzenko, began a continuous human presence that marked the start of a new era in space habitation.

The crew for Expedition 25 was composed of seasoned astronauts from both NASA and Roscosmos. Commander Douglas H. Wheelock, who was on his second spaceflight, led the mission. His team included Flight Engineer Shannon Walker from NASA, on her first spaceflight; Fyodor Yurchikhin from Roscosmos, undertaking his third mission; Scott J. Kelly from NASA, on his third spaceflight; Alexander Kaleri from Roscosmos, with his fifth spaceflight; and Oleg Skripochka from Roscosmos, on his first mission.

The backup crew for Expedition 25 included a group of experienced astronauts and cosmonauts: Andrei Borisenko as Commander, Paolo Nespoli, Catherine Coleman, Anatoli Ivanishin, Sergei Revin, and Ronald J. Garan Jr.

Expedition 25, a milestone in the ongoing research aboard the International Space Station (ISS), was distinguished by its extensive array of scientific experiments and logistical operations. The mission began with the undocking of Soyuz TMA-18 on September 25, 2010, followed by the arrival of Soyuz TMA-01M on October 10, 2010. bringing new crew members Scott Kelly, Alexander Kaleri, and Oleg Skripochka. They joined the existing Douglas Wheelock, Fyodor Yurchikhin, and Shannon Walker team, making up a full six-member crew.

On October 7, 2010, the Soyuz TMA-01M mission embarked on its journey from the Baikonur Cosmodrome, marking a significant milestone in the history of spaceflight. This mission was the 107th flight of a Soyuz spacecraft and the inaugural mission of the upgraded Soyuz TMA-M series. Designed and built by RKK Energia, the TMA-M series represented a substantial evolution from the previous Soyuz-TMA models, reflecting technological and design advancements.

The Soyuz TMA-01M was equipped with several enhancements over its predecessors. Outdated components were replaced with 19 new-generation devices, and the spacecraft's overall mass was reduced by 70 kilograms (154 pounds). Notably, the traditional Argon-16 computer control system, a fixture in Soyuz spacecraft for over three decades, was replaced with the more advanced TsVM-101 digital onboard computer. This upgrade improved power efficiency and simplified the spacecraft's manufacturing process by replacing the magnesium alloy in the instrument module frame with a lighter aluminum alloy.

NASA astronaut Scott Kelly (left), Expedition 25 flight engineer and Expedition 26 commander; along with Russian cosmonauts Alexander Kaleri (center), Expedition 25 commander and Expedition 26 flight engineer and Oleg Skripochka, Expedition 25/26 flight engineer,

NASA astronaut Scott Kelly, a Soyuz TMA-01M crew member, lauded the spacecraft's new display systems, which he noted made operations more intuitive and less demanding for the crew. The TMA-M series was designed to enhance operational capabilities and serve as a testbed for new technologies intended for Russia's next-generation crewed spacecraft.

The Soyuz TMA-01M crew comprised three seasoned astronauts: Commander Aleksandr Kaleri from Russia, Flight Engineer Oleg Skripochka from Russia, and Flight Engineer Scott Kelly from the United States. Kaleri was on his fifth and final spaceflight, Skripochka was making his first voyage into space, and Kelly was undertaking his third mission. The backup crew included Commander Sergey Volkov, Flight Engineer Oleg Kononenko, and Flight Engineer Ronald Garan.

The spacecraft lifted off aboard a Soyuz-FG rocket at 23:10:55 UTC on October 7, 2010. Within nine minutes, Soyuz TMA-01M achieved low Earth orbit, with parameters including a minimum altitude of 199.85 kilometers and a maximum altitude of 258.77 kilometers. Following its orbital insertion, the spacecraft deployed its solar panels and antennas and began approaching the International Space Station (ISS).

Soyuz TMA-01M docked with the ISS at 00:01 UTC on October 10, 2010, precisely one minute ahead of schedule. The docking occurred at the Poisk module while the spacecraft was approximately 354 kilometers above the southern Pacific Ocean, off the coast of Chile. After the docking hooks and latches engaged, extensive leak checks were performed before the hatches were opened at 03:09 UTC, allowing the crew to enter the station.

During Expeditions 25 and 26, the International Space Station (ISS) served as a bustling center of scientific exploration and technological advancement, marked by a comprehensive array of experiments. The Russian Federal Space Agency (Roscosmos) spearheaded an ambitious research program, announcing 504 experimental sessions across 41 distinct studies. This included continuing 34 experiments from previous missions and introducing seven new ones, reflecting a broad and deep commitment to advancing knowledge in various fields.

Among the new experiments, Molniya-Gamma stood out for its focus on understanding gamma rays and optical radiation associated with terrestrial lightning and thunder. By measuring these emissions, Molniya-Gamma aimed to enhance our understanding of terrestrial gamma-ray flashes and the dynamics of upper-atmospheric lightning, providing valuable insights into both space and atmospheric sciences.

Sprut-2, another notable addition, delved into the biological effects of space on living organisms. This experiment and others such as Sonocard, Pilot, Vzaimodeystviye, Tipologia, Pneumocard, and Biorisk comprised a suite of life science studies. The Biorisk experiment was particularly important, as it investigated the interactions of microbial bacteria and fungi with structural materials used in spacecraft construction, an essential factor in ensuring the longevity and integrity of space missions.

In remote sensing, the Rusalka experiment tested methods for remotely determining the concentration of methane and carbon dioxide in Earth's atmosphere. This was part of a broader suite of remote sensing studies that included MW-radiometry, Zeiner, and Econ. These experiments aimed to improve our understanding of atmospheric composition and contribute to environmental monitoring efforts from space.

Space biotechnology was another area of focus, with experiments like Lactolen, Biotrek, Biodegradatsia, Zhinseng-2, Structure, and Constanta exploring the effects of space on biological systems and materials. These studies were crucial for understanding how space environments impact living organisms and biological processes, advancing our knowledge of life sciences in extraterrestrial contexts.

Technical research during this period was extensive and varied. Experiments such as Vektor-T, Izgib, Identification, Veterok, SLS, Sreda-MKS, Contur, VIRU, Bar, Test, and RadioSkaf addressed numerous aspects of space technology and material science. These studies aimed to overcome technical challenges and enhance the technological capabilities necessary for future space exploration.

A significant highlight of the missions was the EXPOSE-R payload, an initiative by the European Space Agency (ESA). This experiment involved exposing organic materials to the harsh conditions of space to study their resilience, contributing to our understanding of how life might survive in extraterrestrial environments.

The study of cosmic rays was advanced through experiments like BTN-Neutron and Matryoshka-R, which sought to improve knowledge about cosmic radiation and its interactions with both human bodies and spacecraft materials.

Educational and humanitarian projects also featured prominently in the mission's agenda. Projects such as MAI-75 and Colon Crystal aimed to provide educational benefits and explored crystallization processes in space, broadening the scope of scientific inquiry.

During Expedition 25 of the International Space Station (ISS), one of the pivotal events was the handling of resupply missions that ensured the station's continued operation and scientific progress. On October 25, 2010, the Progress M-05M spacecraft, which had been docked with the ISS since May 2010, was undocked to make way for its successor, Progress M-08M. The Progress

M-08M, loaded with essential supplies, was launched from Baikonur Cosmodrome's Gagarin's launch pad aboard a Soyuz-U carrier rocket on October 27, 2010. This mission, designated by NASA as Progress 40 or 40P, carried approximately 2.5 tons of cargo, including water, air, fuel, and specialized hardware necessary for ongoing scientific experiments.

After a three-day journey through space, Progress M-08M arrived at the ISS on October 30, 2010. It docked with the Pirs module's nadir port at 16:36 UTC. However, this docking was not without complications. As the spacecraft approached the ISS, it encountered difficulties that led mission controllers in Moscow to instruct cosmonaut Alexander Kaleri to switch from the spacecraft's automated KURS docking system to manual control using the TORU system. Kaleri, from the Zvezda module, expertly piloted Progress M-08M using television views and joysticks, guiding it successfully to its docking port.

Expedition 25 also featured significant spacewalks, which were crucial for station maintenance and scientific experiments. On November 15, 2010, cosmonauts Fyodor Yurchikhin and Oleg Skripochka embarked on a spacewalk, known as EVA 1, that lasted 6 hours and 27 minutes. Their mission involved several critical tasks within the Russian segment of the ISS. They began by installing a portable multipurpose workstation in Plane IV of the Zvezda module and setting up structural struts connecting the Poisk module to both the Zvezda and Zarya modules.

A major part of their work involved conducting the Test experiment, aimed at detecting microorganisms or contamination beneath the insulation in the Russian segment. They photographed, installed protective covers, and removed various scientific instruments, including the Plasma Pulse Injector Science hardware and the Kontur science hardware (ROKVISS), which were cleaned and removed from the Zvezda module. Additionally, they replaced the Expose-R scientific experiment, a European Space Agency project designed to assess space's effects on organic materials.

Further, Yurchikhin and Skripochka installed a handrail on the Pirs docking module and mounted the SKK #1-M2 cassette on the Poisk module. They also attempted to relocate a television camera

from the Rassvet module; however, this effort was unsuccessful due to interference with insulation at the planned installation site.

Expedition 25 commander Douglas Wheelock in the Cupola.

STS-130, designated as ISS assembly flight 20A, was a pivotal NASA Space Shuttle mission to expand the International Space Station (ISS). The mission was launched on February 8, 2010, at 04:14 EST (09:14 UTC) aboard Space Shuttle Endeavour, with its primary objectives being the delivery and installation of the Tranquility module and the Cupola. The Tranquility module, known as Node 3, was manufactured at the Thales Alenia Space factory in Turin, Italy, and transported to the Kennedy Space Center in Florida on May 21, 2009. The Cupola, a versatile robotic control station with six side windows and a central window providing a 360-degree view, was a key mission component.

Endeavour's successful launch began a significant phase for the ISS. As the shuttle ascended into orbit, the crew, under the command of George D. Zamka, conducted essential tasks, including opening the payload bay doors, activating the radiators, and deploying the Ku band antenna. Astronauts Nick Patrick and Kathryn Hire performed a thorough checkout of the Shuttle Robotic Arm (SRMS) and surveyed the payload bay. They also downlinked imagery and video of the external tank back to mission control.

On February 9, the crew focused on inspecting the shuttle's thermal protection system (TPS). This involved a meticulous review of the TPS, with all six crew members participating. After completing this task, astronauts Bob Behnken and Nick Patrick began preparing the spacesuits for the upcoming spacewalks. In contrast, Stephen Robinson and Kathryn Hire, joined by Behnken, prepared the tools for the shuttle's rendezvous with the ISS. These tools included a hand-held LIDAR gun and the Orbiter Docking System (ODS).

STS-130 astronauts George Zamka (right), commander; and Terry Virts, pilot. From the left (standing) are astronaut Nicholas Patrick, Robert Behnken, Kathryn Hire and Stephen Robinson, all mission specialists.

The following day, February 10, marked the shuttle's rendezvous with the ISS. Endeavour executed a series of burns to align its approach and performed the Rendezvous Pitch Maneuver (RPM). ISS Commander Jeff Williams and Flight Engineer Oleg Kotov photographed the shuttle's TPS during this maneuver. The docking occurred at 05:26 UTC (00:06 EST), followed by a successful opening of the hatches at 06:26 UTC (01:26 EST). The crew then conducted a welcome ceremony and safety briefing before transferring the spacesuits and tools needed for the spacewalks.

On February 11, the crew prepared for the first spacewalk scheduled for the next day. Nick Patrick and Bob Behnken organized the necessary tools while Commander Zamka and ISS Flight Engineer Soichi Noguchi addressed an issue with Behnken's suit. The team also conducted a PAO event, engaging with television and radio stations across the United States before reviewing spacewalk procedures and preparing the Quest Airlock for the upcoming EVA.

The first spacewalk of the mission took place on February 12. Nick Patrick and Bob Behnken began their EVA at 02:17 UTC, performing tasks such as moving the payload bay of Endeavour and releasing launch locks on the Tranquility module and Cupola. After successfully relocating

Tranquility to the port side of the Unity node, they connected a temporary heater and data cables. The spacewalk lasted six and a half hours and concluded with additional transfer activities.

February 13 marked the opening of the Tranquility module's hatches for the first time. The crew, including George Zamka, Terry Virts, Stephen Robinson, and Kathryn Hire, conducted the module's initial outfitting and prepared the Cupola for its installation. Spacewalkers Behnken and Patrick also prepared their tools for the second EVA. The day featured PAO events, including a Q&A session with Capcom Mike Massimino and interviews with various news outlets.

The second spacewalk occurred on February 14. Bob Behnken and Nick Patrick completed all their scheduled tasks, including installing ammonia coolant loops and thermal blankets and outfitting the Earth-facing port of Tranquility for the Cupola. This spacewalk lasted 5 hours and 54 minutes, significantly advancing the mission's objectives.

STS-130, culminating with Endeavour's landing on February 21, 2010, at 22:22 EST at the Kennedy Space Center's Shuttle Landing Facility, marked a successful chapter in the ISS's assembly and operational history.

The Cupola, an observatory module built by the European Space Agency (ESA), was a distinctive feature of the International Space Station (ISS). Its name, derived from the Italian word "cupola," meaning "dome," aptly reflects its design and purpose. This module has seven windows, each strategically positioned to facilitate a range of functions, including scientific experiments, spacecraft docking, and Earth observations.

The Cupola was launched aboard Space Shuttle mission STS-130 on February 8, 2010. Upon arrival, it was meticulously attached to the Tranquility (Node 3) module of the ISS. The installation of the Cupola marked a significant milestone in the ISS assembly process, bringing the station's overall construction to 85 percent completion.

One of the most notable features of the Cupola was its central window, which has a diameter of 80 centimeters (31 inches). This expansive window provides an unparalleled view of space and Earth, enhancing the station's observational capabilities and contributing to its scientific research efforts.

The Cupola's strategic placement and advanced design make it a critical asset for both operational and scientific activities aboard the ISS.

The Cupola's windows with shutters open. The Cupola houses controls for the station robotics and will be a location where crew members could operate the robotic arms and monitor other exterior activities.

Preflight preparations for the new crew involved rigorous activities at the Baikonur Cosmodrome in Kazakhstan. From September 26 to October 4, 2010, Soyuz Commander Alexander Kaleri, NASA Flight Engineer Scott Kelly, and Russian Flight Engineer Oleg Skripochka engaged in various preparatory tasks. These included arriving in Baikonur, conducting fit checks both suited and unsuited in their Soyuz spacecraft, and participating in traditional ceremonies, such as raising flags outside their Cosmonaut Hotel quarters. The Soyuz TMA-01M spacecraft was meticulously prepared for its mission, with its final assembly and integration occurring on October 5, 2010. The spacecraft and its booster were transported to Launch Complex 5 at the Baikonur Cosmodrome for final preparations before the scheduled launch.

On September 22, 2010, NASA astronaut Douglas Wheelock took over as the commander of Expedition 25, succeeding Aleksandr Skvortsov. Under Wheelock's leadership, Expedition 25 continued its important work aboard the International Space Station until November 25, 2010. On this date, Wheelock, along with astronauts Shannon Walker and Fyodor Yurchikhin, concluded their time on the station and departed, marking the end of Expedition 25.

The mission of Soyuz TMA-01M, which had played a crucial role as an emergency escape vehicle during Expedition 25, came to a close on March 16, 2011. At 4:27 GMT, the spacecraft undocked from the ISS, initiating the final phase of its journey back to Earth. The spacecraft began its descent with a four-minute and 17-second rocket burn, designed to slow its re-entry and ensure a controlled landing. As the spacecraft re-entered the atmosphere, the modules separated as planned, and the central module, carrying the crew, touched down at 7:54 GMT near Arkalyk in Kazakhstan.

The landing site presented severe challenges, characterized by high winds and frigid temperatures, conditions that NASA spokesman Rob Navias described as reminiscent of the North Pole. Despite the harsh environment, the recovery teams swiftly retrieved the crew from the capsule. Kaleri, Kelly, and Skripochka were immediately placed in reclining chairs to facilitate their recovery after the demanding mission.

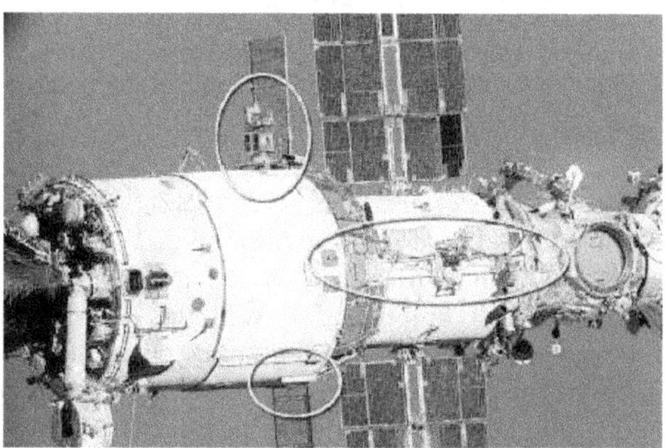

Russian EVA 26 worksites.

Multipurpose workstation on Zvezda module.

Kaleri, Skripochka and Kelly sit in chairs after they landed.

The Soyuz TMA-01M spacecraft as it lands with NASA astronaut Scott Kelly, Expedition 26 commander; and Russian cosmonauts Oleg Skripochka and Alexander Kaleri, both flight engineers, near the town of Arkalyk, Kazakhstan on March 16, 2011. Kelly, Skripochka and Kaleri are returning from almost six months onboard the International Space Station where they served as members of the Expedition 25 and 26 crews.

Expedition 26

Expedition 26 crew members - NASA astronaut Scott Kelly, commander; NASA astronaut Catherine Coleman, Russian cosmonauts Dmitry Kondratyev, Oleg Skripochka, Alexander Kaleri and European Space Agency (ESA) astronaut Paolo Nespoli, all flight engineers.

Expedition 26, a crucial chapter in the history of the International Space Station (ISS), was characterized by a series of high-profile missions and pivotal transitions that contributed to the station's ongoing success.

The expedition began with the launch of the Expedition 26 crew on December 15, 2010, aboard the Soyuz TMA-01M spacecraft. This crew, comprising NASA astronaut Scott Kelly, ESA astronaut Alexander Gerst, and Roscosmos cosmonaut Dmitry Kondratyev, embarked on their mission to join the ISS and carry out their scientific duties.

Their arrival at the ISS on December 17, 2010, marked the beginning of Expedition 26. This mission was marked by numerous significant events aimed at maintaining and enhancing the ISS's capabilities.

Among these events was the successful launch of the Progress 41 spacecraft on January 27, 2011, followed by its docking with the ISS on January 29, 2011. The Progress spacecraft delivered essential supplies and equipment to the station, ensuring the continued smooth operation of the station's systems.

The HTV-2 Kounotori, a Japanese cargo spacecraft, was another key player during this period. It launched on January 22, 2011, and docked with the ISS on January 27, 2011. The HTV-2 delivered additional supplies and scientific experiments to the station. Its docking was closely followed by its relocation on February 18, 2011, to optimize the station's cargo handling capabilities.

The Automated Transfer Vehicle (ATV-2), Johannes Kepler, launched on February 16, 2011, and docked with the ISS on February 24, 2011. This European spacecraft contributed further to the station's resupply operations, bringing substantial cargo, including scientific experiments and provisions.

STS-133 crew - NASA astronauts Steve Lindsey (center right) and Eric Boe (center left), commander and pilot, respectively; along with astronauts (from the left) Alvin Drew, Nicole Stott, Michael Barratt and Steve Bowen, all mission specialists.

During the mission, the crew conducted two significant spacewalks. The first, known as Russian EVA-27, took place on January 21, 2011. Russian EVA-28 followed this on February 16, 2011. Both spacewalks were carried out by cosmonauts Oleg Skripochka and Dmitri Kondratyev, contributing to critical maintenance and upgrades of the ISS's exterior.

Space Shuttle Discovery approaches the ISS on 26 February 2011.

The mission designated STS-133, initially set to launch on November 3, 2010, encountered a delay and was ultimately rescheduled to February 3, 2011. Despite this shift, the arrival of the Space Shuttle Discovery did not impede the progress of Expedition 25. The crew aboard the International Space Station (ISS) continued their crucial tasks, contributing significantly to the station's ongoing scientific and operational endeavors.

NASA officially announced the STS-133 crew on September 18, 2009, marking the beginning of their intensive training regimen in October of the same year. The original lineup included Commander Steven Lindsey, Pilot Eric Boe, and Mission Specialists Alvin Drew, Timothy Kopra, Michael Barratt, and Nicole Stott. However, in a late change, Stephen Bowen replaced Tim Kopra, who was unfortunately injured in a bicycle accident on January 19, 2011, just weeks before the launch.

Each crew member had prior spaceflight experience, except for Commander Lindsey. Notably, five crew members—Drew, Boe, Barratt, Stott, and Kopra—were part of NASA's Astronaut Group 18, selected in 2000. Lindsey, then serving as Chief of the Astronaut Office, relinquished this role to Peggy Whitson to lead the mission, highlighting the significance of STS-133. Remarkably, this mission was the first time that two astronauts—Nicole Stott and Michael Barratt—were already aboard the ISS when their crew assignment was announced, as they were part of Expedition 20.

Several notable firsts and historical markers distinguished STS-133. Alvin Drew became the last African-American astronaut to fly on the Space

Shuttle, as no African-Americans were among the crews of the subsequent STS-134 and STS-135 missions. Stephen Bowen, who had previously flown on Atlantis' STS-132 mission, set a precedent as the first NASA astronaut to be launched on two consecutive missions until Doug Hurley achieved this distinction in May 2020 with the Crew Dragon Demo-2 mission.

A central component of the STS-133 mission was the Permanent Multipurpose Module (PMM), named Leonardo in honor of the Italian Renaissance inventor Leonardo da Vinci. Leonardo's role was pivotal as one of the three Multi-Purpose Logistics Modules (MPLMs). Originally delivered to the Kennedy Space Center (KSC) in August 1998 and debuting on Space Shuttle Discovery during STS-102 in March 2001, Leonardo was affixed permanently to the ISS during this mission. Its primary function was to provide essential additional storage space, addressing the growing demands for scientific and operational support aboard the station.

Before its permanent installation, Leonardo underwent significant modifications. After returning from the STS-131 mission, it was reconfigured for its new role at the Space Station Processing Facility at KSC. This involved removing certain equipment to reduce its weight by 178.1 pounds and enhancing its protection with upgraded multi-layer insulation and micro-meteoroid orbital debris shielding. Additionally, a Planar Reflector was installed at the Japanese Space Agency's (JAXA) request to aid in the module's functionality. Upon docking with the ISS, the contents of Leonardo were redistributed within the station, and the module's launch hardware was deorbited and destroyed in Earth's atmosphere by JAXA's Kounotori 2 (HTV-2), marking the end of its journey in space.

The mission also delivered the Express Logistics Carrier 4 (ELC-4), a robust steel platform designed to support external payloads mounted to the space station's trusses. ELC-4, which weighs approximately 8,235 pounds, was positioned on the starboard 3 (S3) truss' lower inboard passive attachment system (PAS). It carried several Orbital Replacement Units (ORUs), including a Heat Rejection System Radiator (HRSR) Flight Support Equipment (FSE), which was an essential spare for the station's external thermal control system, and the ExPRESS Pallet Controller Avionics 4 (ExPCA #4).

Another significant addition to the ISS during STS-133 was Robonaut2 (R2), the first humanoid robot in space. Designed for both Earth-based and space applications, R2 was upgraded with materials and systems suitable for the harsh conditions of space. It features a nickel-plated carbon fiber and aluminum construction, weighs 300 pounds, and stands at 3 feet 3.7 inches tall with a shoulder width of 2 feet 7.4 inches. Equipped with 54 servo motors and 42 degrees of freedom, R2 was tasked with initial operational testing inside the Destiny module and holds the potential to assist astronauts in future spacewalks and scientific operations.

NASA's Robonaut 2, or R2 for short, R2 was flanked by NASA astronauts Tim Kopra and Nicole Stott, both mission specialists; Eric Boe, pilot; Michael Barratt and Alvin Drew, both mission specialists, and Steve Lindsey, commander.

STS-133 also carried the DragonEye sensor, developed by Advanced Scientific Concepts, Inc. This 3D Flash LiDAR sensor, mounted on Discovery's trajectory control system carrier assembly, provided a three-dimensional image of targets using pulsed laser technology. It was a crucial step in the development and testing of SpaceX's Dragon spacecraft, providing data to enhance the spacecraft's docking capabilities.

In addition to these technical components, STS-133 carried a collection of symbolic items, including the signatures of over 500,000 students who participated in the Student Signatures in Space program. The mission also transported various flags, bookmarks, patches, and personal mementos from the astronauts, reflecting the diverse and personal connections that space exploration fosters. These items included Lego Space Shuttles, medallions, and even a William Shakespeare "action figure," embodying the spirit and collaboration inherent in the space program.

On February 26, 2011, Flight Day 3 of the STS-133 mission marked a significant milestone as the Space Shuttle Discovery docked with the International Space Station (ISS) for the thirteenth time. The docking was executed precisely at 19:14 UTC. However, the hard mate, or the final docking procedure, was delayed by approximately 40 minutes due to relative motion between the shuttle and the station. This delay set the crew behind schedule for the day's tasks. The hatches were eventually opened at 21:16 UTC, and the Expedition 26 crew welcomed the incoming STS-133 astronauts. After a cordial welcome ceremony and a comprehensive safety briefing, the crew's main task was to transfer the ExPRESS Logistics Carrier 4 (ELC-4) to the ISS.

The ELC-4, housed in Discovery's payload bay, was carefully extracted using the Space Station Remote Manipulator System (SSRMS), commonly known as Canadarm2. Astronauts Nicole Stott and Michael Barratt conducted this operation. The SSRMS then handed off ELC-4 to the Shuttle Remote Manipulator System (SRMS), operated by astronauts Stephen Bowen and Alvin Drew. Due to clearance issues, the SRMS grappled the ELC-4 and positioned it onto the Mobile Base System (MBS). From there, the SSRMS reattached ELC-4 to its designated location on the S3 truss of the station. By 03:22 UTC on February 27, the ELC-4 was successfully installed. Meanwhile, Bowen and Lindsey busied themselves with preparing items needed for Flight Day 4 and the upcoming spacewalk on Flight Day 5.

On February 27, Flight Day 4, Stott and Barratt used Canadarm2 to remove the Orbiter Boom Sensor System (OBSS) from the starboard sill of Discovery's payload bay. Due to its size and the constraints of clearance, the OBSS was transferred to the SSRMS for further handling. This transfer was necessary to clear the payload bay for the removal of the Permanent Multipurpose Module (PMM). Later, the STS-133 crew, joined by ISS Expedition 26 commander Scott Kelly and flight engineer Paolo Nespoli, participated in a series of media interviews with various outlets including the Weather Channel and local stations in Boston, Atlanta, and Charlotte. Throughout the day, Drew and Bowen continued their preparations for the upcoming spacewalk. The evening was dedicated to the standard pre-spacewalk campout in the Quest airlock, where the airlock pressure was lowered to 10.2 psi to help the spacewalkers purge nitrogen from their blood and mitigate the risk of decompression sickness.

February 28, Flight Day 5, was marked by the first extra-vehicular activity (EVA) of the mission. Astronauts Steve Bowen and Alvin Drew initiated their spacewalk at 10:46 EST. Their tasks included installing a power cable between the Unity and Tranquility modules to ensure a contingency power source, relocating a failed ammonia pump module to the External Stowage Platform 2, and conducting several other minor tasks. Despite delays caused by technical issues with the robotic control station in the Cupola module, the EVA concluded after six hours and 34 minutes at 17:20 EST.

The following day, March 1, Flight Day 6, focused on the installation of the Leonardo Permanent Multipurpose Module (PMM) to the nadir port of the Unity module. With the PMM securely in place, efforts shifted to its external outfitting to integrate it fully into the ISS. Bowen and Drew reviewed the mission's second spacewalk procedures and prepared for their pre-EVA campout in the Quest airlock.

On March 2, Flight Day 7, the crew conducted their second EVA. Drew removed thermal insulation from a platform, while Bowen replaced an attachment bracket on the Columbus module and installed a camera assembly on the Dextre robot. Other tasks included repairing insulation and installing a light on a cargo cart. Meanwhile, the ISS and shuttle crews began the internal outfitting of the Leonardo PMM.

March 3, Flight Day 8, was dedicated to transferring cargo into the Leonardo PMM and afforded the crew some off-duty time. On March 4, Flight Day 9, equipment from the spacewalks was reconfigured, and a joint crew news conference was held via satellite. The day also gave additional off-duty time to the astronauts.

Flight Day 10, March 5, continued with the internal outfitting of the Leonardo PMM. Although a photo shoot of the ISS with multiple spacecraft was considered, it was ultimately not pursued. On March 6, Flight Day 11, the crew continued outfitting the PMM and performed a checkout of Discovery's rendezvous tools. The shuttle crew then bid farewell to their ISS counterparts, exited the station, and sealed the hatch.

On March 7, Flight Day 12, Discovery undocked from the ISS for the final time, followed by a fly-around and a late inspection of the shuttle's Thermal Protection System using the OBSS. The OBSS was then berthed for the final time.

On the final day, March 9, Flight Day 14, the crew conducted final deorbit preparations and closed the shuttle's payload bay doors. Discovery completed a successful deorbit burn and re-entry, landing at Kennedy Space Center's Shuttle Landing Facility at 11:58:14 EST. This landing marked the end of Discovery's storied career and was the final shuttle landing in daylight, as subsequent missions would land at night.

In 2011, significant developments in global space exploration continued with new astronaut selections and advancements across different programs and initiatives.

In January and February, Roscosmos, the Russian space agency, enrolled a united squad of astronauts, marking a consolidation of talent within the Russian space program. This group included Oleg Artemyev, Andrei Babkin, Ivan Vagner, Andrei Borisenko, Sergei Zhukov, Oleg Kononenko, Mikhail Kornienko, Sergey Kud-Sverchkov, Svyatoslav Morozov, Sergei Revin, Sergey Ryazansky, Yelena Serova, and Nikolai Tikhonov. This move aimed to streamline training and preparation processes at the Y. A. Gagarin Cosmonaut Training Center, emphasizing Russia's commitment to maintaining and expanding its human spaceflight capabilities.

On February 28th, the Association of Spaceflight Professionals announced Group 3, comprising Christopher Altman, Jon-Erik Dahlin, Melania Guerra, Mindy Howard, Kris Lehnhardt, Abhishek Tripathi, Cosan Unuvar, Pavel Zagadailov, and Luis Zea. This group highlighted the growing diversity and specialization within the commercial space sector, focusing on scientific research, technology development, and expanding human presence beyond Earth's orbit.

In October, Virgin Galactic's Astronaut Pilots Group welcomed Keith Colmer, reinforcing Virgin Galactic's role in advancing private sector involvement in space exploration, paving the way for future missions and space tourism endeavors.

In 2011, the realm of space exploration witnessed significant developments and operations involving two prominent space stations: the International Space Station (ISS) and the Chinese Tiangong-1. Throughout this pivotal year, the ISS continued to serve as a hub for international collaboration and scientific advancement. The station hosted a succession of expeditions, with each crew contributing substantially to research and station upkeep.

Meanwhile, on September 29, 2011, China achieved a notable milestone in its space program with the successful launch of the Tiangong-1 space station. This development marked China's significant progress in space technology, particularly in demonstrating docking capabilities and the ability to conduct experiments in space. The Tiangong-1 station received two unmanned spacecraft: the Tiangong-1 target vehicle and the Shenzhou 8 spacecraft. In a notable achievement, the Shenzhou 8 docked with Tiangong-1, and the two vehicles remained connected for twelve days. During this period, the crew conducted a series of experiments, furthering China's capabilities in space operations.

Notable events in spaceflight history also marked the year 2011. In July, NASA retired the Space Shuttle program, concluding its era with the final flight of the shuttle. Concurrently, China's space ambitions were underscored by the launch of its first space station module, Tiangong-1. Globally, there were 84 orbital launches, achieving a success rate of 78. Russia, China, and the United States led the way, with 35, 19, and 18 launches, respectively. Remarkably, 2011 was the first year China outperformed the United States in successful launches. Additionally, seven crewed missions were launched, transporting 28 astronauts to the ISS.

2011 also saw notable advancements in rocket technology. The maiden flights of the Zenit-3F, Long March 2F/T, and Long March 2F/G rockets were significant, while the Delta II Heavy completed its final launch.

In addition to these space missions, in 2011, DARPA and the US Navy flew the Northrop Grumman X-47C UCAV, a significant advancement in unmanned aerial vehicle technology for naval use.

Expedition 27

Expedition 27 crew from the right are Russian cosmonaut Dmitry Kondratyev, commander; Russian cosmonaut Andre Borisenko, NASA astronaut Catherine Coleman, Russian cosmonaut Alexander Samokutyayev, European Space Agency (ESA) astronaut Paolo Nespoli and NASA astronaut Ron Garan, all flight engineers.

Expedition 27 of the International Space Station (ISS) commenced on March 16, 2011, with the new crew's arrival, following a series of essential activities and milestones. Meticulous preparations and key spacecraft operations characterized the initial phase of Expedition 27. On March 28, 2011, the Kounotori 2 spacecraft, also known as HTV-2, was undocked from the ISS's Harmony module using the Canadarm2 robotic arm. This operation, which occurred at 15:46 UTC, marked the end of the spacecraft's resupply mission. Kounotori 2, Japan's second H-II Transfer Vehicle, had been a crucial component in maintaining the station's supply chain since its launch in January 2011. The spacecraft reentered Earth's atmosphere on March 30 at approximately 03:09 UTC, concluding its mission.

The arrival of the Soyuz TMA-21 spacecraft on April 6, 2011, marked a significant moment for Expedition 27. Commanded by Russian astronaut Aleksandr Samokutyayev, Soyuz TMA-21 carried two fellow crew members: Flight Engineer Andrei Borisenko from Russia and NASA astronaut Ronald J. Garan. The spacecraft lifted off smoothly, with Samokutyayev in the center seat, Borisenko to his left, and Garan to his right. The mission's timing was significant as it commemorated the milestone of human spaceflight that began with Gagarin's journey on April 12, 1961.

At the Baikonur Cosmodrome in Kazakhstan, NASA astronaut Ron Garan (left), Expedition 27 flight engineer; along with Russian cosmonauts Alexander Samokutyaev (center), Soyuz commander; and Andrey Borisenko, flight engineer outside their Soyuz TMA-21 spacecraft dubbed "Gagarin" and bearing the likeness of cosmonaut Yuri Gagarin, the first human in space, scheduled for launch on April 5 (April 4, U.S. time), just one week shy of the 50th anniversary of Gagarin's historic journey into space from the same launch pad that the Expedition 27 crew began their mission from.

Upon reaching orbit, Soyuz TMA-21 docked with the International Space Station (ISS) on April 6, 2011, at 23:09 UTC. The docking took place at the Poisk module while the spacecraft was positioned over the Andes Mountains in Chile. The hatches between Soyuz TMA-21 and the ISS opened at 2:13 UTC on April 7, allowing the crew to float into the station. They were greeted by Expedition 27 Commander Dmitry Kondratyev and Flight Engineers Catherine Coleman and Paolo Nespoli, who conducted a crew welcome ceremony and a mandatory safety orientation.

Soyuz TMA-21's mission extended beyond its docking; the spacecraft served as a lifeboat throughout the remainder of Expedition 27 and continued to provide critical support during Expedition 28. The return journey, initially scheduled for September 8, 2011, faced a delay due to the crash of the Progress M-12M cargo spacecraft on August 24. Consequently, Soyuz TMA-21 undocked from the ISS on September 16, 2011, at 00:38 UTC.

The descent was not without its challenges. Shortly after the deorbit burn, the mission experienced a communications malfunction, resulting in a brief loss of voice contact with the crew. Despite the tense moments on the ground, the reentry and landing proceeded without further complications. The descent module safely touched down in central Kazakhstan at 03:59 UTC on September 16.

Russian search and rescue teams, accompanied by NASA flight surgeons and space station program managers, were prepared for the recovery operation. A substantial team, including three Antonov airplanes, 14 Mil Mi-8 helicopters, and seven rescue vehicles, participated in locating the capsule. Commander Samokutyayev was extracted first, followed by Garan and Borisenko. All three astronauts appeared in good health and were seen resting in recliners near the descent module.

Following medical evaluations conducted in a nearby tent, the crew was transported to Karaganda for an official welcome home ceremony. Samokutyayev and Borisenko continued their journey to Star City near Moscow. At the same time, Garan boarded a NASA jet to return to the Johnson Space Center in Houston, concluding the successful Soyuz TMA-21 mission.

The transition to Expedition 27 saw the ISS Progress M-09M cargo spacecraft undock from the Pirs module on April 22, 2011. Since January 28, 2011, Progress M-09 M has played a crucial role in station operations, providing vital supplies, including dry cargo, propellant, oxygen, and water. After its undocking, the spacecraft was repurposed for scientific experiments before being deorbited and reentering over the South Pacific Ocean.

The arrival of the Progress M-10M cargo spacecraft on April 29, 2011, further underscored the expedition's significance. Two days earlier, Progress M-10M had launched autonomously from Baikonur and docked with the Pirs module at 14:19 UTC, delivering essential supplies and equipment to support ongoing missions.

A major highlight of Expedition 27 was the docking of the Space Shuttle Endeavour on May 18, 2011, as part of mission STS-134.

STS-134, designated as the sixth assembly flight (ULF6) for the International Space Station (ISS), marked a pivotal moment in the history of NASA's Space Shuttle program. As the penultimate mission, it was the 25th and final spaceflight of the Space Shuttle Endeavour. This mission's primary objectives included delivering the Alpha Magnetic Spectrometer (AMS-02), a cutting-edge particle physics experiment designed to search for dark matter and antimatter, as well as an ExPRESS Logistics Carrier (ELC) to the ISS.

The mission, commanded by Mark Kelly, was initially slated to be the final Space Shuttle flight unless Congress approved funding for an additional mission, STS-135. However, by February 2011, NASA confirmed that STS-135 would proceed regardless of funding, ensuring the Shuttle program's legacy would extend a bit further. STS-135, flown by Space Shuttle Atlantis, was prepared as a contingency mission in the event that the STS-134 crew became stranded in orbit. This preparation was part of the Launch on Need (LON) strategy, designed to safeguard the Shuttle program and its astronauts.

The mission's launch was delayed by issues with the main payload, AMS-02, as well as scheduling setbacks related to STS-133. The initial launch attempt on April 29, 2011, was aborted due to technical difficulties involving two heaters in one of the orbiter's auxiliary power units (APU). Following these setbacks, Endeavour successfully launched on May 16, 2011, at 08:56:28 EDT (12:56:28 UTC). The Shuttle touched down for the final time on June 1, 2011, completing its distinguished service.

The crew for STS-134 was announced by NASA on August 10, 2009. The team was composed of experienced astronauts, each bringing their expertise to the mission:

Commander Mark Kelly, representing the United States, led his fourth and final spaceflight with this mission.

Pilot Gregory H. Johnson, also from the United States, completed his second and final spaceflight.

Mission Specialist 1 Michael Fincke, a seasoned astronaut from the United States, embarked on his third spaceflight.

Mission Specialist 2 and Flight Engineer Roberto Vittori of Italy, representing the European Space Agency (ESA), undertook his third and final spaceflight with STS-134. Vittori was the last non-U.S. astronaut to fly with the Shuttle.

Mission Specialist 3 Andrew J. Feustel, from the United States, was on his second spaceflight.

Mission Specialist 4 Gregory Chamitoff, also from the United States, completed his second and final spaceflight with this mission.

STS-134 crew portrait. Pictured clockwise are NASA astronauts Mark Kelly (bottom center), commander; Gregory H. Johnson, pilot; Michael Fincke, Greg Chamitoff, Andrew Feustel and European Space Agency's Roberto Vittori, all mission specialists.

In light of the tragic shooting in Tucson on January 13, 2011, which critically wounded Congresswoman Gabrielle Giffords, Mark Kelly's wife, NASA appointed Frederick W. Sturckow as backup commander for STS-134 to ensure the mission's preparedness under all circumstances.

STS-134 stands out not only as the concluding chapter of Endeavour's storied career but also as the final Space Shuttle mission with a six-person, all-male crew. The subsequent mission, STS-135, would introduce the first female crew member since the Shuttle program's inception, marking a significant milestone in the Shuttle's final flights.

The Space Shuttle Program faced a pivotal moment in its history as the retirement of the shuttle fleet approached, initially scheduled to culminate with the completion of STS-133. However, controversy emerged surrounding the cancellation of key International Space Station (ISS) components, notably the Alpha Magnetic Spectrometer (AMS), which was intended to be a crucial addition to the station's scientific capabilities. This controversy led to substantial debate over the implications of the shuttle's retirement and its impact on the ISS's scientific objectives.

On June 19, 2008, the U.S. House of Representatives addressed this issue by passing the NASA Authorization Act of 2008. This legislation provided NASA with funding for an additional Space Shuttle mission aimed specifically at delivering critical science experiments to the ISS. The Senate followed suit with similar provisions in their version of the NASA Authorization Act, which was unanimously approved by the Senate Committee on Commerce, Science, and Transportation on June 25, 2008. The full Senate passed the amended version on September 25, 2008, and the House approved it two days later. President George W. Bush signed the bill into law on October 15, 2008. Although initially resistant to additional shuttle missions due to concerns about delaying the transition to Project Constellation, Bush's administration ultimately acquiesced.

With the new mission authorized, the Obama Administration included funds for STS-134 in its proposed 2010 NASA budget, officially integrating the mission into the shuttle's final flight schedule. STS-134 was initially slated to be the last scheduled mission of the Space Shuttle Program. However, passing an appropriations bill in 2011, which converted the backup STS-335 mission to STS-135, extended the shuttle's operations beyond STS-134.

STS-134's timeline was also affected by delays in the shuttle's launch schedule, which pushed the mission past Expedition 26. Had STS-134 launched as originally planned during Expedition 26, it would have created a unique milestone: astronauts Mark Kelly and his twin brother, Scott Kelly, would have become the first siblings and twins to fly in space simultaneously.

On May 16, 2011, the anticipated launch of STS-134 was a significant event not only for the space community but also for Mark Kelly's wife, U.S. Representative Gabby Giffords. Giffords, who had been recovering in Houston from a severe injury sustained during a shooting in Tucson in January 2011, traveled to the Kennedy Space Center (KSC) to witness the launch attempt. The launch was described by The New York Times as "one of the most anticipated in years."

Adding to the mission's high profile, President Barack Obama scheduled a visit to Kennedy Space Center on April 29, 2011, to view the launch. Despite the launch being postponed, Obama toured the Orbiter Processing Facility at Launch Complex 39 and met with Giffords and the six-member crew of STS-134. This visit underscored the mission's importance and the broader significance of the Space Shuttle Program's final flights.

STS-134 and Expedition 27 crew members in the Harmony node of the International Space Station shortly after space shuttle Endeavour and the space station docked in space and the hatches were opened. Pictured from the left are European Space Agency astronaut Roberto Vittori, STS-134 mission specialist; NASA astronaut Mark Kelly, STS-134 commander; Russian cosmonaut Dmitry Kondratyev, Expedition 27 commander; NASA astronaut Cady Coleman, Expedition 27 flight engineer; and NASA astronaut Greg H. Johnson, STS-134 pilot.

The STS-134 mission of the Space Shuttle Endeavour was a significant undertaking, marking a critical juncture in the Shuttle Program. Its goal was to deliver and install essential components to the International Space Station (ISS).

Among the key payloads was the Alpha Magnetic Spectrometer 2 (AMS-02), a sophisticated particle physics detector designed to advance our understanding of fundamental cosmic phenomena. Housed in Endeavour's payload bay, the AMS-02 was mounted on the ISS's S3 truss segment. This instrument, featuring a large permanent magnet, was tasked with searching for antimatter and investigating the origins and structure of dark matter. Originally designed with a cryogenic superconducting magnet system to achieve the high sensitivity necessary for its mission, the AMS-02 experienced unexpected issues with anomalous heating. As a result, Samuel C. C. Ting, the experiment leader, opted to use the permanent magnet from the earlier AMS-01 model, a decision that maintained the mission's scientific integrity.

Another critical component carried by STS-134 was the ExPRESS Logistics Carrier 3 (ELC-3), which accommodated several Orbital Replacement Units (ORUs). These included a High-Pressure Gas Tank (HPGT), an Ammonia

Tank Assembly (ATA), multiple S-band Antenna Sub-System Assemblies (SASA), and a Special-Purpose Dextrous Manipulator (SPDM) Arm. The ELC-3 also carried a Space Test Program payload and a spare ELC pallet controller avionics box, which were crucial for maintaining and enhancing the ISS's operational capabilities.

The mission also delivered the Materials on International Space Station Experiment 8 (MISSE-8), returning the completed MISSE-7 experiments to Earth. MISSE-7 had been sent to the ISS on STS-129 in 2009, and these experiments were critical for studying material degradation in space.

A notable part of STS-134 was the Orion Rel-nav Sensor Test. The sensor, mounted on the Orbiter Docking System (ODS) and an Adaptive Payload Carrier, was designed to test its performance during the shuttle's rendezvous and proximity operations with the ISS. This included approach, docking, and separation maneuvers, providing valuable data for future missions.

The Glacier Freezer Module was another payload carried by STS-134. This module, along with two older Glacier units returned to Earth, was used to store and return scientific samples from space.

Additionally, STS-134 left behind the Orbiter Boom Sensor System (OBSS) on the ISS. The OBSS was a valuable tool for accessing areas on the station that the Canadarm2 could not reach. Its utility was demonstrated during previous missions, including STS-120 and STS-123, where it proved effective in maintaining and repairing the ISS.

The mission also featured an educational outreach component, with 13 Lego kits delivered to the ISS. Astronauts built Lego models in microgravity as part of the Lego Bricks in Space program, and the results were shared with schools to inspire students in the fields of science, technology, engineering, and mathematics (STEM).

Among other payloads was the Shuttle LIFE precursor mission, a Planetary Society project designed to test technology for a future Mars mission. The mission was deemed fully successful during its stage rehearsal.

Finally, STS-134 included STEM Bars, specialized nutrition bars developed by high school students Mikayla and Shannon Diesch. These bars were crafted to meet NASA's nutritional standards for spaceflight and were part of an outreach effort to promote STEM education.

A figurine of the Little Mole, a beloved character in Czech culture, was also returned to Earth by astronaut Andrew Feustel. The figurine was later presented to its creator, Zdeněk Miler, and used to promote space science education, particularly in the Czech Republic.

The final mission of the Space Shuttle Endeavour, designated STS-134, commenced with a series of significant experiments and achievements. During this mission, Endeavour conducted four Department of Defense payloads of opportunity: MAUI, SEITI, RAMBO-2, and SIMPLEX. Each of these experiments necessitated the use of the shuttle's engines and thrusters and was contingent on the availability of sufficient propellant on board.

This historic flight was notable for several key milestones. It marked the 165th NASA crewed space flight and was the 134th Shuttle mission since the inception of STS-1. Endeavour's mission was particularly significant as it was the 25th and final flight of the shuttle. The mission also represented the 36th Shuttle mission to the International Space Station (ISS), as well as the 109th post-Challenger mission and the 21st post-Columbia mission. Notably, it featured the last non-U.S. astronaut on a Space Shuttle mission, Colonel Roberto Vittori from Italy. Additionally, it was the first instance of a Papal blessing and call to astronauts in space.

Endeavour launched from Kennedy Space Center's Launch Complex 39 at 08:56 EDT on May 16, 2011. The mission progressed with the shuttle's docking to the ISS on May 18, 2011, at 10:14 UTC. The docking involved the shuttle approaching and connecting to the Pressurized Mating Adapter (PMA) 2 on the station. Following the docking, the six astronauts aboard STS-134 joined the Expedition 27 crew aboard the ISS. The crews performed essential leak checks and opened the hatches at 11:38 UTC, marking the beginning of their collaboration on the station.

ISS Starboard truss with the newly installed AMS-02

The initial tasks included the unberthing and installation of the Express Logistics Carrier (ELC) 3, which was placed on the ISS's Port 3 (P3) truss segment. The ELC 3 was maneuvered out of Endeavour's payload bay using the Shuttle Remote Manipulator System (SRMS) and then handed over to the Space Station Remote Manipulator System (SSRMS), also known as Canadarm2. This critical installation was completed at 16:18 UTC. Additionally, astronaut Mark Kelly began transferring oxygen from Endeavour to the ISS, while Mike Fincke and Drew Feustel prepared the Extravehicular Mobility Units (EMUs) for the mission's scheduled spacewalks.

On May 19, during flight day 4, the crew installed the Alpha Magnetic Spectrometer-2 (AMS-2). AMS-2 was lifted from Endeavour's payload bay by the Canadarm, operated by Drew Feustel and Roberto Vittori, and transferred to Canadarm2, operated by Greg Chamitoff and Greg Johnson. The installation on the S3 truss segment was completed at 09:46 UTC. This milestone marked the completion of the U.S. Orbital Segment of the ISS. The crew also prepared for upcoming spacewalks by setting up the EMUs and tools required for Extravehicular Activities (EVAs). The day concluded with an EVA procedures review, followed by a campout in the Quest Airlock to minimize decompression sickness risks for the astronauts.

The first spacewalk, conducted on May 20, flight day 5, saw Drew Feustel and Greg Chamitoff engaging in several critical tasks. They installed a new set of Materials International Space Station Experiments (MISSE) and began installing a wireless video system. However, the spacewalk was interrupted by a CO2 sensor malfunction in Chamitoff's suit, leading to task adjustments. The astronauts installed an ammonia jumper between the Port 3 and Port 6 truss segments and made additional installations on the Crew Equipment Translation Aid (CETA) cart and the starboard Solar Alpha Rotary Joint (SARJ). The spacewalk, which lasted 6 hours and 19 minutes, was a precursor to further equipment transfers between Endeavour and the ISS.

On May 21, flight day 6, Endeavour's crew conducted a focused inspection of the orbiter's thermal protection tiles, which had sustained damage during launch. The inspection, using the Orbiter Boom Sensor System (OBSS), was carried out to ensure re-entry safety. The inspection, performed by the shuttle's Canadarm and controlled by pilot Greg Johnson and mission specialists Mike Fincke and Roberto Vittori, confirmed the Thermal Protection System (TPS) was intact and cleared for entry. This day also featured a historic event as Pope Benedict XVI spoke with the crew and offered a blessing, marking the first Papal communication with astronauts in space.

May 22, flight day 7, was highlighted by the second EVA of the mission. Drew Feustel and Mike Fincke conducted this spacewalk, which became the sixth-longest in spaceflight history at the time, lasting 8 hours and 7 minutes. Their tasks included transferring ammonia, lubricating the Solar Alpha Rotary Joint, and installing a stowage beam. During the spacewalk, some thermal blanket bolts were lost. While the EVA was in progress, the rest of the crew continued with equipment transfers and the Expedition 27 crew prepared for their departure. This day also saw the ceremonial change of command on the ISS, with Dmitri Kondratyev handing over command to Andrei Borisenko.

On May 23, flight day 8, Endeavour's crew had some off-duty time. Mark Kelly and Mike Fincke conducted an in-flight interview with students from Mesa Verde Elementary School, and Roberto Vittori and Paolo Nespoli answered questions from Italian President Giorgio Napolitano. The day concluded with the departure of Expedition 27 commander Dmitri Kondratyev, flight engineers Paolo Nespoli and Cady Coleman aboard Soyuz TMA-20, marking the transition to Expedition 28.

This image of the International Space Station and the docked space shuttle Endeavour, flying at an altitude of approximately 220 miles, was taken by Expedition 27 crew member Paolo Nespoli from the Soyuz TMA-20 following its undocking on May 23, 2011 (USA time). The pictures are the first taken of a shuttle docked to the International Space Station from the perspective of a Russian Soyuz spacecraft. Onboard the Soyuz were Russian cosmonaut and Expedition 27 commander Dmitry Kondratyev; Nespoli, a European Space Agency astronaut; and NASA astronaut Cady Coleman. Coleman and Nespoli were both flight engineers. The three landed in Kazakhstan later that day, completing 159 days in space.

The Soyuz TMA-21 spacecraft on the launch pad at the Baikonur Cosmodrome in Kazakhstan.

Expedition 27 crew members Russian cosmonaut Dmitry Kondratyev (bottom center), commander; NASA astronaut Cady Coleman, Russian cosmonauts Alexander Samokutyaev (center) and Andrey Borisenko (top left), NASA astronaut Ron Garan and European Space Agency astronaut Paolo Nespoli (right), in the Zvezda Service Module honoring of the 50th anniversary of the spaceflight of Yuri Gagarin, the first human launched in space

Expedition 28

Expedition 28 crew from the right (front row) are Russian cosmonaut Andre Borisenko, commander; Russian cosmonaut Alexander Samokutyaev and NASA astronaut Mike Fossum, both flight engineers. Pictured from the left (back row) are Japan Aerospace Exploration Agency (JAXA) astronaut Satoshi Furukawa, NASA astronaut Ron Garan and Russian cosmonaut Sergei Volkov, all flight engineers.

On May 23, 2011, the undocking of Soyuz TMA-20 marked the conclusion of Expedition 27 and the commencement of Expedition 28 aboard the International Space Station (ISS). At 21:35 UTC, Soyuz TMA-20 departed, carrying the outgoing crew members Dmitri Kondratyev, Catherine Coleman, and Paolo Nespoli. As the spacecraft exited, it performed a fly-around of the ISS, capturing detailed photographs of the station's exterior. This maneuver provided a valuable opportunity to document the Space Shuttle Endeavour, which was docked to the ISS at the time on its final mission, STS-134. The departing crew landed safely in Kazakhstan at 02:27 UTC on May 24, 2011, concluding their mission and leaving the station in the capable hands of the new Expedition 28 crew.

The transition to Expedition 28 was facilitated by the arrival of the Soyuz TMA-21 spacecraft, which docked with the ISS on April 4, 2011. This initial crew, consisting of Commander Andrey Babkin from the Russian space agency Roscosmos and Flight Engineers Ron Garan from NASA and Alexander Samokutyayev from Roscosmos, set the stage for ongoing scientific research and station maintenance.

Further bolstering Expedition 28's crew, the Soyuz TMA-02M spacecraft launched from the Baikonur Cosmodrome in Kazakhstan on June 7, 2011. This mission aimed to bring additional personnel to the ISS, ensuring the continuation of critical scientific and operational tasks. By June 9, 2011, Soyuz TMA-02M had successfully docked with the ISS, adding to the Expedition 28 team and reinforcing the station's operational capacity. This crew's arrival was pivotal for maintaining the ISS's functionality and advancing its scientific endeavors.

The Soyuz TMA-02M prime and backup crews conduct their ceremonial tour on 16 May 2011. From left to right are NASA astronaut Mike Fossum, prime crew flight engineer; prime Soyuz commander Sergei Volkov, Russian cosmonaut; Japan Aerospace astronaut Satoshi Furukawa, prime crew flight engineer, NASA astronaut Don Pettit, backup flight engineer; backup Soyuz commander Oleg Kononenko, Russian cosmonaut; and European Space Agency astronaut Andre Kuipers, backup flight engineer.

STS-135, the final mission of the American Space Shuttle program, represented a significant moment in spaceflight history. Launched on July 8, 2011, and concluding on July 21, 2011, STS-135 marked the end of an era with the orbiter Atlantis completing its final flight. This mission utilized hardware originally intended for the contingency mission STS-335, which had not flown. Despite the mission's critical nature, it faced funding uncertainties, initially lacking appropriations in NASA's budget, raising doubts about its execution.

On January 20, 2011, program managers officially designated the mission as STS-135, replacing the STS-335 mission on the flight manifest. This change allowed for necessary training and preparations. By February 13, 2011, NASA had committed to proceeding with STS-135, irrespective of the funding challenges, through a continuing resolution. Until this point, the mission had not been publicly documented.

During a speech at the Marshall Space Flight Center on November 16, 2010, NASA Administrator Charles Bolden emphasized the importance of STS-135. He indicated that the shuttle mission was essential due to potential delays in developing commercial rockets and spacecraft intended for cargo transport to the International Space Station (ISS). Bolden noted that the mission would mitigate risks associated with these delays, stating the need for a third shuttle mission in addition to STS-133 and STS-134.

STS-135 was included in NASA's 2011 authorization, signed into law on October 11, 2010. However, its funding depended on a subsequent appropriations bill. By April 2011, the federal budget allocated $5.5 billion to NASA's space operations, including shuttle and space station programs, resolving funding concerns for the mission.

The mission's primary cargo comprised the Multi-Purpose Logistics Module (MPLM) Raffaello and a Lightweight Multi-Purpose Carrier (LMC), both of which were delivered to the ISS. Notably, Raffaello was the only MPLM flown by Atlantis. The mission crew, the smallest of any shuttle mission since STS-6 in April 1983, included only four astronauts. The absence of other shuttles for potential rescue missions necessitated this reduced crew size as Discovery and Endeavour had been retired. In the event of serious damage to the shuttle, the crew would have relocated to the ISS and returned to Earth using Russian Soyuz capsules over a year. Consequently, all STS-135 crew members were equipped with custom-fitted Russian Sokol space suits and Soyuz seat liners.

The crew of STS-135 consisted of Commander Christopher Ferguson, Pilot Douglas Hurley, Mission Specialist Sandra Magnus, and Mission Specialist/Flight Engineer Rex Walheim. This mission was the third and final spaceflight for each crew member. With a mission duration of 12 days, 18 hours, 28 minutes, and 50 seconds, the crew traveled a total distance of 8,505,161 kilometers (5,284,862 miles). Their journey not only concluded the shuttle program but also underscored the transition towards future space exploration endeavors.

STS-135 NASA astronauts Chris Ferguson (center right), commander; Doug Hurley (center left), pilot; Rex Walheim and Sandy Magnus, both mission specialists.

On July 8, 2011, Space Shuttle Atlantis embarked on its final mission, STS-135, the 135th and concluding flight of the American Space Shuttle program. Atlantis docked with the International Space Station (ISS) at 15:07 UTC, as the two spacecraft orbited 220 miles (350 km) over the South Pacific Ocean east of New Zealand. This docking, facilitated by Atlantis's 19th visit to a Space Station, was confirmed by Pilot Douglas Hurley with the transmission, "Houston, station, Atlantis, capture confirmed and we see free drift." In response, ISS Commander Ron Garan welcomed Atlantis with the phrase, "Welcome to the International Space Station for the last time," following the ceremonial ringing of the station's bell.

After docking, the crew conducted leak checks on both sides of the hatches, which were opened at 16:47 UTC. The astronauts then floated into the Harmony module of the ISS at 16:55 UTC, greeted by the station's crew and received a safety briefing.

On July 11, the fourth day of the mission, the primary objective was the installation of the Raffaello Multi-Purpose Logistics Module (MPLM) on the nadir port of the Harmony module. The crew's day began with the song "Tubthumping" by Chumbawamba playing as their wake-up call. Mission Specialist Sandra Magnus and Pilot Doug Hurley used the Canadarm2 to remove the Raffaello module from Atlantis's

payload bay at 09:09 UTC and successfully installed it at 10:46 UTC. Following leak checks, the hatches between Raffaello and the ISS were opened before noon. The spacewalk associated with this task, completed at 19:53 UTC, was notable as the 160th spacewalk in support of ISS assembly and maintenance and the 249th by U.S. astronauts. The crew began transferring cargo from the Raffaello MPLM, which carried 9,403 pounds of supplies with plans to return 5,666 pounds upon Atlantis's return. The cargo was intended to sustain the ISS through 2012.

On June 20, 2011, the European Space Agency's Johannes Kepler Automated Transfer Vehicle (ATV) undocked from the International Space Station (ISS) after a successful mission that began in February 2011. The ATV's departure marked the end of its operational period, and the following day, on June 21, 2011, it was deorbited. As the spacecraft re-entered Earth's atmosphere, it burned up over the southern Pacific Ocean at approximately 22:44 CET, completing its mission of resupplying and supporting the ISS.

The ISS soon welcomed a new cargo shipment with the arrival of the Russian Progress M-11M spacecraft, also known as Progress 42 by NASA. Launched on June 21, 2011, this cargo ship carried over 2.5 tons of essential supplies, including food, water, scientific equipment, and propellant. The Progress M-11M successfully docked with the Zvezda service module of the ISS on June 23, 2011, at 16:37 UTC, approximately 245 miles above eastern Kazakhstan.

Seven astronauts – six from NASA and one from the Japan Aerospace Exploration Agency (JAXA) – and three Russian cosmonauts participate on July 14, 2011 in a special meal on the Space Shuttle Atlantis' middeck. One of the final meals shared between shuttle and station crews has been called "The All-American Meal." The STS-135 crew consists of NASA astronauts Chris Ferguson, Doug Hurley, Sandy Magnus and Rex Walheim; the Expedition 28 or station crew members are JAXA astronaut Satoshi Furukawa, NASA astronauts Ron Garan and Mike Fossum, and Russian cosmonauts Andrey Borisenko, Alexander Samokutyaev and Sergei Volkov.

On July 19, Flight Day 12, Atlantis undocked from the ISS, marking the end of shuttle visits to the station. Undocking occurred at 06:28 UTC as Atlantis and the ISS moved through orbital night above the Pacific Ocean. Following the undocking, Flight Engineer Ron Garan rang the station's bell and announced, "Atlantis, departing the International Space Station for the last time."

Atlantis then moved to a station-keeping point approximately 600 feet (180 meters) ahead of the ISS. The crew conducted a half-lap flyaround, capturing images and video of the station's areas not previously documented. This maneuver, which began around 07:30 UTC and lasted approximately 25 minutes, provided valuable information on the ISS's condition.

As the mission concluded, shuttle and station flight control teams in Houston completed their final shifts. Commander Christopher Ferguson and the control room teams expressed their mutual appreciation for the collaborative effort. Ferguson emphasized the importance of making memories, while CAPCOM Daniel Tani reflected on the shuttle program's significance in the ISS's development.

At the end of the flyaround, Atlantis executed two Terminal Initiation (TI) separation burns, with the second occurring at 08:18 UTC, to distance itself from the ISS. This marked a poignant end to the Space Shuttle program's critical role in the assembly and maintenance of the International Space Station.

On July 12, 2011, the International Space Station (ISS) was a hive of activity as astronauts Michael Fossum and Ronald Garan undertook a crucial spacewalk that marked the start of Expedition 28. This mission was pivotal for the ISS's scientific research and operational capacity.

The spacewalk began precisely at 13:22 UTC when Fossum and Garan emerged from the airlock, stepping into the vastness of space. Their primary task was the installation of the Alpha Magnetic Spectrometer (AMS-02), a sophisticated particle physics experiment designed to explore the enigmatic realms of antimatter and dark matter. The AMS-02, an advanced instrument poised to unravel cosmic mysteries, was to be mounted on the ISS's S3 truss. As the astronauts floated against the backdrop of Earth, they meticulously secured the AMS-02, a task that would significantly enhance the station's ability to study cosmic rays and fundamental particles.

In addition to this landmark installation, the spacewalk included critical maintenance work on the station's cooling system. Fossum and Garan addressed several essential tasks, including removing and replacing thermal covers, ensuring the temperature control system functioned optimally. Their precision and attention to detail underscored the rigorous planning required to maintain the ISS's operational integrity.

Throughout the extravehicular activity (EVA), Fossum and Garan conducted safety checks and documented their progress with photographs. These images captured the details of the AMS-02 installation and the condition of the station's exterior, aiding in the ongoing assessment and maintenance of the ISS.

As the sun set on July 12, the spacewalk concluded at 19:53 UTC. The astronauts returned to the safety of the ISS, their mission accomplished and their tasks completed with commendable efficiency.

Less than a month later, on August 3, 2011, Russian cosmonauts Sergey Volkov and Aleksandr

Samokutyayev embarked on the 29th Russian spacewalk, Russian EVA #29. This EVA commenced at 14:50 UTC, with Volkov and Samokutyayev donning their spacesuits and venturing into space to perform a series of critical tasks.

Their focus was the Rassvet module, an essential component of the ISS's Russian segment. The cosmonauts worked diligently to install and test external hardware, deploy scientific instruments, and conduct maintenance on the module's exterior systems. These efforts were vital in ensuring the Rassvet module's functionality and its readiness for continued research.

In addition to their installation work, Volkov and Samokutyayev conducted a thorough photographic survey of the station's exterior. This survey provided valuable visual data on the condition of the modules and their attached experiments, which was crucial for future maintenance planning and ensuring the station's structural integrity.

The spacewalk also served as a platform for testing new tools and techniques for future EVAs. The cosmonauts experimented with various methods for securing and positioning equipment, refining their approach to space station maintenance. Throughout the EVA, safety and communication remained paramount, with the cosmonauts maintaining constant contact with ground control to ensure the safe and efficient execution of their tasks.

As the sun set on August 3, the spacewalk concluded at 21:13 UTC. Volkov and Samokutyayev returned to the ISS, their mission complete and their contributions to the station's maintenance and enhancement invaluable.

The culmination of Expedition 28 came on September 16, 2011, with the undocking of the Soyuz TMA-21 spacecraft. Commanded by Aleksandr Samokutyayev, with Flight Engineers Andrei Borisenko and Ronald Garan, the spacecraft departed from the ISS at 00:38 UTC. It safely landed in central Kazakhstan at 03:59 UTC, marking the end of Expedition 28 and the beginning of Expedition 29. These spacewalks and the successful conclusion of Expedition 28 exemplify the dedication and precision required to sustain the ISS's functionality and advance its scientific mission.

Expedition 29

Expedition 29 crew on the front row are NASA astronauts Mike Fossum (left), commander; and Dan Burbank, flight engineer. From the left (back row) are Japan Aerospace Exploration Agency (JAXA) astronaut Satoshi Furukawa along with Russian cosmonauts Sergei Volkov, Anatoly Ivanishin and Anton Shkaplerov, all flight engineers.

On April 4, 2011, at 23:18:20 UTC, Soyuz TMA-21, a mission commemorating the 50th anniversary of Yuri Gagarin's pioneering spaceflight, lifted off from the Baikonur Cosmodrome's Gagarin's Start launch pad in Kazakhstan. This mission marked the 109th flight of a Soyuz spacecraft since its inaugural launch in 1967. Commanded by Russian astronaut Aleksandr Samokutyayev, Soyuz TMA-21 carried Flight Engineers Andrei Borisenko from Russia and Ronald J. Garan from the United States. The spacecraft was tasked with transporting three Expedition 27 crew members to the International Space Station (ISS).

The Soyuz-FG rocket performed a nominal ascent, achieving orbit 8 minutes and 45 seconds after launch. Upon reaching space, the spacecraft deployed its solar panels and communications antennas as planned. Inside the spacecraft, Samokutyayev, who had brought a small stuffed dog as a personal memento, shared a moment of weightlessness as the toy floated alongside the crew. Samokutyayev described the launch as "great," a sentiment echoed by Mission Control in Moscow, celebrating the flight's historical significance.

Soyuz TMA-21 docked with the ISS on April 6, 2011, at 23:09 UTC, connecting with the Poisk module while the station was orbiting over the Andes Mountains. The hatches between the Soyuz and the ISS opened at 2:13 UTC on April 7, allowing the crew to float into the station. They were greeted by Expedition 27 Commander Dmitry Kondratyev and Flight Engineers Catherine Coleman and Paolo Nespoli, who conducted a crew welcome ceremony and a mandatory safety orientation.

At the Baikonur Cosmodrome in Kazakhstan, NASA astronaut Ron Garan (left), Expedition 27 flight engineer; along with Russian cosmonauts Alexander Samokutyaev (center), Soyuz commander; and Andrey Borisenko, flight engineer outside their Soyuz TMA-21 spacecraft during a check of its systems March 22, 2011. The Soyuz, which has been dubbed "Gagarin" and which bears the likeness of cosmonaut Yuri Gagarin, the first human in space launched on April 4 (April 4, U.S. time), just one week shy of the 50th anniversary of Gagarin's historic journey into space from the same launch pad that the Expedition 27 crew will begin their mission from.

The return of Soyuz TMA-21 was originally scheduled for September 8, 2011. However, following the crash of the Progress M-12M resupply vehicle on August 24, the mission was delayed. Soyuz TMA-21 undocked from the ISS on September 16, 2011, at 00:38 UTC. Although there was a brief loss of voice communications from the crew due to a malfunction, the reentry and landing proceeded smoothly. The spacecraft landed in central Kazakhstan at 03:59 UTC.

Immediately following the landing, Russian search and rescue teams were deployed by NASA flight surgeons and space station program managers. The recovery operation involved three Antonov airplanes, 14 Mil Mi-8 helicopters, and seven rescue vehicles. Samokutyayev was the first to be extracted from the descent module, followed by Garan and Borisenko. The crew appeared in good health as they rested in recliners near the module. After quick medical checks, they were flown to Karaganda for an official welcome home ceremony. Samokutyayev and Borisenko subsequently traveled to Star City near Moscow, while Garan returned to the Johnson Space Center in Houston.

Expedition 29 began on September 16, 2011, marking a new phase for the ISS with the departure of Soyuz TMA-21 and the conclusion of Expedition 28. The initial crew of Expedition 29 included Commander Michael Fossum from NASA, Flight Engineer Satoshi Furukawa from JAXA, and Flight Engineer Sergey Volkov from Roscosmos. They continued their work on the station until November 21, 2011, when Soyuz TMA-02M undocked, bringing their mission to a close. This spacecraft, which had served as their transport back to Earth, landed safely in Kazakhstan at 02:26 GMT on November 22, 2011.

NASA astronaut Dan Burbank (from left) and Russian cosmonauts Anton Shkaplerov and Anatoly Ivanishin are set to launch aboard a Soyuz TMA-22 spacecraft.

The arrival of Soyuz TMA-22, initially planned for September 2011 but delayed due to the Progress M-12M launch failure, took place on November 14, 2011, at 04:14 UTC. The spacecraft, carrying Flight Engineers Anton Shkaplerov, Anatoli Ivanishin, and Daniel Burbank, docked with the ISS on November 16, 2011, at 05:24 GMT. This successful docking ensured the continuation of ISS operations and research activities.

The transition to Expedition 29 was marked by the arrival of the new crew members from Soyuz TMA-22. As Commander Michael Fossum, Flight Engineer Satoshi Furukawa, and Flight Engineer Sergey Volkov concluded their mission, the ISS welcomed Shkaplerov, Ivanishin, and Burbank, who continued the essential work aboard the station, maintaining its research and operational activities.

View of Earth taken 16 September 2011 during ISS Expedition 29 from a height of 211 nautical miles (391 km)

Expedition 29 launches aboard the Soyuz TMA-22 spacecraft amid snowy conditions at the launch pad.

Russian support personnel work to help get Expedition 29 crew members out of the Soyuz TMA-02M spacecraft shortly after the capsule landed with Expedition 29 Commander Mike Fossum and Flight Engineers Sergei Volkov and Satoshi Furukawa in a remote area outside of the town of Arkalyk, Kazakhstan, Nov. 22 (Nov. 21 in the United States). NASA astronaut Fossum, Russian cosmonaut Volkov and Japan Aerospace Exploration Agency astronaut Furukawa are returning from more than five months onboard the International Space Station where they served as members of the Expedition 28 and 29 crews.

The descent back to Earth, however, presented a challenge. Soyuz TMA-02M touched down in Kazakhstan at 02:26 AM GMT on November 22, 2011, but not without incident. The spacecraft landed on its side, necessitating a prompt and careful recovery operation. Despite this unusual landing position, the crew emerged safely, concluding their mission with a successful return to Earth.

Expedition 29 crew members, from left to right, Commander Mike Fossum, Flight Engineers Sergei Volkov and Satoshi Furukawa, sit in chairs outside the Soyuz TMA-02M capsule just minutes after they landed in a remote area outside the town of Arkalyk, Kazakhstan, on Nov. 22 (Nov. 21 in the United States). NASA astronaut Fossum, Russian cosmonaut Volkov and Japan Aerospace Exploration Agency astronaut Furukawa are returning from more than five months onboard the International Space Station where they served as members of the Expedition 28 and 29 crews.

Expedition 30

Expedition 30, the thirtieth long-duration mission to the International Space Station (ISS), began its journey with the arrival of its initial crew members on November 16, 2011. This first team included Commander Dan Burbank from NASA, Flight Engineers Anton Shkaplerov, and Anatoli Ivanishin from Roscosmos, who arrived aboard the Soyuz TMA-22 spacecraft. Their arrival marked the final phase of Expedition 29. The formal start of Expedition 30 was signified by the departure of the Soyuz TMA-02M on November 21, 2011, which had carried the previous crew members away from the station.

Expedition 30 was characterized by a seamless transition of crew and supplies, essential for maintaining the ISS's operational capacity and advancing its scientific research. On December 21, 2011, the Soyuz TMA-03M spacecraft launched, bringing additional members to the crew and marking the official beginning of Expedition 30. The spacecraft docked with the ISS on December 23, 2011, ensuring the station continued functioning effectively with a full team. The arrival of this new crew was crucial for the ongoing scientific experiments and maintenance tasks aboard the ISS.

Expedition 30 crew front row are NASA astronaut Dan Burbank, commander; and Russian cosmonaut Oleg Kononenko, flight engineer. Pictured from the left (back row) are Russian cosmonauts Anton Shkaplerov and Anatoly Ivanishin; along with European Space Agency astronaut Andre Kuipers and NASA astronaut Don Pettit, all flight engineers.

In 2012, the ISS received vital supplies and equipment to support its operations. On January 25, 2012, the Progress 46 spacecraft launched, delivering a fresh cargo of provisions and scientific materials. It docked with the ISS on January 27, 2012, following the undocking of Progress 45 on January 23, 2012. This timely resupply ensured the crew had the necessary resources for their ongoing work.

Another significant event during Expedition 30 occurred on March 28, 2012, with the arrival of the Automated Transfer Vehicle (ATV-3). Launched on March 23, 2012, the ATV-3 brought additional cargo and scientific equipment to the ISS, further supporting the station's research capabilities and crew needs. The Progress 46 spacecraft undocked on April 19, 2012, making way for Progress 47, which launched on April 20, 2012, and docked with the ISS on April 22, 2012. This new cargo delivery continued the critical role of maintaining the station's operations.

The transition to Expedition 31 was marked by the departure of Burbank, Shkaplerov, and Ivanishin aboard Soyuz TMA-22 on April 27, 2012. This changeover from Expedition 30 to Expedition 31 highlighted the ongoing cycle of crew rotation that ensures the ISS remains a hub of international cooperation and scientific advancement.

Throughout Expedition 30, the crew witnessed and documented significant celestial events, including the appearance of Comet C/2011 W3 Lovejoy. On December 21, 2011, Commander Dan Burbank and his team observed the comet's close approach to the Sun, initially feared to be a perilous encounter. Despite its close approach, within 140,000 kilometers (87,000 miles) of the Sun's surface, the comet survived this intense solar encounter, providing valuable data for scientists studying celestial bodies and their behaviors.

The first spacewalk of 2012 took place on February 16, involving Russian cosmonauts Oleg Kononenko and Anton Shkaplerov. During this extravehicular activity (EVA), the astronauts moved one of the ISS's Strela cranes from the Pirs module to the Poisk module. In addition to this significant hardware relocation, they installed new debris shields and conducted various material experiments on the station's exterior. This EVA was a critical component of the station's maintenance and enhancement efforts, contributing to the ongoing safety and functionality of the ISS.

Comet Lovejoy was visible near Earth's horizon in this nighttime image photographed by NASA astronaut Dan Burbank, Expedition 30 commander, onboard the International Space Station .

On February 20, 2012, the ISS crew commemorated a landmark event in space history—the fiftieth anniversary of John Glenn's pioneering orbital flight aboard the Project Mercury spacecraft, Friendship 7. In a heartfelt tribute, the crew surprised the 90-year-old Glenn with a video link while he was on stage with NASA Administrator Charlie Bolden at Ohio State University. This gesture highlighted the enduring legacy of Glenn's historic mission and the continued spirit of exploration and achievement in spaceflight.

The European Space Agency's third Automated Transfer Vehicle (ATV), named Edoardo Amaldi, launched on March 23, 2012, and arrived at the ISS on March 28. The ATV carried approximately 6,595 kilograms (14,539 pounds) of cargo, including propellants, water, and dry supplies for the Expedition 30 crew. Additionally, the ATV's thrusters was crucial in boosting the station's orbit, ensuring it remained at its optimal altitude. Edoardo Amaldi remained docked until September 2012, when it was deorbited and re-entered Earth's atmosphere, burning up as planned.

On April 20, 2012, the Progress M-15M spacecraft, a Russian unmanned resupply vehicle, launched toward the ISS from Baikonur Cosmodrome. It docked with the station on April 22, delivering essential supplies and equipment. In preparation for Progress M-15M's arrival, its predecessor, Progress M-14M, undocked on April 19, 2012, having been docked since January 28, 2012. This transition was crucial to maintaining the station's supply chain and operational readiness.

The culmination of Expedition 30 occurred on April 27, 2012, with the departure of Soyuz TMA-22 from the ISS. The spacecraft carried astronauts Dan Burbank, Anton Shkaplerov, and Anatoli Ivanishin back to Earth. The departure marked the end of their mission, and they landed safely in Kazakhstan at 11:45 AM GMT. As Soyuz TMA-22 departed, the ISS was left in the capable hands of the remaining crew members, including Oleg Kononenko, André Kuipers, and Donald Pettit, who began their new roles in Expedition 31.

European Space Agency's "Edoardo Amaldi" Automated Transfer Vehicle-3 (ATV-3) docked to the space station delivering 220 pounds of oxygen, 628 pounds of water, 4.5 tons of propellant, and nearly 2.5 tons of dry cargo

The year 2012 marked significant advancements and achievements in spaceflight. One notable event was initiating the Commercial Orbital Transportation Services (COTS) program, which saw the first private companies conduct resupply missions to the International Space Station (ISS). SpaceX's Dragon spacecraft became the first privately developed spacecraft to dock with the ISS, establishing a milestone in the commercialization of space exploration.

China also made remarkable progress with its manned space program in 2012. The Shenzhou 9 mission, carrying astronauts Jing Haipeng, Liu Wang, and Liu Yang, achieved the country's first crewed space station docking. The mission involved the rendezvous and docking of the Shenzhou 9 spacecraft with the Tiangong 1 space station, demonstrating China's increasing capabilities in human spaceflight.

The ISS remained a focal point for human spaceflight activities throughout the year, hosting various expeditions and missions. The crew of Soyuz TMA-04M, consisting of Gennady Padalka, Sergei Revin, and Joseph M. Acaba, facilitated ISS crew rotation. Similarly, Yuri Malenchenko, Sunita Williams, and Akihiko Hoshide formed the crew of Soyuz TMA-05M, continuing the ISS crew rotation. Oleg Novitskiy, Evgeny Tarelkin, and Kevin A. Ford constituted the crew of Soyuz TMA-06M, and Roman Romanenko, Chris Hadfield, and Thomas Marshburn formed the crew of Soyuz TMA-07M, all contributing to the ongoing ISS crew rotation.

The Salyut program, Russia's precursor to the ISS, achieved notable milestones. The Mir Core Module ("DOS-7") became the first multi-module space station, accumulating an impressive occupancy of 4,592 days. Similarly, the Zvezda module ("DOS-8") within the ISS surpassed 4,310 days of occupancy as of August 21, 2012. These modules showcased the continued legacy of the Salyut program and its contributions to human space exploration.

The year 2012 witnessed 77 attempted orbital launches, with 72 successful launches. Among these launches were five crewed orbital missions that placed 15 individuals into orbit. Russia, China, and the United States led most of these launches, reflecting their continued dominance in spaceflight endeavors. Furthermore, the USSR's Almaz program played a role in developing the Functional Cargo Block modules, with the Zarya module remaining in operation as part of the ISS, alongside the Zvezda module.

Expedition 31

Expedition 31 crew front row are Russian cosmonauts Oleg Kononenko (right), commander; and Gennady Padalka, flight engineer. Ftom the left (back row) are NASA astronaut Joe Acaba, Russian cosmonaut Sergei Revin, European Space Agency astronaut Andre Kuipers and NASA astronaut Don Pettit, all flight engineers.

Expedition 31 of the International Space Station (ISS) began on April 27, 2012 with the departure of the Soyuz TMA-22 spacecraft, which had previously hosted the Expedition 30 crew. Commander Dan Burbank and Flight Engineers Anton Shkaplerov and Anatoli Ivanishin returning to Earth, concluding their successful mission aboard the ISS.

The transition to Expedition 31 was carefully orchestrated to ensure a seamless continuation of operations aboard the ISS. Before this, on December 23, 2011, the station had been joined by the crew of Soyuz TMA-03M, including Commander Oleg Kononenko of Russia, Flight Engineer André Kuipers of the Netherlands, and

Flight Engineer Don Pettit of the United States. Their arrival marked the beginning of their tenure, which would see them overseeing critical research and maintenance until the crew rotation in May 2012.

Soyuz TMA-04M was a pivotal mission in spaceflight history, transporting three members of the Expedition 31 crew to the International Space Station (ISS). Launched on May 15, 2012, from the Baikonur Cosmodrome in Kazakhstan, the spacecraft represented the 113th flight of the Soyuz program, which began in 1967, and was the fourth mission of the improved Soyuz TMA-M series, first introduced on October 7, 2010.

Several key milestones marked the Soyuz TMA-04M mission. The spacecraft lifted off on May 15, 2012, at 03:01:23 UTC, initiating its journey to Low Earth orbit. It docked with the ISS on May 17, 2012, at 04:36 UTC, successfully linking with the Poisk docking module. This docking integrated the new crew members into the ongoing operations aboard the ISS and ensured the spacecraft's role as an emergency escape vehicle throughout Expedition 31.

The crew aboard Soyuz TMA-04M included Commander Gennady Padalka from Russia, Flight Engineer Sergei Revin from Russia, and Flight Engineer Joseph M. Acaba from the United States. This mission marked Padalka's fourth spaceflight, Revin's sole spaceflight, and Acaba's second. Their arrival at the ISS was a significant event, enhancing the station's capabilities and continuity of operations.

Soyuz TMA-04M's role extended beyond the initial crew transport. After the departure of Soyuz TMA-03M on June 1, 2012, the crew of Soyuz TMA-04M continued their work as part of Expedition 31 until the arrival of Soyuz TMA-05M in mid-July, which marked the beginning of the second portion of Expedition 32.

The mission culminated with Soyuz TMA-04M undocking from the ISS on September 16, 2012, at 23:09 UTC. It safely landed in Kazakhstan on September 17, 2012, at 02:53 UTC. The spacecraft's successful return concluded its operational role and underscored the reliability and effectiveness of the Soyuz spacecraft in supporting long-duration missions aboard the ISS. The Soyuz TMA-04M mission highlighted advancements in spacecraft design and continued the tradition of international cooperation in space exploration.

Russian cosmonaut Gennady Padalka (center), Expedition 31 flight engineer and Expedition 32 commander; along with NASA astronaut Joe Acaba (left) and Russian cosmonaut Sergei Revin, both Expedition 31/32 flight engineers at NASA's Johnson Space Center.

SpaceX's unmanned Dragon spacecraft approaches the ISS on 25 May 2012.

Another landmark achievement during Expedition 31 was the debut test mission of SpaceX's Dragon spacecraft. As part of NASA's Commercial Orbital Transportation Services (COTS) program, Dragon pioneered commercial spaceflight. After overcoming several delays, Dragon launched on 22 May 2012. The spacecraft conducted a series of precise orbital maneuvers before successfully docking with the ISS on 25 May. Dragon delivered approximately 460 kilograms (1,010 pounds) of cargo, including

essential supplies such as food, clothing, a laptop computer, and 15 student experiments designed to advance scientific understanding.

Following its mission, Dragon was loaded with 660 kilograms (1,460 pounds) of return cargo, which included completed experiments and surplus equipment. The spacecraft undocked from the ISS on 31 May 2012, and its descent concluded with a successful landing in the Pacific Ocean. Dragon's recovery marked a significant milestone for SpaceX, paving the way for future cargo missions. This success set the stage for the first regular logistics flight, CRS SpX-1, launched in October 2012.

The Expedition 31 crew wave goodbye before the launch on 15 May 2012.

The final transition for Expedition 31 occurred on 1 July 2012, with the departure of Soyuz TMA-03M. This mission ended Expedition 31 and ushered in Expedition 32. Soyuz TMA-03M carried the departing crew members—Andrey Kononenko, André Kuipers, and Donald Pettit—back to Earth, marking the successful conclusion of their tenure aboard the ISS. Their departure facilitated the arrival of a new crew, continuing the ISS's legacy as a hub for international cooperation and scientific advancement.

The Soyuz TMA-04M rocket launches from the Baikonur Cosmodrome in Kazakhstan on Tuesday, May 15, 2012 carrying Expedition 31 Soyuz Commander Gennady Padalka, NASA Flight Engineer Joseph Acaba and Flight Engineer Sergei Revin to the International Space Station.

Expedition 32

Expedition 32 crew from the left are Japan Aerospace Exploration Agency (JAXA) astronaut Akihiko Hoshide, Russian cosmonaut Yuri Malenchenko, NASA astronaut Sunita Williams, NASA astronaut Joe Acaba, all flight engineers; Russian cosmonaut Gennady Padalka, commander; and Russian cosmonaut Sergei Revin, flight

Expedition 32, the 32nd long-duration mission aboard the International Space Station (ISS), commenced with the arrival of a new crew on July 17, 2012. This mission was marked by significant logistical and operational activities that ensured the ISS's continuous functioning and advancement of its scientific objectives.

The Expedition 32 crew, consisting of Russian cosmonaut Anatoly Ivanishin, NASA astronaut Kevin Ford, and European Space Agency astronaut Luca Parmitano, arrived as part of a meticulously planned crew rotation system. This system was designed to maintain the ISS's continuous staffing with highly trained professionals. The rotation began with the launch of the Soyuz TMA-05M spacecraft on July 15, 2012, three days prior to the crew's arrival. This timely launch was essential in ensuring that the ISS's operations and scientific experiments remained uninterrupted during the transition.

During the summer of 2012, the ISS experienced a period of intense logistical activity. On July 20, 2012, the Japanese cargo spacecraft HTV-3 Kounotori was launched, playing a crucial role in the station's supply chain. The HTV-3 successfully docked with the ISS on July 27, delivering essential supplies and scientific materials. Following the completion of its mission, HTV-3 Kounotori was released from the station on September 11, 2012, after transferring its cargo.

At the Baikonur Cosmodrome in Kazakhstan, Expedition 32 Flight Engineer Aki Hoshide (left) of the Japan Aerospace Exploration Agency, Soyuz Commander Yuri Malenchenko (center) and NASA Flight Engineer Sunita Williams walk by the upper stage spacecraft fairing of their Soyuz booster rocket in the Integration Facility July 3, 2012 as they completed a suited "fit check" of the Soyuz TMA-05M spacecraft before launching to the International Space Station.

In addition to the Kounotori mission, the Progress 47P spacecraft, which had previously docked with the ISS, underwent a series of routine docking maneuvers. Progress 47P undocked for the first time on July 22, 2012, redocked on July 28, and undocked again on July 30. These maneuvers were part of the standard cargo and supply management operations that ensured the ISS was well-supplied and that its operational integrity was maintained.

The Soyuz TMA-04M spacecraft, which had transported the Expedition 31 crew to the ISS, concluded its mission with a return to Earth on September 17, 2012. The spacecraft, launched from the Baikonur Cosmodrome in Kazakhstan on May 15, 2012, had docked with the ISS on May 17, carrying a crew of three: Gennady Padalka, Sergei Revin, and Joseph Acaba. After completing its mission, Soyuz TMA-04M undocked from the ISS on September 16 and landed safely in Kazakhstan on September 17 at 2:53 UTC. Following the departure of Soyuz TMA-04M, Expedition 32's crew was completed with the arrival of Soyuz TMA-05M on July 17, 2012. Soyuz TMA-05M was launched on July 15 atop a Soyuz FG rocket and successfully docked with the ISS on July 17. The spacecraft carried Commander Yuri Malenchenko, Flight Engineer Sunita Williams, and Flight Engineer Akihiko Hoshide. This mission was notable for coinciding with the 37th anniversary of the Apollo-Soyuz Test Project. Soyuz TMA-05M remained docked to the ISS throughout Expedition 32, serving as an emergency escape vehicle.

The crew of Soyuz TMA-05M conducted the latter portion of Expedition 32 until their departure on November 18, 2012. The spacecraft undocked from the ISS at 10:26 PM GMT and landed safely at 1:53:30 AM GMT the following day, approximately 4.7 kilometers from the target location in Kazakhstan. The landing marked the conclusion of Expedition 32 and the transition to Expedition 33.

The Soyuz TMA-05M spacecraft was carried by train towards the launch pad at the Baikonur Cosmodrome in Kazakhstan on 12 July 2012.

At the Baikonur Cosmodrome in Kazakhstan, the Soyuz TMA-05M spacecraft was readied for its encapsulation into the upper stage of its Soyuz booster July 8, 2012 in advance of its rollout to the launch pad July 12 and the scheduled launch of its occupants, Expedition 32 Soyuz Commander Yuri Malenchenko, NASA Flight Engineer Sunita Williams and Flight Engineer Aki Hoshide of the Japan Aerospace Exploration Agency, July 15 (Kazakhstan time) for a four-month mission on the International Space Station.

NASA astronaut Sunita Williams appears to touch the bright sun during Expedition 32's third spacewalk on Sept. 5, 2012. Williams and Japanese astronaut Akihiko Hoshide (visible in the reflection of Williams' helmet visor) fixed a key power unit on the exterior of the orbiting outpost during the 6 1/2-hour outing.
(Image credit: NASA)

Expedition 33

Expedition 33 crew from the left are NASA astronaut Sunita Williams, commander; along with Russian cosmonaut Yuri Malenchenko, Japan Aerospace Exploration Agency (JAXA) astronaut Akihiko Hoshide, Russian cosmonaut Evgeny Tarelkin, Russian cosmonaut Oleg Novitskiy and NASA astronaut Kevin Ford, all flight engineers.

On October 21, 2012, the Soyuz-FG rocket, carrying the Soyuz TMA-06M spacecraft, was transported to pad 6 at Site 31 by train and erected for launch. Site 31, known as the Tereshkova pad, was chosen because maintenance and upgrades were being carried out on the usual pad 5 at Site 1. The crew's journey to the pad, which took approximately an hour and ten minutes, was longer than the typical trip to pad 5.

On October 23, 2012, the Soyuz TMA-06M spacecraft embarked on its journey to the ISS, marking the 115th flight of a Soyuz spacecraft since its introduction in 1967. This mission was historically significant, representing the first crewed launch from the remote Site 31 launch pad at Baikonur Cosmodrome since July 1984. The Soyuz TMA-06M was launched atop a Soyuz-FG rocket at 10:51:11 GMT from the Baikonur Cosmodrome in Kazakhstan. This launch was particularly notable because it was the first human liftoff from Site 31 since the Soyuz T-12 mission to the Salyut 7 space station. The rocket achieved a flawless launch, with the spacecraft reaching orbit just nine minutes later and beginning its 34-orbit journey to the ISS.

Expedition 33 marked the 33rd long-duration mission to the International Space Station (ISS), beginning on September 16, 2012. This mission commenced with the departure of the Soyuz TMA-04M spacecraft, which concluded the stay of the Expedition 32 crew and heralded the arrival of a new team dedicated to maintaining the ISS's continuous human presence in space.

The Soyuz TMA-06M crew comprised Commander Oleg Novitskiy, Flight Engineer Evgeny Tarelkin, and NASA astronaut Kevin A. Ford. Novitskiy, representing Roscosmos, was on his inaugural spaceflight, while Tarelkin was also on his first mission. Ford, from NASA, was on his second and final spaceflight. The backup crew included Commander Pavel Vinogradov, Flight Engineer Aleksandr Misurkin, and Flight Engineer Christopher Cassidy, all from Roscosmos and NASA.

As Soyuz TMA-06M approached the ISS, it performed a series of orbital maneuvers to prepare for docking. The spacecraft executed its initial burns on launch day, with the final burn occurring on the second day of flight. The automated Rendezvous Operations commenced at 10:11 UTC on October 25, 2012. Following a successful Flyaround and station-keeping operations, Soyuz TMA-06M docked with the ISS at 12:29 GMT, approximately six minutes ahead of schedule. The docking occurred at the MRM-2 Poisk module while the ISS was above southern Ukraine.

Upon docking, the crew conducted a standard one-hour leak check before opening the hatches and entering the ISS at approximately 15:15 GMT. Expedition 33 members Sunita Williams, Yuri Malenchenko, and Akihiko Hoshide welcomed them. Soyuz TMA-06M remained attached to the ISS throughout the Expedition 33 increment, serving as an emergency escape vehicle. It returned to Earth on March 15, 2013, undocking from the ISS at 11:43 PM GMT. The spacecraft landed safely at 3:06:30 AM GMT the following day, north of Arkalyk.

Expedition 33's crew comprised six international astronauts, reflecting a broad collaboration among spacefaring nations. Sunita Williams of NASA served as the commander, marking her second spaceflight. Williams played a significant role in overseeing the station's operations and ensuring the success of various scientific experiments. Yuri Malenchenko from the Russian Space Agency (RSA) served as Flight Engineer 1, contributing his extensive spaceflight experience, including this being his fifth mission. Akihiko Hoshide from the Japan Aerospace Exploration Agency (JAXA) was Flight Engineer 2 on his second spaceflight, providing expertise for various research projects. Flight Engineer 3 was Kevin A. Ford of NASA, on his second and final spaceflight, where he managed the station's systems and conducted scientific research. Oleg Novitskiy, representing Roscosmos, was Flight Engineer 4 on his first spaceflight, learning the intricacies of life aboard the ISS while supporting the mission's scientific goals. Evgeny Tarelkin, also from RSA, was completing the team and making his only spaceflight with Expedition 33 and adapting to the challenges of living and working in space.

In addition to carrying the crew, Soyuz TMA-06M also transported 32 Medaka fish, housed in an Aquatic Habitat (AQH) in the Japanese Kibo Lab Module. These fish were part of an experiment to study the effects of space conditions on living organisms, contributing valuable data to the mission's scientific endeavors.

During the expedition, the crew engaged in a range of scientific activities. A notable achievement was the successful experimentation with the Delay-Tolerant Networking (DTN) protocol. DTN was designed to improve communication systems in space by addressing challenges posed by long communication delays and intermittent connections, providing crucial insights into its effectiveness for future deep-space missions. The crew also remotely operated a Lego robot on Earth from the ISS, demonstrating the station's role as a platform for advancing remote operation and robotics.

A standout moment of Expedition 33 was the deployment of CubeSats on October 4, 2012. These compact, cube-shaped satellites, each about the size of a loaf of bread, represented a significant advancement in space technology. CubeSats are versatile and cost-effective and designed for various scientific and technological purposes. Their deployment during this mission aimed to gather important data and test new technologies, pushing the boundaries of satellite capabilities and enhancing our understanding of their potential applications.

Expedition 33 officially commenced at 23:09 UTC on September 16, 2012, and concluded on November 18, 2012. The mission, with its international crew and array of scientific experiments, continued the ISS's legacy of international cooperation and exploration, building on the achievements of previous expeditions while preparing the station for future endeavors.

NASA astronaut Sunita Williams (front left) takes command of the International Space Station from cosmonaut Gennady Padalka (front right) during a ceremony marking the start of the Expedition 33 increment aboard the space station on Sept. 15, 2012.

Family of the newly arrived International Station Expedition 33 crew members, Russian cosmonaut Oleg Novitskiy, front left, NASA astronaut Kevin Ford, front center, and Russian cosmonaut Evgeny Tarelkin, front right, talk via phone to the crew from the Russian Mission Control Center in Korolev, Russia shortly after the three joined Flight Engineer Aki Hoshide of the Japan Aerospace Exploration Agency, back left, Expedition 33 Commander Sunita Williams of NASA, back center, and Yuri Malenchenko of the Russian Federal Space Agency on Oct. 25, 2012.

The Soyuz rocket was rolled out to the launch pad by train on Oct. 21, 2012, at the Baikonur Cosmodrome in Kazakhstan to send Expedition 33 Flight Engineer Kevin Ford of NASA, Soyuz Commander Oleg Novitskiy and Flight Engineer Evgeny Tarelkin of Roscosmos on a five-month mission aboard the International Space Station.

Several tiny CubeSat satellites are shown in this image photographed by an Expedition 33 crew member on the International Space Station on 4 October 2012. The satellites were released outside the Kibo laboratory using a Small Satellite Orbital Deployer attached to the Japanese module's robotic arm. Japan Aerospace Exploration Agency astronaut Aki Hoshide, flight engineer, set up the satellite deployment gear inside the laboratory and placed it in the Kibo airlock. The Japanese robotic arm then grappled the deployment system and its satellites from the airlock for deployment. A portion of the station's solar array panels and a blue and white part of the earth provide the backdrop for the scene.

Expedition 34

Expedition 34 crew front row are NASA astronaut Kevin Ford (left), commander; and Canadian Space Agency astronaut Chris Hadfield, flight engineer. Pictured from the left (back row) are Russian cosmonauts Oleg Novitskiy, Evgeny Tarelkin, Roman Romanenko and NASA astronaut Tom Marshburn, all flight engineers.

On December 19, 2012, the Soyuz TMA-07M mission embarked on a crucial journey from the Baikonur Cosmodrome, delivering three members of Expedition 34 to the International Space Station (ISS). This mission, integral to the ISS's ongoing operations, included Roman Romanenko from the Russian space agency Roscosmos, Canadian astronaut Chris Hadfield, and American astronaut Thomas Marshburn. Besides transporting the crew, Soyuz TMA-07M also functioned as a vital emergency escape vehicle for the subsequent Expedition 35.

The pre-launch period began with ceremonial events on November 29, 2012, when the Soyuz TMA-07M crew participated in traditional activities at Red Square. The spacecraft itself was transported to the launch site on December 17, braving the harsh pre-dawn temperatures that dropped to –30 °C. This launch was historic, being the first from the newly upgraded Site 1/5, a significant advancement from the previous Site 31/6.

On the morning of December 19, 2012, the Soyuz TMA-07M crew engaged in their final preparations. After a traditional pre-flight blessing by a Russian Orthodox Priest, the astronauts donned their Sokol launch and entry suits at Site 254. They proceeded to the launch pad approximately two hours and 33 minutes before liftoff. Romanenko occupied the central seat of the spacecraft, flanked by Hadfield and Marshburn.

The Soyuz FG rocket, carrying Soyuz TMA-07M, lifted off precisely at 12:12:35 UTC. Despite the freezing weather conditions, the launch proceeded smoothly. Two minutes and 35 seconds after liftoff, the launch escape system and payload shroud were jettisoned, and the core stage burned until it was shut down at 4 minutes and 45 seconds. The third stage then ignited, propelling the spacecraft into its intended orbit of 200 by 242 kilometers with an inclination of 51.67 degrees. Upon reaching orbit, Soyuz TMA-07M deployed its solar arrays and communication antennas, setting the stage for its approach to the ISS.

The docking sequence commenced on December 21, 2012. Soyuz TMA-07M executed a series of precise orbital maneuvers to align with the ISS. The spacecraft performed several burns, adjusting its trajectory to rendezvous with the station. By 13:50 UTC, Soyuz TMA-07M was 400 meters from the ISS, initiating a flyaround maneuver to align with the Rassvet module's nadir docking port. After a seven-minute orbital lap and a brief period of stationkeeping, the spacecraft closed in for docking, which occurred at 14:09 UTC, three minutes ahead of schedule.

The Soyuz TMA-07M crew—Hadfield, Marshburn, and Romanenko—returned to Earth on May 13, 2013, undocking from the ISS at 23:08 UTC. The capsule successfully landed in Kazakhstan on May 14, 2013, at 02:31 UTC, marking the end of a significant mission phase.

Expedition 34, which began on November 18, 2012, with the departure of Soyuz TMA-05M, was a pivotal period for the ISS. The transition to Expedition 34 was marked by the arrival of the new crew aboard Soyuz TMA-07M. This team, led by Commander Kevin Ford from the United States and joined by Flight Engineers Oleg Novitskiy and Evgeny Tarelkin from Russia, embarked on a new chapter focused on advancing scientific research and maintaining station operations.

Throughout Expedition 34, several critical cargo missions were undertaken. On February 9, 2013, the ISS Progress 48 undocked, clearing the way for Progress 50, which launched on February 11, 2013. Progress 50 was essential in resupplying the ISS with scientific experiments, crew supplies, and critical equipment, ensuring the station's operational readiness.

Further enhancing the station's capabilities, SpaceX CRS-2 launched on March 1, 2013, as part of NASA's Commercial Resupply Services program. This mission was crucial for the station's logistics, delivering a variety of scientific experiments, hardware, and supplies. CRS-2 successfully docked with the ISS on March 3, 2013, bolstering the station's capacity to support ongoing research and operational needs.

Expedition 34's crew, which included Commander Kevin A. Ford, Flight Engineers Oleg Novitskiy, and Evgeny Tarelkin, was later joined by Flight Engineers Thomas Marshburn, Chris Hadfield, and Roman Romanenko. Their work focused on diverse scientific objectives. Research included studying the human cardiovascular system's adaptation to microgravity, essential for long-duration space missions and future deep-space travel. This research aimed to understand physiological changes such as blood flow alterations and heart function, providing insights into maintaining cardiovascular health in space.

Another area of research examined fish behavior in microgravity, exploring how aquatic life adapts to the ISS environment. This study provided valuable data on sensory perception and balance, with implications for understanding both aquatic and human responses to gravity changes.

Additionally, Expedition 34 investigated solar radiation's effects on Earth's climate. Solar radiation drives climate processes, and this research aimed to improve climate models and predict atmospheric changes by providing detailed data on solar radiation interactions with the Earth's climate system.

A notable aspect of Expedition 34 was the continued testing of Robonaut, a humanoid robot designed to assist astronauts with various tasks aboard the ISS. The ongoing testing aimed to refine Robonaut's functionality and explore its potential applications in future missions, where robotic assistance could significantly support human activities and extend mission capabilities.

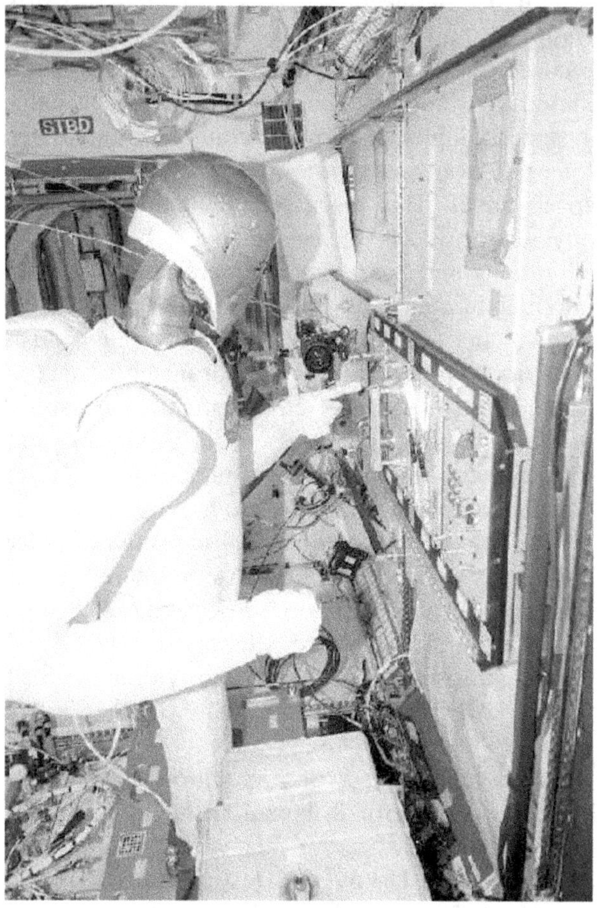

In 2013, the realm of human spaceflight was predominantly marked by the International Space Station (ISS), a focal point for numerous expeditions and missions. Throughout the year, the ISS continued to serve as a hub for scientific discovery and international cooperation. Astronauts and cosmonauts aboard the ISS engaged in diverse experiments spanning human physiology, biology, physics, and Earth observation. Their responsibilities extended beyond research, encompassing essential maintenance tasks and spacewalks necessary for the station's upkeep and functionality.

In March, the ISS saw a change in its crew with the departure of Expedition 34 and the arrival of Expedition 35. This rotation facilitated the continuation of ongoing scientific experiments and maintenance work. By May, the second operational mission of SpaceX's Dragon spacecraft was completed, delivering crucial supplies, experiments, and equipment to the ISS. The Dragon's successful return to Earth with scientific samples highlighted the growing role of commercial cargo vehicles in supporting space missions, alongside the Orbital Sciences Cygnus spacecraft, which also visited the ISS that year.

The rhythm of crew rotations continued with the arrival of Expedition 36 in July and the subsequent departure of Expedition 35. The new crew engaged in further research and welcomed the European Space Agency's ATV-4 spacecraft, which delivered additional supplies and experiments to the ISS. By November, Expedition 38 took over from Expedition 37, undertaking further spacewalks and installations of new equipment while continuing the station's research agenda.

Concurrent with these ISS activities, China made significant strides with its Tiangong-1 space station program. June 2013 saw the successful launch and docking of the Shenzhou 10 spacecraft with Tiangong-1, marking China's second manned space mission to its space station. The mission's crew conducted a series of experiments over their 15-day stay, demonstrating China's growing capabilities in manned spaceflight and space station operations.

On the broader stage of space exploration, 2013 was marked by notable advancements and challenges. 81 orbital launches were attempted globally, with 77 achieving success. The year saw Russia, the United States, and China leading most of these missions. Among the highlights was the debut of Russia's Soyuz-2-1v rocket, a development indicative of the country's efforts to advance its launch capabilities. However, the year also brought to light concerns regarding missing payload numbers and mysterious "add-on" payloads, raising questions about potential covert activities in space.

The year also witnessed organizational changes to address challenges within the Russian space sector. In August, the Russian government established the United Rocket and Space Corporation to tackle reliability issues and a series of launch failures. The reorganization aimed to streamline the industry through a unified command structure and eliminating redundant capabilities, signaling a substantial restructuring with the potential for significant workforce reductions.

The Shenzhou 10 mission, launched on June 11, 2013, featured Nie Haisheng, Zhang Xiaoguang, and Wang Yaping. Their mission was a significant milestone for China, marking its fifth human spaceflight and second crewed docking with Tiangong. The crew conducted various experiments during their 15-day stay before returning to Earth on June 26, 2013.

Expedition 35

Expedition 35, which succeeded Expedition 34, brought a fresh wave of activity and scientific focus to the International Space Station (ISS). This period was marked by a series of critical missions and crew changes that ensured the station continued to advance its research goals and operational capabilities.

On March 28, 2013, the Soyuz TMA-08M spacecraft, also known as Soyuz 34 or 34S by NASA, embarked on a pivotal journey to the International Space Station (ISS). This mission was noteworthy as it marked the 117th flight of a Soyuz spacecraft, a lineage that began with the first launch in 1967. Soyuz TMA-08M transported three members of the Expedition 35 crew to the ISS, including Commander Pavel Vinogradov,

Flight Engineer Alexander Misurkin, and Flight Engineer Christopher Cassidy.

Expedition 35 crew front row are Canadian Space Agency astronaut Chris Hadfield (right), commander; and Russian cosmonaut Pavel Vinogradov, flight engineer. Left (back row) are Russian cosmonaut Alexander Misurkin, NASA astronaut Chris Cassidy, Russian cosmonaut Roman Romanenko and NASA astronaut Tom Marshburn, all flight engineers.

The mission utilized the new 6-hour fast rendezvous profile developed by the Russian Federal Space Agency (RKA). This innovative approach, previously tested on Progress M-16M and M-17M, significantly shortened the time between launch and docking compared to the traditional two-day rendezvous. This advancement allowed the crew to transition from ground facilities to the ISS in a timeframe comparable to a transatlantic flight.

The Soyuz TMA-08M mission began with the rollout of the Soyuz FG Rocket from Site 1/5 at the Baikonur Cosmodrome on March 26, 2013. The rocket, carrying the Soyuz spacecraft, was erected at the launch pad and prepared for its countdown. Final cargo items, including time-sensitive experiment payloads for the Russian segment of the ISS, were loaded into the spacecraft just before launch.

At 20:43 GMT, the Soyuz FG Rocket launched from the Baikonur Cosmodrome in Kazakhstan. The rocket performed flawlessly, and within nine minutes, it delivered the Soyuz TMA-08M crew into orbit. Commander Pavel Vinogradov occupied the center seat, while Flight Engineer Misurkin and

Flight Engineer Cassidy took the left and right seats, respectively.

Immediately following orbital insertion, Soyuz TMA-08M commenced its rendezvous operations. During its first orbit, the spacecraft executed two programmed engine burns. On its second orbit, it transmitted actual orbital parameters from a Russian ground site. Using these parameters, Soyuz conducted eight additional rendezvous burns over the next five hours. The spacecraft docked with the ISS's MRM-2 Poisk module after just four orbits and less than six hours post-launch. This achievement set a new record for the fastest crewed docking to the ISS.

Upon docking at 2:28 GMT on March 29, 2013, the Soyuz TMA-08M crew was welcomed aboard the ISS by Expedition 35 Commander Chris Hadfield and Flight Engineers Thomas Marshburn and Roman Romanenko. A ceremonial welcome was held with family members and mission officials at the Russian Mission Control Center near Moscow.

Expedition 35 generated significant media attention, largely due to Commander Hadfield's remarkable presence. Leveraging the power of social media, Hadfield became a global sensation, captivating the public with his engaging updates and videos from space. One of the most notable moments was the release of the "first music video recorded in space," a rendition of David Bowie's 1969 song "Space Oddity," which garnered millions of views on YouTube and showcased Hadfield's musical talent and creativity in orbit.

In addition to his musical endeavors, Hadfield participated in the unveiling of the Bank of Canada's new $5 note, part of the Frontier Series of polymer bills released in 2013. This event was broadcast via video from the ISS, adding a unique space-based element to the note's introduction.

The mission also included significant logistical achievements. The SpaceX CRS-2 mission delivered essential supplies to the ISS. It returned cargo to Earth, marking SpaceX's second contracted cargo flight and its first to utilize the unpressurized trunk section of the Dragon spacecraft.

On May 11, 2013, Expedition 35 encountered a critical situation that required an unplanned spacewalk. Astronauts Christopher Cassidy and Thomas Marshburn conducted the spacewalk to replace a faulty pump controller box suspected of causing an ammonia coolant leak. This urgent repair ensured the continued safe operation of the station.

Expedition 35 was a milestone mission that advanced scientific research and international collaboration and brought the ISS into the global spotlight through Commander Hadfield's extraordinary efforts and achievements.

During Expedition 35 aboard the International Space Station (ISS), the crew engaged in a range of significant robotic activities, reflecting the mission's focus on advancing space technology and enhancing the station's operational capabilities.

A notable highlight of Expedition 35's robotic operations was the ongoing use of the station's sophisticated robotic systems, such as the Canadian Space Agency's Canadarm2 and the European Space Agency's (ESA) European Robotic Arm (ERA). These versatile robotic arms are integral to the ISS and used for a variety of tasks, including docking spacecraft, performing maintenance, and conducting scientific experiments.

One of the key robotic activities during this expedition involved the Canadarm2, crucial in supporting the arrival and integration of the Dragon cargo spacecraft. This spacecraft, operated by SpaceX, delivered essential supplies and scientific payloads to the ISS. The Canadarm2's precise control and maneuverability were essential for capturing and berthing the Dragon capsule, achieved with remarkable accuracy by the onboard crew using the station's robotics systems.

Additionally, Expedition 35 saw the continuation of the assembly and maintenance tasks facilitated by the robotic arms. The crew used the Canadarm2 for various maintenance operations, including installing new equipment and repositioning scientific instruments. These tasks are vital for maintaining the station's functionality and supporting ongoing research.

While not actively deployed in every mission, the ERA remained a key component of the ISS's robotic suite. The arm's role in future missions involves assisting with manipulating external payloads and maintaining the station's structure, underscoring the continued importance of robotic technology in space operations.

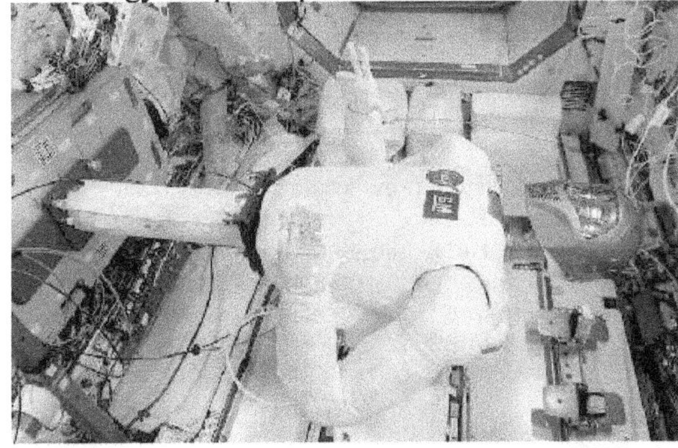

Elsewhere in international space activies, in 2013, significant developments unfolded across the aerospace landscape, reflecting space exploration ventures' diverse and dynamic nature. Among these milestones, the Virgin Galactic Astronaut Pilots Group, based in the UK, saw former NASA astronaut Frederick W. Sturckow and Michael "Sooch" Masucci join its ranks, marking a pivotal step in the burgeoning field of commercial space travel.

Simultaneously, the Association of Spaceflight Professionals welcomed its Group 4 members, including David Ballinger, Jessica Cherry, Michael Gallagher, Jamie Guined, Tanya Markow-Estes, and Aaron Persad. These individuals, each contributing unique expertise, highlighted the increasing specialization within the aerospace industry and its global reach.

Meanwhile, on June 17th, NASA introduced Group 21, affectionately dubbed the 8-Balls, underscoring the agency's ongoing commitment to fostering diverse talent pools. Comprising Josh A. Cassada, Victor J. Glover, Tyler N. Hague, Christina M. Hammock, Nicole Aunapu Mann, Anne C. McClain, Jessica U. Meir, and Andrew R. Morgan, this cohort symbolized NASA's dedication to advancing scientific discovery and space exploration capabilities.

These developments underscored a transformative period in space history, characterized by collaborative efforts between commercial entities and governmental agencies alike.

The mission concluded on September 10, 2013, with the Soyuz TMA-08M spacecraft undocking from the Poisk Module at 23:37 GMT. The spacecraft initiated its Deorbit Burn at 2:05 GMT on September 11, followed by a 4-minute 46-second retrograde burn of its SKD Main Propulsion System. At 2:32 GMT, the three modules of the Soyuz separated at an altitude of 140 kilometers. As the Entry Module re-entered the atmosphere, it experienced the first traces of atmospheric entry, leading to heat build-up on the spacecraft's thermal protection system. During the peak stress of reentry, communications with the Soyuz were temporarily lost.

As the Entry Module descended, it slowed to approximately 240 meters per second. At an altitude of 10.6 kilometers, the parachute deployment sequence began. Three chutes were deployed sequentially: the Pilot Chute, the Drogue Chute, and finally the Main Chute, which deployed at around 7.5 kilometers, slowing the spacecraft to 6 meters per second.

The Soyuz TMA-08M spacecraft landed on the steppe of Kazakhstan, southeast of the town of Dzhezkazgan, at 2:58 GMT. Recovery personnel quickly surrounded the spacecraft, preparing to open the hatch. Commander Vinogradov was the first to be extracted, followed by NASA astronaut Cassidy, with Flight Engineer Misurkin being the last to leave the Entry Module. Due to adverse weather conditions, the crew was swiftly transported to a nearby medical tent.

Post-landing, Cassidy and Misurkin underwent physiological assessments to gauge their condition after the mission. These tests were part of an ongoing effort to understand how astronauts might feel after landing on Mars. Meanwhile, Vinogradov and Misurkin were flown to Star City near Moscow, and Cassidy boarded a NASA Gulfstream III aircraft for his return to Houston.

The Soyuz TMA-08M spacecraft departs from the International Space Station's Poisk Mini-Research Module 2 (MRM2) and heads toward a landing in a remote area near the town of Zhezkazgan, Kazakhstan, on Sept. 11, 2013 (Kazakhstan time). Russian cosmonaut Pavel Vinogradov, Expedition 36 commander; along with NASA astronaut Chris Cassidy and Russian cosmonaut Alexander Misurkin, both flight engineers, are ending a five-and-a-half month stay at the space station where they served as members of the Expedition 35 and 36 crews.

Expedition 36

Expedition 36, which spanned from May 13, 2013, to September 10, 2013, represented the 36th long-duration mission to the International Space Station (ISS). This mission underscored the ISS's pivotal role in fostering scientific research and international collaboration in space.

The crew of Expedition 36 was a testament to international cooperation, comprising astronauts from the United States, Russia, and Italy. Leading the mission was Commander Pavel Vinogradov from Russia, a seasoned astronaut whose extensive experience was crucial in steering the team through their objectives. Vinogradov's leadership was instrumental in guiding the crew through a series of complex tasks and experiments.

Expedition 36 crew front row are Russian cosmonauts Pavel Vinogradov (left), commander; and Fyodor Yurchikhin, flight engineer. Left (back row) are Russian cosmonaut Alexander Misurkin, NASA astronaut Chris Cassidy, European Space Agency astronaut Luca Parmitano and NASA astronaut Karen Nyberg, all flight engineers.

The flight engineers on this mission were a blend of experienced space travelers and first-timers. Aleksandr Misurkin, also from Russia, embarked on his inaugural spaceflight as Flight Engineer 1. Chris Cassidy from the United States, on his second space mission, served as Flight Engineer 2, bringing valuable experience from his previous spaceflight. Karen L. Nyberg, representing NASA, was on her second and final spaceflight as Flight Engineer 3. Notably, on June 16, 2013, Nyberg, along with Wang Yaping from the Chinese space module Tiangong-1, commemorated the 50th anniversary of Valentina Tereshkova's historic Vostok 6 flight—the first spaceflight by a woman. At that time, Nyberg and Yaping were the only two women in space.

Another key member of the crew was Fyodor Yurchikhin, also from Russia, who was undertaking his fourth spaceflight as Flight Engineer 4. Italian astronaut Luca Parmitano, participating in his first space mission as Flight

Engineer 5, was involved in a significant event during his tenure.

A critical incident occurred on July 16, 2013, during Extravehicular Activity (EVA) 23. Parmitano encountered a serious problem when water leaked into his helmet, leading to an immediate abort of the spacewalk. Parmitano, accompanied by Chris Cassidy, returned to the Quest airlock after an hour and 32 minutes outside the station. The intrusion of water into Parmitano's suit impaired his vision, hearing, and speech. Commander Vinogradov and Yurchikhin promptly assisted Parmitano by removing his helmet and using towels to absorb the water. Despite the distressing situation, Parmitano emerged unscathed and in good spirits.

Throughout Expedition 36, the crew engaged in numerous experiments and maintenance tasks, ensuring the continued functionality of the ISS. The mission saw the overlap of the outgoing Expedition 35/36 and incoming Expedition 36/37 crews, facilitating a seamless transition and continuity of operations.

Soyuz TMA-09M played a crucial role in the mission, transporting astronauts to the ISS. Launched from the Baikonur Cosmodrome in Kazakhstan on May 28, 2013, the Soyuz-FG rocket propelled the spacecraft into low Earth orbit. The spacecraft separated from the rocket's upper stage nine minutes after liftoff and docked with the ISS on May 29, 2013, at 02:10 UTC, at the nadir port of the Rassvet module. The swift rendezvous minimized the crew's confinement within the Soyuz spacecraft, enhancing their comfort.

The Soyuz TMA-09M crew consisted of Fyodor Yurchikhin as the mission commander, Luca Parmitano from the European Space Agency (ESA), and Karen L. Nyberg from NASA. Yurchikhin's extensive experience guided the mission, while Parmitano and Nyberg contributed their expertise. Throughout Expeditions 36 and 37, Soyuz TMA-09M remained docked as an emergency escape vehicle, ensuring the crew's safety.

The mission lasted a total of 166 days, 6 hours, and 18 minutes. Soyuz TMA-09M undocked from the ISS on November 1, 2013, at 23:26 UTC, briefly relocated to the Zvezda aft docking port, and departed the station on November 10, 2013, at 08:54 UTC. The spacecraft landed on November 11, 2013, at 02:49 UTC, returning the same trio of astronauts who had embarked on the mission.

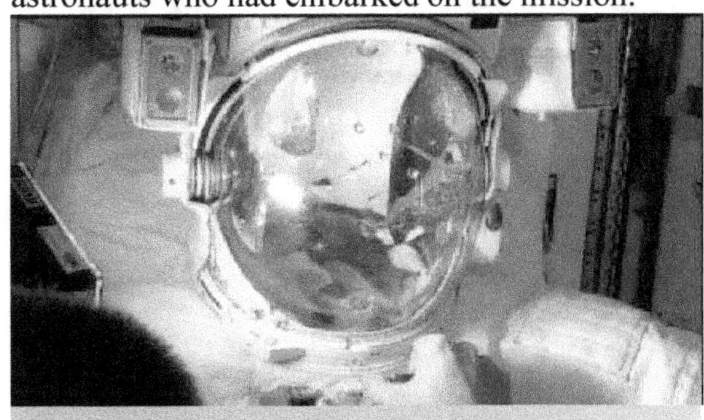

With famed St. Basil's Cathedral serving as a backdrop, Expedition 36/37 Flight Engineer Karen Nyberg of NASA (left), Soyuz Commander Fyodor Yurchikhin (center) and Flight Engineer Luca Parmitano of the European Space Agency pose for pictures May 8 during a ceremonial tour of Red Square in Moscow. Nyberg, Yurchikhin and Parmitano are preparing for their launch May 29, Kazakh time, in their Soyuz TMA-09M spacecraft from the Baikonur Cosmodrome in Kazakhstan for a six-month mission on the International Space Station.

Expedition 37

Expedition 37 crew front row are Russian cosmonauts Fyodor Yurchikhin (left), commander; and Oleg Kotov, flight engineer. Left (back row) are NASA astronaut Karen Nyberg, European Space Agency astronaut Luca Parmitano, NASA astronaut Michael Hopkins and Russian cosmonaut Sergey Ryazanskiy, all flight engineers.

Expedition 37, an important chapter in the history of the International Space Station (ISS), unfolded from September 2013 to November 2013. It highlighted the intricate coordination and international cooperation essential for modern space missions.

The expedition commenced on September 18, 2013, with the launch of the Cygnus spacecraft from Cape Canaveral. Developed by Orbital Sciences Corporation, the Cygnus was a critical component of NASA's Commercial Resupply Services program. Its primary mission was to deliver vital supplies and scientific experiments to the ISS, ensuring the station's continuous operation and support for its crew.

Expedition 37 itself began on September 10, 2013, with a diverse and experienced team. Fyodor Yurchikhin of Roscosmos led the mission as the commander, bringing his extensive experience from four previous spaceflights. On her second and final spaceflight, flight Engineer Karen L. Nyberg from NASA contributed significantly to the mission's scientific and technical aspects. Representing Italy, Luca Parmitano of the European Space Agency (ESA) was on his first spaceflight, infusing the team with fresh expertise.

Joining them was Oleg Kotov from Roscosmos, marking his third and final spaceflight, and Sergey Ryazansky, also from Roscosmos, on his inaugural spaceflight. NASA's Michael S. Hopkins completed the team, supporting the mission's objectives.

Upon its arrival at the ISS on September 29, 2013, the Cygnus spacecraft garnered significant attention within the space community. The Expedition 37 crew, including Commander Fyodor Yurchikhin and Flight Engineers Luca Parmitano, Karen Nyberg, and Michael Hopkins, meticulously executed the docking operation. Their precise handling of the spacecraft underscored their expertise and the robustness of the docking procedures.

Just a few weeks later, on October 22, 2013, the Cygnus spacecraft was undocked from the ISS, marking the end of its resupply mission. This operation demonstrated the careful planning and execution required to safely and efficiently complete such missions.

Following this, on October 28, 2013, the Automated Transfer Vehicle-4 (ATV-4), a European Space Agency (ESA) resupply spacecraft, was undocked. The ATV-4 had been crucial in delivering additional supplies and experiments, enhancing the station's capabilities and contributing to its scientific endeavors.

During their time aboard, the crew conducted a range of scientific experiments in fields such as biology, physics, and materials science. These experiments leveraged the ISS's unique microgravity environment to investigate fundamental processes, including fluid behavior and combustion in space, which are crucial for future space missions.

A significant achievement of Expedition 37 was installing and activating the Express Logistics Carrier-3 (ELC-3). This large, rectangular aluminum structure plays a vital role in expanding the station's external payload capabilities. The ELC-3, delivered by the Space Shuttle Endeavour during its final flight on STS-135, measures approximately 6.5 meters in length and 4.3 meters in width, weighing around 6.5 metric tons. It provides multiple mounting points for various payloads and equipment, facilitating the integration and deployment of new technologies.

The ELC-3 was installed on the ISS's S3 truss using the Canadarm2, a robotic arm, showcasing the sophisticated technology and coordination involved in maintaining and upgrading the station. The crew also performed routine maintenance and upgrades to ensure the ISS remained in optimal condition, including system inspections, repairs, and preventive maintenance.

Expedition 37 was characterized by its international collaboration, involving NASA, Roscosmos, ESA, and the Japanese Aerospace Exploration Agency (JAXA).

The expedition also saw a significant maneuver on November 1, 2013, when the Soyuz spacecraft was relocated. This routine operation was part of the crew rotation and logistics processes necessary to ensure the ISS remained well-staffed and operational.

Shortly after that, on November 7, 2013, the Soyuz TMA-11M spacecraft was launched, carrying a new crew to join the ISS. This mission was essential for maintaining the station's human presence and ensuring the continuity of its operations.

On November 10, 2013, the crew of Expedition 36, who had been aboard the ISS since May, returned to Earth, concluding their mission. Their handover to the new crew was a testament to the rigorous training and preparation that characterize ISS operations.

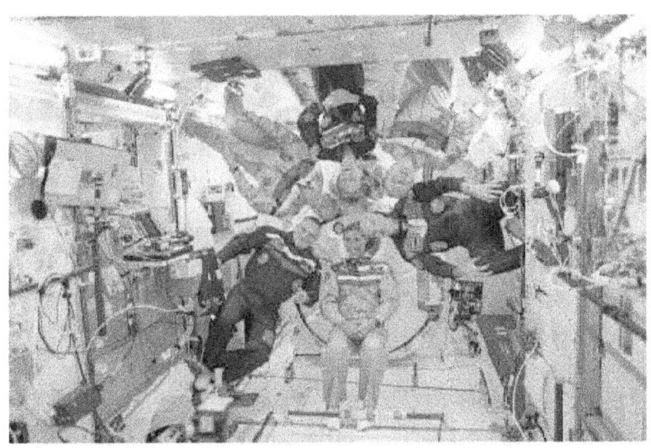

Expedition 37 crew members pose for an inflight crew portrait in the Kibo laboratory of the International Space Station. Pictured (clockwise from lower left) are Russian cosmonaut Fyodor Yurchikhin, commander; Russian cosmonauts Sergey Ryazanskiy, NASA astronaut Karen Nyberg, Russian cosmonaut Oleg Kotov, ESA astronaut Luca Parmitano and NASA astronaut Michael Hopkins, all flight engineers.

The year 2014 was a notable period for space exploration, marked by significant advancements and activities from various spacefaring nations. Central to human spaceflight during this year was the International Space Station (ISS), which continued to be a hub of scientific research and international cooperation.

The year began with an intriguing development from the Soviet Union, now Russia. On May 6, 2014, the country launched the Kosmos-2495 satellite, an evolution of the Yantar-4K2 high-resolution imaging reconnaissance satellite first deployed in 1981. Initially launched as part of the Kobalt-M series, Kosmos-2495 drew international attention due to suspicions surrounding its potential military capabilities. The satellite, which replaced its predecessors' ion-electric engines with conventional chemical thrusters, was noted for carrying an amateur radio payload designated RS-46. The U.S. remained cautious as Kosmos-2495, along with the subsequent launch of Kosmos-2499, exhibited unusual orbital maneuvering capabilities that suggested secretive military applications. The nature and purpose of these satellites, often referred to as "ghost" satellites, remained ambiguous due to the lack of official disclosures from Russia.

The global landscape of intercontinental ballistic missiles (ICBMs) also saw developments in 2014. The Russian Federation was actively advancing the RS-28 Sarmat, a liquid-fueled ICBM designed to carry multiple independently targetable reentry vehicles (MIRVs). This missile was intended to be equipped with a mix of heavy and lighter warheads, hypersonic glide vehicles, and countermeasures to evade anti-missile systems, representing a response to the U.S. Prompt Global Strike initiative. Similarly, in July 2014, China unveiled its Dongfeng-41 (DF-41), a new generation ICBM capable of reaching up to 7,500 miles (12,000 km) and equipped with MIRV technology. This announcement was accompanied by a successful non-destructive test of an anti-satellite weapon, showcasing China's growing capabilities in space and missile technology.

Amidst these developments, the ISS continued to serve as a focal point for human spaceflight. The station hosted a series of expeditions throughout the year, each contributing to a diverse range of scientific experiments and operational tasks. In March, the Expedition 39 crew, consisting of Russian cosmonauts Aleksandr Skvortsov and Oleg Artemyev, along with NASA astronaut Steven Swanson, concluded their mission and were succeeded by the Expedition 40 crew. The arrival of the Expedition 40 team marked the continuation of research in human physiology, biology, and Earth observation. The SpaceX Dragon and Orbital Sciences Cygnus spacecraft facilitated the delivery of crucial supplies and experiments to the ISS.

The following months saw the arrival of Expedition 41 in May, with astronauts engaging in research, maintenance, and receiving supplies from the Japanese HTV-5 spacecraft. In September, Expedition 42 took over, continuing the station's operations, performing maintenance tasks, and conducting experiments. The year closed with the arrival of Expedition 43 in November, as the Soyuz TMA-15M spacecraft brought Anton Shkaplerov, Samantha Cristoforetti, and Terry W. Virts to the ISS. Their mission involved ongoing maintenance, experiments, and the installation of new equipment to support the station's activities.

China also made strides in its space program with the successful launch of the unmanned Tianzhou-1 cargo spacecraft in 2014, which delivered supplies and equipment for their Tiangong-1 space station program. This mission was a step towards establishing a permanently manned space station, reflecting China's growing ambitions in space exploration.

Throughout 2014, the ISS witnessed 92 attempted orbital launches, with 88 achieving success. The year saw seven spacewalks conducted by Russia, the United States, and China astronauts. These activities underscored the ongoing commitment to space exploration and the collaborative efforts of various space agencies.

Expedition 38

Expedition 38 crew front row are Koichi Wakata (left), flight engineer; and Russian cosmonaut Oleg Kotov, commander. Left (back row) are Russian cosmonaut Mikhail Tyurin, NASA astronaut Rick Mastracchio, Russian cosmonaut Sergey Ryazanskiy and NASA astronaut Michael Hopkins, all flight engineers.

Expedition 38 of the International Space Station (ISS) unfolded between November 2013 and March 2014, a period marked by significant milestones and advancements in space research. This mission began with the arrival of Soyuz TMA-11M, which launched on November 7, 2013. Soyuz TMA-11M, the 120th flight of the Soyuz spacecraft series since its inception in 1967, carried three members of the Expedition 38 crew to the ISS. The successful docking of Soyuz TMA-11M on November 7, 2013, was particularly notable as it marked the first time since October 2009 that nine astronauts were aboard the ISS simultaneously, without the presence of a Space Shuttle.

The mission's rocket and spacecraft featured Olympic symbols, and a notable event occurred when Russian cosmonauts Oleg Kotov and Sergey Ryazansky passed the Olympic torch in space. This symbolic gesture was the first time an Olympic torch had been passed in the vacuum of space, reflecting the mission's role in promoting international cooperation and unity.

The Expedition 38 crew comprised six members, each bringing a wealth of experience and expertise. Mikhail Tyurin from the Russian space agency Roscosmos commanded the mission, marking his third and final spaceflight. As Commander, Tyurin was responsible for overseeing the mission and ensuring the successful coordination of crew activities.

The team included Richard Mastracchio from NASA, who served as Flight Engineer 1 and was on his fourth and final spaceflight. His extensive experience was crucial in managing complex experiments and maintaining the station's systems. Koichi Wakata of JAXA, Japan's space agency, was Flight Engineer 2 on his fourth spaceflight, contributing significantly to daily operations and scientific research. Oleg Kotov of Roscosmos, who later became the Commander of Expedition 38, was on his third spaceflight, bringing valuable leadership and oversight to the mission. Sergey Ryazansky, also from Roscosmos, was on his first spaceflight and served as Flight Engineer 1, contributing fresh perspectives and energy to the team.

The mission saw a series of critical events and logistical support activities. On November 30, 2013, the ISS Progress 53 spacecraft docked with the station, delivering essential supplies and scientific equipment. This was followed by the launch of Orbital Sciences' Cygnus spacecraft, Orbital-1, on January 9, 2014. This mission began the private sector's involvement in resupplying the ISS under NASA's Commercial Resupply Services program. The Cygnus spacecraft arrived at the ISS on January 12, 2014, delivering scientific experiments and supplies.

Expedition 38 also saw the successful deployment and activation of the Express Logistics Carrier-4 (ELC-4), a component designed to expand the ISS's ability to support additional external experiments and equipment. Delivered and installed by Space Shuttle Atlantis during its final mission, STS-135, the ELC-4 enhanced the station's external payload capacity and supported new experiments and technology demonstrations.

On February 3, 2014, the ISS Progress 52 spacecraft undocked as part of routine spacecraft management. Shortly after, the ISS Progress 54 launched on February 5, 2014, carrying additional cargo to support the mission. The Cygnus spacecraft was released from the ISS on February 18, 2014, concluding its successful cargo mission.

Expedition 39

Expedition 39 crew front row are Japan Aerospace Exploration Agency (JAXA) astronaut Koichi Wakata (right), commander; and NASA astronaut Steve Swanson, flight engineer. Left (back row) are Russian cosmonauts Oleg Artemyev, Alexander Skvortsov, Mikhail Tyurin and NASA astronaut Rick Mastracchio, all flight engineers.

On March 25, 2014, Soyuz TMA-12M embarked on its mission to the International Space Station (ISS), marking a milestone as the 121st flight of a Soyuz spacecraft since its first launch in 1967. This mission, the 38th Soyuz mission dedicated to the ISS, was integral in transporting three members of the Expedition 39 crew to the space station. The launch, executed aboard a Soyuz-FG rocket from the Baikonur Cosmodrome in Kazakhstan at 21:17 UTC, was timed precisely as the ISS passed over Baikonur, allowing NASA astronaut Richard Mastracchio to capture a striking photograph of the spacecraft's ascent.

Initially, Soyuz TMA-12M was scheduled to dock with the ISS's Poisk module on March 26, 2014, at 03:04 UTC, with the hatch opening planned for 04:45 UTC. However, an unexpected issue arose during the critical trajectory refinement phase. A problem with the spacecraft's attitude control prevented one of the orbital burns necessary for path adjustment. This anomaly left the spacecraft in an incorrect orientation, necessitating a shift to the previously used two-day rendezvous profile instead of the more recent six-hour trajectory. Consequently, the docking was rescheduled to March 27, 2014, at 23:58 UTC. Despite these challenges, the spacecraft docked successfully, arriving five minutes ahead of the revised schedule at 23:53 UTC. The hatch between Soyuz TMA-12M and the ISS opened at 02:35 UTC on March 28, 2014. Upon arrival, the crew, comprising Commander Aleksandr Skvortsov, Flight Engineer Oleg Artemyev, and Flight Engineer Steven R. Swanson, joined the Expedition 39 team, marking the beginning of their mission aboard the station.

Expedition 39, which officially commenced with the arrival of Soyuz TMA-12M, was a period of intense activity and significant milestones for the ISS. The mission began with the launch of Soyuz TMA-12M on March 25, 2014, carrying Russian Commander Aleksandr Skvortsov, NASA's Steve Swanson, and European Space Agency (ESA) astronaut Alexander Gerst. Their arrival signaled a key transition in crew operations, integrating new personnel into the station's ongoing activities.

Following their arrival on March 27, 2014, the Expedition 39 crew, which included Commander Koichi Wakata of JAXA, NASA astronaut Rick Mastracchio, and Roscosmos astronaut Mikhail Tyurin, joined forces with Skvortsov, Swanson, and Gerst. This integration was crucial for continuing the station's scientific research and maintaining operational integrity. The crew undertook a series of significant tasks, including the installation of new experiments and the management of critical systems aboard the ISS.

Throughout April 2014, a series of resupply missions ensured the station remained well-equipped. On April 7, 2014, the ISS Progress 54 spacecraft undocked, making way for the arrival of ISS Progress 55, which launched on April 9, 2014, and docked with the ISS shortly thereafter, delivering essential supplies. The SpaceX CRS-3 mission further bolstered the station's inventory when the SpaceX Dragon spacecraft launched from Cape Canaveral on April 18, 2014, and docked with the ISS on April 20, 2014, providing scientific payloads and general cargo.

Expedition 39 was distinguished by its focus on advancing space-based agriculture. On May 7, 2014, NASA astronauts Steve Swanson and Rick Mastracchio installed the Veggie experiment in the Columbus module of the ISS. This project aimed to cultivate plants in microgravity, contributing valuable insights into food production for long-duration space missions. The Veggie experiment involved the use of red, blue, and green LED lights to simulate natural sunlight, promoting plant growth. The chamber was equipped with six plant "pillows" containing seeds of 'Outredgeous 'red romaine lettuce, and the growth media included a controlled-release fertilizer and calcined clay to support root development.

NASA astronaut Mike Hopkins tending multiple plant experiments growing lettuce, radishes, mustard greens, and pak choi.

Beyond the Veggie project, Expedition 39 encompassed various scientific experiments and routine maintenance. The crew worked on upgrading the station's communication systems, conducting material science and human physiology experiments, and ensuring operational efficiency. This period also featured a redocking procedure for the ISS Progress 53 spacecraft on April 25, 2014, following an earlier undocking on April 23, 2014, to optimize docking ports.

Expedition 39's leadership was notable for its international representation, with Koichi Wakata becoming the first Japanese astronaut to command the ISS. This marked only the third instance of non-NASA or non-Russian leadership of the station, following Japanese astronauts in Expedition 21 in 2009 and Expedition 35 in 2013. Wakata's command underscored the global cooperation that defines ISS operations.

Expedition 39 concluded on May 13, 2014, when the outgoing crew returned to Earth, marking the successful integration of new members and the continuation of vital ISS operations. This period exemplified the ongoing efforts required to sustain human presence in space, highlighting the international collaboration and logistical coordination essential for advancing scientific knowledge and exploration aboard the ISS.

The Soyuz TMA-12M spacecraft departs the ISS to land near Zhezkazgan, Kazakhstan. NASA astronaut Steve Swanson, Expedition 40 commander; Russian cosmonaut Alexander Skvortsov, Soyuz commander and flight engineer; and Russian cosmonaut Oleg Artemyev, flight engineer, after five months of the Expedition 39 and 40 crews

Expedition 40

The six Expedition 40 crew from left are cosmonaut Alexander Skvortsov of Roscosmos, NASA astronaut Steve Swanson and cosmonaut Oleg Artemyev; ESA astronaut Alexander Gerst, cosmonaut Maxim Suraev and NASA astronaut Reid Wiseman. The 38S crew was composed of Swanson, Skvortsov and Artemyev. The 39S crew includes Suraev, Gerst and

Soyuz TMA-13M was a significant mission in the annals of space exploration. It marked the 122nd flight of a Soyuz spacecraft and the 39th mission to the International Space Station (ISS). Launched on May 28, 2014, from the Baikonur Cosmodrome in Kazakhstan, Soyuz TMA-13M carried three astronauts to the ISS, transitioning into a pivotal phase of Expedition 40 and later contributing to Expedition 41.

The mission began with the Soyuz FG rocket, a towering 49.5 meters in height, being rolled out to the launch pad at Site 1/5 of the Baikonur Cosmodrome. This event, occurring under clear skies on May 26, 2014, was witnessed by the backup crew members: Anton Shkaplerov, Samantha Cristoforetti, and Terry Virts. Superstition dictated that the prime crew, who were to be aboard the mission, abstained from attending this public occasion to avoid any ill fortune. Once the rocket was positioned vertically, it was enclosed by its service structure, which safeguarded the vehicle and provided access for technicians.

The launch of Soyuz TMA-13M took place at precisely 19:57 UTC on May 28, 2014. After achieving orbit roughly nine minutes later, the spacecraft embarked on a four-orbit rendezvous with the ISS. The docking occurred successfully at 1:44 UTC on May 29, with Soyuz TMA-13M connecting to the Rassvet module of the space station. The hatches between the Soyuz and the ISS were opened at 3:52 UTC, allowing the crew to enter the station and join their colleagues already in orbit.

The crew of Soyuz TMA-13M comprised Commander Maksim Surayev from Roscosmos, Flight Engineer Gregory R. Wiseman from NASA, and Flight Engineer Alexander Gerst from ESA. Surayev, on his second spaceflight, brought valuable experience to the mission, while Wiseman and Gerst were embarking on their first journeys into space. This mission was crucial for continuing the scientific work aboard the ISS and the ongoing management and maintenance of the station's systems.

During Expedition 40, the ISS saw a flurry of activity and international cooperation. The transition from Expedition 39 to Expedition 40 involved the arrival and departure of several spacecraft. The SpaceX Dragon spacecraft, arriving on March 25, 2014, brought essential supplies and experiments. On June 9, 2014, the ISS Progress 53 spacecraft undocked, making way for new cargo. The Orbital Sciences Cygnus spacecraft, known as Orb-2, launched on July 13, 2014, and successfully docked on July 16, delivering additional cargo and experiments. The undocking of ISS Progress 55 on July 21, 2014, was soon followed by the arrival of ISS Progress 56 on July 23, which continued to support the station with essential supplies.

A notable milestone occurred on August 12, 2014, with the docking of the Automated Transfer Vehicle (ATV-5), named "Georges Lemaitre." This European Space Agency cargo spacecraft further bolstered the station's supplies and scientific resources. The Orb-2 spacecraft was subsequently released on August 15, 2014, concluding its mission.

The mission of Soyuz TMA-13M came to a close with its undocking from the ISS on November 10, 2014, at 00:31 UTC. The spacecraft re-entered Earth's atmosphere and landed safely northwest of Arkalyk, Kazakhstan, at 03:58 UTC.

Expedition 40, which ran from May to November 2014, saw a seamless integration of crew transitions and the continuation of significant scientific and operational work aboard the ISS. This period demonstrated the effectiveness of international cooperation and the smooth handover of responsibilities between different crews.

The advancements during this mission also highlighted the capabilities of the Mobile Servicing System (MSS) on the ISS. The MSS, comprising the Canadarm2, the Special Purpose Dexterous Manipulator (SPDM), and the Mobile Base System (MBS), played a critical role in the station's assembly, maintenance, and operations. The Canadarm2, a highly versatile robotic arm mounted on the station's main truss, performed various tasks, including capturing and berthing visiting spacecraft and manipulating external payloads. The SPDM, known as Dextre, executed delicate maintenance and repair operations challenging for human astronauts. The MBS provided the necessary mobility for the robotic arms to perform their functions effectively across the station's truss structure.

The Soyuz TMA-13M rocket was launched with Expedition 40 Soyuz Commander Maxim Suraev, of the Russian Federal Space Agency, Roscosmos, Flight Engineer Alexander Gerst, of the European Space Agency, ESA, and Flight Engineer Reid Wiseman of NASA, Thursday, May 29, 2014 at the Baikonur Cosmodrome in Kazakhstan. Suraev, Gerst, and Wiseman will spend the next five and a half months aboard the International Space Station.

Dextre and Canadarm2 docked side by side on Power Data Grapple Fixtures

Soyuz TMA-13M spacecraft on 5-28-14 at a distance of 25m on final approach, astronaut's hand could be seen in the window. This was during a test of the new Russian ALHAT automated docking system

Expedition 41

Expedition 41 crew front row are Russian cosmonaut Maxim Suraev (left), commander; and NASA astronaut Barry Wilmore, flight engineer. Left (back row) are NASA astronaut Reid Wiseman, European Space Agency astronaut Alexander Gerst, Russian cosmonauts Alexander Samokutyaev and Elena Serova, all flight engineers.

Expedition 41 marked a significant chapter in the ongoing exploration and utilization of the International Space Station (ISS). Commencing on September 10, 2014, the expedition began with the undocking of Soyuz TMA-12M, which safely transported the Expedition 40 crew back to Earth. This marked the beginning of Expedition 41, which would see a series of critical milestones in space exploration.

The core crew of Expedition 41, initially assembled for the mission, included several astronauts from various space agencies. The commander for the first part of the expedition, which ran from September to November 2014, was Maksim Surayev of the Russian space agency (RSA). Surayev, on his second and final spaceflight, was joined by Flight Engineer Gregory R. Wiseman from NASA, embarking on his inaugural mission, and Flight Engineer Alexander Gerst from the European Space Agency (ESA), also on his first spaceflight. They were complemented by Russian Flight Engineers Aleksandr Samokutyayev, who was on his second and last spaceflight, and Yelena Serova, making her only spaceflight.

The mission saw a pivotal moment with the arrival of Soyuz TMA-14M, which launched on September 25, 2014. This spacecraft, part of the 123rd Soyuz flight, lifted off from the Baikonur Cosmodrome aboard a Soyuz-FG rocket. Despite a brief malfunction where the spacecraft's port solar array failed to deploy initially, it successfully achieved low Earth orbit within nine minutes. Following its orbit insertion, the solar array eventually deployed after docking with the ISS, and the issue did not jeopardize the mission's success.

Soyuz TMA-14M, after a swift four-orbit rendezvous, docked with the Poisk module of the ISS just under six hours post-launch. The docking occurred at 02:11 UTC on September 26, 2014, and the hatches between the Soyuz and the ISS were opened at 04:06 UTC, allowing the TMA-14M crew to join Expedition 41. The arrival of Samokutyayev, Serova, and Wilmore marked the enhancement of Expedition 41's team, where they integrated into the station's operations and continued their work until November 2014.

With the departure of Soyuz TMA-13M on November 10, 2014, the remaining crew members of Expedition 41, Samokutyayev, Serova, and Wilmore, transitioned to Expedition 42. Soyuz TMA-14M remained docked to the ISS as a critical emergency escape vehicle until its scheduled undocking on March 11, 2015. After a successful undocking at 22:44 UTC, the spacecraft re-entered Earth's atmosphere, with its descent module and crew landing safely at 02:07 UTC on March 12, 2015.

The mission concluded with the crew's successful return and the Soyuz spacecraft's safe landing, marking the end of Expedition 41 and the transition into the subsequent expedition.

Expedition 42

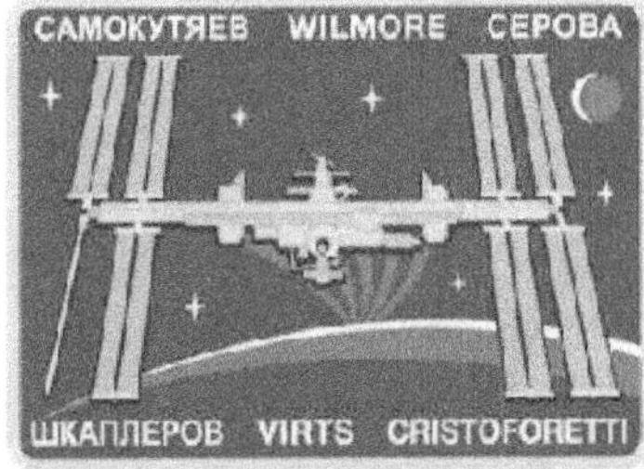

Expedition 42 crew front row are NASA astronauts Barry Wilmore (left), commander; and Terry Virts, flight engineer. Left (back row) are Russian cosmonauts Elena Serova, Alexander Samoukutyaev and Anton Shkaplerov and European Space Agency astronaut Samantha Cristoforetti, all flight engineers.

Expedition 42 began with the launch of the Soyuz TMA-15M spacecraft on November 23, 2014, heralding the arrival of a new crew to the International Space Station (ISS). This marked the commencement of their stay aboard the orbital laboratory, which was pivotal for advancing the station's scientific research and operational capabilities. The crew of Expedition 42 included Russian cosmonauts Anton Shkaplerov and Elena Serova, along with NASA astronaut Barry Wilmore. Their mission was to carry forward the extensive scientific research and maintenance operations critical to the ISS's ongoing objectives.

The arrival of the Soyuz TMA-15M on November 11, 2014, initiated Expedition 42's tenure aboard the ISS, marking the end of Expedition 41. The transition was meticulously planned to ensure a smooth handover, with the new crew quickly settling into their roles. Their responsibilities encompassed a range of activities, from conducting scientific experiments to performing routine maintenance and system upgrades that were essential for the station's operational integrity.

Soon after their arrival, significant spacecraft activities unfolded. On January 10, 2015, SpaceX launched its CRS-5 mission, which carried vital supplies and scientific experiments to the ISS. The cargo spacecraft successfully docked with the station on January 12, 2015, and remained attached until its undocking on February 10, 2015, after completing its mission. This delivery was crucial for supporting the ISS's research and daily operations.

On February 14, 2015, the Automated Transfer Vehicle (ATV-5), named Jules Verne in honor of the renowned French scientist, undocked from the ISS. The ATV-5 had previously fulfilled its role by delivering cargo and supplies to the station. Its departure marked the conclusion of its operational mission.

Adding to the momentum of space activities, the ISS Progress 58 spacecraft launched on February 17, 2015. This uncrewed cargo vehicle provided essential supplies and equipment necessary for the station's continuous operations and research endeavors.

Expedition 42's mission culminated on March 11, 2015, with the undocking of the Soyuz TMA-14M spacecraft. This event marked the expedition's end and its crew's return to Earth. The Soyuz TMA-15M, part of the crew rotation, also departed as part of the crew transition process.

During Expedition 42, the ISS continued its comprehensive array of scientific and technological research. The crew engaged in a variety of experiments across multiple disciplines, including biology, physics, astronomy, and materials science. These experiments were designed to leverage the ISS's unique microgravity environment, providing insights unattainable on Earth.

Elsewhere in the global space race, in 2015, pivotal developments continued to shape the landscape of space exploration, highlighting significant advancements and international collaborations. On January 23rd, Mark Stucky joined the Virgin Galactic Astronaut Pilots Group in the UK, adding to the roster of commercial astronauts poised to push the boundaries of space tourism and research.

Later in the year, on July 9th, NASA's Commercial Crew Program made strides with the inclusion of Robert Behnken, Sunita Williams, Eric Boe, and Douglas Hurley. This marked a crucial step toward NASA's goal of fostering commercial partnerships to transport astronauts to and from the International Space Station (ISS), enhancing accessibility to space and scientific research capabilities.

Simultaneously, the European Space Agency (ESA) expanded its Astronaut Corps with the addition of Matthias Maurer, reflecting Europe's commitment to advancing space exploration and scientific discovery.

In Denmark, Copenhagen Suborbitals made headlines in the commercial astronautics arena. Mads Stenfatt, Anna Olsen, and Carsten Olsen contributed to innovative suborbital missions, underscoring the diverse global efforts to explore space and push technological boundaries.

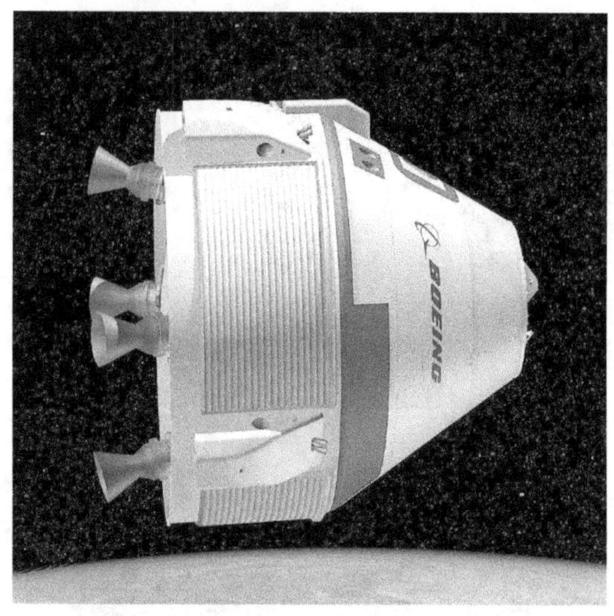

In 2011, NASA selected Boeing during Commercial Crew Development Round 2 (CCDev2) activities to mature the design and development of a crew transportation system with the overall goal of accelerating a United States-led capability to the International Space Station. The goal of CCP was to drive down the cost of space travel as well as open up space to more people than ever before by balancing industry's own innovative capabilities with NASA's 50 years of human spaceflight experience. Six other aerospace companies also matures launch vehicle and spacecraft designs under CCDev2, including Alliant Techsystems Inc. (ATK), Blue Origin, Excalibur Almaz Inc., Sierra Nevada Corp., Space Exploration Technologies (SpaceX), and United Launch Alliance (ULA).

New docking ports for Dragon capsule The SpaceX Dragon commercial cargo craft May 25, 2012 for grapple and berthing. Expedition 31 Flight Engineers Don Pettit and Andre Kuipers, becoming the first commercially developed space vehicle to be launched to the station to join Russian, European and Japanese resupply craft that service the complex while restoring a U.S. capability to deliver cargo to the orbital laboratory. Dragon spent about a week docked with the station before returning to Earth on May 31 for retrieval.

In 2015, the realm of space exploration continued to evolve with significant developments and missions spanning several nations. Russia, in particular, made notable strides in its space endeavors.

In March, Russia launched Kosmos 2504, a maneuverable satellite intended to replace Kosmos 2491. Kosmos 2491 had met with an unfortunate end due to a suspected breakup, possibly caused by residual fuel in its propulsion system. Kosmos 2504 was part of Russia's ongoing efforts to advance its satellite technology, particularly in military applications.

May 7, 2015, marked a significant moment for Russia as it celebrated the 70th anniversary of its victory over Nazi Germany with a grand display of military prowess on Moscow's Red Square. The highlight was the presentation of the Yars RS-24 intercontinental ballistic missile system, also known in NATO circles as the SS-18 Satan. This new missile, which succeeded the older Voevoda models, demonstrated enhanced capabilities with its superior speed and virtually unlimited range. President Vladimir Putin made bold claims about the destructive power of the Sarmat missiles, asserting that a few of these could potentially obliterate the entire east coast of the United States. In his 2018 State of the Nation address, Putin elaborated on the RS-28 Sarmat missile's specifications, noting its length of 35.3 meters (116 feet), weight of approximately 200 metric tons, and its ability to travel at Mach 5 speeds. This missile could deliver at least ten warheads across a distance of 7,800 miles, bridging the gap between Moscow and Washington, D.C. By 2022, the Sarmat missile was put into combat service.

In a significant development in military space operations, Russia integrated space assets into its military campaign in Syria throughout 2015. The Russian Ministry of Defense revealed the use of ten spacecraft, including Resurs-P and Kanopus satellites, initially intended for civilian remote sensing. These satellites were repurposed to adjust orbits and enhance coverage of conflict zones. General Valery Gerasimov, Chief of the General Staff, confirmed that these space assets were instrumental in target identification and improving the precision of coordinates.

December 3, 2015, saw the Russian Ministry of Defense release satellite and aerial imagery to support claims of Turkey's involvement in illegal oil trade with terrorist groups in Syria and Iraq. The imagery revealed oil trucks traveling from terrorist-controlled regions to Turkish ports, with around 1,720 trucks observed in makeshift parking areas. Approximately 3,200 trucks were also tracked moving from Iraq to a Turkish oil refinery. The photographs, taken on October 18 and November 14, 2015, highlighted the role of space reconnaissance in these investigations, though specific satellite details were not disclosed.

Internal power struggles and shifting priorities marked the early years of the Russian Space Agency. The Energia design bureau, for instance, decided to extend the Mir space station's operational life beyond 1999, contrary to the agency's initial plans. Similarly, developing the

Angara rocket stemmed more from Khrunichev's resource acquisition rather than a deliberate long-term agency strategy. By 2015, these dynamics culminated in the merger of the Federal Space Agency with the United Rocket and Space Corporation, resulting in the establishment of Roscosmos as the central authority for Russia's space endeavors.

On the International Space Station (ISS), 2015 continued to be a year of dynamic activity and international cooperation. In March, the Expedition 42 crew concluded their mission, making way for the Expedition 43 team. This crew conducted a broad range of experiments across human physiology, biology, physics, and Earth observation, while also performing essential maintenance tasks, including equipment replacements and spacewalks.

April saw the sixth operational mission of the SpaceX Dragon spacecraft to the ISS, delivering crucial supplies and experiments while returning with scientific samples and equipment. This mission coincided with the Orbital Sciences Cygnus spacecraft's fourth commercial cargo delivery to the station.

By June, the ISS welcomed the Expedition 44 crew, who replaced the departing Expedition 43 team. This period also saw the arrival of the Japanese HTV-5 spacecraft, which brought additional supplies and experiments to the station. September marked the transition to Expedition 45, with the crew conducting multiple spacewalks for maintenance and equipment installation, alongside continuing their scientific research.

China also made significant strides in its space program in 2015. The successful launch of the unmanned cargo spacecraft Tianzhou-1 in September, which docked with the Tiangong-1 space station, demonstrated China's growing capabilities and ambitions for a permanent manned space station.

December closed out the year with NASA achieving a major milestone. The unmanned test flight of the Orion spacecraft, designed for deep space exploration beyond low Earth orbit, completed two orbits around Earth before re-entering the atmosphere and landing in the Pacific Ocean. This test was crucial to NASA's broader goal of sending humans to Mars in the 2030s.

Throughout 2015, the ISS remained a focal point for international cooperation in space, with regular resupply missions and ongoing scientific research. The year also saw a notable rotation of ISS crews, including significant contributions from astronauts and cosmonauts such as Gennady Padalka, Mikhail Korniyenko, Scott Kelly, and Kjell N. Lindgren. These crew members played vital roles in maintaining and advancing the ISS's mission, contributing to both scientific research and station operations.

Additionally, in November, Russia tested its direct ascent anti-satellite missile, the PL-19 Nudol. China confirmed its development of the DN-3 satellite-killing missile, highlighting the ongoing advancements and competitive nature of space technology.

Expedition 43

Expedition 43 of the International Space Station (ISS) began with the launch of the Soyuz TMA-16M spacecraft on March 27, 2015. This mission marked the arrival of a new crew aboard the ISS, consisting of Russian cosmonauts Gennady Padalka and Mikhail Kornienko, alongside NASA astronaut Scott Kelly. Their objectives were to advance ongoing scientific research, maintain operational integrity, and contribute to the station's continuous functionality.

The crew's first major milestone came on April 14, 2015, with the launch of SpaceX's CRS-6 mission. This cargo resupply mission was critical for delivering essential supplies and scientific experiments to the ISS. On April 17, 2015, the CRS-6 spacecraft successfully docked with the station, offloading its cargo, which included groundbreaking experiments aimed at improving human health and exploring fundamental physics in space.

However, the expedition encountered a challenge on April 29, 2015, when the ISS Progress 59 spacecraft, an uncrewed cargo vehicle, experienced an anomaly. This incident underscored the difficulties in maintaining the ISS and ensuring a steady flow of necessary resources to the station.

The successful completion of the CRS-6 mission was marked by the spacecraft's splashdown on May 21, 2015. The safe return of its cargo highlighted the critical role of resupply missions in supporting the ISS's ongoing research and operational needs.

Expedition 43, which spanned from March 11, 2015, to September 11, 2015, was a pivotal period in the ISS's journey of scientific exploration and international collaboration. The crew during this expedition included Commander Terry Virts of NASA, Flight Engineers Samantha Cristoforetti from the European Space Agency (ESA), and Anton Shkaplerov from Roscosmos. They were joined later by Scott Kelly and Mikhail Kornienko, whose expertise further enriched the mission.

A key focus of Expedition 43 was to advance the ISS's scientific research capabilities and maintain operational efficiency. The crew conducted a diverse array of experiments, exploring fluid physics, material science, and biological studies. These investigations were crucial for understanding phenomena in microgravity and for preparing for future long-duration space missions beyond low Earth orbit.

In addition to their research efforts, the crew engaged in significant maintenance and upgrade tasks. This included the installation of new equipment and the repair of existing systems to ensure the station's long-term functionality. The astronauts performed multiple spacewalks (extravehicular activities, or EVAs) to install and maintain various components, including upgrades to the station's external hardware and systems.

Expedition 43 also continued to strengthen international partnerships. The collaborative efforts of NASA, ESA, and Roscosmos were central to the mission's success, reflecting a global commitment to space exploration. The crew's work demonstrated the importance of international cooperation in overcoming the challenges of operating a complex space laboratory and advancing human spaceflight.

A notable milestone of this expedition was the arrival and integration of the Dragon cargo spacecraft. Developed by the American private space transportation company SpaceX, Dragon delivered essential supplies and scientific equipment to the ISS. Dragon, initially known as Dragon 1, was designed to transport cargo rather than astronauts and flew 23 missions to the ISS between 2010 and 2020 before its retirement. Funded by NASA through the Commercial Orbital Transportation Services program and contracted under the Commercial Resupply Services (CRS) program, Dragon played a vital role in sustaining the station's operations and research endeavors.

Expedition 43 NASA Astronaut Scott Kelly, left, and Russian Cosmonauts Gennady Padalka, center, and Mikhail Kornienko of the Russian Federal Space Agency (Roscosmos)

The SpaceX Dragon resupply ship approaching the International Space Station as both spacecraft were soaring 267 miles above the African nation of Namibia.

Soyuz TMA-16M undocked from the ISS at 21:29 UTC on 11 September 2015, containing Gennady Padalka of Roscosmos and visiting crew members Andreas Mogensen of ESA (European Space Agency) and Aidyn Aimbetov of the Kazakh Space Agency. Following a deorbit burn, the Soyuz spacecraft's descent module reentered the Earth's atmosphere. The crew landed safely in Kazakhstan at 00:51 UTC on 12 September 2015, just over three hours after departing the ISS.

Expedition 44

In the summer of 2015, the International Space Station (ISS) witnessed a series of pivotal events, showcasing both the advancements and challenges of space exploration. The period began on June 28, 2015, when the SpaceX CRS-7 mission, part of NASA's Commercial Resupply Services program, faced a dramatic setback. Intended to deliver crucial supplies and scientific experiments to the ISS, the mission suffered a catastrophic failure shortly after launch, resulting in the loss of the spacecraft.

In response to this setback, the Russian space agency swiftly moved to ensure the station's operational continuity. On July 3, 2015, the ISS Progress 60 spacecraft was launched, carrying essential cargo and scientific equipment. It docked with the ISS on July 5, replenishing the station's supplies and supporting ongoing research efforts.

Expedition 44 crew portrait Soyuz 42 (Gennady Padalka, Mikhail Kornienko, Scott Kelly) and Soyuz 43 (Oleg Kononenko, Kimiya Yui and Kjell Lindgren)

A significant crew rotation followed on July 22, 2015, with the launch of Expedition 44/45. This mission transported a new team of astronauts to the ISS, including American astronaut Scott Kelly, Russian cosmonauts Mikhail Kornienko and Sergey Prokopyev, and ESA astronaut Samantha Cristoforetti. Their primary objective was to continue the vital work of conducting scientific experiments and advancing research in the unique microgravity environment of the ISS.

August 2015 saw further critical activities. On August 19, the Japanese cargo spacecraft HTV-5, also known as Kounotori 5, was launched. Designed to deliver crucial supplies, HTV-5 was successfully captured by the ISS's robotic arm on August 24 and securely attached, allowing the crew to unload its cargo. Concurrently, the ISS Progress 58 spacecraft undocked on August 14, completing its mission and clearing the way for new cargo and crew operations.

On September 1, 2015, the Soyuz Taxi Flight was launched, a mission integral to ensuring safe crew transport to and from the ISS. The Soyuz spacecraft docked with the station on September 4 and concluded its mission with a successful landing on September 11, marking the end of its operational cycle.

Expedition 44, from September 11 to December 11, 2015, was a particularly significant chapter in the ISS's history. This expedition, featuring a diverse international crew, focused on understanding the effects of extended spaceflight on the human body. NASA commander Scott Kelly, alongside flight engineers Mikhail Kornienko, Kjell N. Lindgren, Oleg Kononenko, Sergey Revin, and Kimiya Yui, embarked on this mission with crucial objectives.

One of the primary research areas was the study of bone density changes experienced by astronauts in microgravity. Without the gravitational forces present on Earth, astronauts experience a reduction in bone mass similar to osteoporosis. Understanding these changes was essential for developing countermeasures to protect astronaut health during long-duration missions.

Muscle atrophy was another critical focus. In microgravity, the lack of gravitational force leads to muscle degradation as the muscles do not need to support the body's weight. The crew engaged in rigorous exercise routines to counteract this effect, and the data collected provided insights into muscle loss and the effectiveness of various exercise protocols.

Cardiovascular health was also examined during Expedition 44. The absence of gravity affects blood distribution and heart function in unique ways. Studying these effects was vital for ensuring astronauts' cardiovascular well-being on extended missions and upon their return to Earth.

In addition to human health studies, Expedition 44 continued a broad range of scientific experiments. Biological experiments investigated how microgravity impacts living organisms, enhancing our understanding of fundamental biological processes. Physics experiments explored material and fluid behavior without gravity, offering valuable insights with potential technological applications. Astronomy experiments leveraged the ISS's vantage point to observe celestial phenomena, advancing our knowledge of the universe. In July 2015, Kelly and Lindgren became the first Americans ever to eat food grown entirely in space.

Maintenance work was another essential component of the expedition. The crew undertook several spacewalks to upgrade and repair critical systems, including life support and power systems, ensuring the ISS's continued operational stability.

The successful completion of Expedition 44's objectives contributed to the ISS's ongoing role as a premier laboratory for scientific discovery and international collaboration.

In 2016, the landscape of human space exploration saw significant developments, marked by a blend of international cooperation and advancements in space technology. That year witnessed notable milestones both on Earth and in orbit, underscoring the ongoing progress of global space programs.

China, making strides in its ambitious space endeavors, launched the Tiangong-2 space laboratory in September 2016. This mission, a key component of the country's Project 921-2 space station program, was designed to pave the way for China's future space station. Tiangong-2 was to serve as a testbed for the technology and systems necessary for a permanent space habitat. By October 2016, China sent a manned mission to Tiangong-2, where the crew conducted a variety of experiments and tests over 33 days. This marked a significant leap forward in China's space program, showcasing its growing capabilities in long-term space missions. However, on July 19, 2017, as planned, China intentionally deorbited Tiangong-2, concluding its mission and demonstrating its commitment to evolving its space station technology.

Meanwhile, the International Space Station (ISS) remained a vital hub for scientific research and international collaboration throughout 2016. In March, the Expedition 46 crew concluded their tenure and was succeeded by the Expedition 47 crew. The latter group continued a broad range of scientific experiments and maintenance tasks. Their research spanned various disciplines, including human physiology, biology, physics, and Earth observation, reflecting the ISS's role as a versatile laboratory orbiting Earth.

April saw the arrival of critical supplies and experiments delivered by the SpaceX Dragon spacecraft, which also returned to Earth with scientific samples and equipment. Additionally, the Orbital ATK Cygnus spacecraft contributed by delivering its fifth commercial cargo to the ISS, highlighting the growing reliance on commercial partnerships for resupply missions.

The summer months brought further activity to the ISS with the arrival of the Expedition 48 crew in July. They were greeted by the Japanese HTV-6 spacecraft, which delivered additional supplies and experiments. The crew engaged in ongoing research and maintenance, ensuring the station's smooth operation.

October was a particularly busy month for the ISS. The Expedition 49 crew arrived, undertaking several spacewalks to install new equipment and conduct experiments related to biology, physics, and human physiology. These missions were crucial for the continued advancement of scientific knowledge and the enhancement of the ISS's capabilities.

Parallel to these activities, NASA demonstrated its commitment to advancing aerospace technology. The agency flew the ESAero/Tecnam X-57 Maxwell, a low-emission aircraft powered entirely by electric motors, as part of its Scalable Convergent Electric Propulsion Technology Operations Research (SCEPTOR) project. Furthermore, the Kratos XQ-58 Valkyrie was designated as an X-58, showcasing the progress in unmanned aerial systems.

China's Shenzhou 11 spacecraft, launched on October 16, 2016, carried Jing Haipeng and Chen Dong to Tiangong-2. This mission provided valuable experience for the Chinese space program, allowing them to test life-support systems and conduct an extended stay in a space laboratory. After a successful mission, Jing Haipeng and Chen Dong returned to Earth on November 18, 2016.

The year 2016 also saw significant events on Earth. On September 9, North Korea announced a successful nuclear weapon test, claiming to have developed a warhead capable of being mounted on a missile or long-range rocket. The test, accompanied by a 5.3 magnitude earthquake, marked a notable development in the region's geopolitical landscape.

In November, the United States experienced a major political shift with the election of Donald Trump as president, a surprising outcome in the political arena. Additionally, on December 15, the US Army Space and Missile Defense Command tested a US Army Zombie Pathfinder rocket at White Sands Missile Range, part of NASA's sounding rocket program for anti-ballistic missile scenarios.

The Japan Aerospace Exploration Agency (JAXA) Kounotori 5 H-II Transfer Vehicle (HTV-5) berthed to the International Space Station. The external CALET experiment, which will search for signatures of dark matter, being extracted from the unpressurized section by the station's robotic arm, Canadarm2. The unpiloted cargo craft, named "Kounotori," Japanese for "white stork," was loaded with more than four-and-a-half tons of research and supplies, including water, spare parts and experiment hardware.
NASA

The Soyuz TMA-17M spacecraft after it landed with Expedition 45 crew members Oleg Kononenko of the Russian Federal Space Agency (Roscosmos), Kjell Lindgren of NASA and Kimiya Yui of the Japan Aerospace Exploration Agency (JAXA) near the town of Zhezkazgan, Kazakhstan on Friday, Dec. 11, 2015. Kononenko, Lindgren, and Yui returned after 141 days in space where they served as members of the Expedition 44 and 45 crews onboard the International Space Station.

Expedition 45 crew portrait with (from left) Commander Scott Kelly and Flight Engineers Sergey Volkov, Mikhail Kornienko, Kjell Lindgren, Oleg Kononenko and Kimiya Yui.

Earlier in the year, on September 28, the HTV-5 spacecraft, also known as Kounotori 5, was successfully detached from the ISS. Having fulfilled its mission of transporting vital supplies and experimental equipment, the HTV-5's departure marked a transition, clearing the way for new cargo deliveries and maintaining the station's operational schedule.

December 2015 brought further vital support to the ISS with the launch of the Orbital ATK CRS-4 spacecraft on December 6. This mission, part of NASA's Commercial Resupply Services program, was designed to replenish the station's supplies and deliver scientific materials. By December 9, the spacecraft had arrived at the ISS, where the station's robotic arm expertly captured its cargo. This resupply operation was crucial, ensuring that the crew had all they needed for their research and daily activities.

The end of 2015 also saw the conclusion of Expedition 44/45, with the landing of the crew on December 11. This mission marked the end of the tenure for astronauts Scott Kelly, Mikhail Kornienko, Sergey Prokopyev, and Samantha Cristoforetti. Their time aboard the ISS was significant for its focus on long-duration spaceflight research, contributing valuable data to the understanding of human adaptability to extended space missions. Their return to Earth completed a successful chapter, providing insights that would benefit future space exploration endeavors.

Following their departure, Expedition 45 commenced on December 11, 2015, and continued until March 26, 2016. This expedition was a critical period for the ISS, emphasizing scientific research and technical maintenance. Led by Commander Scott Kelly of NASA and Flight Engineers Mikhail Kornienko of Roscosmos and Kjell N. Lindgren of NASA, the crew also included Flight Engineers Timothy Peake of ESA and Yuri Malenchenko and Sergey Prokopyev of Roscosmos. The diverse and international composition of Expedition 45 was integral to achieving the mission's objectives.

During this period, the crew focused on advancing various research projects, including studies on fluid dynamics, material science, and the effects of microgravity on biological systems. Their work enhanced our understanding of these scientific areas and laid the groundwork for future deep-space missions.

Expedition 45 also involved crucial maintenance tasks. The crew updated station systems and conducted multiple spacewalks to address hardware issues and upgrade essential components.

-

Expedition 46

Expedition 46 crew portrait with (from left) Commander Scott Kelly and Flight Engineers Sergey Volkov, Mikhail Kornienko, Timothy Kopra, Timothy Peake and Yuri Malenchenko.

Expedition 46, which began in December 2015, was a pivotal period for the International Space Station (ISS), marked by significant events and collaborative efforts that highlighted the ongoing operations and advancements in space research.

The mission commenced with the launch of the ISS Progress 62 spacecraft on December 21, 2015. This uncrewed Russian cargo spacecraft played a critical role by delivering vital supplies, including fuel, water, and scientific payloads necessary for the ISS's daily functions and ongoing research. Two days later, on December 23, Progress 62 successfully docked with the ISS, ensuring that the crew had the essential resources needed for their activities.

As the year transitioned into 2016, Expedition 46 continued to advance with various key milestones. On February 19, 2016, the Orbital ATK CRS-4 spacecraft, which had been part of

NASA's Commercial Resupply Services program, was released from the ISS. This spacecraft had fulfilled its mission of delivering supplies and experiments, and its departure marked the completion of its resupply role.

A notable event occurred on March 1, 2016, when the first crew of Expedition 46 concluded their mission and returned to Earth. This marked the end of their significant contributions, which included advancing scientific experiments and maintaining the ISS's functionality.

Expedition 46 officially ran from March 26, 2016, to September 6, 2016, and was characterized by a focus on scientific research and station maintenance. The crew, led by Commander Tim Kopra of NASA, included Flight Engineers Timothy Peake of ESA, Yuri Malenchenko of Roscosmos, Jeff Williams of NASA, and Oleg Skripochka of Roscosmos. Their collective efforts were instrumental in achieving the mission's objectives.

Scientific research during Expedition 46 was extensive, with the crew conducting fluid mechanics, combustion, and life sciences experiments. These studies were crucial for understanding how various processes operate in microgravity and for developing technologies essential for future space missions. Maintenance work was equally important, involving several spacewalks to install new equipment and repair existing systems, thereby ensuring the ISS's continued smooth operation.

Expedition 46 also witnessed several notable events and activities. On March 2, 2016, the Soyuz TMA-18M spacecraft landed near Zhezkazgan, Kazakhstan, bringing Expedition 46 Commander Scott Kelly and Russian cosmonauts Mikhail Kornienko and Sergey Volkov back to Earth. Their return concluded their significant contributions to the ISS mission.

Following their departure, the crew of Soyuz TMA-19M arrived at the ISS, marking the beginning of a new mission phase. The Progress M-28M spacecraft, launched on July 3, 2015, was used to resupply the ISS and was succeeded by the Progress MS-1, which launched on December 21, 2015. The Progress MS-1 delivery was crucial for replenishing supplies and supporting ongoing operations.

A key event during this period was a spacewalk conducted on December 21, 2015, by Commander Scott Kelly and astronaut Timothy Kopra. This contingency extravehicular activity (EVA) addressed issues with the Mobile Base System, ensuring its continued functionality. Progress MS-1 docked with the ISS on December 23, delivering essential cargo.

Expedition 46 also included an inspiring outreach event on January 8, 2016, when students from Sandringham School in St Albans made the first amateur radio call to British astronaut Tim Peake, highlighting the ISS's role in education and international engagement. On January 15, 2016, Tim Peake and Timothy Kopra conducted a notable EVA, during which Peake became the first British astronaut to perform a spacewalk. Unfortunately, the EVA was cut short due to water found in Kopra's helmet, echoing a similar issue experienced by astronaut Luca Parmitano in 2013.

In a lighter moment, on February 23, 2016, Scott Kelly humorously donned a gorilla suit and playfully chased Tim Peake through the ISS, with a video of the antics set to "Yakety Sax." Tim Peake's congratulatory video message to Adele at the 2016 Brit Awards further demonstrated the crew's engagement with the public.

Typhoon Soudelor photographed from the International Space Station on Aug. 5, 2015 while the storm was traveling in the western Pacific. The Soyuz TMA-17M (bottom left) and the Progress 60 (top left) cargo craft are visible.

The Soyuz TMA-18M spacecraft as it lands with Expedition 46 Commander Scott Kelly of NASA and Russian cosmonauts Mikhail Kornienko and Sergey Volkov of Roscosmos near the town of Zhezkazgan, Kazakhstan on Wednesday, March 2, 2016 (Kazakh time). Kelly and Kornienko completed an International Space Station record year-long mission to collect valuable data on the effect of long duration weightlessness on the human body that will be used to formulate a human mission to Mars. Volkov returned after spending six months on the station.

The ISS Progress 60 (60P) cargo craft shortly after undocking from the International Space Station. The uncrewed Russian resupply ship spent 166 days attached to the Pirs docking compartment and was packed with trash and other disposable items prior to departure.

Official Expedition 47 crew portrait with 45S crew (Yuri Malenchenko/Tim Kopra/Tim Peake) and the 46S crew (Jeff Williams, Oleg Skripochka, Aleksei Ovchinin)

Expedition 47 was a pivotal chapter in the history of the International Space Station (ISS), unfolding between March and June 2016 and continuing into the subsequent year. This expedition marked a period of significant advancements and operational milestones for the ISS, featuring notable crew rotations, critical cargo missions, and groundbreaking scientific experiments.

The journey began on March 18, 2016, with the launch of Expedition 47/48, ushering in a new phase for the ISS. The incoming crew, comprising Commander Timothy Kopra of NASA, Flight Engineer Timothy Peake of ESA, and Flight Engineer Yuri Malenchenko of Roscosmos, was set to join the existing team aboard the ISS. Their mission was to advance scientific research, maintain station operations, and contribute to various research projects essential for the station's ongoing success.

The ISS Progress 61 spacecraft, launched on March 31, 2016, played a critical role during this period. It docked with the ISS on April 2, delivering a substantial cargo load that included essential supplies, scientific experiments, and equipment necessary for the ISS crew's daily operations and research endeavors. This resupply mission was pivotal for sustaining the station's activities.

Adding to the support of the ISS's operations, two other significant cargo missions followed. On March 22, 2016, the Orbital ATK CRS-6 mission was launched and successfully docked with the station on March 26, bringing a variety of scientific experiments, hardware, and supplies. Shortly thereafter, on April 8, 2016, the SpaceX CRS-8 mission launched and docked with the ISS on April 10. This mission, part of NASA's Commercial Resupply Services, was notable for delivering the Bigelow Expandable Activity Module (BEAM), a prototype inflatable habitat designed to expand once in space. The BEAM was installed on the ISS's Tranquility module and, as of May 2022, has continued to serve as a testbed for expandable habitat technology.

As Expedition 47 transitioned into its operational phase, the focus shifted towards scientific experimentation and station maintenance. The crew, now including Flight Engineers Peggy Whitson of NASA and Andrey Babkin of Roscosmos, engaged in a wide range of research activities. Their studies encompassed materials science, fluid dynamics, and biological research aimed at understanding the effects of microgravity and preparing for future deep-space missions.

Significant maintenance and upgrades were carried out on the ISS during this period. The crew conducted several spacewalks to address hardware issues, replace components, and install new systems, ensuring the station's functionality and safety. These efforts were crucial for the continued operation of the ISS and for supporting its complex scientific objectives.

In April 2017, the SpaceX Dragon spacecraft embarked on its tenth operational mission to the ISS, delivering essential supplies and scientific experiments while also returning valuable samples and equipment. Concurrently, the Orbital ATK Cygnus spacecraft completed its seventh commercial cargo delivery, further supporting the ISS's operational needs.

The arrival of Expedition 52 in June 2017 marked another transition in crew composition. The team focused on maintaining ongoing research activities and conducting necessary maintenance tasks. The Russian Progress MS-06 spacecraft delivered supplies and experiments during this period.

By September 2017, the Expedition 53 crew took over, conducting various maintenance activities and scientific experiments. This period also included notable spacewalks to install new equipment and perform upgrades. China's space program made strides with successfully launching the Tianzhou-1 cargo spacecraft to the Tiangong-2 space station, advancing towards a permanently manned space station.

Throughout 2017, the ISS remained a central hub for international cooperation and scientific advancement. Crew rotations ensured a continuous presence of astronauts and cosmonauts, each contributing to the station's mission with their expertise and dedication. Notable figures included Peggy Whitson, who set the record for the most cumulative days in space by an American astronaut, and Thomas Pesquet, who conducted significant experiments and captured stunning images of Earth.

The year concluded with the launch of the Soyuz MS-07 spacecraft on December 17, bringing Anton Shkaplerov, Scott D. Tingle, and Norishige Kanai to the ISS. Their mission involved scientific research and station maintenance, continuing the legacy of exploration and cooperation that defines the ISS.

NASA and Russian cosmonauts arrive at the international space station Mar 19, 2016. NASA astronaut Jeff Williams and Russian cosmonauts Alexey Ovchinin and Oleg Skripochka docked to the International Space Station in their Soyuz TMA-20M spacecraft. They will spend 6 months onboard the station carrying out scientific experiments, performing maintenance and other duties.

Expedition 48

Expedition 48 crew portrait with 46S crew (Jeff Williams, Oleg Skripochka, Aleksei Ovchinin) and 47S crew (Anatoli Ivanishin, Kate Rubins, Takuya Onishi).

Expedition 48 to the International Space Station (ISS) encompassed a dynamic and pivotal period in the summer of 2016, marked by significant operational transitions and the continuation of crucial scientific work. The mission, which began on July 1, 2016, saw the undocking of the Progress 62 spacecraft. This maneuver was critical for managing cargo and maintaining the station's operational efficiency. The Progress 62, a Russian cargo vehicle, had been essential in resupplying the ISS with vital supplies and scientific equipment. Its departure was followed by a final undocking on July 3, 2016, clearing the way for new spacecraft to arrive and sustain the ISS's ongoing research and daily functions.

On July 6, 2016, the Expedition 48/49 mission was launched aboard a Soyuz spacecraft. This mission was instrumental in rotating crew members and exchanging scientific experiments. The spacecraft docked with the ISS on July 9, 2016, bringing aboard Commander Jeffrey Williams from NASA, Flight Engineers Oleg Skripochka and Alexey Ovchinin from Roscosmos, and Kate Rubins, also from NASA. The arrival of this new crew was crucial for advancing the research objectives and operational tasks on the station.

The ISS continued to receive vital supplies with the launch of Progress 64 on July 16, 2016. This Russian resupply spacecraft docked with the ISS on July 18, delivering additional materials necessary for the station's operations and research. On the same day, SpaceX CRS-9 was launched, a commercial resupply mission part of NASA's Commercial Resupply Services program. This mission aimed to deliver cargo, including scientific experiments and equipment. SpaceX CRS-9 docked with the ISS on July 20, further supporting the station's research endeavors and operational needs.

Expedition 47's culmination occurred on September 6, 2016, with the landing of the previous expedition's crew. This marked the end of their mission and allowed the astronauts to return to Earth, concluding their time aboard the ISS.

At the Gagarin Cosmonaut Training Center in Star City, Russia, Expedition 46-47 backup crew members Anatoly Ivanishin of the Russian Federal Space Agency (Roscosmos, left), Kate Rubins of NASA (center) and Takuya Onishi of the Japan Aerospace Exploration Agency

Expedition 48, which officially spanned from March 30, 2017, to September 2, 2017, continued to build on the ISS's scientific and operational achievements. Led by Commander Peggy Whitson of NASA, alongside Flight Engineers Jack Fischer of NASA and Fyodor Yurchikhin of Roscosmos, with Sergey Ryzhikov and Andrey Baskin of Roscosmos joining them, the crew focused on advancing scientific research and maintaining the station's operational efficiency.

During this expedition, the crew conducted a range of scientific experiments across various disciplines, including fluid mechanics, biology, and materials science. These experiments were critical in generating data that would inform future space missions and contribute to technological advancements.

One notable event was the spacewalk on August 19, 2016, conducted by astronauts Jeff Williams and Kate Rubins. This extravehicular activity (EVA) lasted nearly six hours and was pivotal for installing the International Docking Adapter (IDA) on Pressurized Mating Adapter-2 (PMA-2). The IDA was designed to facilitate the docking of future spacecraft with the ISS, thereby enhancing the station's capability to receive a variety of visiting vehicles. During this EVA, the crew also prepared for the installation of a second IDA by routing and securing cables.

On September 1, 2016, Williams and Rubins embarked on another significant EVA, lasting six hours and forty-eight minutes. This spacewalk involved retracting a thermal radiator from the Integrated Truss Structure's segment P1. Initially deployed during Expedition 33 to isolate a coolant leak, the radiator's retraction had been deferred from a previous spacewalk in Expedition 45. This EVA also included the installation of high-definition video cameras to improve visual monitoring of the ISS, additional torque application to the Solar Array Rotary Joint (SARJ) bolts, and maintenance tasks such as securing the Crew and Equipment Translation Aid (CETA) Cart brake handles to ensure the smooth operation of the station's components.

Sunrise panorama taken by Jeff Williams during Expedition 48.

Expedition 49

Expedition 49 official crew portrait with (from left) Sergei Ryzhikov, Shane Kimbrough, Andrei Borisenko, Kate Rubins, Anatoli Ivanishin, and Takuya Onishi.

In the annals of spaceflight history, Soyuz MS-02 stands out as a pivotal mission. Originally slated for launch on September 23, 2016, this mission experienced a delay due to technical issues, with the revised launch date set for October 19, 2016. Soyuz MS-02 was instrumental in ferrying three members of the Expedition 49 crew to the International Space Station (ISS). This mission marked the 131st flight of a Soyuz spacecraft, underscoring its critical role in ensuring the operational continuity of the ISS.

The Soyuz MS-02 spacecraft, launched from Baikonur Cosmodrome atop a Soyuz-FG rocket, successfully docked with the ISS on October 21, 2016. It attached to the Poisk (MRM-2) module, bringing aboard Commander Sergey Ryzhikov and Flight Engineers Andrei Borisenko, both of Roscosmos, and Shane Kimbrough from NASA. For Ryzhikov and Borisenko, this mission was their first spaceflight, while Kimbrough was on his second mission. The backup crew comprised Alexander Misurkin, Nikolai Tikhonov, and Mark T. Vande Hei.

The Soyuz MS-02 mission lasted 173 days and concluded with the spacecraft's return to Earth on April 10, 2017. However, the descent was marred by a technical anomaly. About eight kilometers above the ground, a partial depressurization occurred when a buckle from the parachute deployment system struck a welding seam. Despite this malfunction, the crew's safety was ensured by their altitude and the capsule's design. Russian officials attributed the issue to a packing error in the parachute system.

Throughout its mission, Soyuz MS-02 completed 2,768 orbits around Earth, covering approximately 118 million kilometers. The spacecraft, built by RKK Energia and modeled as Soyuz MS 11F732A48, demonstrated the resilience of Soyuz spacecraft in handling extended missions and overcoming unforeseen challenges. The mission's total duration was 173 days, 3 hours, 16 minutes, and 21 seconds. Upon undocking from the ISS at 07:57 UTC on April 10, 2017, Soyuz MS-02 safely landed on the steppes of Kazakhstan at 11:20 UTC, marking a successful conclusion to the mission.

The Soyuz MS-02 spacecraft was lowered from a vertical position prior to being encapsulated in its fairing on Thursday, Sept. 15, 2016 at the Baikonur Cosmodrome in Kazakhstan. Expedition 49 flight engineer Shane Kimbrough of NASA, flight engineer Andrey Borisenko of Roscosmos, and Soyuz commander Sergey Ryzhikov of Roscosmos are scheduled to launch to the International Space Station aboard the Soyuz MS-02 spacecraft from the Baikonur Cosmodrome on September 24 Kazakh time.

Expedition 49, which commenced with the arrival of Soyuz MS-02, ushered in a period of significant activity aboard the ISS during the fall of 2016. On October 14, 2016, the ISS Progress 63 spacecraft undocked after delivering essential supplies and equipment, making way for new missions. The critical Soyuz launch on October 19, 2016, was followed by the spacecraft's docking with the ISS on October 21, 2016. This docking marked the official beginning of Expedition 49, which included Commander Shane Kimbrough of NASA, Flight Engineers Sergey Ryzhikov and Andrey Babkin from Roscosmos, and Thomas

Pesquet from the European Space Agency (ESA). Their arrival heralded a new phase of scientific research and station operations.

In parallel, Orbital ATK CRS-5, a commercial resupply mission, launched on October 17, 2016, and docked with the ISS on October 23, 2016. This mission, part of NASA's Commercial Resupply Services program, was crucial for delivering vital supplies and scientific experiments necessary for the ISS's ongoing research and logistical needs. The transition of responsibilities from Expedition 48 to Expedition 49 was marked by the return of the previous crew on October 29, 2016, signaling the successful handover and continuation of ISS operations.

Expedition 49 itself, running from September 2, 2017, to December 13, 2017, was a period of significant scientific and technical advancement. The crew, comprising Commander Randy Bresnik and Flight Engineers Mark T. Vande Hei, Alexander Gerst, Sergey Prokopyev, Oleg Kononenko, and Paolo Nespoli, worked collaboratively to achieve key mission objectives. Their efforts focused on experiments in human health, material science, and physics, aimed at understanding the effects of microgravity on biological systems and physical processes critical for future long-duration space missions. Maintenance activities were also integral, with several spacewalks conducted to address hardware issues and upgrade systems, ensuring the ISS's continued functionality.

The year 2017 also witnessed notable advancements in astronaut programs worldwide. On June 7, NASA introduced Group 22, known as The Turtles, a diverse cohort of new astronaut candidates, including Kayla Barron, Zena Cardman, Raja Chari, Matthew Dominick, Robert Hines, Warren Hoburg, Jonny Kim, Jasmin Moghbeli, Loral O'Hara, Francisco Rubio, and Jessica Watkins. Their selection highlighted NASA's commitment to nurturing a new generation of explorers. Robb Kulin, initially part of Group 22, resigned in August 2018, underscoring the rigorous demands of astronaut training.

The Canadian Space Agency (CSA) also expanded its astronaut corps, announcing Jennifer Sidey and Joshua Kutryk as new recruits on July 1, 2017. This selection emphasized Canada's ongoing contributions to space exploration. Meanwhile, Germany's 2017 Die Astronautin Selection program, which initially selected Nicola Baumann and later replaced her with Suzanna Randall, aimed to advance women's participation in European space missions, promoting inclusivity and diversity in astronaut training and space exploration.

The Soyuz MS-02 spacecraft was lowered from a vertical position prior to being encapsulated in its fairing at the Baikonur Cosmodrome in Kazakhstan for. Expedition 49 flight engineer Shane Kimbrough of NASA, flight engineer Andrey Borisenko of Roscosmos, and Soyuz commander Sergey Ryzhikov of Roscosmos to the International Space Station aboard the Soyuz MS-02

Expedition 50

Expedition 50 official crew portrait with (from left) Andrei Borisenko, Shane Kimbrough, Sergei Ryzhikov, Thomas Pesquet, Peggy Whitson and Oleg Novitsky.

In the latter part of 2016 and early 2017, the International Space Station (ISS) witnessed a series of significant events, underscoring the continued collaboration and progress in space exploration.

On November 17, 2016, Soyuz MS-03 was launched from the Baikonur Cosmodrome in Kazakhstan, marking a pivotal moment for the ISS. Soyuz MS-03 was the 132nd flight of a Soyuz spacecraft and played a crucial role in transporting members of the Expedition 50 crew to the space station. The mission's crew comprised Russian commander Oleg Novitsky, French astronaut Thomas Pesquet, and American flight engineer Peggy Whitson. This launch was particularly notable as Peggy Whitson, at 56 years old, became the oldest woman to fly into space.

The Soyuz MS-03 spacecraft docked with the ISS on November 19, 2016, precisely at 21:58 UTC, attaching to the Rassvet module. This successful docking facilitated the crew and equipment transfer, allowing them to begin their mission of scientific research and station maintenance. The crew's tasks included conducting fluid dynamics, human physiology, and materials science experiments—essential for understanding microgravity effects and advancing technologies for future space exploration.

During this period, the ISS experienced a technical anomaly on December 1, 2016, involving the ISS Progress 65 spacecraft. While details of the anomaly were not extensively documented, it highlighted the complexities and challenges in managing the space station's sophisticated systems.

Throughout this time, the ISS also received critical resupply missions. On December 9, 2016, the HTV-6 Kounotori, a Japanese cargo spacecraft, launched and subsequently docked with the station on December 13, delivering essential supplies, scientific experiments, and equipment. The Orbital ATK Cygnus spacecraft from the CRS-5 mission had already been released from the ISS on November 21, 2016, following the completion of its resupply objectives.

A unique celebration on December 25, 2016 marked the holiday season aboard the ISS. The crew, experiencing the enchantment of microgravity, exchanged Christmas presents that had been delivered by a Japanese cargo spacecraft. In a festive touch, one astronaut donned a Santa hat, adding a whimsical element to the holiday observance. French astronaut Thomas Pesquet enhanced the festivities by sharing traditional French cuisine with his fellow crew members, creating a special culinary experience in orbit. Additionally, Pesquet produced a Christmas-themed video for the European Space Agency (ESA), capturing the spirit of the season and the camaraderie among the astronauts.

In early 2017, additional cargo operations were carried out. After fulfilling its mission, the HTV-6 Kounotori was undocked from the ISS on January 27, 2017. This was followed by the undocking of ISS Progress 64 on January 31, 2017, which cleared the docking port for future spacecraft.

The SpaceX CRS-10 mission, launched on February 19, 2017, carried crucial supplies and scientific experiments to the ISS. Shortly after that, ISS Progress 66 was launched on February 22, 2017, and successfully docked with the ISS on February 24, 2017, further supporting the station's operational needs.

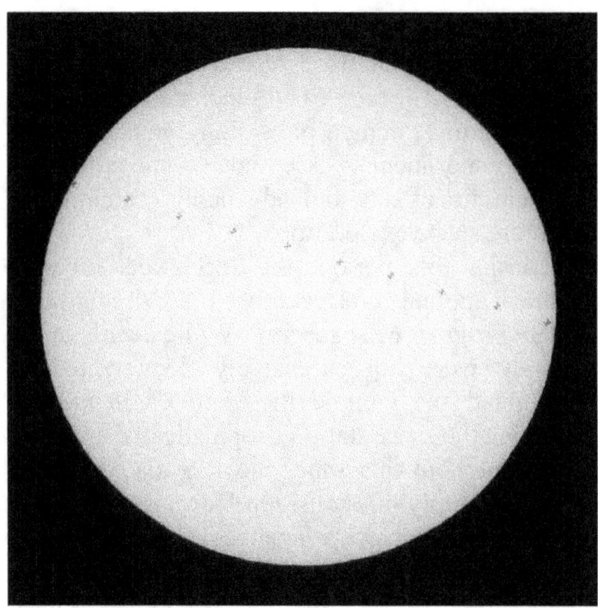

This composite image, made from ten frames, shows the International Space Station, with a crew of six onboard, in silhouette as it transits the sun at roughly five miles per second, Saturday, Dec. 17, 2016, from Newbury Park, California. Onboard as part of Expedition 50 are: NASA astronauts Shane Kimbrough and Peggy Whitson: Russian cosmonauts Andrey Borisenko, Sergey Ryzhikov, and Oleg Novitskiy: and ESA (European Space Agency) astronaut Thomas Pesquet.

The six-person Expedition 50 crew inside the Columbus lab module. (Top row from left) Thomas Pesquet, Peggy Whitson and Oleg Novitskiy. (Bottom row from left) Andrey Borisenko, Kimbrough and Sergey Ryzhikov.

The Expedition 50 crew, consisting of six members, took a moment to document their presence in space with a group portrait inside the Columbus laboratory module. The photograph featured (top row from left) Thomas Pesquet, Peggy Whitson, and Oleg Novitskiy, alongside (bottom row from left) Andrei Borisenko, Robert S. Kimbrough, and Sergey Ryzhikov.

Throughout Expedition 50, the International Space Station continued to host a series of uncrewed resupply missions, ensuring the station was well-equipped and supplied for ongoing research and daily operations. These resupply missions were crucial in maintaining the station's readiness for scientific experiments and crew activities.

The International Space Station (ISS) continued its mission of scientific research and international cooperation during Expedition 49/50. Several spacecraft visited the station, each contributing to its operations and logistics in distinct ways.

The Japanese HTV-6 cargo vehicle grappled by the International Space Station's robotic arm after arrival. HTV-6 launched from the Tanegashima Space Center in southern Japan on Friday, Dec. 9 and arrived at the space station on Tuesday, Dec. 13. The vehicle was loaded with more than 4.5 tons of supplies, water, spare parts and experiment hardware for the six-person station crew, including six new lithium-ion batteries and adapter plates that will replace the nickel-hydrogen batteries currently used on the station to store electrical energy generated by the station's solar arrays.

The Cygnus CRS OA-5 mission, operated by Northrop Grumman, was launched from the United States on October 17, 2016, at 23:45:40 UTC using the Antares 230 rocket. This mission was primarily focused on logistics, delivering essential supplies and equipment to the ISS. The Cygnus spacecraft docked with the station on October 23, 2016, at 11:28 UTC. After spending 29 days and 1 hour docked to the station, the spacecraft undocked on November 21, 2016, at 12:35 UTC. Cygnus CRS OA-5 was deorbited and re-entered Earth's atmosphere on November 27, 2016, at 23:36 UTC.

The Progress MS-04, a Russian logistics mission, was launched on December 1, 2016, at

14:51:45 UTC using the Soyuz-U rocket. Unfortunately, this mission encountered difficulties as the spacecraft separated from the launch vehicle's third stage before reaching orbit, resulting in the loss of the mission. Consequently, Progress MS-04 was unable to dock with the ISS.

Expedition 50 crew members Shane Kimbrough of NASA (left) and Thomas Pesquet of ESA (right) work inside the Cupola module to robotically capture the Japanese HTV-6 cargo craft. HTV-6 launched from the Tanegashima Space Center in southern Japan on Friday, Dec. 9 and arrived at the space station on Tuesday, Dec. 13. The vehicle was loaded with more than 4.5 tons of supplies, water, spare parts and experiment hardware for the six-person station crew, including six new lithium-ion batteries and adapter plates that will replace the nickel-hydrogen batteries currently used on the station to store electrical energy generated by the station's solar arrays.

Japan's Kounotori 6, also known as HTV-6, was launched on December 9, 2016, at 13:26:47 UTC aboard the H-IIB rocket. This mission was crucial for delivering logistics and cargo to the ISS. Kounotori 6 docked with the station on December 13, 2016, at 10:37 UTC. The spacecraft remained docked for 45 days and 5 hours before undocking on January 27, 2017, at 15:45 UTC. Kounotori 6 was deorbited and re-entered Earth's atmosphere on February 5, 2017, at 15:06 UTC.

The SpaceX CRS-10 mission, part of NASA's Commercial Resupply Services program, was launched from the United States on February 19, 2017, at 14:39:00 UTC using a Falcon 9 rocket. This mission was instrumental in supplying the ISS with cargo and scientific experiments. SpaceX CRS-10 docked with the station on February 23, 2017, at 13:12 UTC. The spacecraft spent 23 days, 19 hours, and 54 minutes docked to the ISS before undocking on March 19, 2017, at 09:11 UTC. The

spacecraft was deorbited and re-entered Earth's atmosphere on March 19, 2017, at 14:00 UTC.

Another Russian logistics mission, the Progress MS-05 mission, was launched on February 22, 2017, at 05:58:33 UTC using the Soyuz-U rocket. This spacecraft docked with the ISS on February 24, 2017, at 08:34 UTC. Progress MS-05 remained docked for an extended period of 146 days and 9 hours before undocking on July 20, 2017, at 17:46 UTC. The spacecraft was deorbited and re-entered Earth's atmosphere on July 20, 2017, at 20:58 UTC.

Expedition 49/50 featured a series of spacewalks, or Extravehicular Activities (EVAs), essential for maintaining, upgrading, and enhancing the International Space Station (ISS). These spacewalks involved intricate tasks and showcased the astronauts' skills in managing and upgrading the station's systems.

The first spacewalk of Expedition 49/50 took place on January 6, 2017, and lasted 6 hours and 32 minutes. American astronauts Peggy Whitson and Shane Kimbrough were the spacewalkers for this mission. Their tasks included installing adapter plates and cables for new batteries on the 3A power channel. They also took the opportunity to capture pictures of the Alpha Magnetic Spectrometer (AMS), removed a camera, and routed an Ethernet cable. These activities ensured the continued functionality and efficiency of the ISS's power and communication systems.

The second spacewalk occurred on January 13, 2017, and lasted 5 hours and 58 minutes. Shane Kimbrough, from the United States, and Thomas Pesquet, from France, conducted this EVA. The primary objectives included retrieving Adapter Plates E and F from the Express Pallet (EP) and installing these plates in specific slots. They also relocated Battery 4 to Adapter Plate F and installed Adapter Plate E in Slot 4. Additionally, they retrieved and installed Adapter Plate D in Slot 2, and fastened H1 bolts on Li-Ion batteries in Slots 1 and 5. Beyond these primary tasks, they completed several get-ahead tasks, including temporarily stowing Node 3 Shields Bundle #3, swapping the Mobile Transporter Relay Assembly (MTRA) Camera Light Pan and Tilt Assembly (CLPA), replacing the Latching End Effector (LEE) A Worksite Interface Fixture (WIF) Adapter, photo mapping the Rat's Nest from S0 aft to Z1 forward, securing Solar Array Blanket Boxes (SABB) restraints, and relocating Shields Bundle #2 to Node 3. These actions were integral to maintaining and upgrading the station's systems.

Shane Kimbrough and Thomas Pesquet conducted the third spacewalk on March 24, 2017, lasting 6 hours and 34 minutes. The spacewalk involved preparing the Pressurized Mating Adapter (PMA) for relocation, removing the EXT-2 MDM (Multiplexer/Demultiplexer) and replacing it with a new EPIC MDM (Enhanced Processor Integrated Circuit), and conducting camera work. They also lubricated the Canadarm2 end effector and inspected a radiator valve. These tasks were essential for maintaining the station's equipment and ensuring its operational efficiency.

The final spacewalk of the expedition occurred on March 30, 2017, and lasted 7 hours and 4 minutes. Shane Kimbrough and Peggy Whitson were the astronauts performing this EVA. They focused on removing and replacing the EXT-1 MDM with a new EPIC MDM, installing Node 3 axial shields, and replacing a lost shield with the PMA-3 cover. Additionally, they installed PMA-3 cummerbunds, removed the PMA-3 cover, completed PMA-3 connections, and closed the Node 3 port Common Docking System (CDC). They also inspected and cleaned the Earth-facing berthing port of the Harmony module. These tasks were crucial for maintaining the ISS's structural integrity and operational capabilities.

In the 2018 international space race, significant advancements in global space exploration programs highlighted the diverse and expanding landscape of astronaut training and missions.

On August 10th, Russia's 17th Cosmonaut Group was announced, comprising Konstantin Borisov, Alexander Gorbunov, Alexander Grebenkin, Sergei Mikayev, Kirill Peskov, Oleg Platonov, Yevgeny Prokopyev, and Alexei Zubritsky. Following rigorous training, all members except Yevgeny Prokopyev passed the state exam in December 2020, qualifying them for future spaceflight assignments. Prokopyev, however, did not qualify initially and was reassigned to undergo further basic space training.

Meanwhile, on September 3rd, the United Arab Emirates (UAE) introduced its Emirati Astronaut Group, consisting of Hazza Al Mansouri and Sultan Al Neyadi. Al Mansouri and Al Neyadi were selected as candidates to fly to the International Space Station (ISS) aboard a Soyuz spacecraft. Al Mansouri completed his mission in 2019, while Al Neyadi served as his backup. In 2020, both astronauts were designated to train as full-fledged mission specialist astronauts in Houston, marking a significant milestone for the UAE's space program and integration into international space missions. Al Neyadi became the first Emirati astronaut to embark on a long-duration mission aboard SpaceX Crew-6 in March 2023.

Expedition 51

The five-member Expedition 51 crew consisted of (from left) Jack Fischer, Fyodor Yurchikhin, Thomas Pesquet, Peggy Whitson and Oleg Novitskiy.

In the spring of 2017, the International Space Station (ISS) was a hub of activity, marked by crew transitions and critical resupply missions, which highlighted the ongoing collaboration and operational continuity essential for the station's success.

On April 18, 2017, the Orbital ATK CRS-7 spacecraft, part of NASA's Commercial Resupply

Services program, lifted off from Cape Canaveral Air Force Station in Florida. This resupply mission was pivotal, carrying essential supplies, scientific experiments, and equipment to the ISS. Its successful launch supported the station's scientific objectives and day-to-day operations. After a journey of just under four days, CRS-7 docked with the ISS on April 22, 2017. The docking process was a critical operation, enabling cargo transfer and ensuring that the crew had immediate access to the materials necessary for their ongoing research and maintenance tasks.

As CRS-7 arrived, the ISS was also welcoming a new crew. On April 20, 2017, a new team of astronauts launched from the Baikonur Cosmodrome in Kazakhstan to join Expedition 51. This crew included NASA astronaut Jack Fischer, Russian cosmonaut Fyodor Yurchikhin, and European Space Agency (ESA) astronaut Paolo Nespoli. Their mission was to continue their predecessors' scientific work and maintenance efforts, ensuring a smooth transition and uninterrupted progress in space research.

The arrival of the Expedition 51/52 crew marked the beginning of their mission aboard the ISS, signaling the start of a new phase of exploration and scientific endeavor. Their integration into the station's operations was seamless, setting the stage for the continuation of vital research and international cooperation.

On June 2, 2017, the Expedition 50/51 crew concluded their mission and departed from the ISS. This handover marked a significant transition, as the crew members wrapped up their time on the station and transferred responsibilities to their successors. The departure underscored the cyclical nature of the ISS missions, ensuring that the station's scientific and operational activities continued without interruption.

Expedition 51, which ran from April to June 2017, was notable for several reasons. It began with the departure of the Soyuz MS-02 spacecraft on April 10, 2017, and concluded with the departure of Soyuz MS-03 on June 2, 2017. During this period, the ISS saw the continuation of Peggy Whitson's, Oleg Novitskiy's, and Thomas Pesquet's mission from Expedition 50 into Expedition 51. Notably, Peggy Whitson made history as the first woman to command two separate expeditions to the ISS. Her previous term as commander during Expedition 16 highlighted her significant contributions to space exploration and leadership in the ISS program.

The expedition saw a reduction in the number of participating Russian cosmonauts due to changes in the 2017 mission schedule. Only two cosmonauts were launched on Soyuz MS-04 on April 20, 2017. This adjustment brought the total crew number to five, highlighting a shift in crew composition for this expedition.

The transfer of command from Expedition 51 to Expedition 52 was officially carried out on June 1, 2017. The mission formally concluded on June 2, 2017, at 10:47 UTC, with the undocking of Soyuz MS-03. This event marked the end of a significant chapter in the ISS's ongoing mission to advance scientific knowledge and foster international cooperation in space.

During Expedition 51, the crew composition evolved across its duration, reflecting the dynamic nature of space missions.

Astronaut Shane Kimbrough (far left) handed over station command to Peggy Whitson (far right). In between, from left to right are Sergey Ryzhikov and Andrey Borisenko of Expedition 50 in black and Thomas Pesquet and Oleg Novitiskiy of Expedition 51 in blue.

At the beginning of the expedition in April 2017, the crew was led by Commander Peggy A. Whitson from NASA, marking her third spaceflight. Whitson, renowned for her expertise and leadership, was the only American astronaut serving as the commander during this phase. She was joined by Russian Flight Engineer Oleg Novitsky, embarking on his second spaceflight, and French Flight Engineer Thomas Pesquet, participating in his inaugural mission. This trio formed the initial team of Expedition 51.

As the expedition progressed, the crew was expanded with the arrival of Russian Flight Engineer Fyodor Yurchikhin, who brought extensive experience from his fifth and final spaceflight. Additionally, American Flight Engineer Jack D. Fischer joined the team on his sole spaceflight. This composition underscored the international collaboration and expertise present on the ISS during Expedition 51.

The launch of Soyuz MS-03 marked the beginning of a new chapter for both Expedition 50 and 51. On November 17, 2016, Soyuz MS-03 lifted off, carrying three crew members: Russian cosmonaut Oleg Novitskiy, NASA astronaut Peggy Whitson, and French astronaut Thomas Pesquet. After a successful journey, the spacecraft docked with the Rassvet module of the International Space Station (ISS) on November 19, 2016, establishing a critical link for the upcoming missions.

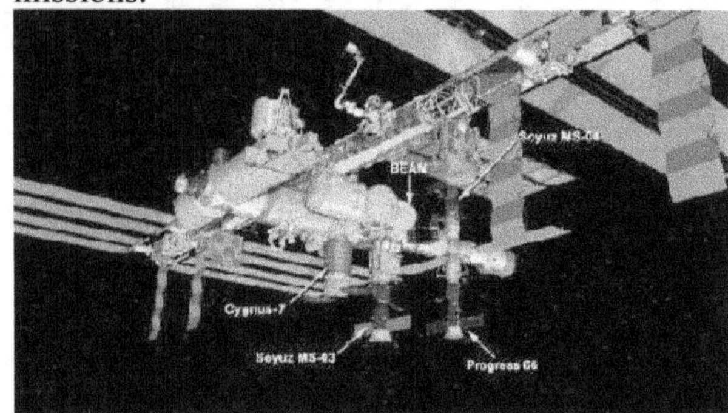

ISS Configuration - April 2017

As Expedition 51 officially began in April 2017, it signaled a transition of command and responsibility aboard the ISS. On April 9, 2017, the handover of station control took place. NASA astronaut Shane Kimbrough, the commander of Expedition 50, transferred command to Peggy Whitson of Expedition 51. The ceremonial handover included a formal exchange among crew members, with Kimbrough and his fellow cosmonauts Sergey Ryzhikov and Andrey Borisenko departing the station aboard Soyuz MS-02. Their spacecraft undocked at 7:57 UTC on April 10 and safely landed southeast of Dzhezkazgan, Kazakhstan, concluding their mission.

With Expedition 50's departure, the crew of Expedition 51, led by Whitson, immediately began their responsibilities. Whitson focused on assessing the impact of newly installed lighting systems on crew health and wellness. Meanwhile, Pesquet prepared for a spacewalk by managing the cooling water and purging gas buildup from the water tanks within the spacesuits.

Throughout Expedition 51, the crew adeptly balanced their roles between conducting scientific research and performing essential maintenance. Their work ranged from investigating the effects of extreme temperatures in microgravity on various materials to managing the water tanks in the Progress M-66 spacecraft. This multifaceted approach underscored the ISS's role as a hub of scientific inquiry and international cooperation, demonstrating the crew's adaptability and commitment to advancing our understanding of space.

Expedition 51 was marked by significant research initiatives aimed at advancing our understanding of space-based agriculture and protein crystallization.

One of the notable projects was the continuation of the Veggie plant growth experiments. The crew cultivated Chinese cabbage and Red Romaine lettuce within the Veggie facility, an integral part of the ISS's plant growth experiments. The Veg-03 mission aimed to validate the effectiveness of the Veggie plant growth chamber and its planting pillows. This research was pivotal for long-duration missions in space, where fresh food supply becomes essential for crew nutrition. Previous studies had explored plant productivity, but the confined space of the ISS had limited large-scale crop production trials. Veg-03 sought to overcome these limitations and enhance our ability to grow food in space, paving the way for future missions with extended durations.

In addition to agricultural experiments, Expedition 51 also focused on the JAXA Protein Crystal Growth #12 experiment. This research, conducted in the Japan Aerospace Exploration Agency's (JAXA) Ryutai rack within the Kibo laboratory module, involved installing and configuring two canisters containing 47 protein samples. Russian and Japanese researchers prepared these samples to grow high-quality proteins in the unique microgravity environment of the ISS. The experiment aimed to maintain a constant temperature for six weeks, facilitating the growth of these proteins to support developing drugs targeting multi-drug-resistant bacteria, Alzheimer's disease, muscular dystrophy, and periodontitis. This research underscores the ISS's role as a critical platform for advancing biomedical sciences and developing treatments for various health conditions.

In April 2017, Expedition 51 witnessed a series of significant arrivals that enhanced the International Space Station's operational capabilities and research potential.

On April 20, 2017, Soyuz MS-04 launched from the Baikonur Cosmodrome in Kazakhstan at 07:13 UTC. This mission carried Russian cosmonaut Fyodor Yurchikhin and American astronaut Jack Fischer to the ISS. After a successful journey, Soyuz MS-04 docked with the Poisk module at 13:18 UTC. The hatches between the spacecraft and the station were opened at 15:25 UTC, completing the Expedition 51 crew complement. This brought the total number of crew members aboard the ISS to five: Peggy Whitson, Oleg Novitskiy, Thomas Pesquet, Fyodor Yurchikhin, and Jack Fischer.

On the same day, the crew also managed the arrival of the Cygnus-7 cargo spacecraft. At 10:05 UTC, Peggy Whitson and Thomas Pesquet utilized the Space Station's robotic arm, Canadarm2, to capture Cygnus-7. The cargo spacecraft was then docked to the Earth-facing port of the Unity module at 12:39 UTC. Cygnus-7 delivered over 7,600 pounds of research equipment, supplies, and other critical materials to support both Expedition 51 and the upcoming Expedition 52.

On April 24, 2017, a significant milestone was achieved in space exploration as Peggy Whitson surpassed the United States record for the most cumulative time spent in space. With her achievement, she exceeded the previous record held by astronaut Jeff Williams, who had accumulated 534 days in space.

This remarkable accomplishment was celebrated with a congratulatory call from then-President Donald Trump. During the call, Whitson and her fellow astronaut Jack Fischer discussed with the President about NASA's ongoing research efforts and the ambitious plans for future space exploration, including the goal of sending humans to Mars by the 2030s.

Whitson (in red), transferred command to Yurchikhin (front row, right). Fischer (left). Behind (from left): Pesquet & Novitskiy.

In May 2017, Expedition 51 continued to address key scientific questions related to the effects of long-term space habitation on the human body. One major area of focus was the phenomenon known as Spaceflight-Associated Neuro-ocular Syndrome (SANS), which involves increased pressure behind the eyes due to the upward shift of bodily fluids in microgravity. Astronauts on the ISS regularly underwent ultrasound scans and vision tests as part of an extensive series of studies designed to help NASA better understand and mitigate these issues for future deep-space missions.

Another critical aspect of space research during this period was the study of bone loss, a common concern for astronauts spending extended periods in space. Peggy Whitson and Thomas Pesquet conducted the OsteoOmics bone study to investigate the molecular mechanisms underlying bone degradation in microgravity. This research was essential for developing strategies to counteract bone loss and maintain astronaut health on long-duration missions.

The crew also engaged in the "Fluid Shifts" study, a joint NASA-Roscosmos experiment designed to explore methods for counteracting the upward flow of fluids in the human body. Fyodor Yurchikhin, Jack Fischer, and Oleg Novitskiy tested a specialized suit designed to address this issue, contributing valuable data to the collaborative effort between the two space agencies.

Additionally, Thomas Pesquet carried out a student-submitted winning proposal for the "Genes in Space" competition. This study aimed to track how space travel impacts astronauts' DNA and immune systems, providing insights into the long-term effects of spaceflight on human genetics and health.

On May 19, 2017, Jack Fischer participated in the NeuroMapping experiment, which involved strapping himself into a device to study brain structure and function changes during spaceflight. This research sought to enhance understanding of how microgravity affects cognitive and neural processes, which was crucial for preparing astronauts for the cognitive challenges of extended missions.

In mid-May 2017, the International Space Station saw significant advancements in both research and infrastructure. On May 15, the Kibo laboratory module, operated by Japan's space agency JAXA, executed a major deployment of Cubesats. These small, versatile satellites were released to investigate the Earth's thermosphere and test new radar technologies. This deployment was part of ongoing efforts to enhance our understanding of space environments and improve satellite technology. Later in the month, on May 21 and 22, NanoRacks, a private company with facilities on the Kibo module, launched an additional 17 Cubesats, further expanding the scope of space research conducted aboard the ISS.

The month also marked a significant milestone in extravehicular activities (EVAs). On May 12, astronauts Peggy Whitson and Jack Fischer undertook a four-hour spacewalk, notable for being the 200th spacewalk in the history of the ISS. During this EVA, they replaced a large avionics box responsible for providing electricity and data connections to various science experiments on the station. Additionally, they installed a new connector to facilitate data routing to the Alpha Magnetic Spectrometer, repaired insulation at a connection point for the Japanese Robotic Arm and fitted a protective shield on the Pressurized Mating Adapter-3. Collectively, these spacewalks had accumulated 1,247 hours and 55 minutes of work outside the ISS.

Another critical EVA occurred on May 23, when Whitson and Fischer completed a 2-hour and 46-minute spacewalk to address a malfunction on the S0 truss. One of the two redundant multiplexer-demultiplexer (MDM) data relay boxes had failed, although this did not pose immediate danger to the crew. During this spacewalk, they replaced the faulty MDM-1 relay box and installed a pair of

antennas designed to enhance wireless communication for future spacewalks.

As June began, Expedition 51 prepared for its transition to the next crew rotation. On June 1, 2017, Peggy Whitson formally transferred command of the ISS to Fyodor Yurchikhin in a traditional Change of Command ceremony. The handover marked the end of Expedition 51 and the beginning of Expedition 52. The transition was completed with the departure of Soyuz MS-03 on June 2, 2017, carrying Oleg Novitskiy and Thomas Pesquet, who had been part of the Expedition 51 crew. This changeover ensured the continued smooth operation of the ISS and the advancement of ongoing scientific research.

Expedition 51 was notable for its significant spacewalks, or extravehicular activities (EVAs) , crucial in maintaining and upgrading the International Space Station (ISS).

The first spacewalk, designated EVA 1, took place on May 12, 2017. Peggy Whitson and Jack Fischer were the astronauts assigned to this mission. They began their spacewalk at 13:08 UTC and concluded at 17:21 UTC, making it 4 hours and 13 minutes. During this EVA, the crew accomplished several important tasks. They replaced the ExPRESS Carrier Avionics (ExPCA), a critical component for managing science experiments on the ISS. They also installed a forward shield on the Pressurized Mating Adapter-3 (PMA-3) to protect it from micrometeoroid impacts. Additionally, they installed a MIL-1553 terminator on the Alpha Magnetic Spectrometer (AMS), which was crucial for its data transmission. The astronauts secured multilayer insulation (MLI) on the Japanese Manipulator System and relocated a portable foot restraint to the PMA-3, facilitating future spacewalks.

The second spacewalk, EVA 2, occurred on May 23, 2017. Whitson and Fischer embarked on this EVA at 12:20 UTC and completed it at 15:06 UTC, for 2 hours and 46 minutes. This spacewalk addressed a hardware issue; the astronauts replaced a failed multiplexer-demultiplexer (MDM) data relay box, essential for managing communication and power distribution within the station's systems. They also installed two new wireless communication antennas to improve the station's communication capabilities for future spacewalks.

In addition to these human activities, Expedition 51 also saw several uncrewed resupply missions that were critical for sustaining the ISS's operations and research activities.

One significant resupply mission was conducted by the Cygnus CRS OA-7 spacecraft. Launched by United Launch Alliance on an Atlas V 401 rocket on April 18, 2017, at 15:11 UTC, Cygnus CRS OA-7 docked with the ISS on April 22, 2017, at 10:16 UTC. It remained docked for 43 days and 49 minutes, delivering essential supplies and scientific cargo to the station. The spacecraft undocked on June 4, 2017, at 11:05 UTC and was deorbited on June 11, 2017, at 17:08 UTC. This mission provided crucial logistics support for both Expedition 51 and the upcoming Expedition 52, ensuring the continuous operation and research activities aboard the ISS.

Expedition 52

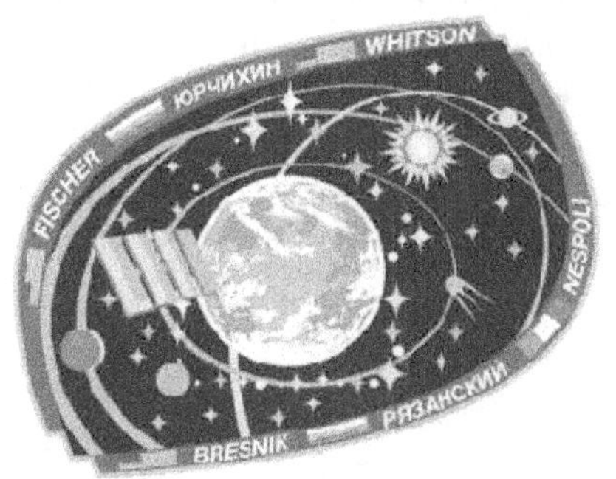

Expedition 52 to the International Space Station (ISS) marked a period of significant activities and transitions, involving a series of cargo missions and crew changes that were crucial for the ongoing research and operations on the station.

The sequence of events began with the launch of SpaceX CRS-11 on June 3, 2017. This mission was part of NASA's Commercial Resupply Services program, designed to deliver supplies and equipment to the ISS. The Dragon spacecraft launched from Kennedy Space Center in Florida, carrying a variety of scientific experiments and cargo.

Following the launch, SpaceX CRS-11 arrived at the ISS on June 5, 2017. The spacecraft was captured and berthed to the station, bringing vital supplies and experiments needed for ongoing research. The successful capture was a key operation, carried out with precision by the crew onboard the ISS.

On June 14, 2017, the ISS welcomed a new Progress spacecraft, Progress 67, which launched from the Baikonur Cosmodrome in Kazakhstan. This cargo mission was designed to deliver additional supplies, including fuel, water, and scientific experiments. Progress 67 docked with the ISS on June 16, 2017, further enhancing the station's capabilities and resources.

Before the arrival of Progress 67, the station experienced the release of Orbital ATK CRS-7 on June 4, 2017. This event marked the end of the mission for the Dragon spacecraft that had been previously berthed at the ISS, completing its cargo delivery and returning to Earth.

The final stages of Expedition 51, which were concluding as Expedition 52 was beginning, saw Progress 66 undock on July 20, 2017. The spacecraft had fulfilled its mission, and its departure made way for the upcoming crew and cargo operations.

Expedition 52, which spanned from June to September 2017, marked the fifty-second crew rotation aboard the International Space Station (ISS). The expedition officially commenced on June 2, 2017, at 10:47 UTC, following the undocking of Soyuz MS-03, which had been carrying the previous crew. The transition of command from Expedition 51 was completed the day before, on June 1, 2017.

In 2017, a strategic reduction in the number of Russian cosmonauts participating in space missions led to a temporary decrease in the ISS crew complement. Initially, only two crew members were launched aboard Soyuz MS-04, which reduced the onboard team to five astronauts. This adjustment was part of a broader plan to streamline crew numbers, reflecting changes in mission requirements and resources.

However, this reduction was short-lived. To maintain a full complement of six astronauts, Peggy Whitson extended her stay on the ISS, transitioning from Expedition 51. Her continued presence ensured the ISS could fully support its scientific and operational objectives. The crew number was bolstered further with the arrival of three new members on Soyuz MS-05, which arrived later in the summer.

During Expedition 52, which ran from June to September 2017, the International Space Station

(ISS) was home to a dynamic team of astronauts, each bringing unique skills and experiences to the mission. The crew was divided into two distinct phases, reflecting the rotation of personnel and the completion of their respective missions.

In the first phase, from June to July 2017, the ISS crew was led by Commander Fyodor Yurchikhin of Russia, embarking on his fifth and final spaceflight. Yurchikhin's extensive experience in spaceflight provided valuable leadership and continuity for the mission. The crew also included Flight Engineer 1, Jack Fischer from the United States, representing NASA. Fischer was on his sole spaceflight, contributing fresh expertise to the team. Flight Engineer 2, Peggy Whitson, also from the United States and representing NASA, was on her third spaceflight. Her extensive background and previous missions added significant depth to the crew's experience. Flight Engineer 3 was Randy Bresnik, another American astronaut from NASA, engaged in his second spaceflight. Rounding out the first phase was Flight Engineer 4, Sergey Ryazansky from Russia, who was on his second and final spaceflight. Lastly, Flight Engineer 5, Paolo Nespoli from Italy, representing ESA, was participating in his third and final space flight.

The Expedition 52 Commander Fyodor Yurchikhin and NASA astronaut Randy Bresnik. In the back row (from left) are, NASA astronauts Jack Fischer and Peggy Whitson, European Space Agency astronaut Paolo Nespoli and Roscosmos cosmonaut Sergey Ryazanskiy.

In the second phase, from July to September 2017, the crew changed to incorporate new members and continue the mission's objectives.

This phase featured a mix of returning and new astronauts, ensuring continuity and fresh perspectives. The transition underscored the dynamic nature of ISS operations, with crew rotations reflecting both planned schedules and the need to maintain a fully operational and effective team.

Expedition 52's commencement in June 2017 was marked by a series of significant events and activities aboard the International Space Station (ISS). On June 1, 2017, NASA astronaut Peggy Whitson formally handed over the ISS's command to Russian cosmonaut Fyodor Yurchikhin in a traditional Change of Command ceremony that began at 15:50 UTC. This ceremony symbolized the transition of leadership and the continuation of the mission's objectives under new command.

The official start of Expedition 52 was heralded by the departure of the Expedition 51 crew members. After a successful 194-day mission aboard the ISS, Oleg Novitskiy of Roscosmos and Thomas Pesquet undocked from the station at 10:47 UTC on June 2. Their Soyuz MS-03 spacecraft landed in Kazakhstan at 14:10 UTC, marking the conclusion of their mission and the beginning of Expedition 52.

SpaceX CRS-11 Dragon approaching the ISS. The river in the background was the Euphrates, running from right to left out of the Hadithah Dam Lake. The body of water at the bottom of the photo was Lake Tharthar. North was to the right of the photo.

In the days following, the ISS underwent a series of logistical updates. On June 3, a SpaceX Falcon 9 rocket, launching from Cape Canaveral's pad 39A, carried the Dragon resupply ship as part of SpaceX mission CRS-11. This mission delivered essential supplies and a new Roll Out Solar Array (ROSA) prototype to the station. The previous Cygnus OA-7 spacecraft, known as "SS John Glenn," was undocked by astronaut Jack Fischer on June 4 to make space for the new Dragon cargo ship. On June 5, Fischer and Whitson utilized the Canadarm2 to grapple the Dragon spacecraft and dock it on the Earth-facing side of the Harmony module.

In parallel, the Roscosmos Progress MS-06 launched from the Baikonur Cosmodrome on June 14, delivering over 3 tons of supplies to the ISS. This spacecraft docked at the aft port of the Zvezda module on June 16.

Scientific research during June was both diverse and impactful. Fischer and Whitson observed mold and bacteria samples as part of student-led biology experiments and conducted protein crystal studies. Whitson took care of rodents in the Rodent Habitat Facility to explore healing mechanisms and the efficacy of osteoinductive drugs in microgravity. She also performed experiments related to cardiac stem cells to understand the effects of space on aging. In addition, Fischer and Whitson initiated a botany study investigating how light and microgravity affect the growth of Arabidopsis thaliana seedlings.

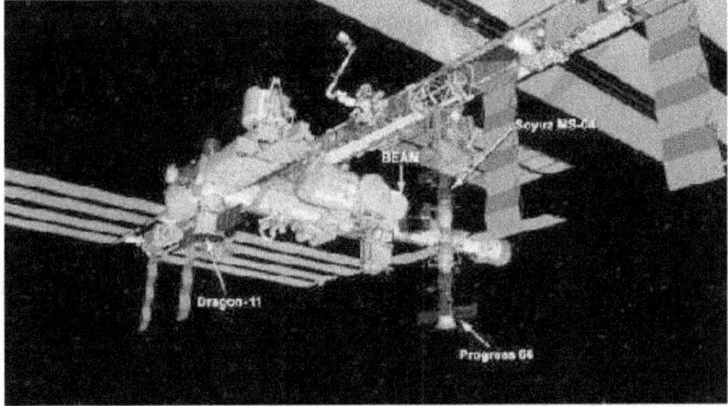

ISS Configuration after swapping Dragon for Cygnus, 5 June 2017

Fischer participated in a Vascular Echo study examining changes in blood vessels and the heart during spaceflight and their subsequent recovery on Earth. Yurchikhin focused on pain sensation research to aid in developing improved healthcare protocols for space missions. He, along with Whitson and Fischer, also conducted body measurements to provide data for comparative studies of in-flight, pre-flight, and post-flight conditions. Whitson began a cancer study to evaluate antibody-drug conjugates designed to enhance chemotherapy's effectiveness while minimizing its side effects.

In July 2017, Expedition 52 welcomed a full crew after the arrival of Soyuz MS-05. On July 3, at 06:41 UTC, NASA astronauts Jack Fischer and Peggy Whitson released the SpaceX Dragon CRS-11 from the International Space Station (ISS). The Dragon spacecraft's departure marked the end of its mission, which had delivered critical supplies and equipment to the ISS.

A series of significant research activities marked the month. Peggy Whitson participated in a Fine Motor Skills study to evaluate how prolonged exposure to microgravity affects fine motor skills. This study aimed to assess the effects of long-term spaceflight on motor skills and the phases of adaptation and recovery upon returning to Earth.

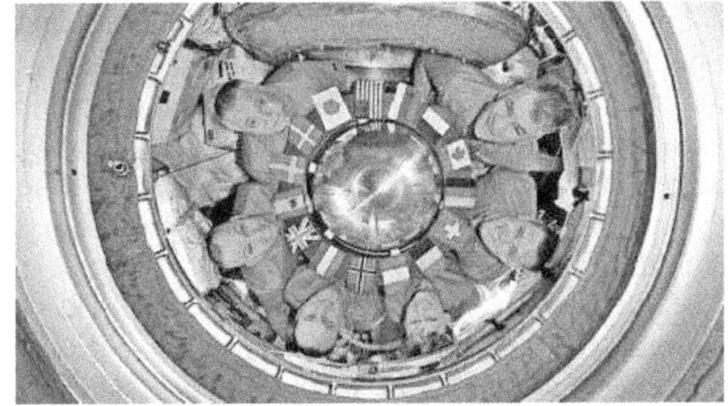

The Expedition 52 crew poses for a unique portrait. Pictured clockwise from top right are, Flight Engineers Paolo Nespoli, Jack Fischer, Peggy Whitson, Sergey Ryazanskiy, Randy Bresnik and Commander Fyodor Yurchikhin.

Jack Fischer concluded his work with the Group Combustion Module experiment, where he analyzed the behavior of decane droplets arranged on a thin-fiber lattice. This experiment was designed to measure flame and temperature distribution as the flame spreads, providing valuable data on combustion processes in microgravity.

Both Fischer and Whitson contributed to the Fluid Shifts experiment by collecting blood, urine, and saliva samples. This study sought to understand how fluids shift from the lower to the upper body in microgravity and its effects on the human eye, contributing to the broader understanding of physiological changes during spaceflight.

Whitson also initiated the Mag3D experiment, which involved magnetizing cells to facilitate their manipulation in microgravity. This innovative approach aimed to enhance the handling of biological samples during space missions. Additionally, she set up equipment for the Two-Phase Flow experiment, which investigated the behavior of perfluorohexane, an electronic coolant, in various conditions to study interfacial phenomena.

Fischer conducted high-intensity, low-volume exercise research using the station's exercise bike. His findings suggested that maximum-intensity exercise was more effective for improving aerobic capacity in microgravity than standard-intensity exercise.

In July 2017, the crew of Expedition 52 undertook crucial maintenance tasks to ensure the continued smooth operation of the International Space Station (ISS). Fyodor Yurchikhin focused on maintaining the life support systems on the Russian side of the station. This involved replacing pumps and hoses as needed and re-pressurizing the cabin to maintain optimal environmental conditions for the crew.

On the U.S. side of the station, Jack Fischer addressed a critical issue by replacing a failed water separator inside the Tranquility module. This component was integral to the Common Cabin Air Assembly, which plays a key role in regulating the station's humidity and temperature. Ensuring the proper function of this system was essential for maintaining a comfortable and safe environment for the astronauts.

In addition to this repair work, Fischer also installed new equipment in the Window Observational Research Facility (WORF). This facility was used for a variety of scientific observations and experiments, and the installation of new equipment was aimed at enhancing its capabilities and supporting ongoing research initiatives.

On July 28, 2017, the International Space Station (ISS) welcomed new crew members as the Soyuz MS-05 spacecraft launched from Baikonur Cosmodrome in Kazakhstan. Aboard were Randy Bresnik of NASA, Sergey Ryazanskiy of Roscosmos, and Paolo Nespoli of the European Space Agency (ESA). The launch marked the beginning of their mission to join Expedition 52.

The Soyuz MS-05 docked with the ISS later that same day, with the docking occurring while both spacecraft were positioned over Germany. This timely arrival added three experienced astronauts to the station's crew, enriching the ISS's ongoing research and daily operations. Adding Bresnik, Ryazanskiy, and Nespoli brought new expertise and perspectives to the mission, contributing to the station's dynamic and collaborative environment.

In August 2017, the International Space Station (ISS) continued its rigorous research and resupply operations schedule.

During this month, astronauts Paolo Nespoli and Randy Bresnik documented their experiences with space headaches. Research into these headaches revealed that they were likely caused by changes in cerebral blood flow and intracranial pressure, rather than the previously considered theory of space motion sickness.

Meanwhile, Jack Fischer and Peggy Whitson studied the effects of a new drug designed to combat mouse bone atrophy. Current therapies had been unable to restore lost bone. Still, this innovative drug, developed by researchers at the University of California, Los Angeles, showed promise in potentially rebuilding bone and preventing further loss. The ongoing research aimed to address the significant issue of bone density loss experienced by astronauts during extended space missions, which remains a critical area of study for long-duration spaceflight.

On August 16, 2017, the International Space Station (ISS) received a vital resupply shipment when SpaceX's CRS-12 Dragon arrived. Astronauts Jack Fischer and Paolo Nespoli played a key role in this operation by capturing the Dragon spacecraft using the station's robotic arm. They then carefully maneuvered and installed the spacecraft onto the ISS's Harmony module.

The Dragon cargo craft brought over 6,400 pounds of supplies, essential for maintaining the station's operations and supporting the crew's research activities. Among the delivered items was a special treat for the astronauts—ice cream, a welcome and rare indulgence for those living and working in space.

On August 17, 2017, Russian cosmonauts Fyodor Yurchikhin and Sergey Ryazanskiy conducted a spacewalk from the Pirs Docking

Compartment. During their extravehicular activity, they manually deployed five nanosatellites into orbit, contributing to small satellite technology and research advancement.

In addition to deploying the nanosatellites, Yurchikhin and Ryazanskiy collected test containers from various locations outside the Russian segment of the International Space Station (ISS). These containers were part of ongoing experiments designed to study the effects of the space environment on different materials and technologies.

During this extensive spacewalk, Yurchikhin and Ryazanskiy undertook a variety of tasks essential to the upkeep and advancement of the International Space Station (ISS). Their activities included retrieving the "Restavratsiya" (Restoration) experiment hardware, crucial for ongoing scientific research. They also launched five nanosatellites, one of a Sputnik satellite named "Zerkalo." This satellite was specifically launched to commemorate the 60th anniversary of the original Sputnik and honor the contributions of rocket scientist Konstantin Tsiolkovsky.

In addition to these significant tasks, the spacewalkers cleaned the windows on the Russian segment of the ISS and installed "Test" containers on the hatches of the Pirs Docking Compartment and the Poisk Module. They retrieved CKK 9M9 cassettes from the Zvezda module and installed various components, including struts, gap spanners, handrails, and ladders on both Zvezda and Poisk. These installations were preparatory measures for the future arrival of the Nauka module.

The astronauts also installed "Impact" trays near the Zvezda thrusters, photographed the aft end of Zvezda and the "OHA" antenna, and captured images of the station's Russian segment.

During Expedition 52, several uncrewed spacecraft visited the International Space Station (ISS), delivering essential supplies and conducting various logistical operations.

On September 3, 2017, the mission concluded with the triumphant return of Expedition 52's crew. The Soyuz MS-04 capsule touched down at 01:21 UTC, southeast of Dzhezkazgan in Kazakhstan. This landing marked the end of a notable chapter for the returning astronauts: Peggy Whitson, Jack Fischer, and Fyodor Yurchikhin.

Peggy Whitson's return was especially historic. Having completed a remarkable 288-day mission during her third extended stay aboard the ISS, Whitson surpassed 665 days in space throughout her career. This achievement set a new record for the longest cumulative time spent in space by a U.S. astronaut and placed her eighth in the history of human spaceflight.

Jack Fischer, who spent 136 days in space, and Fyodor Yurchikhin, who accumulated 673 days across his spaceflights, also made significant contributions—Yurchikhin's total time in space positioned him seventh on the all-time list at that time.

In the Integration Building at the Baikonur Cosmodrome in Kazakhstan, Expedition 51 crewmembers Fyodor Yurchikhin of the Russian Federal Space Agency (Roscosmos, left) and Jack Fischer of NASA (right) pose for pictures April 6 in front of their Soyuz MS-04 spacecraft as part of pre-launch training preparations. Fischer and Yurchikhin launched April 20 on the Soyuz MS-04 spacecraft for a four and a half month mission on the International Space Station.

Expedition 53

The six-member Expedition 53 crew Flight Engineer Alexander Misurkin of Roscosmos and Commander Randy Bresnik of NASA. Standing in the back (from left) are NASA astronauts Joe Acaba and Mark Vande Hei, Sergey Ryazanskiy of Roscosmos and Paolo Nespoli of the European Space Agency.

Expedition 53, from September to December 2017, was a pivotal chapter in the International Space Station's (ISS) history, marked by key transitions, significant missions, and notable international collaboration. This expedition began with the departure of Soyuz MS-04 on September 2, 2017, and concluded with the departure of Soyuz MS-05 on December 14, 2017.

The mission's timeline was punctuated by the arrival and departure of various cargo spacecraft, which were crucial in supporting ISS operations and scientific research. On November 12, 2017, the Orbital ATK CRS-8 cargo spacecraft was launched from Kennedy Space Center, Florida. As part of NASA's Commercial Resupply Services program, CRS-8 delivered essential supplies, scientific experiments, and equipment to the ISS. The spacecraft docked with the ISS on November 14, 2017, and was later unberthed on December 6, 2017, marking the end of its mission. In addition to CRS-8, the ISS Progress 68, a Russian cargo spacecraft, was launched on October 14, 2017, and docked with the ISS two days later, bringing crucial supplies for the station's ongoing operations.

Expedition 53 saw the transition of crew members from Expedition 52 to Expedition 53. On September 13, 2017, Soyuz MS-06 carried astronauts Alexander Misurkin, Mark T. Vande Hei, and Joe Acaba to the ISS, officially marking the start of their mission. Soyuz MS-06 docked with the Poisk module, integrating the new crew into the station's activities. The arrival of this spacecraft was a significant milestone, enabling the incoming astronauts to commence their duties.

The mission was characterized by meticulously coordinating crew activities and spacecraft operations. Commander Randolph Bresnik, Flight Engineer Sergey Ryazansky, and Flight Engineer Paolo Nespoli, who were part of Expedition 52, completed their duties and returned to Earth on December 14, 2017. Their departure was succeeded by the official handover to Expedition 53, which concluded with the undocking of Soyuz MS-05.

Expedition 53 featured a distinguished international crew. In the mission's first phase, from September 2 to September 13, 2017, Commander Randolph J. Bresnik from NASA led the team. Alongside him were Flight Engineer Sergey Ryazansky from the Russian Space Agency (RSA) and Flight Engineer Paolo Nespoli from the European Space Agency (ESA). Ryazansky and Nespoli were undertaking their final spaceflights while Bresnik, an experienced astronaut, commanded the expedition.

The second phase introduced new crew members from September 13 to December 14, 2017. Flight Engineer Alexander Misurkin, returning for his second spaceflight with RSA, and Flight Engineers Mark T. Vande Hei from NASA and Joseph M. Acaba, also from NASA, who were on their first and third spaceflights, respectively, joined the team. This phase marked a period of intense scientific research, maintenance, and operational continuity.

The transition between Expedition 52 and Expedition 53 was meticulously planned and executed. The arrival of Soyuz MS-05 on July 28, 2017, marked the beginning of the new crew's integration into the ISS. The spacecraft docked with the Rassvet module, facilitating the transfer of crew members and ensuring a smooth transition between expeditions.

On September 17, 2017, astronauts Randolph Bresnik and Paolo Nespoli released the SpaceX CRS-12 Dragon cargo spacecraft from the International Space Station (ISS) using the Canadarm2. This maneuver cleared the docking port on the Harmony module, allowing it to return to Earth.

The Dragon spacecraft descended to the Pacific Ocean, landing southwest of Long Beach, California. Onboard, it carried several important scientific experiments and research specimens aboard the ISS. Among these was the Lung Tissue Experiment, which involved growing lung tissue from stem cells in the unique environment of microgravity. This experiment aimed to advance the understanding of lung development and diseases.

Also included was the CASIS PCG7 study, which focused on growing larger versions of the protein LRRK2. This research was crucial for studying Parkinson's disease and its underlying mechanisms. Additionally, the Dragon carried mice from the Rodent Research-9 study, which investigated cartilage loss in hip and knee joints, a critical area of research for understanding joint health and degeneration.

In September 2017, several important research activities were conducted aboard the International Space Station (ISS) as part of Expedition 53 and the transition to Expedition 54.

Paolo Nespoli and Sergey Ryazansky were the subjects of a bone marrow experiment aimed at understanding the negative effects of microgravity on bone marrow and the blood cells it produces. This research was crucial for comprehending how long-term spaceflight affects hematopoiesis and overall bone health.

Randolph Bresnik participated in a study focusing on physiological changes induced by microgravity. As one of the 33 astronauts involved in NASA's Biological Specimen Repository, Bresnik's contributions were pivotal in studying the broader impacts of spaceflight on human physiology.

Mark Vande Hei worked on the Meteor experiment, which involved operating a specialized camera designed to capture high-resolution video and images of the Earth's atmosphere. This experiment provided valuable data on meteor chemical composition, enhancing our understanding of atmospheric phenomena from space.

Joe Acaba was responsible for setting up the Fast Neutron Spectrometer, an advanced instrument designed to measure neutron radiation in mixed radiation fields. This device offered more precise measurements, improving the understanding of radiation environments encountered during space missions.

Additionally, Ryazansky, Nespoli, and Bresnik participated in the Sarcolab-3 study, which examined the deterioration of the calf muscle at the Achilles tendon junction. This research aimed to investigate muscle atrophy and its implications for astronaut health during extended missions.

Bresnik and Nespoli also tested the Miniature Exercise Device-2 (MED-2), a compact exercise device utilizing small robotic actuators to provide motion and resistance. This innovation offered a more space-efficient solution for maintaining physical fitness onboard the ISS.

Acaba set up hardware for the Veggie-3 experiment, which involved growing Extra Dwarf Pak choi, amara mustard, and Red Romaine Lettuce. This experiment was designed to support future long-duration space missions by advancing the technology needed for growing food in space.

Finally, Ryazansky and Acaba installed radiation sensors on the U.S. segment of the ISS to characterize the station's radiation environment.

This data was essential for understanding the levels of radiation that astronauts are exposed to during their missions and for improving safety measures.

In October 2017, the International Space Station (ISS) witnessed significant progress with three extravehicular activities (EVAs) and the arrival of a Progress spacecraft.

The first of these spacewalks occurred on October 5, when astronauts Randolph Bresnik and Mark Vande Hei ventured outside the station for a mission that lasted 6 hours and 55 minutes. During this EVA, the primary objective was the replacement of two Latching End Effectors (LEE) on the Canadarm2, the station's robotic arm. The successful installation of these components was crucial for maintaining the arm's functionality, and essential for various tasks, including docking and capturing spacecraft.

In addition to the primary task, Bresnik and Vande Hei made notable progress by removing insulation from a direct current (DC) switching unit and preparing a flex hose rotary coupler. This preparation work was instrumental for subsequent maintenance and operational tasks on the ISS.

This spacewalk marked Mark Vande Hei's debut in extravehicular activities. At the same time, it was Randolph Bresnik's third spacewalk of his career, showcasing his extensive experience in performing complex tasks outside the station.

On October 10, 2017, Randolph Bresnik and Mark Vande Hei conducted their second extravehicular activity (EVA) within a week, marking a notable achievement in their mission schedule. This spacewalk, lasting 6 hours and 26 minutes, focused on several critical maintenance tasks.

The primary objective of this EVA was to lubricate the Latching End Effectors (LEE) that Bresnik and Vande Hei had installed during their previous spacewalk on October 5. Proper lubrication was essential to ensure the continued smooth operation of the Canadarm2's components.

Additionally, they addressed a malfunction by replacing a faulty camera system and a smudged lens cover, crucial for maintaining the quality of visual data used for station operations and navigation. The spacewalkers also removed two handrails from the exterior of the Tranquility module, a task necessary for preparing the module for future work and modifications.

This spacewalk was Vande Hei's second and final EVA of his mission, highlighting his quick adaptation to the demands of spacewalking and contributing to the successful completion of essential maintenance tasks.

Following a postponed launch attempt on October 12, 2017, the Progress MS-07 resupply spacecraft launched from the Baikonur Cosmodrome on October 14. After a precise and timely journey, the spacecraft docked with the International Space Station (ISS) on October 16, approximately 252 miles above eastern China.

The Progress MS-07 delivered a critical cargo load of three tons, comprising essential supplies, including food, fuel, and other necessary materials for the continued operation and support of the ISS. This resupply mission was vital for sustaining the crew's daily needs and maintaining the station's systems, ensuring the ISS could continue its research and operational activities without interruption.

EVA 3 on October 20, 2017

On October 20, 2017, astronauts Randolph Bresnik and Joe Acaba embarked on the mission's third extravehicular activity (EVA). This spacewalk, which lasted several hours, was focused on a series of critical tasks to enhance the functionality and maintenance of the International Space Station (ISS).

During this EVA, Bresnik and Acaba installed a new camera system on the end of Canadarm2's Latching End Effector (LEE), essential for improving the station's operational capabilities. Additionally, they mounted a high-definition camera on the ISS's starboard truss, further augmenting the station's observational and monitoring systems.

The spacewalk also involved replacing a fuse on the Dextre robotic arm, ensuring its continued reliability for various station tasks. Bresnik took on additional responsibilities, including installing a new radiator grapple bar, preparing one of two spare pumps, and initiating work on the second pump.

This spacewalk was notable for both astronauts: it marked Bresnik's fifth career EVA, extending his total extravehicular hours to 32, while it was Acaba's third spacewalk, increasing his total EVA hours to 19.

In November 2017, the crew aboard the International Space Station (ISS) engaged in a variety of scientific studies and outreach activities, alongside their routine maintenance and research tasks.

Astronauts Mark Vande Hei and Joe Acaba took part in educational outreach by sharing their experiences and insights with Shaker Heights High School students in Cleveland, Ohio. This initiative aimed to inspire the next generation by connecting students with real-world science and technology being conducted in space.

The crew also contributed to the Canadian Space Agency's "At Home in Space" study, which explored the concept of space culture. This research investigated how, despite their diverse backgrounds, astronauts develop a shared culture due to the unique and isolated environment of the ISS. The study sought to understand living in space's psychological and social dynamics.

Vande Hei was involved in setting up equipment to analyze the air within the space station for dust particles. This research was crucial for identifying potential health impacts and ensuring the well-being of the crew members by maintaining a clean and safe environment.

Paolo Nespoli participated in a Fine Motor Skills study examining how microgravity affects fine motor control and dexterity. This research aimed to assess how the unique environment of space influences motor skills, essential for both daily tasks and complex operations.

To address the issue of microgravity-induced vision impairment, astronauts Alexander Misurkin, Sergey Ryazansky, Nespoli, and Vande Hei conducted eye examinations on each other. This study aimed to monitor and understand the visual changes in the space environment, providing valuable data to mitigate potential impacts on astronauts' vision.

In November 2017, the International Space Station (ISS) received a vital resupply mission with the arrival of the Cygnus spacecraft, named in honor of astronaut Gene Cernan. After an initial launch attempt was scrubbed on November 11, the Cygnus resupply vehicle lifted off on November 12 atop an Orbital ATK Antares rocket from the Wallops Flight Facility in Virginia.

The spacecraft docked with the ISS on November 14, delivering approximately 7,400 pounds of essential research materials and supplies. This resupply was crucial for maintaining the station's ongoing scientific experiments and meeting the crew's daily needs.

Later in the month, the Bigelow Expandable Activity Module (BEAM) was prepared for future operations. The crew removed excess equipment, including inflation tanks and sensors, to make way for stowage operations. This process also involved placing the removed gear and other waste into the Cygnus spacecraft for disposal.

On December 5, astronauts Mark Vande Hei and Joe Acaba and ground controllers detached the Cygnus spacecraft from the ISS. This marked the end of a successful resupply mission and the commencement of the spacecraft's journey to re-enter the Earth's atmosphere and burn up safely.

Expedition 53 crew (from left) Nespoli, Ryazanskiy and Bresnik aboard Soyuz MS-05

On December 13, 2017, the International Space Station (ISS) witnessed a formal change of command ceremony, marking the transition from Expedition 53 to Expedition 54. Expedition 53 Commander Randy Bresnik officially handed over the station command responsibilities to Expedition 54 Commander Alexander Misurkin.

The ceremonial handover was a significant moment in the ISS's continuous operation, symbolizing the seamless collaboration between different crew expeditions. The transition was completed the following day when Soyuz MS-05 undocked from the station, signaling the end of Expedition 53 and the commencement of Expedition 54.

After undocking, Soyuz MS-05 touched down later that day southeast of Dzhezkazgan, Kazakhstan, concluding the mission. By the end of Expedition 53, the crew members had accumulated substantial time in space: Randy Bresnik had logged 150 days across two missions, Sergey Ryazansky had spent 306 days over two missions, and Paolo Nespoli had amassed 313 days in space across three missions. This marked the end of a highly productive and significant period aboard the ISS, setting the stage for the new crew's mission and ongoing research.

During Expedition 53, the crew conducted several critical spacewalks, each contributing significantly to the maintenance and enhancement of the International Space Station (ISS).

EVA 1, conducted on October 5, 2017, saw astronauts Randy Bresnik and Mark T. Vande Hei undertake a spacewalk that lasted 6 hours and 55 minutes. They began their extravehicular activity at 12:05 UTC and concluded at 19:00 UTC. This spacewalk focused on replacing the Latching End Effector (LEE-A) on the Canadarm2, a crucial component for the station's robotic operations. Additionally, the astronauts removed multi-layer insulation from a spare direct current switching unit and prepared a flex hose rotary coupler for future use.

On October 10, 2017, Randy Bresnik and Mark T. Vande Hei conducted their second spacewalk of Expedition 53. The spacewalk lasted 6 hours and 26 minutes, beginning at 12:56 UTC and ending at 19:22 UTC.

During this EVA, Bresnik and Vande Hei focused on completing several important tasks. They finished the repairs on the Canadarm2 by adding lubricating oil to its moving parts, ensuring smoother operation for future missions. The astronauts also replaced the station's cameras, crucial for NASA TV broadcasts, and installed new lens covers. Additionally, they closed and locked a latch on the high-pressure gas tanks and prepared a pump module for relocation to the P6 truss on a future spacewalk.

They also updated the degraded Latching End Effector on Canadarm2 by changing the sockets and reinstalling them on the new unit. Furthermore,

they removed handrails on the Tranquility module in anticipation of installing the EWS antennas on a future spacewalk.

On October 20, 2017, Randy Bresnik and Joseph M. Acaba conducted their third spacewalk of Expedition 53. During this spacewalk, Bresnik and Acaba undertook a range of crucial tasks to maintain and upgrade the International Space Station. They completed the repairs to the Canadarm2, adding lubricating oil to all its moving parts and installing a new camera while replacing a degraded one. They also replaced several station cameras used for NASA TV broadcasts and addressed a blown fuse on the Dextre robotic arm.

In addition to these tasks, the astronauts removed multi-layer insulation (MLI) from two Orbital Replacement Units (ORUs) stored on the Express Logistics Carrier (ESP2) in preparation for their movement by Dextre later in the year. The crew was able to perform three "get-ahead" tasks: they removed MLI from the Pump Modules on ESP2, although Bresnik was unable to complete the second module due to time constraints and had to close the flap on it, which will be moved on a subsequent spacewalk. They also installed radiator grapple bars delivered on the SpaceX CRS-12 mission, further enhancing the station's functionality and preparedness for future operations.

During Expedition 53, two uncrewed resupply spacecraft visited the International Space Station (ISS), each contributing significantly to the station's logistics and research capabilities.

The Progress MS-07, designated as ISS flight number 68P, was a Russian resupply mission. Launched atop a Soyuz-2.1a rocket on October 14, 2017, at 08:47 UTC, it docked with the ISS on October 16, 2017, at 00:00 UTC. The spacecraft remained attached to the station for 163 days and 13 hours before undocking on March 28, 2018, at 13:50 UTC. It was deorbited and re-entered Earth's atmosphere on April 26, 2018, at 04:08 UTC.

The Cygnus CRS OA-8E, launched by the United States, was designated CRS OA-8E. The spacecraft was launched aboard an Antares 230 rocket from Wallops Flight Facility on November 12, 2017, at 12:20 UTC. It docked with the ISS on November 14, 2017, at 10:04 UTC. The Cygnus remained attached to the station for 22 days and 3 hours before undocking on December 6, 2017, at

Expedition 54

The six-member Expedition 54 crew Flight Engineers Joe Acaba and Mark Vande Hei of NASA, Commander Alexander Misurkin of Roscosmos and Flight Engineers Anton Shkaplerov of Roscosmos, Scott Tingle of NASA and Norishige Kanai of the Japan Aerospace Exploration Agency.

Expedition 54 to the International Space Station (ISS) marked a significant chapter in the station's ongoing history. This mission began with the departure of Soyuz MS-05 on December 14, 2017, and ended with the undocking of Soyuz MS-06 on February 27, 2018.

During this period, Expedition 54's crew consisted of Alexander Misurkin, Mark Vande Hei, and Joseph Acaba. Alexander Misurkin, who had been part of Expedition 53, assumed the role of commander for this expedition. His leadership was instrumental in guiding the mission through its various phases, ensuring smooth operations and coordination aboard the station.

Expedition 54 was marked by a significant transition as the crew prepared for the handover to Expedition 55. This transfer of command occurred on February 26, 2018, a day before the official conclusion of the expedition. The handover was a carefully orchestrated process, involving detailed briefings and coordination to ensure a seamless transition between the outgoing and incoming teams.

The expedition officially concluded at 23:08 Coordinated Universal Time (UTC) on February 27, 2018, with the undocking of Soyuz MS-06 from the ISS. This marked the end of Expedition 54's tenure and the beginning of the new chapter with Expedition 55.

Expedition 54 to the International Space Station (ISS) was characterized by an evolving crew roster reflecting the continuity and transition inherent to space missions. The expedition unfolded in two distinct phases, each with its own complement of crew members.

In the initial phase, from December 14, 2017, to December 17, 2017, the crew included Alexander Misurkin, serving as the commander. A seasoned cosmonaut with this being his second spaceflight, Misurkin brought extensive experience to his leadership role. Mark T. Vande Hei and Joseph M. Acaba, both Flight Engineers joined him. Vande Hei, embarking on his first spaceflight, and Acaba, on his third, contributed their skills and expertise to the mission. Their collective efforts ensured the successful execution of the mission's early objectives.

The second phase of Expedition 54, from December 17, 2017, to February 27, 2018, saw additional crew members join the ISS. On his third spaceflight, Anton Shkaplerov, a Russian cosmonaut, assumed the role of Flight Engineer. He was accompanied by Scott D. Tingle, an

American astronaut making his first spaceflight, and Norishige Kanai from Japan, also on his first spaceflight. This diverse team brought a range of perspectives and expertise, enriching the mission's collaborative efforts.

During Expedition 54, the crew conducted a notable spacewalk, or Extravehicular Activity (EVA), which showcased their technical expertise and teamwork.

The first and only spacewalk of Expedition 54 took place on January 23, 2018. The spacewalkers, Mark T. Vande Hei and Scott D. Tingle, both from the United States, began their task at 11:49 Coordinated Universal Time (UTC) and completed it at 19:13 UTC. This EVA lasted 7 hours and 24 minutes, underscoring the complexity and duration of the work conducted outside the International Space Station (ISS).

The primary objectives of this spacewalk included several critical maintenance tasks. The crew replaced the Latching End Effector (LEE-B) on the Canadarm2, a key component of the station's robotic arm system, ensuring its continued functionality for future operations. Additionally, they installed a failed LEE onto the External Stowage Platform-2 (ESP2), replaced a LEE camera, and updated the EVA socket.

Expedition 54 featured a significant second spacewalk on February 2, 2018, highlighting the mission's technical demands and the astronauts' capabilities.

Alexander Misurkin and Anton Shkaplerov, both from Russia, conducted this EVA. They commenced their spacewalk at 15:34 Coordinated Universal Time (UTC) and concluded at 23:47 UTC, making it an extensive 8 hours and 13 minutes. This spacewalk's primary focus was removing and replacing an electronics box for a high-gain communications antenna located on the Zvezda service module.

The high-gain antenna was crucial for maintaining reliable communication between the ISS and mission control on Earth. By replacing the malfunctioning electronics box, Misurkin and Shkaplerov ensured that the antenna could continue to perform its essential function, vital for the uninterrupted exchange of data and communications necessary for the smooth operation of the ISS.

Expedition 54's third spacewalk occurred on February 16, 2018, showcasing the crew's continued dedication to maintaining and upgrading the International Space Station's (ISS) systems.

Mark T. Vande Hei and Norishige Kanai were the spacewalkers for this mission. They initiated their EVA at 12:00 Coordinated Universal Time (UTC) and completed it at 17:57 UTC, totaling 5 hours and 57 minutes. This spacewalk involved a series of critical tasks aimed at ensuring the functionality and efficiency of the station's equipment.

The primary objectives included completing the removal and replacement of the Latching End Effector (LEE) on the Payload Orbital Adapter (POA), a key component of the station's robotic arm system. They also replaced a LEE camera to monitor the arm's operations clearly.

Additionally, the spacewalkers installed a ground strap on the Canadarm2, which helps in reducing static electricity that could interfere with the arm's performance. They brought the failed LEE inside the station for further analysis and maintenance. Other tasks included lubricating the Canadarm2 to ensure its smooth operation, relocating the Tool Platform on Dextre, and adjusting the struts on the Flex Hose Rotary Coupler to enhance its functionality.

During Expedition 54, the International Space Station (ISS) welcomed several uncrewed resupply missions, crucial in maintaining the station's operations and supporting its crew.

One notable mission was SpaceX CRS-13, also known as CRS SpX-13. This mission, conducted by the United States, was focused on delivering logistics and supplies to the ISS. The spacecraft was launched aboard a Falcon 9 rocket on December 15, 2017, at 15:36 Coordinated Universal Time (UTC). It docked with the ISS on December 17, 2017, at 10:57 UTC.

The CRS-13 mission remained docked to the station for 26 days, 23 hours, and 1 minute. During this period, it provided essential supplies and scientific experiments. The spacecraft undocked from the ISS on January 13, 2018, at 09:58 UTC, marking the end of its mission. The deorbit burn, which ensured the spacecraft re-entered Earth's atmosphere, occurred on January 13, 2018, at 14:43 UTC.

Progress MS-08, designated as ISS flight 69P, was launched on February 13, 2018, at 08:13:33 Coordinated Universal Time (UTC) aboard a Soyuz-2.1a rocket. This mission aimed to deliver crucial supplies and equipment to the ISS. The spacecraft docked with the station on February 15, 2018, at 10:38 UTC.

The Progress MS-08 spacecraft remained attached to the ISS for 188 days, 15 hours, and 38 minutes. During its time docked, it provided essential logistics support, including cargo delivery, scientific experiments, and other supplies necessary for the smooth operation of the ISS.

The spacecraft undocked from the ISS on August 23, 2018, at 02:16 UTC. Following its departure, Progress MS-08 was deorbited on August 30, 2018, marking the end of its mission

In 2018, the International Space Station (ISS) remained a vibrant hub of human spaceflight and scientific discovery. The year saw a series of missions and crew rotations, each contributing to the ISS's ongoing role in advancing space research and exploration.

The ISS continued its crucial work in various scientific disciplines, including human physiology, biology, physics, and Earth observation. Regular maintenance and repairs were also essential, with crews engaged in equipment replacements and spacewalks to ensure the station'

In the realm of commercial spaceflight, December 13, 2018, marked a milestone with the successful test flight of the VSS Unity VP-03 spacecraft by Virgin Galactic. Piloted by Mark P. Stucky and Frederick W. Sturckow, this test flight reached an apogee of 82.7 kilometers, surpassing the U.S. definition of space, though not the Kármán line. This achievement signaled progress towards the future of space tourism.

2018 also saw notable advancements in aviation technology by NASA and the U.S. Air Force. NASA's Lockheed Martin X-59 QueSST, a prototype for quiet supersonic transport, took flight, demonstrating significant strides in reducing sonic booms. Additionally, the U.S. Air Force's Generation Orbit Launch Services X-60, an air-launched rocket designed for hypersonic research, debuted. NASA's X-57 Maxwell Electric Propulsion Airplane, the first crewed X-plane in two decades, aimed to advance certification for electric aircraft by showcasing its innovative electric propulsion system.

Expedition 55

Expedition 55 commenced with the departure of the Soyuz MS-06 spacecraft on February 27, 2018. As the 55th rotation of crew members, this expedition saw the transfer of astronauts from the previous Expedition 54. Among them were Anton Shkaplerov, Scott D. Tingle, and Norishige Kanai. Shkaplerov, an experienced cosmonaut, assumed this mission's commander role, guiding the crew through their tasks and responsibilities aboard the ISS.

The crew's mission during Expedition 55 involved a range of scientific experiments and station maintenance activities designed to further the understanding of long-term spaceflight effects on both humans and equipment. Their efforts were integral to advancing the ISS's research goals and ensuring its continued operation as a premier platform for space exploration.

The expedition concluded with Soyuz MS-07's departure in June 2018, marking the end of a successful period of operation and research.

Expedition 55 was a distinguished chapter in the International Space Station's history. It showcased a diverse and skilled crew from various space agencies. The expedition was divided into two parts, each featuring a distinct team of astronauts working in orbit.

The six-member Expedition 55 crew: front row (from left) are Scott Tingle of NASA, Commander Anton Shkaplerov of Roscosmos and Norishige Kanai of the Japan Aerospace Exploration Agency. Back row (feft) are NASA astronauts Ricky Arnold and Andrew Feustel and Roscosmos cosmonaut Oleg Artemyev.

The first part of Expedition 55, from February 27, 2018, to March 23, 2018, was commanded by Anton Shkaplerov, a seasoned Russian cosmonaut representing the Russian space agency, Roscosmos. Shkaplerov, marking his third spaceflight, brought extensive experience to his role as commander. Joining him in this initial phase were Scott D. Tingle from NASA, embarking on his first spaceflight as Flight Engineer 1, and Norishige Kanai from JAXA, also on his first space mission, serving as Flight Engineer 2.

Following the departure of the initial crew, the second part of Expedition 55 began on March 23, 2018. This phase saw the arrival of new crew members: Andrew Feustel, representing NASA, took on the role of Flight Engineer 3 for the second part of the expedition, marking his third spaceflight. Oleg Artemyev, another Russian cosmonaut with Roscosmos, joined as Flight Engineer 4, having completed his second spaceflight. Richard R. Arnold, a NASA astronaut and veteran of two space missions, assumed the position of Flight Engineer 5.

During Expedition 55, the crew conducted a noteworthy spacewalk, both a technical achievement and a critical step in preparing the International Space Station (ISS) for upcoming missions. This spacewalk, designated EVA 1, took place on March 29, 2018, and was executed by NASA astronauts Andrew J. Feustel and Richard R. Arnold.

The spacewalk began at 13:33 UTC and concluded at 19:43 UTC, lasting 6 hours and 10 minutes. During this extravehicular activity, the astronauts performed several key tasks. They installed two WiFi antennas on the Node 3 module, a crucial step to facilitate the upcoming arrival of the ECOSTRESS payload aboard SpaceX CRS-15. Additionally, Feustel and Arnold removed ammonia jumpers and inspected two working jumpers on the station's truss, ensuring the integrity and functionality of critical systems. They also replaced cameras and lights broadcasting NASA TV, essential for communication and public engagement.

Expedition 55 featured a second significant spacewalk, EVA 2, conducted on May 16, 2018. NASA astronauts Andrew J. Feustel and Richard R. Arnold carried out this extravehicular activity. The spacewalk commenced at 11:39 UTC and concluded at 18:10 UTC, lasting 6 hours and 31 minutes.

During EVA 2, Feustel and Arnold undertook several critical tasks. They transferred a Pump Flow Control Subassembly (PFCS) to the Dextre robotic system and stowed a failed PFCS on the External Stowage Platform-1 (ESP-1). They also replaced cameras and lights used for filming NASA TV, ensuring continued public engagement and mission communication.

Additionally, the astronauts replaced a Space to Ground Transceiver Controller and performed preparatory work for future installations, including installing handrails on Radiator Grapple Bars on the S1 truss. They removed thermal blankets and Multi-Layer Insulation (MLI) from two Direct Current Switching Units on the External Stowage Platform-2 (ESP-2) and prepared the Flex Hose Rotary Coupler on the S1 truss for replacement.

The spacewalk encountered a brief delay of 7 minutes due to a water leak in the airlock, which formed ice crystals. The astronauts completed their tasks despite this interruption, contributing to the ISS's ongoing maintenance and operational readiness.

During Expedition 55, the International Space Station (ISS) welcomed several uncrewed resupply missions essential for maintaining station operations and supporting scientific research.

One of the key resupply missions was SpaceX CRS-14, designated as CRS SpX-14. This mission, organized by the United States, was launched on April 2, 2018, at 20:30 UTC using a Falcon 9 rocket. The spacecraft docked with the ISS on April 4, 2018, at 13:00 UTC. SpaceX CRS-14 remained docked to the station for approximately 31 days and 23 minutes before undocking on May 5, 2018, at 13:23 UTC. After completing its mission, the spacecraft was deorbited on the same day.

During Expedition 55, the International Space Station also hosted the Cygnus CRS OA-9E, launched by the United States, was part of Northrop Grumman's resupply program for the ISS. The mission utilized an Antares 230 rocket and took off on May 21, 2018, at 08:44 UTC. The Cygnus spacecraft docked with the ISS on May 24, 2018, at 12:13 UTC. It remained attached to the station for 51 days, 22 hours, and 7 minutes before undocking on July 15, 2018, at 10:20 UTC. Following its departure, the spacecraft was deorbited on July 30, 2018.

Expedition 56

The official portrait of the Expedition 56 crew. In the front row from left are astronauts Drew Feustel of NASA and Alexander Gerst of the European Space Agency. In the rear from left are crew members Oleg Artemyev of Roscosmos, Ricky Arnold of NASA, Sergei Prokopev of Roscosmos and Serena Auñón-Chancellor of NASA.

Expedition 56 marked the 56th mission to the International Space Station (ISS), commencing on June 3, 2018, with the departure of the Soyuz MS-07 spacecraft. This mission saw the transition of crew members from Expedition 55 to the new team, led by Andrew Feustel, who assumed the role of commander. Joining Feustel were Russian cosmonaut Oleg Artemyev and NASA astronaut Richard R. Arnold.

The expedition was further bolstered by new crew members, who launched aboard Soyuz MS-09 on June 6, 2018. This crew included European Space Agency astronaut Alexander Gerst, NASA astronaut Serena M. Auñón-Chancellor, and Russian cosmonaut Sergey Prokopyev. Their arrival marked a key moment in maintaining the ISS's operational capabilities and advancing ongoing scientific research. During Expedition 56, the crew's composition was divided into two phases. The first phase, from June 3 to June 6, 2018, included:

Commander: Andrew Feustel, representing the United States and NASA. This was Feustel's third spaceflight.

Flight Engineer 1: Oleg Artemyev from Russia, representing the Russian Space Agency (RSA), on his second spaceflight.

Flight Engineer 2: Richard R. Arnold from the United States, also with NASA, undertaking his second spaceflight.

As the expedition progressed, from June 6 to October 4, 2018, the crew saw additional members join:

Flight Engineer 3: Sergey Prokopyev, representing Russia and RSA, embarking on his first spaceflight.

Flight Engineer 4: Alexander Gerst from Germany, representing the European Space Agency (ESA), on his second spaceflight.

Flight Engineer 5: Serena M. Auñón-Chancellor from the United States, with NASA, undertaking her first spaceflight.

Initially, NASA astronaut Jeanette Epps was assigned as a flight engineer for Expeditions 56 and 57. Epps was set to become the first African American to serve as a space station crew member and the 15th African American to fly in space. However, on January 16, 2018, NASA announced that Epps had been replaced by her backup, Serena M. Auñón-Chancellor. The reasons for this change were not publicly disclosed.

During Expedition 56, the crew conducted one notable spacewalk:

EVA 1 took place on June 14, 2018. This extravehicular activity, which began at 12:06 UTC and concluded at 18:55 UTC, lasted 6 hours and 49 minutes. The spacewalk was carried out by NASA astronauts Andrew Feustel and Richard R. Arnold.

The primary objectives of this spacewalk were to enhance the station's capabilities and prepare for future missions. Feustel and Arnold installed new high-definition cameras near the International Docking System Adapter 2 (IDA 2) on the front end of the Harmony module. These cameras were designed to provide improved views during the final phases of approach and docking for upcoming commercial crew spacecraft, including the SpaceX Crew Dragon and Boeing CST-100 Starliner. The installation of these cameras was crucial for monitoring and guiding the docking process of these spacecraft, set to launch from American soil.

In addition to the camera installation, the astronauts performed several other tasks. They replaced a camera assembly on the station's starboard truss, used for broadcasting NASA TV. They closed an aperture door on the Cloud Aerosol Transport System (CATS) experiment outside the Japanese Kibo module. The CATS experiment was scheduled for disposal on SpaceX CRS-15 and would be replaced by the ECOSTRESS experiment. The spacewalkers also relocated an adjustable grapple bar to the S1 Truss and secured the Flex Hose Rotary Coupler in preparation for its replacement on a future spacewalk.

During this spacewalk, Feustel achieved a significant milestone in his career. By completing this EVA, he surpassed fellow astronauts Jerry Ross, John Grunsfeld, Fyodor Yurchikhin, and Peggy Whitson to become the third astronaut on the all-time list of spacewalkers.

During Expedition 56, the second spacewalk, known as EVA 2, took place on August 15, 2018. The extravehicular activity began at 16:17 UTC and concluded at 00:03 UTC the following day, lasting 7 hours and 46 minutes. Russian cosmonauts Oleg Artemyev and Sergey Prokopyev conducted this spacewalk.

The primary objectives of EVA 2 included deploying four CubeSats and installing the Icarus experiment. CubeSats are small, standardized satellites used for various scientific and technological experiments. Their deployment marked an important step in expanding the ISS's space-based research and communication capability.

However, the spacewalk encountered unexpected challenges. The Icarus experiment, designed to track the migration patterns of animals using radio frequency technology, failed to seat properly, which delayed the completion of the tasks. As a result, the spacewalk fell behind schedule by approximately 90 minutes, necessitating an additional hour to complete the planned activities.

Despite the delay, Artemyev and Prokopyev concluded the spacewalk by retrieving experiments from the Pirs docking compartment and the Poisk module. This task was crucial for maintaining the operational efficiency of these modules and ensuring the continuity of ongoing scientific experiments aboard the station.

View of Earth taken during ISS Expedition 56 on 13 August 2018 at 17:49:12.

On August 29, 2018, flight controllers detected a small leak aboard the International Space Station through a noticeable drop in air pressure. Upon being informed of the situation, the astronauts, upon waking up, investigated the source of the leak and discovered a 2 mm hole in the orbital module of the Soyuz MS-09 spacecraft.

Initially, the crew applied a temporary fix using tape to stem the leakage. Following this, a more permanent repair was conducted using gauze and epoxy, which effectively sealed the hole and restored the integrity of the spacecraft's pressure system. This incident highlighted the resilience and resourcefulness of the ISS crew and mission control in addressing and resolving unexpected challenges in space.

During Expedition 56, several uncrewed resupply missions visited the International Space Station, crucial in sustaining the station's operations and supporting its scientific research. Among these, the SpaceX CRS-15 mission was a notable resupply flight.

SpaceX CRS-15, also known as CRS SpX-15, was a logistics mission conducted by the United States. It was launched aboard a Falcon 9 rocket on June 29, 2018, at 09:42 UTC. The spacecraft docked with the ISS on July 2, 2018, at 13:50 UTC. The mission remained docked to the station for 32 days, 2 hours, and 48 minutes.

The spacecraft undocked from the ISS on August 3, 2018, at 16:38 UTC and completed its deorbiting maneuver later that same day at 22:17 UTC.

During Expedition 56, the Progress MS-09 mission was a significant uncrewed resupply flight from Russia. This logistics mission was launched aboard a Soyuz-2.1a rocket on July 9, 2018, at 21:51:34 UTC. The spacecraft docked with the International Space Station on July 10, 2018, at 01:30:48 UTC.

Progress MS-09 remained attached to the ISS for an extended period of 199 days, 11 hours, and 24 minutes. This duration allowed the spacecraft to provide continuous support and resupply to the station, including essential cargo, scientific equipment, and experiments. The spacecraft undocked from the ISS on January 25, 2019, at 12:55 UTC, and it completed its deorbiting maneuver and re-entered Earth's atmosphere later that day at 16:50 UTC.

During Expedition 56, the Kounotori 7 mission, also known as HTV-7, was an important uncrewed resupply flight from Japan. Launched aboard an H-IIB rocket on September 22, 2018, at 17:52:27 UTC, HTV-7 was designed to deliver cargo, scientific experiments, and supplies to the International Space Station.

The spacecraft docked with the ISS on September 27, 2018, at 14:09 UTC. It remained docked to the station for 40 days, 9 hours, and 23 minutes, providing crucial logistical support and facilitating the ongoing research and operations aboard the ISS.

HTV-7 undocked from the station on November 6, 2018, at 23:32 UTC, and it completed its deorbiting and re-entry procedures on November 10, 2018, at 21:14 UTC.

The mission of Expedition 56 concluded with the departure of Soyuz MS-08 on October 4, 2018. This marked the end of a productive and collaborative period of international research aboard the International Space Station. During Expedition 56, crew members from the United States, Russia, Germany, and other participating nations worked together on a wide range of scientific experiments and technical projects, furthering our understanding of space and advancing the goals of the ISS program.

Expedition 57

Official crew portrait of Expedition 57 crew members (from left) Serena Auñón-Chancellor of NASA, Alexander Gerst of ESA (European Space Agency) and Sergey Prokopyev of Roscosmos.

Expedition 57 marked the fifty-seventh mission to the International Space Station (ISS), commencing on October 4, 2018, with the departure of the Soyuz MS-08 spacecraft. This expedition followed a period of significant uncertainty and adjustment due to an incident involving the Soyuz MS-10 mission.

The Soyuz MS-09 crew, consisting of Commander Alexander Gerst, Flight Engineer Serena M. Auñón-Chancellor, and Flight Engineer Sergey Prokopyev, had been aboard the ISS since June 2018. Their mission was to be complemented by the arrival of Aleksey Ovchinin and Nick Hague, who were scheduled to join them in October 2018. However, the launch of Soyuz MS-10 on October 11 was interrupted by a critical failure of the rocket booster. This malfunction forced the spacecraft to abort its mission in mid-flight. Thankfully, Ovchinin and Hague survived the emergency landing, executed by ballistic descent, and returned safely to Earth.

The failure of Soyuz MS-10 introduced considerable uncertainty regarding the ISS crew schedule. The Soyuz MS-09 crew was expected to return to Earth by mid-December 2018, given the spacecraft's operational lifespan of approximately 200 days. The possibility of extending their stay was limited, with a marginal buffer extending into early January 2019. To prevent a situation where the ISS would be left uncrewed, NASA had contingency plans to manage the station from the ground if necessary.

On October 23, 2018, NASA Administrator Jim Bridenstine announced that Soyuz flights to the ISS were anticipated to resume in December 2018. This forecast proved accurate when the Soyuz MS-11 mission, carrying Commander Oleg Kononenko and Flight Engineers Anne McClain and David Saint-Jacques, launched on December 3, 2018. This flight also marked the 100th orbital launch of the year. Subsequently, the Soyuz MS-09 crew departed the ISS on December 20, 2018, transitioning to Expedition 58, which began as a three-person increment.

During Expedition 57, the crew underwent a notable transition between October 4 and December 3, 2018, and then from December 3 to December 20, 2018.

Initially, the crew was led by Commander Alexander Gerst of Germany, representing the European Space Agency (ESA). This was Gerst's second spaceflight, bringing his prior experience to his role as the station's commander. Flight Engineer Serena M. Auñón-Chancellor joined him from the United States, a NASA astronaut embarking on her first space mission. The team also included Flight Engineer Sergey Prokopyev from Russia, representing the Russian Space Agency (RSA), who was undertaking his first spaceflight.

From December 3, 2018, the crew composition changed with the arrival of new members. Commander Oleg Kononenko from Russia, a seasoned astronaut with four previous spaceflights, joined the team as one of the new flight engineers. He was accompanied by Flight Engineer Anne McClain from the United States, another first-time space traveler with NASA, and Flight Engineer David Saint-Jacques from Canada, who was also on his inaugural space mission with the Canadian Space Agency (CSA).

Originally, NASA Astronaut Jeanette Epps was assigned as a flight engineer for Expeditions 56 and 57, making her the first African American astronaut scheduled to serve on a space station crew and the fifteenth African American to fly in space. However, on January 16, 2018, NASA announced a significant change in crew assignments: Epps was replaced by her backup, Serena M. Auñón-Chancellor. This decision sparked controversy when, on January 20, 2018, Epps' brother, Henry, posted a statement on Facebook alleging that Epps had faced systemic racism and misogyny at NASA, which he believed led to her removal from the mission. The post was subsequently deleted, but the sentiments expressed drew public attention.

Jeanette Epps herself refrained from commenting on her brother's statement or the reasons for reassignment. She clarified that she had no medical conditions or family issues that would prevent her from flying and affirmed that her training had been successful. The Washington Post noted that such last-minute crew changes are common at NASA, reflecting the complexities and fluidity of space mission planning.

In a related development, Cosmonaut Nikolai Tikhonov was originally slated to make his debut spaceflight aboard the Soyuz MS-10 spacecraft. However, he was removed from the mission roster due to delays in the launch of the Russian Nauka module, marking the second instance where Tikhonov was excluded from an ISS crew due to similar delays. These adjustments underscore the challenges faced in space mission scheduling and the need for flexibility in crew assignments.

During Expedition 57, a series of pivotal events took place, reflecting the dynamic and often unpredictable nature of life aboard the International Space Station (ISS).

Global Ecosystem Dynamics Investigation (GEDI): The Global Ecosystem Dynamics Investigation (GEDI) was a pioneering endeavor on the International Space Station dedicated to unraveling the complexities of Earth's forests and topography. As a critical component of space-based environmental research, GEDI employs cutting-edge laser-ranging technology to deliver unprecedented insights into our planet's ecological systems.

Mounted on the Japanese Experiment Module's Exposed Facility (JEM-EF), GEDI's mission was to provide high-resolution observations of forest vertical structure on a global scale. This state-of-the-art instrument meticulously measures the aboveground carbon stored in vegetation, capturing a detailed snapshot of the Earth's carbon dynamics. By monitoring changes in vegetation caused by disturbances and natural recovery processes, GEDI offers a window into the shifting patterns of our forests.

One of GEDI's key objectives was to assess the future potential of forests to sequester carbon. By analyzing how vegetation absorbs and stores carbon, scientists could better predict how forests might contribute to mitigating climate change in the coming years. In addition to its focus on carbon, GEDI also investigates habitat structure, shedding light on how the arrangement and quality of habitats influence biodiversity.

The data collected by GEDI was vital for advancing our understanding of carbon and water cycling processes, which are essential for maintaining the health of our planet's ecosystems. Through its observations, GEDI helps paint a comprehensive picture of how forests function and respond to environmental changes, thus playing a crucial role in global ecological monitoring and management.

In essence, GEDI's mission was about capturing data and forging a deeper connection with our planet's natural processes. Its work from the vantage point of space offers a profound perspective on the Earth's dynamic ecosystems, providing invaluable information that helps shape our approach to environmental stewardship.

Angiex Cancer Therapy: The Endothelial Cells in Microgravity as a Model System for Evaluation of Cancer Therapy Toxicity investigation, often called the Angiex Cancer Therapy study, represented a groundbreaking approach in cancer research. This investigation focused on cultivating endothelial cells (ECs) in the unique space environment to enhance our understanding of cancer therapy toxicity.

Angiex, a leader in innovative cancer treatments, has developed a novel therapy targeting tumor cells and the surrounding vasculature. However, a more sophisticated testing model was needed to refine this treatment. The microgravity environment aboard the International Space

Station offers a unique opportunity to culture ECs under impossible conditions. This experimental setup was designed to improve the accuracy and efficiency of evaluating the safety and effectiveness of vascular-targeted drugs.

By utilizing space-based research, the study aims to create a cost-effective testing model that eliminates the need for animal testing. This advancement could potentially lead to safer and more effective treatments, paving the way for significant progress in cancer therapy.

Cimon: The Crew Interactive MObile companioN (Cimon) pilot study marks an exciting foray into integrating artificial intelligence (AI) within space missions. This technology demonstration and observational study was designed to explore how an AI companion could impact crew support during extended spaceflights.

Space missions often subject astronauts to intense stress and heavy workloads, making every aspect of crew support crucial. Cimon aims to address these challenges by providing operational assistance and enhancing the crew's efficiency and well-being. By introducing AI into this demanding environment, the study seeks to assess the effectiveness and acceptance of such technology in space.

This investigation will be pivotal in understanding how AI could support astronauts, potentially offering significant benefits for future long-term missions. The insights gained from Cimon's deployment will be instrumental in developing AI systems that could alleviate the burdens of space travel and improve overall mission success and crew satisfaction.

On December 11, 2018, Russian cosmonauts Oleg Kononenko and Sergey Prokopyev conducted a significant extravehicular activity (EVA). This spacewalk began at 15:59 UTC and concluded at 23:44 UTC, lasting 7 hours and 45 minutes. The primary goal of this EVA was to address a critical issue with the Soyuz MS-09 spacecraft, which had been compromised by an air leak caused by a power tool incident. The cosmonauts installed a plug and applied thermal insulation to the damaged area, ensuring the spacecraft's integrity. Their work enabled Soyuz MS-09 to be declared fit for re-entry, facilitating the crew's return on December 20, 2018. During this spacewalk, the cosmonauts also performed a

"get-ahead" task by exchanging experiments on the Rassvet module, further enhancing the ISS's research capabilities.

Several uncrewed resupply missions also was crucial during Expedition 57, each contributing to the station's operational and research needs. One such mission was the launch of Progress MS-10, designated ISS flight number 71P. Launched from Russia aboard a Soyuz-FG rocket on November 16, 2018, at 18:14 UTC, Progress MS-10 docked with the ISS on November 18, 2018, at 19:28 UTC. The spacecraft remained attached to the station for 197 days, 13 hours, and 12 minutes, providing essential supplies and conducting various experiments before undocking on June 4, 2019, at 08:40 UTC. It was subsequently deorbited, marking the end of its mission.

Another notable mission was Cygnus NG-10, also known as CRS NG-10E, launched on November 17, 2018, at 09:01 UTC aboard an Antares 230 rocket. Cygnus NG-10 docked with the ISS on November 19, 2018, at 12:31 UTC, delivering critical supplies and scientific experiments. The spacecraft spent 81 days, 3 hours, and 45 minutes attached to the station before undocking on February 8, 2019, at 16:16 UTC. It re-entered Earth's atmosphere on February 25, 2019, at 09:05 UTC, concluding its mission.

The SpaceX CRS-16 mission, also known as CRS SpX-16, was another key uncrewed resupply flight. Launched on December 5, 2018, at 18:16 UTC, this mission used a Falcon 9 Block 5 rocket. The CRS-16 spacecraft docked with the ISS on December 8, 2018, at 15:36 UTC, bringing additional supplies and research materials. It remained attached to the station for 36 days, 7 hours, and 57 minutes, undocking on January 13, 2019, at 23:33 UTC. The spacecraft re-entered Earth's atmosphere and concluded its mission on January 14, 2019.

A series of other significant missions marked Expedition 57's operational rhythm. On October 11, 2018, an anomaly occurred during the launch of Soyuz MS-10, initially planned to ferry additional crew members. The mission experienced a mid-flight abort due to a booster failure, resulting in a safe ballistic descent of the crew. Later, on November 7, 2018, the HTV-7 Kounotori, a Japanese cargo spacecraft, was released from the ISS after completing its resupply mission.

The transition to Expedition 58 was highlighted by the launch of Soyuz MS-11 on December 3, 2018. This spacecraft docked with the ISS on the same day, bringing new crew members aboard and marking the end of Expedition 57, which concluded with the return of its crew to Earth on December 20, 2018.

Expedition 58

The official Expedition crew portrait with (from left) NASA astronaut Anne McClain, Roscosmos cosmonaut Oleg Kononenko and astronaut David Saint-Jacques of the Canadian Space Agency.

Expedition 58, the 58th mission to the International Space Station (ISS), commenced on December 20, 2018, following the departure of the Expedition 57 crew. This expedition was led by Russian cosmonaut Oleg Kononenko, who took command alongside American astronaut Anne McClain and Canadian astronaut David Saint-Jacques. The trio embarked on their journey to the ISS aboard Soyuz MS-11, which lifted off on

December 3, 2018. This launch was a significant milestone, marking the 100th orbital launch of that year.

During their tenure aboard the ISS, Kononenko, McClain, and Saint-Jacques contributed to a variety of scientific experiments and station operations, continuing the station's ongoing research and maintenance tasks. Their mission, however, was relatively brief in the context of ISS expeditions. On March 15, 2019, they handed over their responsibilities to the incoming crew of Expedition 59. This transition occurred with the arrival of Aleksey Ovchinin, Nick Hague, and Christina Koch, who arrived on Soyuz MS-12. The seamless exchange of crews ensured that the ISS continued its vital scientific work uninterrupted, embodying the spirit of international cooperation that was central to the operation of the station.

During the early planning stages for Expedition 58, cosmonaut Nikolai Tikhonov was initially slated to join the mission as a rookie. However, due to delays in the launch of the Russian Nauka module, Tikhonov's assignment was postponed for the second time. He was later reassigned to the Soyuz MS-14 mission, scheduled for late 2019.

As of October 2018, the plan for Expedition 58 envisaged a crew of five. Cosmonaut Aleksey Ovchinin and astronaut Nick Hague were to join the Expedition 57 crew in October 2018 and subsequently transition into Expedition 58. They would have been accompanied by Oleg Kononenko, Anne McClain, and David Saint-Jacques in December 2018. Following their tenure, Ovchinin and Hague were expected to return to Earth in April 2019, with Expedition 59 commencing under Kononenko's command.

However, the mission faced a significant setback when the Soyuz MS-10 spacecraft, carrying Ovchinin and Hague, experienced a launch abort on October 11, 2018. Both crew members were safely returned to Earth. In response to this incident, NASA Administrator Jim Bridenstine announced on October 23, 2018, that Soyuz flights to the ISS were anticipated to resume in December 2018.

Initially, it was thought Expedition 58 would start with a reduced crew of three members, with additional astronauts joining later. However, during the post-launch press conference for Soyuz MS-11, NASA confirmed that the crew of Soyuz MS-12 would serve as the Expedition 59/60 team. As a result, Expedition 58 proceeded as a three-person increment, with Kononenko, McClain, and Saint-Jacques as the sole members.

Expedition 58's crew consisted of three distinguished members representing three different countries. The mission was commanded by Oleg Kononenko from Russia, who brought extensive experience from his previous spaceflights, having embarked on four previous missions.

Anne McClain served as Flight Engineer 1, representing the United States and NASA. This mission marked her first journey into space, as she was a rookie astronaut making her debut aboard the International Space Station.

David Saint-Jacques from Canada, serving as Flight Engineer 2, also made his first spaceflight during Expedition 58. Saint-Jacques became the first Canadian resident on the ISS since Chris Hadfield, who commanded Expedition 35, concluded his term on May 13, 2013, nearly six years earlier. Saint-Jacques's presence at the station highlighted Canada's ongoing contributions to international collaboration in space exploration.

During Expedition 58, several uncrewed resupply missions visited the International Space Station and played crucial roles in maintaining station operations and delivering essential supplies.

One notable mission was the SpaceX Demo-1 (SpX-DM1) flight, representing the United States. This mission, a test flight of the Crew Dragon spacecraft, launched on March 2, 2019, at 07:49 UTC using the Falcon 9 Block 5 rocket. The spacecraft docked with the ISS on March 3, 2019, at 10:51 UTC. It remained docked for approximately four days, undocking on March 8, 2019, at 07:32 UTC. The mission's duration at the station was 4 days, 20 hours, and 41 minutes. Following undocking, the spacecraft re-entered Earth's atmosphere and was deorbited at 13:45 UTC on March 8, 2019.

On December 20, 2019, President Donald Trump inaugurated the United States Space Force (USSF) as an independent branch of the U.S. Armed Forces. This landmark decision created the world's first distinct space-focused military service. As a new arm of the military, the Space Force was originally part of the U.S. Air Force Department, overseen by the Secretary of the Air Force, a civilian appointee who reports to the Secretary of Defense. The top officer within the Space Force, known as the Chief of Space Operations, was a member of the Joint Chiefs of Staff, tasked with supervising the service's various units.

With approximately 8,400 military personnel, the USSF managed an array of crucial space operations, including the operation of 77 spacecraft supporting vital programs such as GPS, military satellite communications, space surveillance, and missile warning systems. The Space Force organized, trained, and equipped space forces for deployment under the U.S. Space Command. Its origins trace back to the early days of the Cold War, with the Army Air Forces initiating space programs as early as 1945. Over the decades, military space operations evolved under various Air Force commands, culminating in establishing the United States Air Force Space Command in 1982. During conflicts like the Gulf War, U.S. space forces played a significant role, providing satellite communications, weather data, and navigation support, marking what some termed the first "space war."

Discussions about creating a dedicated military space service began in earnest in 1958 and were revisited by President Ronald Reagan in 1982. The concept gained momentum through the years, with the 2001 Space Commission recommending the formation of a Space Corps and a subsequent bipartisan proposal establishing the U.S. Space Corps in 2017. The final step came on December 20, 2019, when President Trump signed the US Space Force Act into law, reorganizing the Air Force Space Command and other Air Force space elements into the independent U.S. Space Force. This event marked the creation of the first new military service since the establishment of the U.S. Air Force in 1947.

The Space Force's roots could be traced to 1945, when General Bernard Schriever spearheaded the creation of the Western Development Division, the U.S. Armed Forces' first dedicated space organization. Throughout the Cold War, military space operations were managed by various Air Force commands until the USAF Space Command was officially established on September 1, 1982.

Space Exploration Highlights of 2019

The year 2019 was notable for its significant space missions and advancements in human spaceflight. On November 25, Russia launched its Kosmos 2542 inspector satellite at the Plesetsk Cosmodrome. This satellite, which adjusted its orbit to closely approach the U.S. National Reconnaissance Office's KH-11 image-gathering spy satellite, raised concerns about potential surveillance or even hostile intentions. The exact purpose of the Kosmos 2542 remained unclear. However, its proximity to the U.S. satellite suggested it might have been intended to gather intelligence or, in a more ominous scenario, to act as a "killer satellite" capable of damaging or disabling its target.

Throughout 2019, the International Space Station (ISS) continued to be the epicenter of human spaceflight. In February, the Expedition 58 crew completed their mission, making way for Expedition 59. The latter group focused on a range of experiments in human physiology, biology, physics, and Earth observation, while also performing maintenance tasks and conducting spacewalks to ensure the station's optimal operation.

In May, SpaceX's Dragon spacecraft completed its seventeenth operational mission to the ISS, delivering supplies, experiments, and scientific samples. Shortly after, the Northrop Grumman Cygnus spacecraft carried out its eleventh commercial cargo mission, further supporting the station's needs.

July saw the arrival of the Expedition 60 crew, who replaced the Expedition 59 team. This crew continued research and maintenance activities, and welcomed the Russian Progress MS-12 spacecraft, which brought essential supplies and experiments.

October brought the Expedition 61 crew to the ISS, following the departure of Expedition 60. Expedition 61's goals included conducting maintenance spacewalks, installing new equipment, and continuing scientific experiments in biology, physics, and human physiology.

Meanwhile, China was advancing its own space program. In December, China launched the unmanned cargo spacecraft Tianzhou-3, which docked with the Tiangong-2 space station. This mission delivered critical supplies and equipment and marked the final operational phase of Tiangong-2, with plans to replace it with the larger Tiangong-3.

2019 also featured several significant crew rotations. On February 22, Virgin Galactic's VSS Unity VF-01 embarked on its first test flight with a passenger, Beth Moses, Virgin Galactic's Chief Astronaut Instructor. Moses became the first woman to fly aboard a commercial spacecraft, reaching an apogee of 89.9 km, thereby fulfilling the U.S. definition of space.

On March 14, the Soyuz MS-12 spacecraft launched with Aleksey Ovchinin and Nick Hague, marking the 59th/60th expedition to the ISS. Their mission involved scientific experiments and station duties, and they returned to Earth on October 3, 2019.

Later, on July 20, the Soyuz MS-13 spacecraft carried Aleksandr Skvortsov and Luca Parmitano to the ISS, beginning the 60th/61st expedition. Their mission focused on research and station operations, with their return to Earth scheduled for February 6, 2020.

On February 6, 2020, the Soyuz MS-13 spacecraft embarked on its second mission, adding Christina Koch to the crew. This mission was part of the 59th/60th/61st expedition, with Koch setting a record for the longest continuous spaceflight by a woman at 328 days. The crew concluded their rotation on April 17, 2020.

Finally, the Soyuz MS-15 spacecraft launched on September 25, 2019, carrying Oleg Skripochka and Jessica Meir, marking the 61st/62nd expedition to the ISS. This rotation included Hazza Al Mansouri, the first Emirati in space. The crew returned to Earth on April 17, 2020, completing their ISS rotation.

In 2019, the U.S. Air Force continued to advance aerospace technology with several notable flights. These included the Dynetics X-61 Gremlins, an air-launched, recoverable reconnaissance UAV, and the classified X-42, a liquid-fueled rocket upper stage. The X-39, associated with the Future Aircraft Technology Enhancements (FATE) program, was reserved for a nuclear-powered General Electric X47 turbojet, although it remained unassigned by early 1999.

In December 2019, President Trump formalized the U.S. Space Force under General John Raymond as Chief of Space Operations. The National Reconnaissance Office partnered with the

Space Force's Space and Missile Systems Center to manage the National Security Space Launch (NSSL) program, coordinating with U.S. Space Command and the Space Force to launch critical government payloads. This collaboration underscored the growing importance of space in national security and defense.

Expedition 59

The official Expedition 59 crew portrait with (from left) astronauts David Saint-Jacques of the Canadian Space Agency and Anne McClain of NASA; cosmonauts Oleg Kononenko and Aleksey Ovchinin of Roscosmos; and NASA astronauts Nick Hague and Christina Koch.

Expedition 59 to the International Space Station marked a significant chapter in human space exploration. The expedition commenced with the arrival of the Soyuz MS-12 spacecraft on March 15, 2019, transporting astronauts Aleksey Ovchinin, Nick Hague, and Christina Koch to the station. Their arrival was a pivotal moment, as they joined the existing crew of Expedition 58—Oleg Kononenko, David Saint-Jacques, and Anne McClain—who had been stationed on the ISS since the previous mission.

Aleksey Ovchinin and Nick Hague were originally scheduled to travel to the ISS aboard Soyuz MS-10. However, their mission was interrupted by a critical contingency abort just minutes after launch, resulting in their return to Earth. Despite this setback, they were soon reassigned to Soyuz MS-12, transporting them to the ISS alongside Koch.

The formal conclusion of Expedition 59 occurred on June 24, 2019, when the Soyuz MS-11 spacecraft undocked from the ISS, marking the end of their time aboard. This departure carried Oleg Kononenko, David Saint-Jacques, and Anne McClain back to Earth. At the same time, Aleksey Ovchinin, Nick Hague, and Christina Koch transitioned into Expedition 60, continuing the ongoing cycle of human presence on the station.

Expedition 59 to the International Space Station was led by a distinguished crew, each bringing their unique experience and expertise to the mission.

The commander of expedition 59 was Oleg Kononenko from Russia's space agency, Roscosmos (RSA). Kononenko, a seasoned astronaut on his fourth spaceflight, brought a wealth of experience to commanding the ISS during this mission.

David Saint-Jacques of the Canadian Space Agency (CSA) served as Flight Engineer 1. This mission marked his first spaceflight, and his role was crucial in supporting the daily operations and scientific experiments on board the ISS.

Flight Engineer 2, Anne McClain from NASA, was also on her first spaceflight. Her role involved conducting scientific research and maintenance tasks, and her training and skills contributed to the overall success of the mission.

Aleksey Ovchinin, another Russian astronaut from Roscosmos, held the position of Flight Engineer 3. This was Ovchinin's third spaceflight, and his experience was instrumental in the mission's operations and ensuring the ISS's smooth functioning.

Nick Hague, representing NASA as Flight Engineer 4, was on his second spaceflight. Hague's duties included managing experiments and systems aboard the station, utilizing his previous experience to enhance the mission's effectiveness.

Christina Koch, Flight Engineer 5 from NASA, was on her first spaceflight. Koch's involvement in the mission was vital for conducting scientific investigations and performing technical tasks essential to the success of Expedition 59.

Expedition 59 included a series of significant spacewalks, each contributing to maintaining and enhancing the International Space Station's (ISS) capabilities.

The first spacewalk, designated EVA 1, took place on March 22, 2019. Conducted by Anne McClain and Nick Hague from the United States, this six-hour, thirty-nine-minute extravehicular activity began at 12:01 UTC and concluded at 18:40 UTC. During this spacewalk, McClain and Hague installed adapter plates on the station's external structure. In preparation for the upcoming Cygnus NG-11 resupply mission scheduled for April, they also removed debris from the Unity Module. Additionally, they stowed tools to repair the flex hose rotary coupler and secured tiebacks on the solar array blanket boxes. Originally planned to be executed by European astronaut Alexander Gerst and Hague as part of Expedition 57, this task was delayed due to the Soyuz MS-10 launch abort.

The second spacewalk, EVA 2, occurred on March 29, 2019. Nick Hague and Christina Koch, both from the United States, performed this six-hour, forty-five-minute spacewalk starting at 11:42 UTC and ending at 18:27 UTC. They completed the work from the first spacewalk by installing the remaining three adapter plates on the P4 Truss. Additional tasks included transferring tools and installing a grapple bar on the flex hose rotary coupler. The robotic arm Dextre was used to change batteries between the spacewalks, and an exposed pallet was prepared for disposal via HTV-8. A malfunctioning battery was removed and will be returned to Earth on SpaceX CRS-17 for repair. This spacewalk marked Christina Koch's first-ever spacewalk, making her the 14th woman to walk in space. McClain and Koch initially planned to conduct the spacewalk; however, due to issues with Koch's suit and the unavailability of a properly sized spare, McClain had to be replaced by Hague for this activity.

The third spacewalk, EVA 3, was carried out on April 8, 2019, by Anne McClain and David Saint-Jacques. This spacewalk lasted six hours and thirty minutes from 11:31 UTC to 18:00 UTC and focused on routing cables to establish a redundant power supply for the Canadarm2. The crew also installed studs on the Columbus Module in preparation for future installations. They replaced a set of old batteries with spares to address a malfunctioning battery from the previous spacewalk. David Saint-Jacques became the first Canadian astronaut to walk in space, joining the ranks of Canadian spacewalkers since Chris Hadfield.

The fourth and final spacewalk of Expedition 59 took place on May 29, 2019, involving Russian astronauts Oleg Kononenko and Aleksey Ovchinin. This spacewalk, lasting six hours and one minute from 15:42 UTC to 21:43 UTC, included the removal of experiments from the Pirs docking compartment and cleaning of its windows. The crew also installed a handrail connecting the Zarya module to Poisk and repositioned the Plume Measuring Unit. In addition, they removed and jettisoned the Plasma Monitoring Units from the Zvesda Service Module. As a tribute, Kononenko and Ovchinin sent a video message singing "Happy Birthday" in Russian to Alexei Leonov, the first human to walk in space, on his 85th birthday.

During Expedition 59, several uncrewed spacecraft carried out resupply missions to the International Space Station (ISS), each contributing vital supplies and equipment to the station's maintenance and scientific research.

Progress MS-11 was a Russian logistics spacecraft designated as ISS flight number 72P. It launched on April 4, 2019, at 11:01:35 UTC aboard a Soyuz-2.1a rocket. The spacecraft docked with the ISS at 14:22 UTC the same day. Progress MS-11 remained docked at the station for approximately 115 days, 20 hours, and 22 minutes, undocking on July 29, 2019, at 10:44 UTC. It was deorbited and re-entered Earth's atmosphere, burning up upon re-entry at 13:50 UTC on the same day. This mission was critical for delivering essential supplies, including fuel, equipment, and scientific experiments.

Cygnus NG-11 was an American logistics spacecraft designated CRS NG-11 for this mission. It launched on April 17, 2019, at 20:46:07 UTC using an Antares 230 rocket. After reaching orbit, it docked with the ISS on April 19, 2019, at 09:28 UTC. The spacecraft stayed attached to the station for 109 days, 6 hours, and 47 minutes before undocking on August 6, 2019, at 16:15 UTC. The deorbit burn, which led to its re-entry and disposal in the Earth's atmosphere, occurred on December

6, 2019, at 15:28 UTC. Cygnus NG-11 delivered a variety of cargo, including scientific experiments, crew supplies, and station hardware.

SpaceX CRS-17, an American logistics mission operated by SpaceX, was launched on May 4, 2019, at 06:48 UTC aboard a Falcon 9 rocket. It docked with the ISS on May 6, 2019, at 13:33 UTC. This mission was relatively short at the station, lasting 28 days, 2 hours, and 28 minutes. CRS-17 undocked from the ISS on June 3, 2019, at 16:01 UTC, and its deorbit burn occurred later that day, at 20:56 UTC. The spacecraft provided crucial supplies and scientific experiments, continuing SpaceX's role in supporting the ISS's operations and research.

During Expedition 59, diverse scientific research and experimentation was conducted aboard the International Space Station, leveraging the unique microgravity environment to advance our understanding of various fields.

One notable focus of the expedition was the investigation of tissue chips, which are small devices that contain living human cells organized into micro-engineered tissue structures. These chips allow researchers to study the effects of microgravity on aging and disease processes, offering insights into how the human body might react to long-term space travel. The microgravity environment of the ISS provides a unique opportunity to replicate and examine these effects in ways that are impossible on Earth.

Additionally, Expedition 59 included experiments involving regolith simulants, which are materials designed to mimic the surface of celestial bodies like the Moon or Mars. By studying these simulants, researchers aim to better understand how such materials interact with various tools and systems, which was crucial for future exploration missions and developing technologies for extraterrestrial habitats.

The expedition also focused on Earth's atmospheric carbon cycle, investigating how carbon dioxide and other greenhouse gases behave in the space environment. This research was vital for understanding the impacts of these gases on Earth's climate and developing strategies to address climate change.

Another key research component involved Astrobee robots, autonomous robotic systems designed to perform routine chores and tasks aboard the ISS. These robots help manage daily operations and contribute to the station's efficiency by performing inventory management and equipment checks, allowing the crew to focus more on scientific research and other critical activities.

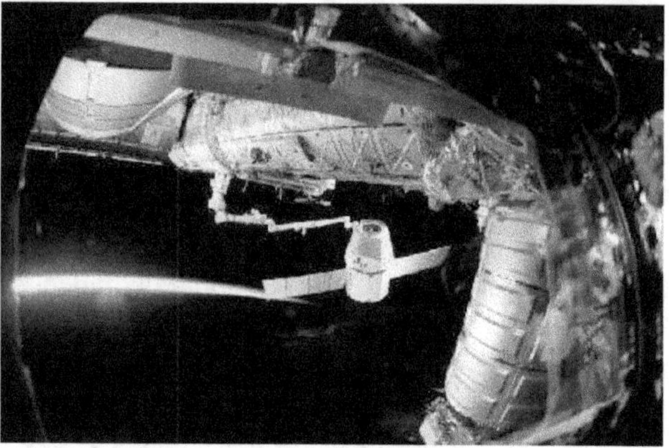

The SpaceX Dragon cargo craft on its 17th contracted mission to resupply mission to the International Space Station moments before being released from the Canadarm2 robotic arm.

Expedition 59 crew members Christina Koch of NASA (left), Alexey Ovchinin of Roscosmos (center) and Nick Hague of NASA (right) pose for pictures in front of their Soyuz MS-12 spacecraft

Expedition 60

The official Expedition 60 crew portrait with (clockwise from top right) astronauts Nick Hague of NASA and Luca Parmitano of ESA (European Space Agency), Roscosmos cosmonaut Alexander Skvortsov, NASA astronauts Drew Morgan and Christina Koch and Roscosmos cosmonaut Alexey Ovchinin.

Soyuz MS-13, also known as ISS flight 59S, was a pivotal mission in the International Space Station (ISS) history. Launched on July 20, 2019—the 50th anniversary of the Apollo 11 Moon landing—this mission was notable for its crew's diverse international composition and its role in maintaining ISS operations amidst evolving spaceflight programs.

The Soyuz MS-13 spacecraft, the 142nd flight of the Soyuz series, carried a crew of three to the ISS: Russian commander Aleksandr Skvortsov, Italian flight engineer Luca Parmitano, and American flight engineer Andrew R. Morgan. This mission represented a significant point in the Soyuz program as it was initially expected to be the last Soyuz flight contracted by NASA, with future astronaut transport anticipated to come from the Commercial Crew Program. However, delays in the Commercial Crew Program led NASA to secure additional Soyuz seats to ensure continuity in crew transportation.

Upon launch, Soyuz MS-13 was propelled into space aboard a Soyuz-FG rocket from the Baikonur Cosmodrome in Kazakhstan. The spacecraft docked with the ISS's Zvezda module at 22:47 UTC on the same day, marking the beginning of its mission. This docking was crucial for integrating the new crew into Expedition 60, which had commenced on June 24, 2019, with the undocking of Soyuz MS-11.

During their stay, the crew of Soyuz MS-13 contributed to the station's scientific research and daily operations. On August 26, 2019, the spacecraft was manually relocated from the Zvezda aft port to the Poisk module's zenith port. This relocation was necessary to clear the Zvezda module for the arrival of Soyuz MS-14, which had experienced a failed docking attempt on August 24 due to a faulty signal amplifier. The relocation of Soyuz MS-13 marked the first time a Soyuz spacecraft had been manually docked in this manner since Soyuz TMA-16M in August 2015.

The Soyuz MS-13 mission concluded on February 6, 2020, when the spacecraft undocked from the Poisk module and landed in the steppes of Kazakhstan. The mission was 200 days, 16 hours, and 44 minutes, during which the crew completed 3,216 orbits of Earth. This mission saw a temporary increase in ISS crew size to nine members with the arrival of Soyuz MS-15, surpassing the standard six-person crew for the first time since September 2015.

Expedition 61

The official Expedition 61 crew portrait with (from left) NASA astronaut Andrew Morgan, Roscosmos cosmonaut Alexander Skvortsov, astronaut Luca Parmitano of ESA (European Space Agency), Roscosmos cosmonaut Oleg Skripochka, and NASA astronauts Jessica Meir and Christina Koch.

Expedition 61 commenced on October 3, 2019, marking the sixty-first mission to the International Space Station (ISS). This milestone was notable for the undocking of the Soyuz MS-12 spacecraft, signaling the transition from Expedition 60.

The mission was led by European Space Agency (ESA) astronaut Luca Parmitano, who made history as both the third European and the first Italian to command the ISS. His role was a significant achievement in space exploration, underscoring Europe's growing leadership in international space missions.

Parmitano was accompanied by his Soyuz MS-13 colleagues, Aleksandr Skvortsov and Andrew Morgan, and Christina Koch from the Soyuz MS-12 mission. These astronauts seamlessly continued their duties from the previous expedition, ensuring a smooth transition in ongoing operations aboard the station.

Joining them were Oleg Skripochka and Jessica Meir, who arrived aboard Soyuz MS-15 on September 25, 2019. Their addition to the crew brought new expertise and perspective to the ISS, enriching the mission's scientific and operational objectives. This blend of experienced astronauts and new arrivals marked a dynamic period for the ISS, characterized by collaborative international efforts and groundbreaking research.

Expedition 61's crew was a distinguished assembly of astronauts representing a blend of nations and space agencies, each bringing their unique expertise to the International Space Station.

Luca Parmitano from Italy, representing the European Space Agency (ESA), served as the mission commander. Parmitano's leadership was marked by his historic achievement as the first Italian astronaut to lead an ISS expedition, further enhancing Italy's prominent role in space exploration. This was his second spaceflight.

Jessica Meir (left) and Christina Koch (right) before their spacewalk, next to (left to right) Aleksandr Skvortsov, Luca Parmitano (Commander), Oleg Skripochka, and Andrew Morgan

Aleksandr Skvortsov from Russia, a veteran astronaut with the Russian Space Agency (RSA), took on the role of Flight Engineer 1. His participation was notable as this was his third and final spaceflight, reflecting his extensive experience and contribution to space missions.

Andrew Morgan, representing NASA, was Flight Engineer 2. As a first-time space traveler, Morgan brought fresh perspectives to the ISS and contributed to various scientific and operational tasks during his inaugural mission.

Christina Koch, also from NASA, served as Flight Engineer 3. Like Morgan, Koch was embarking on her first spaceflight, and her fresh insights and contributions added to the mission's dynamic.

Flight Engineer 4 was Oleg Skripochka, another seasoned astronaut from RSA. This was Skripochka's third spaceflight, underscoring his extensive experience and previous contributions to space missions.

Jessica Meir, representing NASA as Flight Engineer 5, was on her first spaceflight. Her participation brought new energy and expertise to the crew, complementing the diverse skill set of the mission team.

In the planning stages for Expedition 61, there were ambitious plans to integrate two significant Commercial Crew Development missions into the schedule, as outlined in a Flight Planning Integration Panel (FPIP) document obtained by NASAspaceflight.com in June 2019. These missions , intended to enhance the station's operational capabilities and demonstrate new spacecraft technologies, ultimately faced delays that prevented their execution as initially planned.

One of these missions was the Crew Dragon Demo-2, a groundbreaking initiative by SpaceX. The mission was initially slated for launch on November 15, 2019, with a return scheduled for November 22, 2019. American astronauts Bob Behnken and Douglas Hurley were to be the primary crew for this mission. The Crew Dragon Demo-2 was designed to conduct a brief stay aboard the ISS, which was a pivotal step in validating the Crew Dragon spacecraft's performance and compatibility with station operations.

The second planned mission was the Boeing Crewed Flight Test of the Starliner capsule. This mission was scheduled to launch on November 30, 2019, with American astronauts Mike Finke, Nicole Mann, and Chris Ferguson planned to dock with the ISS on December 1, 2019. The Starliner crew was expected to remain on the station until May 2020. There was some ambiguity regarding their status—whether they would be officially considered part of the Expedition 61 crew, be recognized as ISS visitors, or herald the commencement of Expedition 62.

These planned missions were integral to advancing NASA's Commercial Crew Program, which aimed to enhance the ISS's crew transport capabilities and validate new spacecraft designs. However, these visits did not occur during Expedition 61 due to schedule adjustments, impacting the mission's original operational expectations.

During Expedition 61, the crew conducted an unprecedented nine spacewalks, marking the highest number of extravehicular activities (EVAs) in the history of the International Space Station (ISS) up to that point. This remarkable achievement underscored the mission's intensive focus on maintaining and upgrading critical systems aboard the station.

A significant portion of these spacewalks was dedicated to the repair of the Alpha Magnetic Spectrometer (AMS-02), a crucial particle physics experiment mounted on the ISS. The AMS-02, designed to detect cosmic rays and contribute to our understanding of the universe, had encountered technical issues that necessitated extensive repairs. ESA astronaut Luca Parmitano and NASA astronaut Andrew Morgan undertook four challenging spacewalks to address these issues. Their efforts were supported by NASA astronauts Christina Koch and Jessica Meir, who operated the Canadarm2 robotic arm from inside the station, precisely facilitating the repair tasks. These spacewalks were regarded as among the most demanding since the historic Hubble Space Telescope repairs, highlighting the complexity and critical nature of the repairs.

In addition to the AMS-02 repairs, the crew conducted several spacewalks focused on upgrading and replacing the ISS's battery systems. This work was crucial for maintaining the station's power generation capabilities and ensuring operational efficiency.

Notably, one of these spacewalks on October 18, 2019, achieved a historic milestone as Christina Koch and Jessica Meir undertook the first all-female spacewalk. This landmark event was a significant moment in space exploration history, showcasing the advancing role of women in the field and the diverse capabilities of the ISS crew.

The year 2020 marked a pivotal chapter in the history of the International Space Station (ISS), as it continued to serve as a crucial hub for human spaceflight and scientific research. Throughout the year, the ISS hosted a succession of expeditions and missions, each contributing to our understanding of space and advancing human presence in orbit.

In May, a historic milestone was achieved with the launch of SpaceX's Crew Dragon spacecraft, named Endeavour. This mission was notable for being the first crewed spacecraft built by a commercial entity to transport astronauts to the ISS. NASA astronauts Bob Behnken and Doug Hurley were the first to board this groundbreaking spacecraft, which docked with the ISS on May 31. This event not only marked a significant achievement for SpaceX but also represented the first crewed launch from U.S. soil since the end of the Space Shuttle program in 2011.

July brought a transition of crew members with the arrival of the Expedition 63 team, who took over from the Expedition 62 crew. During their time aboard the ISS, the Expedition 63 astronauts engaged in a variety of research and maintenance tasks. They also welcomed the Russian Progress MS-14 spacecraft, which delivered essential supplies and experiments to the station, ensuring its continued functionality and research capabilities.

As October approached, the Expedition 64 crew arrived to replace their predecessors. This new team was tasked with carrying out a series of spacewalks to perform critical maintenance and install new equipment. Their activities included conducting experiments across various scientific disciplines such as biology, physics, and human physiology, all vital to the ongoing mission of the ISS.

Amidst these developments, China's space program made significant strides in advancing its Tiangong space station project. In May, China launched the Long March 5B rocket, carrying the Tianhe module, the first component of its new space station, Tiangong-3. The Tianhe module, which translates to "Heavenly Harmony," was designed to serve as the living quarters for future Chinese astronauts, known as taikonauts. This launch marked the beginning of China's ambitious plan to establish a space station of its own.

The ISS continued to be a focal point of space exploration as the year progressed. On April 9, the Soyuz MS-16 spacecraft launched with a new crew, including Anatoli Ivanishin, Ivan Vagner, and Christopher Cassidy. Their mission involved conducting scientific experiments and performing station duties. The Soyuz MS-16 returned these crew members to Earth on October 22, concluding their rotation.

Another key event occurred on May 30 with the launch of the Crew Dragon Demo-2 spacecraft.

On October 14, 2020, the Soyuz MS-17 spacecraft embarked on its mission, marking the 145th crewed flight of a Soyuz vehicle. This mission carried three astronauts—Russian Commander Sergey Ryzhikov, Russian Flight Engineer Sergey Kud-Sverchkov, and American Flight Engineer Kathleen Rubins—into space, where they would join the International Space Station (ISS) as part of Expedition 63/64.

Soyuz MS-17 was notable for implementing a new "ultrafast" two-orbit rendezvous flight plan, which allowed the spacecraft to reach the ISS within approximately three hours of its launch. This approach significantly shortened the time required for docking compared to previous methods.

Upon arrival, the spacecraft docked with the ISS at the Rassvet module on October 14, 2020, at 08:48 UTC. The crew settled into their roles and contributed to various scientific and operational activities on the station.

On March 19, 2021, the Soyuz MS-17 crew performed a relocation maneuver, moving their spacecraft from the Rassvet module to the Poisk module. This maneuver was necessary to accommodate the arrival of Soyuz MS-18, which launched on April 9, 2021, with cosmonauts Oleg Novitsky and Pyotr Dubrov, and NASA astronaut Mark T. Vande Hei. The handover period allowed both crews to work together on the ISS, ensuring the station's operations continuity.

Another significant event was the SpaceX Crew-1 mission, which began on November 15, 2020. The Resilience spacecraft transported astronauts Michael Hopkins, Victor Glover, Soichi Noguchi, and Shannon Walker to the ISS as part of the 64th/65th expedition. This crew contributed to a wide range of scientific experiments and played a vital role in the station's operations. They safely returned to Earth on May 2, 2021.

Throughout the year, several notable astronauts and cosmonauts contributed to the ISS missions. Jessica Meir and Andrew R. Morgan, both from NASA, continued their participation in space expeditions. Chris Cassidy, also from NASA, served as commander of Expedition 63. Russian cosmonauts Anatoli Ivanishin and Ivan Vagner completed their missions on the Soyuz MS-16, while Kate Rubins, another NASA astronaut, joined the Soyuz MS-17 mission.

In addition to the ISS activities, the year witnessed significant developments in global space exploration. In October, an H-6 bomber at Neixiang Air Base was observed landing with a missile resembling the DF-17. In November, the U.S. demonstrated its missile defense capabilities by intercepting and destroying a mock intercontinental ballistic missile launched from Kwajalein Atoll toward Hawaii.

In 2021, the International Space Station (ISS) stood as a cornerstone of human spaceflight, playing host to a series of pivotal missions and crew rotations that underscored its role as a hub for scientific research and international collaboration. This year marked a dynamic period of activity both aboard the ISS and within the broader realm of space exploration.

The year began with the continued presence of astronauts from previous missions, who diligently carried out their scientific and operational tasks. The arrival of the SpaceX Crew-2 mission in April was a notable highlight. This mission, which utilized the Endeavour spacecraft, transported a diverse crew comprising Shane Kimbrough, K. Megan McArthur, Akihiko Hoshide, and Thomas Pesquet. Their six-month stay aboard the ISS involved conducting extensive scientific experiments, performing critical station maintenance, and contributing to the ongoing research in human physiology, biology, and physics. The Crew-2 mission marked a significant achievement in commercial spaceflight and exemplified the advancing capabilities of private space enterprises. On November 9, 2021, this team completed their mission and safely returned to Earth.

May 2021 saw a major milestone in space exploration as China made substantial progress with its space station program. The successful launch of the Tianhe module, the first component of the Tiangong space station, aboard the Long March 5B rocket represented a crucial advancement for China's ambitions in space. Tianhe, intended to serve as the living quarters for future Chinese astronauts, laid the foundation for the country's long-term goals of establishing a permanent presence in low Earth orbit.

The summer months brought further changes to the ISS crew. In June, the Expedition 65 crew took over from Expedition 64, continuing the station's work in scientific research and maintenance. The departure of the Russian Pirs module in August, which had been an integral part of the ISS since 2001, marked the end of its role as an airlock and docking port, making way for new additions to the station.

The private sector made notable strides in space exploration during 2021. The Unity 21 mission in July, conducted by Virgin Galactic, represented the first human spaceflight from New Mexico, reaching 89.24 km. Although this mission met the US definition of space, it did not satisfy the Fédération Aéronautique Internationale (FAI) criteria. Later in the year, the Unity 22 mission, which included notable figures such as Richard Branson, reached 86 km, crossing the US space boundary but falling short of the FAI's definition. Meanwhile, Blue Origin's NS-16 mission, which featured Jeff Bezos and other crew members, achieved 107 km, meeting the FAI's definition of space and marking a significant moment for private space travel.

In September, the Inspiration 4 mission, led by Jared Isaacman, launched aboard the Resilience spacecraft. This mission aimed to raise funds and awareness for St. Jude Children's Research Hospital and was notable for its all-civilian crew. The mission was a testament to the growing role of private ventures in space exploration and concluded with the crew's return to Earth on September 18, 2021.

Throughout the year, the ISS continued to receive regular resupply and crew missions via Soyuz spacecraft. The Soyuz MS-18 mission, which began in April, brought Oleg Novitsky and his team to the ISS as part of the 64th/65th expedition. Their mission involved conducting scientific research and performing station operations before returning to Earth in October. Similarly, Soyuz MS-19, carrying Piotr Dubrov and Mark T. Vande Hei, joined the ongoing expedition, contributing to the station's activities until their return in 2022.

China's space efforts continued with the Shenzhou 12 mission in June, to developing the first crewed mission to visit the Tiangong space station. The crew's activities, though not fully detailed, contributed to China's space station program. Shenzhou 13, which launched in October, brought additional crew to Tiangong, further advancing the station's development.

The year closed with the arrival of the SpaceX Crew-3 mission in November. The Endurance spacecraft delivered Raja Chari, Thomas Marshburn, Kayla Barron, and Matthias Maurer to the ISS for the 66th/67th expedition. Their mission focused on scientific research and operational duties aboard the station. The crew returned to Earth on May 6, 2022.

Throughout 2021, the ISS remained a vibrant center for scientific discovery and international cooperation. The year also saw advancements in private spaceflight and significant progress in China's space station program. These developments enriched our understanding of space and highlighted the growing role of commercial enterprises in the realm of human space exploration.

Additionally, the year saw advancements in aerospace technology on Earth, such as the USAF's Lockheed Martin/Calspan X-62 VISTA (Variable In-flight Simulator Test Aircraft). The X-62A, an upgrade from its previous iteration as the NF-16D, continued its role at the USAF Test Pilot School, contributing to research in advanced flight systems and simulation technologies.

In 2022, space exploration witnessed a series of significant and groundbreaking events that underscored space technology's rapid advancements and increasing complexities.

On February 2, the United States launched USA-326, an advanced American spy satellite, into orbit. Positioned approximately 500 kilometers above the Earth's surface, this state-of-the-art satellite was designed to capture highly detailed imagery, boasting capabilities so precise that it could discern car license plates from space. This launch underscored the growing sophistication of satellite technology and its pivotal role in global surveillance and reconnaissance.

Later in the year, August brought an alarming development. The US Space Command, while monitoring the ongoing conflict in Ukraine, discovered that Russia had deployed a new satellite, Kosmos-2558. This satellite's presence raised serious concerns within the US defense community. There was suspicion that Kosmos-2558 might possess offensive capabilities, potentially including a projectile designed to target American satellites. Such an action, if executed, could breach NATO's Article 5, which regards an attack on one member as an attack on all. This situation highlighted a critical gap in NATO's space defense, revealing that the US was the sole nation with an operational anti-satellite system, while NATO's capabilities in this domain were markedly underdeveloped.

Amid these geopolitical tensions, advancements in Intercontinental Ballistic Missile (ICBM) technology continued to evolve. developing hypersonic glide vehicles, such as those carried by the RS-28 Sarmat missile, marked a significant leap in missile technology. The Sarmat, alongside other land-based ICBMs from countries like Russia, the United States, China, North Korea, and India, represented cutting-edge strategic missile systems. Although Israel had also tested ICBMs, it kept details of their deployment confidential.

In the US, the operational landscape of ICBMs underwent a strategic shift. By 2022, the US Air Force maintained 405 ICBMs at three bases, exclusively deploying the LGM-30G Minuteman-III. This shift marked the end of the Minuteman II missiles and the Peacekeeper missiles, which had been decommissioned in 2005. The transition underscored a streamlined strategic deterrence approach, reflecting technological and tactical advancements.

Manned space missions continued to push the boundaries of exploration throughout the year. On March 18, the Soyuz MS-21 mission, featuring astronauts Oleg Artemyev, Denis Matveev, and Sergey Korsakov, docked with the International Space Station (ISS). This mission was part of the 66th/67th ISS expedition and was critical in maintaining the station's operations. The crew later returned on September 29 as part of a routine crew rotation.

Another notable event was the NS 20 mission on March 31, which saw astronauts Marty Allen, Sharon Hagle, Marc Hagle, Jim Kitchen, George Nield, and Gary Lai exceed 100 kilometers in altitude, thereby crossing the threshold of space as defined by the Fédération Aéronautique Internationale (FAI). This mission marked a significant achievement in private space travel.

The Axiom Mission 1, launched on April 8, carried astronauts Michael López-Alegría, Larry Connor, Mark Pathy, and Eytan Stibbe aboard the Endeavour spacecraft to the ISS. This mission not only contributed to scientific research but also advanced the ISS's operational capabilities. The crew returned to Earth on April 25, concluding their mission with notable achievements in space research.

Following this, the SpaceX Crew-4 mission took off on April 27, featuring astronauts Kjell N.

Lindgren, Robert Hines, Samantha Cristoforetti, and Jessica Watkins aboard the Freedom spacecraft. Their mission involved joining the 67th/68th ISS expedition, where they engaged in a range of scientific experiments and station duties. The mission concluded on October 14 with the crew's return to Earth.

The NS 21 mission on June 4 saw a new crew, including Hamish Harding, Victor Correa Hespanha, Evan Dick, Katya Echazarreta, Jaison Robinson, and Victor Vescovo, ascend beyond 100 kilometers, entering space as per the FAI definition. This mission continued the trend of significant private space ventures.

On June 5, the Shenzhou 14 mission launched with astronauts Chen Dong, Liu Yang, and Cai Xuzhe. This crew, the third to visit the Tiangong Space Station, carried out their mission with a focus on advancing China's space station program. They returned to Earth on December 4, marking the end of their successful mission.

The NS 22 mission began on August 4 with astronauts Coby Cotton, Mário Ferreira, Vanessa O'Brien, Clint Kelly III, Sara Sabry, and Steve Young, reaching altitudes exceeding 100 kilometers and affirming their place within the realm of space.

The Soyuz MS-22 mission, which launched on September 21, included astronauts Sergey Prokopyev and Dmitry Petelin, joining the ISS as part of the 67th/68th expedition. Their presence was vital for the ongoing ISS crew rotation.

On October 5, the SpaceX Crew-5 mission began with Nicole Aunapu Mann, Josh A. Cassada, Koichi Wakata, and Anna Kikina aboard the Endurance spacecraft. On October 5, 2022, SpaceX Crew-5, the fifth operational mission of NASA's Commercial Crew Program, was launched, marking the eighth crewed orbital flight for Crew Dragon. This mission was a significant milestone as it transported four astronauts to the International Space Station (ISS), docking with the station on October 6, 2022, at 21:01 UTC.

The Crew-5 mission's crew was a diverse group, including two NASA astronauts, one JAXA astronaut, and one Russian cosmonaut. Commander Nicole Mann and Pilot Josh Cassada were reassigned to Crew-5 from Boeing's Starliner program due to delays with Boeing's Starliner-1 mission. Mission Specialist Koichi Wakata also joined the team, transferring from the Boeing Starliner-1 mission. Anna Kikina, a Russian cosmonaut from Roscosmos, was reassigned from Soyuz MS-22 to participate in this historic mission.

SpaceX Crew-5 Official Crew Portrait - Anna Kikina, Josh Cassada, Nicole Mann and Koichi Wakata

This mission was notable for several reasons. It was the first time a Russian cosmonaut, Anna Kikina, flew aboard a Crew Dragon spacecraft. This inclusion was part of the Soyuz-Dragon crew swap system, which ensures that each crew rotation mission aboard the ISS includes at least one NASA astronaut and one Roscosmos cosmonaut. This arrangement allows both space agencies to maintain a continuous presence on the ISS and ensures backup crew capabilities if either Soyuz or commercial crew vehicles experience grounding or delays.

The inclusion of Kikina also marked the first time a Russian cosmonaut had flown on a U.S. spacecraft since STS-113 in 2002, when Nikolai Budarin flew aboard the Space Shuttle Endeavour. Additionally, it was the first launch of a Russian cosmonaut aboard a U.S. space capsule, a historic first in the collaboration between NASA and Roscosmos.

Roscosmos cosmonaut Anna Kikina, left, NASA astronaut Josh Cassada, second from left, NASA astronaut Nicole Mann, second from right, and Japan Aerospace Exploration Agency (JAXA) astronaut Koichi Wakata, right, are seen as they prepare to depart the Neil A. Armstrong Operations and Checkout Building for Launch Complex 39A to board the SpaceX Crew Dragon spacecraft for the Crew-5 mission launch, Wednesday, Oct. 5, 2022, at NASA's Kennedy Space Center in Florida. NASA's SpaceX Crew-5 mission was the fifth crew rotation mission of the SpaceX Crew Dragon spacecraft and Falcon 9 rocket to the International Space Station as part of the agency's Commercial Crew Program. Photo Credit: (NASA/Joel Kowsky)

In June 2022, the Russian authorities approved the seat exchange between American and Russian astronauts, facilitating this unprecedented collaboration.

However, the mission was not without challenges. Soyuz MS-22, which had been docked at the ISS with a crew that included Francisco Rubio, experienced a critical issue when a micrometeorite punctured a 0.8 mm hole in the radiator of its service module. This damage raised concerns about the safety of returning the Soyuz MS-22 crew to Earth.

In response to this issue, Soyuz MS-23 was launched uncrewed on February 24, 2023, to replace Soyuz MS-22. Until MS-23 docked with the ISS on February 26, 2023, SpaceX Crew-5 was considered a potential emergency option for returning the MS-22 crew if necessary. Crew Dragon was designed to accommodate up to seven astronauts, which provided flexibility for emergency situations.

To enhance safety, the International Space Station mission management team decided to transfer NASA astronaut Francisco Rubio's Soyuz seat liner from the Soyuz MS-22 spacecraft to the Crew Dragon Endurance. This transfer, carried out on January 17, 2023, was the first instance of a seat liner being swapped from a Soyuz spacecraft to a Crew Dragon. This adjustment was intended to reduce the heat load inside Soyuz MS-22 and provide additional protection for the cosmonauts Sergey Prokopyev and Dmitry Petelin in the event of an emergency return.

Following the arrival of Soyuz MS-23 at the ISS, Rubio's seat liner was moved to the new Soyuz on March 6, 2023, while the seat liners for Prokopyev and Petelin were transferred from MS-22 to MS-23 on March 2, 2023, in preparation for their return. This logistical maneuver ensured that the crew had a safe and functional spacecraft for their journey back to Earth.

Finally, on November 29, the Shenzhou 15 mission launched with Fei Junlong, Deng Qingming, and Zhang Lu, marking the fourth crew to visit the Tiangong Space Station. Their mission continued the advancement of China's space station program and its international collaborations.

In addition to these space missions, NASA made strides in supersonic air travel with the X-59 QueSST aircraft. This innovative vehicle aimed to revolutionize commercial supersonic flight by minimizing sonic booms—a significant step toward more practical and environmentally friendly supersonic travel. The X-59's design aimed to produce a gentle thump rather than a disruptive sonic boom, signaling a potential new era in high-speed aviation.

In the years 2023 and 2024, the landscape of space exploration was marked by a series of groundbreaking missions and technological advancements that underscored the rapid progress and international collaboration in the field.

As 2023 unfolded, the Artemis I mission continued to capture the world's attention. This ambitious endeavor, launched in November 2021, marked a pivotal moment for NASA as it tested the Space Launch System (SLS) rocket and the Orion spacecraft in an uncrewed flight. The mission aimed to evaluate the spacecraft's systems, laying the crucial groundwork for future crewed lunar missions. The data gathered during this mission would play a vital role in shaping humanity's return to the Moon.

Meanwhile, the James Webb Space Telescope (JWST), a collaboration between NASA, the European Space Agency (ESA), and the Canadian Space Agency (CSA), continued its extraordinary journey through space. Since its launch in December 2021, the JWST had been offering unprecedented glimpses into the universe's most distant reaches. Its infrared observations revealed new details about distant galaxies, star formation, and the atmospheres of exoplanets, significantly advancing our understanding of the cosmos.

SpaceX remained at the forefront of space exploration with several notable missions. On March 2, 2023, the Crew-6 mission transported four astronauts to the International Space Station (ISS) aboard the Crew Dragon spacecraft. Their presence aboard the ISS was vital for conducting scientific research and maintaining the station's operations.

In April, SpaceX launched the Transporter-8 mission, deploying a constellation of small satellites into orbit using the Falcon 9 rocket. This mission highlighted SpaceX's ability to deliver diverse payloads to space, continuing its role as a key player in satellite deployment.

China's space program also made significant strides with the Shenzhou 15 mission, which launched in late November 2022 and continued into 2023. The crewed mission, part of China's Tiangong space station program, saw astronauts conducting important scientific experiments and performing maintenance tasks. This mission was a testament to China's growing expertise in long-duration spaceflight and space station operations.

The Lunar Reconnaissance Orbiter (LRO), which had been mapping the Moon's surface since its launch in 2009, continued its critical work throughout 2023. The LRO's detailed lunar maps and data supported future lunar exploration efforts and deepened our understanding of the Moon.

Looking ahead to 2024, the excitement continued with the anticipated Artemis II mission, scheduled for April. This mission would mark NASA's first crewed flight of the Space Launch System (SLS) rocket and Orion spacecraft. The crew would orbit the Moon, preparing for subsequent lunar landings and advancing humanity's return to lunar exploration.

The ExoMars rover, a collaborative effort between the European Space Agency (ESA) and the Russian space agency Roscosmos, was also set to make headlines in March 2024. This mission aimed to search for signs of past life on Mars, with the rover analyzing soil samples and investigating the Martian environment, promising to deliver new insights into the Red Planet's history.

SpaceX's activities were far from over. The Crew-7 mission, slated for July 2024, would send another group of astronauts to the ISS for crew rotation, ensuring the continuation of critical research and station upkeep. Additionally, the Transporter-9 mission, planned for September, would carry a new batch of small satellites into orbit, further demonstrating SpaceX's ongoing commitment to satellite deployment.

China's Tianwen-1 mission, which included the orbiter, lander, and rover sent to Mars, continued to deliver valuable data in 2024. The rover, Zhurong, explored the Martian surface, contributing to our understanding of Mars through its surface investigations and sending data back to Earth.

Finally, various international space agencies prepared for missions aimed at exploring the Moon's polar regions. Scheduled for 2024, these missions sought to investigate the presence of water ice and other resources that could support future lunar exploration and habitation.

Expedition 62

Expedition 62 crew portrait with NASA astronaut Andrew Morgan, Roscosmos cosmonaut Oleg Skripochka and NASA astronaut Jessica Meir.

On April 9, 2020, the Soyuz MS-16 mission launched, marking a significant milestone as it became the first crewed flight to use the Soyuz 2.1a launch vehicle. This mission also stood out as the first crewed Russian mission not to depart from Gagarin's Start, which was undergoing modernization. Since Soyuz MS-02 in 2016, no crewed mission had taken off from a different launch pad.

The Soyuz MS-16 crew consisted of three astronauts: Anatoli Ivanishin, Ivan Vagner, and Christopher Cassidy. Ivanishin, a seasoned cosmonaut making his third spaceflight, commanded the mission. Vagner, a rookie astronaut, was embarking on his first space journey. Cassidy, representing NASA, was also on his third spaceflight. Notably, this mission would have included Nikolai Tikhonov, a new cosmonaut, but he was replaced due to a medical issue. Tikhonov's eye injury prompted Russian officials to substitute him and his backup, Andrei Babkin, with Ivanishin and Vagner.

The mission was originally planned to include Tikhonov, Babkin, and Japanese astronaut Akihiko Hoshide. However, in late 2019, it was decided that Cassidy would join the crew instead of Hoshide to ensure a continuous American presence on the ISS, especially in light of delays in NASA's Commercial Crew Program. Consequently, Soyuz MS-16's crew became part of Expedition 62.

Soyuz MS-16 launched at 08:05 UTC from the Baikonur Cosmodrome. The Soyuz 2.1a rocket's core and first stage engines ignited flawlessly, propelling the spacecraft into orbit. The rocket climbed into the orbital plane as the International Space Station (ISS) passed over the launch site. After six orbits, Soyuz MS-16 docked with the Poisk module of the ISS at 14:13 UTC.

The unprecedented impact of the COVID-19 pandemic characterized the mission's duration. Traditionally, crewed launches are marked by pre-launch ceremonies and the presence of families and media. However, due to the pandemic, these customary traditions were cancelled, and the usual launch viewing was not permitted.

The Soyuz MS-16 mission, which lasted for 195 days, 18 hours, and 49 minutes, concluded with the spacecraft undocking from the ISS on October 21, 2020, at 23:32 UTC. It safely landed in the Kazakh steppes at 02:54 UTC the following day, approximately 150 kilometers southeast of Zhezkazgan.

Expedition 62, which began on February 6, 2020, with the undocking of Soyuz MS-13, was initially composed of Russian commander Oleg Skripochka and American flight engineers Jessica Meir and Andrew Morgan. The second part of Expedition 62, from April 9 to April 17, 2020, included the Soyuz MS-16 crew. This expedition marked the return of a three-member Russian crew on the ISS for the first time since Expedition 50, making it one of the shortest expeditions in the program's history.

Elsewhere in international space activities, Looking ahead to 2024, Hungary announced its first astronaut mission, HUNOR 1, scheduled for May 27th. Tibor Kapu was selected as the primary astronaut, with Gyula Cserényi serving as the reserve astronaut. This mission marked Hungary's entry into crewed space missions, symbolizing its aspirations in scientific discovery and space exploration.

These developments in 2023 and 2024 underscored a global commitment to expanding human presence in space, fostering international cooperation, and pushing the boundaries of scientific knowledge beyond Earth's atmosphere.

The commercial space sector has evolved significantly, driven by a burgeoning market that has surpassed $330 billion and was projected to approach $3 trillion in the coming decades. Among the sectors poised for substantial growth, human spaceflight stands out prominently. Commercial astronauts are set to play a pivotal role in this transformative phase of space exploration.

The foundation for commercial astronauts was laid with the inception of the Ansari X PRIZE in 2004, which aimed to spur developing the first privately-built, reusable crewed spacecraft. This milestone initiative saw the selection of pioneering individuals who would shape the future of commercial space travel.

Notable figures among the early commercial astronauts include Steve Bennett and Matt Shewbridge from Starchaser Industries, alongside esteemed former NASA astronauts such as John Bennett Herrington of Pioneer Rocketplane, Richard Searfoss, and pilot Dick Rutan of XCOR Aerospace. Canadian engineer Brian Feeney represented the da Vinci Project, contributing to the advancement of private space initiatives. Additionally, Wally Funk, a trailblazer from the Mercury 13 program, continued her legacy with Interorbital Systems, further enriching the commercial space landscape.

Boeing's commitment to advancing space exploration was underscored by the significant addition of former NASA astronaut Chris Ferguson to its Space Exploration Team. Ferguson's extensive experience and expertise from his NASA tenure make him a valuable asset to Boeing's ambitious space initiatives.

Boeing's astronaut corps included various candidates, ranging from former NASA astronauts to commercial scientist astronauts and test pilots who have yet to experience spaceflight. This eclectic mix highlights Boeing's strategic approach to assembling a team equipped to tackle the challenges and opportunities of the evolving space industry.

SpaceX, a commercial space exploration pioneer, has integrated former NASA astronauts into its team while maintaining distinct criteria for crewing its commercial vehicles to the International Space Station (ISS). However, specific citations are required to verify this information.

An example of SpaceX's integration with NASA was Anil Menon, formerly the medical director at SpaceX. He transitioned into a NASA astronaut, selected in 2021 as part of NASA Astronaut Group 23. This transition exemplifies SpaceX and NASA's collaborative synergy, contributing to advancements in human spaceflight and exploration efforts.

The Association of Spaceflight Professionals (ASP) emerged as a pioneering initiative in commercial space exploration, purportedly forming the world's first commercial astronaut corps. This organization reportedly secured funding through the NASA Flight Opportunities Program in March 2012, marking a significant milestone in its efforts to conduct crewed spaceflight missions.

ASP's notable endeavors include several million dollars reportedly allocated for detailed spectroscopic analysis of high-altitude noctilucent cloud formations during suborbital flights. These missions utilize rapidly reusable, task-and-deploy spaceplanes, highlighting ASP's innovative approach to scientific research in space.

ASP's selection process for commercial astronauts mirrors that of NASA's prestigious Astronaut Corps, emphasizing rigorous training and qualification standards. Some ASP members are reported to serve as astronaut trainers themselves and have participated as finalists in National Space Agency astronaut candidate selection campaigns.

One prominent ASP member mentioned was Yi So-yeon, known for her historic orbital mission to the International Space Station. Her involvement underscores ASP's international reach and collaborative efforts in advancing commercial space ventures.

Virgin Galactic and its subsidiary Scaled Composites have been pivotal in advancing commercial spaceflight, and a roster of astronauts and key figures has contributed to their missions and developments.

Among the notable individuals associated with Scaled Composites and Virgin Galactic are:

Michael Alsbury, tragically killed in the 2014 Virgin Galactic crash, was instrumental in the development and testing phases.

Rob Bendall, representing Canada, a key figure in Virgin Galactic's early test flights and operations.

Richard Branson, the founder of Virgin Galactic, who played a significant role in promoting and supporting commercial space tourism.

Peter Kalogiannis, an integral member of the engineering team at Virgin Galactic.

Niki Lauda, the renowned Austrian Formula 1 champion and aviation enthusiast.

Brian Maisler, Clint Nichols, and Wes Persall contributed to various aspects of Virgin Galactic's operations.

Burt Rutan, the legendary aerospace engineer and founder of Scaled Composites, was known for his innovative design and contributions to spacecraft development.

Key pilots involved in test flights and operational missions are Peter Seiffert and Peter Siebold.

Mark Stucky and Dave Mackay are both renowned pilots who have flown Virgin Galactic's spacecraft on numerous test and operational flights.

The Teachers in Space program, launched in 2005, aimed to pioneer educational opportunities in space exploration. In 2012, the United States Rocket Academy announced a significant expansion, broadening the initiative to include a wider array of participants and rebranding it as Citizens in Space.

Under the Citizens in Space banner, the program's initial phase focused on selecting and training ten citizen astronaut candidates to serve as payload operators. This cohort included four astronaut candidates already undergoing training: Maureen Adams, Steve Heck, Michael Johnson, and Edward Wright. Among the notable participants was Gregory Kennedy, an informal educator and aerospace historian who contributed his expertise to the program's mission.

Copenhagen Suborbitals, founded in 2008 in Denmark, has set an ambitious goal to achieve human spaceflight beyond the Kármán line, aiming to make Denmark the fourth nation to accomplish this milestone. The organization represents a grassroots effort in amateur space exploration, driven by the passion and dedication of its members to push the boundaries of space technology.

Mars One, an ambitious private initiative announced in May 2012 by Dutch entrepreneur Bas Lansdorp, aimed to establish the first permanent human colony on Mars by 2023. The project garnered significant attention with its astronaut selection process beginning in April 2013, attracting over 200,000 applicants from around the globe by August of that year. Round Two of selections, announced in December 2013, identified 1,058 finalists from 107 countries eager to embark on the mission.

Despite early enthusiasm, Mars One faced substantial skepticism and criticism regarding its medical, technical, and financial viability. Concerns ranged from the feasibility of sustaining life on Mars to doubts about the project's funding model. Criticism intensified with unverified allegations suggesting Mars One might have been a fundraising scheme rather than a genuine space mission.

In a notable development, Mars One declared bankruptcy in a Swiss court on January 15, 2019, leading to the permanent dissolution of the company. This marked the end of the Mars One venture, highlighting the challenges and uncertainties inherent in pioneering efforts toward human settlement on Mars.

Expedition 63

The official Expedition 63 crew portrait with (clockwise from bottom left) Doug Hurley of NASA, Anatoly Ivanishin of Roscosmos, Chris Cassidy of NASA, Ivan Vagner of Roscosmos and Bob Behnken of NASA.

Expedition 63 marked the sixty-third long-duration mission to the International Space Station (ISS). It spanned an unusually extended period from April 17, 2020, when the Soyuz MS-15 spacecraft was undocked, to October 21, 2020, when Soyuz MS-16 departed. This double-length increment was notable for its extended duration, reflecting the ISS's crew rotation schedule adjustments.

The initial crew of Expedition 63 included American astronaut Chris Cassidy, who served as the mission commander. Cassidy's leadership guided the station's operations and scientific endeavors during this phase. He was joined by Russian flight engineers Anatoli Ivanishin and Ivan Vagner, who brought their expertise and contributed to various mission objectives and daily activities aboard the ISS.

A significant development occurred on May 31, 2020, when the crew of Crew Dragon Demo-2 arrived at the station. This marked the first crewed flight of SpaceX's Crew Dragon spacecraft, named Endeavour in honor of the Space Shuttle vehicle. The Demo-2 mission represented a milestone in commercial spaceflight and demonstrated the viability of private spacecraft in ferrying astronauts to and from the ISS. The crew for this historic flight comprised astronauts Doug Hurley and Bob Behnken. Their presence was instrumental in advancing the mission's research objectives and operations.

During their stay, Hurley and Behnken contributed to a range of scientific activities and conducted several spacewalks outside the station , critical for maintenance and upgrades. Their work helped bolster ISS research efforts, ensuring that the station remained at the forefront of space exploration and scientific discovery.

Expedition 63, spanning from April 17 to October 21, 2020, involved a dynamic crew rotation across several phases, reflecting both the station's operational needs and the integration of new spacecraft technology.

The first part of Expedition 63, from April 17 to May 31, 2020, was led by American astronaut Chris Cassidy, representing NASA. Cassidy, a spaceflight veteran, undertook his third and final mission as commander of the ISS. He was joined by Russian flight engineers Anatoli Ivanishin and Ivan Vagner, who contributed significantly to the mission's operations. Ivanishin was on his third and final spaceflight, while Vagner, on his inaugural spaceflight, provided a fresh perspective to the team.

The Crew Dragon Demo-2 mission, officially known as Crew Demo-2, SpaceX Demo-2, or Demonstration Mission-2, marked a significant milestone in spaceflight history as it became the first crewed test flight of the Crew Dragon spacecraft. Launched on May 30, 2020, the spacecraft, named Endeavour, lifted off atop a Falcon 9 rocket. This mission was groundbreaking for several reasons: it was the first crewed orbital spaceflight from the United States since the final Space Shuttle mission in 2011, and notably, it was the first crewed mission operated by a commercial provider, SpaceX. Moreover, it was the first two-person orbital spaceflight from the U.S. since STS-4 in 1982.

The Demo-2 mission was a crucial step in validating crewed spaceflight operations using SpaceX's hardware. The spacecraft received human-rating certification, which confirmed its suitability for carrying astronauts. This certification included comprehensive testing of Crew Dragon's capabilities in orbit.

Astronauts Doug Hurley and Bob Behnken of NASA's Commercial Crew Program were aboard the SpaceX Crew Dragon as it approached the International Space Station. The Crew Dragon's nose cone was open revealing the spacecraft's docking mechanism that would connect to the Harmony module's forward International Docking Adapter.

Docking with the International Space Station (ISS) was autonomously managed by the Crew Dragon, though the flight crew remained vigilant to intervene if necessary manually. On May 31, 2020, Endeavour achieved a soft docking with the ISS's pressurized mating adapter PMA-2 on the Harmony module approximately nineteen hours after launch. This was followed by a hard capture, completed eleven minutes later through the engagement of twelve docking hooks. Astronauts Doug Hurley and Bob Behnken then joined Expedition 63 crew, which included NASA astronaut Christopher Cassidy and Russian cosmonauts Anatoly Ivanishin and Ivan Vagner.

During their 62-day mission aboard the ISS, Hurley and Behnken conducted extensive scientific experiments, accumulating over 100 hours of research and traversing approximately 27 million miles during 1,024 orbits of Earth. Behnken also participated in four spacewalks with Cassidy to replace batteries brought up by a Japanese cargo vehicle.

The return phase of the mission was equally historic. Endeavour undocked from the ISS on August 1, 2020, after spending 62 days, 9 hours, and 8 minutes attached to the station. Weighing approximately 12,520 kg (27,600 lb) at undocking, the spacecraft executed a series of burns to transition from its ISS orbit to re-entry trajectory. The crew's final day was marked by personalized voice messages from their families, and a maneuver was performed to separate the "claw" umbilical mechanism from the trunk.

The deorbit burn, lasting 11 minutes and 22 seconds, was followed by the closure of the nose cone and the deployment of drogue chutes at 18:44 UTC. Main parachutes followed shortly thereafter, with the spacecraft experiencing a maximum acceleration of 4 g during re-entry. Endeavour splashed down in the Gulf of Mexico off the coast of Pensacola, Florida, on August 2, 2020, at 18:48 UTC. This was the first splashdown of NASA astronauts since 1975 and the first crewed spacecraft splashdown in the Gulf of Mexico.

The recovery operation saw a fast boat crew approach the capsule to assess for any traces of hypergolic propellants, and another crew collected the parachutes. Despite initial safety warnings, some private boats entered the hazardous area, requiring a coordinated response to ensure safety. SpaceX and NASA later agreed to implement a 16 km (9.9 mi) enforceable keep-out zone for future missions to enhance safety during recovery operations.

Upon recovery, the MV GO Navigator, assisted by a fast boat, retrieved the capsule and transported it aboard the ship for a 30-minute purge of the service section, which had detected unusually high levels of dinitrogen tetroxide. Hurley and Behnken were then assisted out of the capsule and into the ship's medical facility before being flown back to land by helicopter.

Post-flight analysis revealed that the Dragon's heat shield experienced higher-than-expected erosion at four attachment points, likely due to unforeseen airflow phenomena. SpaceX addressed these issues by modifying the heat shield design to incorporate more erosion-resistant materials. Additionally, the parachutes had deployed slightly lower than anticipated, prompting a replacement of the barometric pressure instrument used for altitude measurement.

From August 1 to October 14, 2020, the third part of the expedition saw the transition of the station's crew. With the departure of Crew Dragon Demo-2, the station's new arrivals were onboard Soyuz MS-17. The Soyuz MS-17 spacecraft remained docked at the Poisk module until April 17, 2021, when it undocked and returned to Earth. It landed safely in the Kazakh Steppe, Kazakhstan, at 04:55 UTC. The entire mission lasted 184 days, 23 hours, and 10 minutes.

NASA's SpaceX Demo-2 mission will return U.S human spaceflight to the International Space Station from U.S. soil with astronauts Robert Behnken and Douglas Hurley on an American rocket and spacecraft for the first time since 2011. In March 2020, at a SpaceX processing facility on Cape Canaveral Air Force Station in Florida, SpaceX successfully completed a fully integrated test of critical crew flight hardware ahead of Crew Dragon's second demonstration mission to the International Space Station for NASA's Commercial Crew Program; the first flight test with astronauts onboard the spacecraft. Behnken and Hurley participated in the test, which included flight suit leak checks, spacecraft sound verification, display panel and cargo bin inspections, seat hardware rotations, and more

The SpaceX Crew Dragon Endeavour spacecraft landing with NASA astronauts Robert Behnken and Douglas Hurley onboard in the Gulf of Mexico off the coast of Pensacola, Florida, Aug. 2, 2020. The Demo-2 test flight for NASA's Commercial Crew Program was the first to deliver astronauts to the International Space Station and return them safely to Earth onboard a commercially built and operated spacecraft. Behnken and Hurley returned after spending 64 days in space. Photo Credit: (NASA)

Support teams and curious recreational boaters arrive at the SpaceX Crew Dragon Endeavour spacecraft shortly after it landed with NASA astronauts Robert Behnken and Douglas Hurley onboard in the Gulf of Mexico off the coast of Pensacola, Florida, Sunday, Aug. 2, 2020. The Demo-2 test flight for NASA's Commercial Crew Program was the first to deliver astronauts to the International Space Station and return them safely to Earth onboard a commercially built and operated spacecraft. Behnken and Hurley returned after spending 64 days in space. Photo Credit: (NASA/Bill Ingalls)

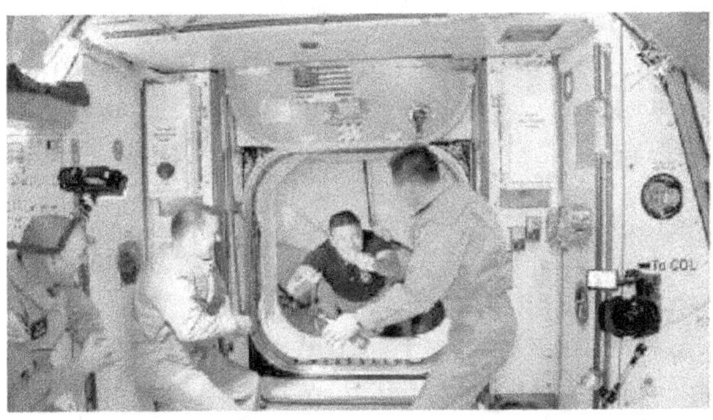

Bob Behnken enters the ISS shortly after the Crew Dragon hatch opened.

The mission's crew included:

Commander Sergey Ryzhikov: A veteran cosmonaut with previous spaceflight experience.

Flight Engineer Sergey Kud-Sverchkov: A first-time space traveler who participated in scientific research and station maintenance.

Flight Engineer Kathleen Rubins: An experienced NASA astronaut with a background in biological research.

Backup crew members prepared for the mission in case of unforeseen issues. They included:

Commander Oleg Novitsky
Flight Engineer Pyotr Dubrov
Flight Engineer Mark T. Vande Hei

Expedition 64 NASA astronaut Kate Rubins, left, and Russian cosmonauts Sergey Ryzhikov, center, and Sergey Kud-Sverchkov, right, of Roscosmos take a moment during the Soyuz MS-17 spacecraft fit check to pose for a photograph, Monday, Sept. 28, 2020, at the Baikonur Cosmodrome in Kazakhstan. The trio are preparing for launch to the International Space Station in their Soyuz MS-17 spacecraft from the Baikonur Cosmodrome in Kazakhstan on October 14, Baikonur time.

In response to the COVID-19 pandemic, Roscosmos also designated a two-cosmonaut reserve crew to ensure mission continuity in the event of illness affecting both the prime and backup crews. The reserve crew consisted of Anton Shkaplerov and Andrei Babkin, with the latter remaining unflown at the time.

The final part of Expedition 63, from October 14 to October 21, 2020, saw the transition to the new Soyuz MS-16 crew. The expedition concluded with the return of Chris Cassidy, Anatoli Ivanishin, and Ivan Vagner, who completed their respective missions.

The Crew Dragon Demo-2 mission, conducted by SpaceX, was a landmark event in spaceflight history. Launched on May 30, 2020, this mission marked the first crewed test flight of SpaceX's Crew Dragon spacecraft, designated C206 and later named Endeavour in tribute to the Space Shuttle Endeavour. This name honored the Space Shuttle orbiter that replaced the Space Shuttle Challenger, tragically destroyed in 1986.

The spacecraft docked with the International Space Station (ISS) on May 31, 2020, at 14:27 UTC, approximately 19 hours after its launch. The mission, originally planned as a short two-week test flight, was extended due to delays, allowing the astronauts to remain aboard the ISS for 64 days.

The extension was necessary to ensure comprehensive testing and integration of the Crew Dragon's systems with the ISS.

NASA astronauts Douglas Hurley and Robert Behnken were the crew aboard this historic flight. Their stay aboard the ISS, which lasted until August 1, 2020, provided valuable insights and data on the Crew Dragon's performance in a real space environment. Their work on the station contributed to various scientific experiments and maintenance tasks.

The readiness of Crew-1, the next scheduled crewed mission influenced the mission's duration and scope. Crew-1 was planned to launch approximately three months after Demo-2's landing, with the potential to join either Expedition 63 or Expedition 64. Crew-1 launched in November 2020, continuing the momentum of commercial crewed spaceflight and reinforcing the role of private spacecraft in supporting the ISS program.

Extravehicular activities (EVAs) during Expedition 63 were a notable aspect of the mission, highlighting the crew's international collaboration and preparedness. On June 26, 2020, astronauts Robert Behnken and Chris Cassidy undertook the first of four planned spacewalks. This initial EVA was a significant event, marking the continuation of critical maintenance and research tasks on the International Space Station (ISS).

Due to potential delays in the NASA Commercial Crew Program, there was a possibility that Cassidy could have been the sole crew member on the US Orbital Segment (USOS) for an extended period. Russian cosmonaut Anatoli Ivanishin underwent extensive training on American Extravehicular Mobility Unit (EMU) spacesuits to address this contingency. In the unlikely event of an unscheduled EVA before additional USOS crew members arrived, Ivanishin would have been the first Russian cosmonaut to use an EMU since Yuri Malenchenko's EVA with NASA astronaut Peggy Whitson during Expedition 16 in 2007.

Additionally, Ivan Vagner trained on the USOS Robotic Arm, known as Canadarm2, to support any spacewalks performed by Cassidy and Ivanishin from a robotic perspective. This training ensured that Vagner could operate the arm to assist in the spacewalk operations if necessary.

The Expedition 62 and 63 crews pose together moments after Oleg Skripochka handed over ISS command to Chris Cassidy.

The extension of Crew Dragon Demo-2's mission to approximately 65 days further underscored the importance of having a well-prepared crew. NASA astronauts Doug Hurley and Bob Behnken received training to conduct EVAs alongside Cassidy if needed. During these excursions, Cassidy and Behnken would perform the spacewalks, while Hurley would provide robotic support from within the station, ensuring seamless execution of tasks and maintaining the station's operational integrity

During Expedition 63, several spacewalks were planned to address critical tasks related to the International Space Station's scientific and power systems. Among these tasks was the activation of the Bartolomeo scientific payload, which had been delivered to the ISS on the CRS-20 mission earlier in the year. Bartolomeo, mounted on the exterior of the Columbus laboratory module, represented a significant addition to the station's scientific capabilities, and its installation required precise and coordinated spacewalk activities.

On May 19, 2020, following the confirmation of the Crew Dragon Demo-2 mission, NASA outlined an ambitious schedule for the mission's extravehicular activities (EVAs). The plan included up to five spacewalks by astronauts Chris Cassidy and Robert Behnken. These EVAs were intended to achieve several key objectives: installing the Bartolomeo module and the replacement of the remaining nickel-hydrogen batteries on the S6 Truss with more advanced lithium-ion batteries.

The battery replacement was a crucial task aimed at enhancing the station's power systems.

The transition to lithium-ion batteries was part of a broader effort to modernize and extend the operational life of the ISS's power infrastructure. Each spacewalk was meticulously planned to ensure the successful execution of these complex tasks, highlighting the ongoing commitment to maintaining and upgrading the ISS's scientific and operational capabilities.

On June 26, 2020, the first spacewalk of Expedition 63, designated American spacewalk 65, commenced at 11:32 UTC. NASA astronauts Chris Cassidy and Robert Behnken carried out the six-hour and seven-minute EVA, concluding their tasks at 17:39 UTC. This spacewalk, the first of four scheduled for the expedition, was focused on critical upgrades to the International Space Station's power systems.

On 26 June 2020, spacewalkers Behnken and Cassidy completed the first of four scheduled spacewalks.

During this spacewalk, Cassidy and Behnken replaced batteries supplying power to the station's solar arrays on the starboard truss. Their work included replacing aging nickel-hydrogen batteries with new lithium-ion batteries designed to offer improved efficiency and extended power capacity. Specifically, the astronauts removed five of the six old nickel-hydrogen batteries from one of the power channels of the S6 truss. They then installed

293

two of the three new lithium-ion batteries and two of the three necessary adapter plates that complete the power circuit for these new batteries.

Installing these new batteries significantly enhanced the station's power management. Mission control confirmed that the two newly installed batteries were operational, supporting the station's power systems and contributing to its continued functionality and efficiency. This spacewalk also included completing some tasks originally scheduled for the second planned spacewalk on July 1, 2020, showcasing the crew's effective management and adaptation of their work schedule.

On July 1, 2020, American spacewalk 66 commenced at 11:13 UTC with NASA astronauts Chris Cassidy and Robert Behnken. The spacewalk, which lasted six hours and one minute, concluded at 17:14 UTC. This EVA was part of the ongoing efforts to upgrade the International Space Station's power systems, following the earlier work performed on June 26.

During this spacewalk, Cassidy and Behnken focused on enhancing the power capabilities of one of the station's solar array channels. They installed and connected one new lithium-ion battery, completing the circuit for this battery. Additionally, the astronauts relocated one nickel-hydrogen battery to an external platform, where it would be prepared for future disposal. They also made significant progress on the remaining work by loosening the bolts on other nickel-hydrogen batteries scheduled for replacement. This task was essential to finalize the power system upgrades on the far starboard truss.

The work performed on July 1 continued the battery replacement initiative that began in January 2017. The successful completion of these tasks was crucial for ensuring the station's power systems remained robust and reliable, further extending the operational life of the ISS.

On July 16, 2020, American spacewalk 67 commenced at 11:10 UTC with NASA astronauts Chris Cassidy and Robert Behnken. The spacewalk, which lasted six hours, concluded at 17:10 UTC. This EVA marked a significant milestone in the ongoing effort to upgrade the International Space Station's power systems.

During this spacewalk, Cassidy and Behnken completed the final tasks needed to replace the batteries that power the station's solar arrays on the starboard truss. They removed six aging nickel-hydrogen batteries from the second of two power channels on the starboard 6 (S6) truss and installed three new lithium-ion batteries, along with the three associated adapter plates necessary to complete the power circuit. This work was a key part of a comprehensive 3.5-year initiative to enhance the station's power infrastructure.

The battery replacement effort aimed to replace 48 old nickel-hydrogen batteries with 24 new lithium-ion batteries and their corresponding adapter plates. This upgrade was critical for ensuring the continued efficiency and reliability of the station's power systems.

Notably, in April 2019, a lithium-ion battery installed on the near port truss experienced a fuse blowout, requiring the temporary reinstallation of two nickel-hydrogen batteries. A replacement lithium-ion battery arrived at the ISS in January 2020 aboard SpaceX's 19th Commercial Resupply Services mission and was currently stowed on the station's truss. This battery was installed during a future spacewalk later in the year to complete the power system upgrades.

On July 21, 2020, NASA's Spacewalk 68 commenced at 11:12 UTC with astronauts Chris Cassidy and Robert Behnken stepping out of the International Space Station (ISS). Their extravehicular activity lasted five hours and 29 minutes and concluded at 16:41 UTC. During this operation, Cassidy and Behnken undertook a series of critical tasks to enhance the ISS's functionality and prepare for future missions.

One of their primary objectives was installing a protective storage unit equipped with two Robotic External Leak Locator (RELL) units. These RELL devices, which will be utilized by the Canadian Space Agency's Dextre robot, are designed to detect leaks of ammonia, a crucial component in the station's cooling system. The astronauts also removed two "H-fixtures" from the base of the solar arrays located on the station's near port truss, or backbone. These fixtures, used during the ground processing of the solar arrays before their launch, were no longer needed, and their removal was an essential step in maintaining the array's functionality.

In preparation for the upcoming arrival of the Nanoracks commercial airlock, scheduled to be delivered later that year by a SpaceX cargo mission, Cassidy and Behnken completed several preparatory tasks on the exterior of the Tranquility module. Once installed, the new airlock will facilitate the deployment of commercial and government-sponsored experiments into space. The astronauts also routed Ethernet cables and removed a lens filter cover from an external camera, further contributing to the station's operational readiness.

This spacewalk marked the tenth for both Cassidy and Behnken, tying them with astronauts Michael Lopez-Alegria and Peggy Whitson as the only U.S. astronauts to achieve this milestone. Behnken now was the U.S. astronaut with the third most total spacewalking hours, amassing 61 hours and 10 minutes, surpassed only by Lopez-Alegria and Andrew Feustel. Cassidy has accumulated 54 hours and 51 minutes of spacewalking time, placing him ninth on the worldwide list for total spacewalking duration.

To date, space station crews have conducted 231 spacewalks to support the assembly and maintenance of the ISS. These spacewalks have accumulated 60 days, 12 hours, and 3 minutes of work outside the station, underscoring the extensive efforts invested in sustaining and advancing the orbiting laboratory.

On July 13, 2020, the Kibo Remote Manipulator System (RMS) performed a crucial operation aboard the International Space Station (ISS) by removing the Nanoracks NRCSD-18 deployer from the Kibo airlock. This deployer was responsible for releasing two distinct CubeSats into orbit.

At 13:40:25 UTC, the deployer ejected the Deformable Mirror experiment (DeMi), a 6U CubeSat developed for the Defense Advanced Research Projects Agency (DARPA) and Aurora Flight Sciences. DeMi was designed to advance the technology of deformable mirrors, which are instrumental in improving the performance of space-based telescopes and other optical systems.

Later in the day, at 16:55:25 UTC, the TechEdSat-10 CubeSat was released. This 6U CubeSat, developed by NASA's Ames Research Center in collaboration with San Jose State University, was focused on testing controlled reentry technology. The experiment aims to refine methods for the safe disposal of spacecraft and satellites at the end of their operational lives, thereby mitigating space debris and enhancing the sustainability of space operations.

The Crew Dragon Demo-2 mission concluded with the undocking of Hurley and Behnken from the ISS on August 1, 2020. Their return marked the end of a successful demonstration mission and highlighted commercial spacecraft's growing capabilities and role in supporting the ISS program.

This oblique view from the port side of the International Space Station's truss structure shows JAXA's (Japan Aerospace Exploration Agency) and the Canadian Space Agency's (CSA) contribution to the orbiting lab. Included in this photograph was a major portion of the Kibo laboratory module, the H-II Transfer Vehicle-9 (HTV-9) resupply ship and the Canadarm2 robotic arm.

Expedition 64

Expedition 64, the sixty-fourth long-duration mission aboard the International Space Station (ISS), began on October 21, 2020, following the departure of Soyuz MS-16. This expedition marked a significant transition, with the arrival of Soyuz MS-17, which launched on October 14, 2020. The initial crew included Russian cosmonauts Sergey Ryzhikov and Sergey Kud-Sverchkov, along with NASA astronaut Kathleen Rubins. This team spent several weeks aboard the

ISS, performing critical tasks and preparing for the arrival of additional crew members.

The official Expedition 64 crew portrait with (from left) NASA astronauts Kathleen Rubins and Victor Glover, JAXA astronaut Soichi Noguchi, Roscosmos cosmonaut Sergey Ryzhikov, NASA astronauts Michael Hopkins and Shannon Walker as well as Roscosmos cosmonaut Sergey Kud-Sverchkov. The expanded seven-member Expedition 64 crew with (from left) Flight Engineers Kate Rubins, Victor Glover and Soichi Noguchi, Commander Sergey Ryzhikov and Flight Engineers Michael Hopkins, Shannon Walker and Sergey Kud-Sverchkov. Rubins, Glover, Hopkins and Walker are all NASA astronauts. Noguchi was a JAXA (Japan Aerospace Exploration Agency) astronaut. Ryzhikov and Kud-Sverchkov are cosmonauts representing Roscosmos.

The pivotal moment of Expedition 64 occurred on November 17, 2020, when SpaceX Crew-1 docked with the ISS. Launched on November 16, 2020, aboard the Crew Dragon spacecraft Resilience, Crew-1 was the first operational flight of NASA's Commercial Crew Program (CCP) and the second crewed orbital flight overall. The Crew-1 mission featured NASA astronauts Michael S. Hopkins, Victor J. Glover, and Shannon Walker, as well as Japanese astronaut Soichi Noguchi. This event marked the first time the ISS hosted a crew of seven, highlighting the growing international collaboration and technological advancements in space exploration.

SpaceX Crew-1

Throughout Expedition 64, the ISS continued to serve as an orbiting laboratory, providing a unique microgravity environment for scientific experiments and technological advancements. The crew undertook diverse research projects that span disciplines from biology and human physiology to material science and astronomy. One of the key focuses of this expedition was to investigate the effects of long-duration spaceflight on the human body, a crucial aspect for future deep-space missions.

In addition to scientific research, Expedition 64 marked a significant step forward in expanding the ISS's capabilities. The mission saw the installation of the new International Docking System Standard (IDSS) docking port, which facilitates future spacecraft docking. This upgrade was essential for accommodating upcoming missions and ensuring the station's continued operation as a hub for international space exploration.

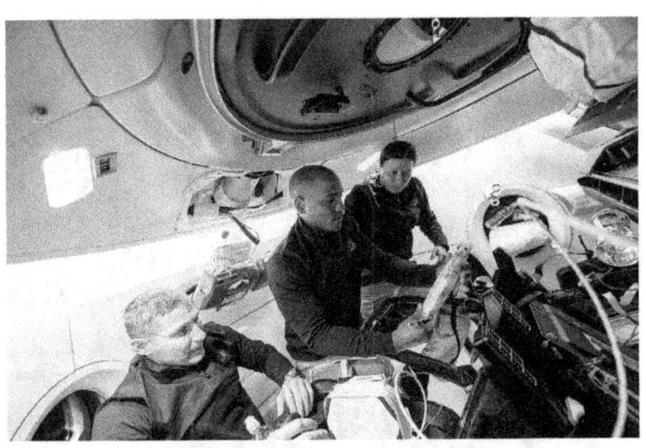

NASA astronauts (from left) Michael Hopkins, Victor Glover and Shannon Walker work inside the SpaceX Crew Dragon spacecraft as they get ready for their approach and rendezvous with the International Space Station. Out of frame was their SpaceX Crew-1 crewmate Soichi Noguchi of JAXA (Japan Aerospace Exploration Agency).

During Expedition 63, several spacewalks were scheduled to address critical tasks involving the scientific and power systems of the International Space Station (ISS). Initially, delays in the NASA Commercial Crew Program resulted in Chris Cassidy being the sole crew member aboard the U.S. Orbital Segment (USOS) for an extended period. The arrival of the Crew Dragon Demo-2 mission in May 2020 facilitated four extravehicular activities (EVAs) by Cassidy and Robert Behnken, focused primarily on replacing the remaining nickel-hydrogen batteries on the S6 Truss with new lithium-ion batteries, marking a significant upgrade to the station's power systems.

The activation of the Bartolomeo scientific package, delivered to the ISS on SpaceX CRS-20, was initially planned for Expedition 63 but was postponed until Expedition 64.

On November 18, 2020, astronauts Sergey Ryzhikov and Sergey Kud-Sverchkov undertook a spacewalk to prepare for the upcoming replacement of the Pirs docking compartment with the Nauka laboratory module. This EVA lasted 6 hours and 48 minutes and was notable as the first to be conducted from the Poisk airlock. Designated "Russian Spacewalk #47" by NASA, the activity began at 14:30 UTC and extended for over six hours.

In early 2021, NASA conducted five spacewalks to advance station operations further. On January 27, 2021, astronauts Michael Hopkins and Victor Glover embarked on a spacewalk starting at 12:28 UTC, lasting 6 hours and 56 minutes. Their tasks included installing a Ka-band antenna on the Columbus module in preparation for Bartolomeo's activation, replacing a pin on the Quest Joint Airlock, and removing a grapple fixture from the P4 Truss as part of a series of experimental solar array wing upgrades.

On February 1, 2021, Hopkins and Glover returned for another spacewalk, starting at 12:56 UTC and lasting 5 hours and 20 minutes. This spacewalk completed a four-year effort, initiated by Shane Kimbrough and Peggy Whitson during Expedition 50, to replace batteries on the Integrated Truss Structure. Additionally, Hopkins and Glover upgraded several cameras on the starboard truss, the Destiny laboratory, and the Kibo robotic arm.

Another significant spacewalk occurred on February 28, 2021, when astronauts Kate Rubins and Victor Glover ventured out at 11:12 UTC for a 7-hour and 4-minute EVA. They installed brackets on the P6 Truss to support experimental solar array upgrades, with the main materials for this project launching aboard SpaceX CRS-22 in June 2021.

The subsequent spacewalk on March 5, 2021, involved Rubins and Soichi Noguchi, starting at 11:37 UTC and lasting 6 hours and 56 minutes. Their primary focus was continuing the bracket installation work. They had planned additional tasks, such as deploying a new airlock cover to reinforce Quest, replacing a wireless video transceiver on the Unity node, routing more cables on Bartolomeo, and managing ammonia hoses. However, they encountered difficulties with several bolts during the bracket installation, leading them to defer these additional tasks.

On March 13, 2021, Hopkins and Glover conducted a final spacewalk to complete the outstanding work from the previous EVA. This spacewalk, which began at 13:14 UTC and lasted 6 hours and 47 minutes, concluded the tasks left unfinished by Rubins and Noguchi. However, they chose to defer the installation of clamps on Bartolomeo to a future spacewalk.

The mission concluded in April 2021, with the crew returning to Earth after a successful stint aboard the ISS.

A SpaceX Falcon 9 rocket and Crew Dragon Resilience for NASA SpaceX's Crew-1 mission are seen inside the SpaceX Hangar at NASA's Kennedy Space Center in Florida.The Crew Dragon capsule will launch atop a Falcon 9 rocket from Launch Complex 39A carrying NASA astronauts Michael Hopkins, Victor Glover, Shannon Walker and Japan Aerospace Exploration Agency (JAXA) astronaut Soichi Noguchi to the space station for a six-months.

SpaceX Crew Dragon Resilience spacecraft shortly after it landed with NASA astronauts Mike Hopkins, Shannon Walker, and Victor Glover, and Japan Aerospace Exploration Agency (JAXA) astronaut Soichi Noguchi

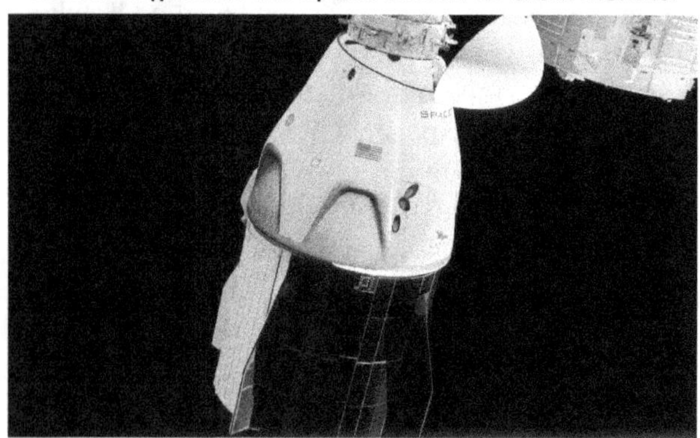

The SpaceX Crew Dragon Resilience shortly before undocking from the Harmony module's space-facing international docking adapter. Resilience would parachute to a splashdown in the Gulf of Mexico several hours later returning four SpaceX Crew-1 crew members back to Earth after 168 days in space.

A view of Earth from the Dragon 2 capsule during Expedition 64

Expedition 65

The seven-member Expedition 65 crew with (from left) Flight Engineers Pyotr Dubrov, Shane Kimbrough, Megan McArthur and Thomas Pesquet, with Commander Akihiko Hoshide and Flight Engineers Oleg Novitskiy and Mark Vande Hei. Kimbrough, McArthur and Vande Hei are all NASA astronauts. Pesquet was an astronaut representing ESA (European Space Agency). Hoshide was a JAXA (Japan Aerospace Exploration Agency) astronaut. Dubrov and Novitskiy are cosmonauts representing Roscosmos.

Expedition 65, the 65th extended mission to the International Space Station (ISS), began on April 17, 2021. This mission was notable for its frequent crew changes, reflecting the ongoing evolution of the station's international and scientific activities.

The first phase of Expedition 65, from April 17 to April 24, 2021, featured a diverse crew from three space agencies. Oleg Novitsky of Roscosmos, Russia, joined the ISS as a Flight Engineer for his third spaceflight aboard Soyuz MS-18. Pyotr Dubrov, also from Roscosmos and embarking on his inaugural spaceflight, served alongside Novitsky. On his second space mission, Mark T. Vande Hei of NASA was also part of the Soyuz MS-18 team.

From April 24 to May 2, 2021, the crew shifted with the arrival of SpaceX Crew-1. Michael S. Hopkins of NASA, making his second spaceflight, served as a Flight Engineer. He was joined by Victor J. Glover, also from NASA, on his first space mission. Soichi Noguchi of JAXA, Japan, was in his third and final spaceflight, while Shannon Walker, also of NASA and serving as Commander for her second mission, led the SpaceX Crew-1 team.

The next phase, from May 2 to October 4, 2021, saw further changes with the arrival of SpaceX Crew-2. On his third and final spaceflight, Shane Kimbrough of NASA became a Flight Engineer. Megan McArthur, also of NASA, joined as a Flight Engineer on her second spaceflight. Akihiko Hoshide of JAXA, in his third spaceflight, initially joined SpaceX Crew-2 as a Flight Engineer before taking over as Commander. Thomas Pesquet of ESA, France, completed his second spaceflight as a Flight Engineer and subsequently became the Commander of Expedition 65.

From October 4 to October 5, 2021, Akihiko Hoshide served as the Commander of the ISS. Following this brief period, from October 5 to October 17, 2021, Anton Shkaplerov of Roscosmos, on his fourth and final spaceflight, joined the ISS as a Flight Engineer aboard Soyuz MS-19.

This dynamic crew lineup reflected the crucial rotation and continuity of personnel required for the ISS's operations and scientific research. Several notable transitions in leadership marked Expedition 65: Shannon Walker became the third female commander of the ISS, Akihiko Hoshide succeeded her as commander, and Thomas Pesquet became the fourth European and the first French astronaut to command the ISS.

The expedition unfolded during a particularly dynamic period for the ISS, featuring a busy schedule of visiting modules, spacecraft, and crews. It hosted two SpaceX Crew Dragon flights—Crew-1 and Crew-2—as well as two Soyuz flights—Soyuz MS-18 and Soyuz MS-19.

The mission commenced with the undocking of Soyuz MS-17 in April 2021. On April 9, 2021, at precisely 07:42:41 UTC, Soyuz MS-18, designated "Y. A. Gagarin," embarked on its journey to the International Space Station (ISS). This mission marked the 146th crewed flight of a Soyuz spacecraft, continuing the longstanding legacy of these reliable space vehicles. The spacecraft, named in honor of Yuri Gagarin, the first human in space, launched from the Baikonur Cosmodrome atop a Soyuz-2.1a rocket. Roscosmos operated the mission and was the 748th Soyuz MS spacecraft, built by the renowned RSC Energia.

The Soyuz MS-18 crew consisted of Russian Commander Oleg Novitsky, Flight Engineer Pyotr Dubrov, and American Flight Engineer Mark Vande Hei. Novitsky, a veteran of spaceflight, was leading his third mission. Dubrov, on his first spaceflight, joined him along with Vande Hei, who was on his second space mission. Notably, this mission also carried Russian film director Klim Shipenko and actress Yulia Peresild, who were participating in a unique endeavor to film the movie Vyzov (The Challenge) during their twelve-day stay aboard the ISS.

Expedition 64 backup crew members NASA astronaut Mark Vande Hei, left, Russian cosmonaut Oleg Novitskiy of Roscosmos, center, and Russian cosmonaut Petr Dubrov of Roscosmos pose for a photo during qualification exams Tuesday, Sept. 22, 2020 at the Gagarin Cosmonaut Training Center (GCTC) in Star City, Russia.

The flight plan for Soyuz MS-18 was adjusted in early March 2021, at NASA's request, to include astronaut Mark Vande Hei in place of Sergei Korsakov and astronaut Anne McClain as the backup instead of Dmitriy Petelin. This adjustment facilitated an extended presence of NASA astronauts on Soyuz missions, with the arrangement being a mutual agreement between NASA and Roscosmos, involving no financial transaction.

Upon arrival at the ISS, Soyuz MS-18 docked with the Rassvet module at 11:05 UTC on April 9, 2021. The mission had a total duration of 190 days, 20 hours, and 53 minutes. During this period, the spacecraft's crew contributed to various scientific experiments and station operations. The film crew, Shipenko and Peresild, conducted their filming as part of a groundbreaking project to bring space exploration to a broader audience.

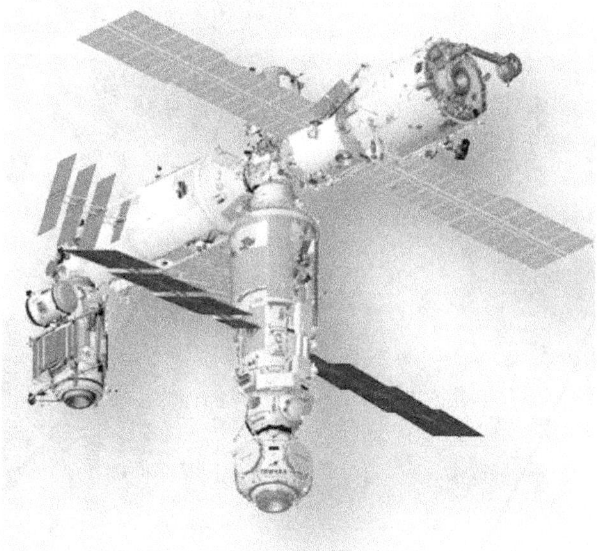

ISS Russian Orbital Segment before and and after the undocking of Progress M-UM from the Prichal module.

On September 28, 2021, the Soyuz MS-18 crew moved the spacecraft to a new docking port, the Nauka module, to make way for the arrival of Soyuz MS-19. The relocation was completed at 13:04 UTC, and Soyuz MS-18 remained docked at this new position for an additional 18 days, 12 hours, and 10 minutes. The mission concluded with the spacecraft's departure from the ISS and reentry into Earth's atmosphere.

On October 17, 2021, Soyuz MS-18 safely landed in the Kazakh Steppe at 04:35:44 UTC. The landing marked the end of a successful mission and a milestone in the ongoing collaboration between international space agencies and the film industry. The return journey included Shipenko and Peresild, who were joined by Commander Novitsky. Cosmonaut Pyotr Dubrov and astronaut Mark Vande Hei, who had arrived on Soyuz MS-18, were part of the crew on Soyuz MS-19's return, completing their own part of the mission.

At that time, the ISS hosted two active spacecraft: Crew Dragon Resilience, which had delivered the SpaceX Crew-1 team, and Soyuz MS-18.

SpaceX Crew-2 marked a significant milestone as the second operational flight of the Crew Dragon spacecraft and the third overall crewed orbital mission of NASA's Commercial Crew Program. The mission commenced with a launch on April 23, 2021, at 09:49:02 UTC, utilizing a Falcon 9 rocket. This mission employed the Crew Dragon capsule previously used for Crew Dragon Demo-2, known as Endeavour, and was launched atop the same Falcon 9 booster as SpaceX Crew-1, designated B1061.1.

The Crew-2 spacecraft docked with the International Space Station (ISS) on April 24, 2021, at 09:08 UTC, attaching to the International Docking Adapter (IDA) at the forward port of the Harmony module. Notably, this was the first crewed mission to use a previously flown booster, marking a significant achievement in the reuse of spaceflight hardware.

The crew for this mission was a distinguished group of seasoned astronauts. Shane Kimbrough of NASA served as the Commander, marking his third and final spaceflight. Kimbrough was joined by Pilot K. Megan McArthur, also from NASA, who was embarking on her second spaceflight and making her debut visit to the ISS. McArthur's previous spaceflight experience included the STS-125 mission to the Hubble Space Telescope, and she occupied the same seat on Crew Dragon Endeavour that her husband, Bob Behnken, had used on the Demo-2 mission.

Mission Specialists included Akihiko Hoshide of JAXA, who was on his third spaceflight and served as the second Japanese ISS Commander during this mission, and Thomas Pesquet of ESA, who was on his second visit to the ISS. This mission was named "Alpha" in reference to Alpha Centauri, the closest star system to Earth, symbolizing the forward-looking spirit of the mission.

The Crew Dragon Endeavour's mission was not without its logistical complexities. To prepare for the arrival of a Starliner spacecraft, Endeavour was undocked from the Harmony forward port on July 21, 2021, at 10:45 UTC and subsequently relocated to the Harmony zenith port at 11:36 UTC.

From left, ESA astronaut Thomas Pesquet, NASA astronaut Megan McArthur, NASA astronaut Shane Kimbrough and JAXA astronaut Akihiko Hoshide enter the elevator inside the Neil Armstrong Operations and Checkout Building at NASA's Kennedy Space Center in Florida, on April 23, 2021. The four astronauts will head to the International Space Station on NASA's SpaceX Crew-2 mission. SpaceX's Crew Dragon Endeavour launched on the company's Falcon 9 rocket from Kennedy's Launch Complex 39A that day at 5:49 a.m. EDT.

The Crew-2 mission also witnessed a notable period where three Dragon spacecraft were simultaneously in space. From September 16 to 18, 2021, the ISS hosted not only Crew-2 but also CRS-23 (Cargo Dragon C208) and the Inspiration4 mission (Resilience).

As the mission neared its end, delays caused by weather and a minor health issue with one of the Crew-3 astronauts prompted NASA to bring the Crew-2 astronauts back to Earth before launching Crew-3. This marked the first indirect handover of space station crews by Crew Dragon. The Crew Dragon Endeavour undocked from the ISS on November 8, 2021, at 19:05 UTC and made its return to Earth, splashing down off the coast of Florida at 03:33 UTC on November 9, 2021. The return was complicated by one of the spacecraft's four parachutes deploying slower than the others, though the landing was otherwise successful.

SpaceX Crew-2 set a new record for the longest spaceflight by a U.S. crewed spacecraft, spending 199 days in orbit. This mission demonstrated Crew Dragon spacecraft's continued advancement and reliability and solidified its role in the ongoing exploration and utilization of the International Space Station.

The International Space Station pictured from the SpaceX Crew Dragon Endeavour during a fly around of the orbiting lab that took place following its undocking from the Harmony module's space-facing port on Nov. 8, 2021.

Shannon Walker, a mission specialist from Crew-1, initially served as station commander. Her tenure and that of fellow Crew—1 members Mike Hopkins, Victor Glover, and Soichi Noguchi ended with their departure aboard Crew Dragon Endeavour, which delivered the SpaceX Crew-2 team to the ISS in April 2021.

With Crew-1's departure, Akihiko Hoshide of Crew-2 assumed command of the ISS, marking him as the second Japanese astronaut to lead the station. The station's activities continued with the arrival of Soyuz MS-19 on October 5, 2021. This mission brought Russian cosmonaut Anton Shkaplerov, film director Klim Shipenko, and actress Yulia Peresild to the ISS. Shipenko and Peresild spent 12 days onboard as part of a unique film project, highlighting the ISS's role in scientific research and international cultural endeavors.

Soyuz MS-19, a significant mission in the history of crewed spaceflights, was launched on October 5, 2021, at 08:55:02 UTC. This flight marked the 147th crewed Soyuz mission and was notable for being the first mission to the International Space Station (ISS) with an entirely Russian crew in over two decades.

The crew of Soyuz MS-19 included Russian commander Anton Shkaplerov, Russian film director Klim Shipenko, and Russian actress Yulia Peresild. This mission was remarkable not only for its all-Russian crew but also for its unique objective: Shipenko and Peresild spent approximately twelve days aboard the ISS filming the movie Vyzov (Russian: Вызов, translated as "The Challenge"). Their presence on the station was part of an ambitious project to bring filmmaking to the microgravity environment of space.

Following its launch, Soyuz MS-19 executed a precise docking maneuver, reaching the ISS and docking with the Rassvet module on October 5, 2021, at 12:22:31 UTC. The docking process was achieved through a manual system operated by Commander Anton Shkaplerov, utilizing a three-hour, two-orbit rendezvous profile.

Soyuz MS-19 crew

This mission also played a critical role in expanding the Russian Orbital Segment (ROS) of the ISS. Scheduled to dock the Prichal module to the Nauka module on November 24, 2021, Prichal was designed with a hybrid docking system, including both active and passive ports. These ports are essential for future docking of various spacecraft and modules, enhancing the ROS's capability to function independently after 2024. To integrate Prichal into the ROS, cosmonauts Anton Shkaplerov and Pyotr Dubrov performed a spacewalk on January 19, 2022, to lay cables between Nauka and Prichal. This task was part of a series of planned spacewalks to complete the integration process throughout 2022.

The return phase of Soyuz MS-19 was equally notable. On October 17, 2021, the crew members, including Shipenko and Peresild, returned to Earth aboard Soyuz MS-18, under the command of Oleg Novitskiy. The spacecraft successfully landed on March 30, 2022. Notably, the return of Shipenko and Peresild was financed by Konstantin Ernst,

CEO of Channel One, highlighting the collaboration between the film industry and space agencies. Klim Shipenko shot about 35–40 minutes of film on the ISS and took on the positions of director, operator, art director, and makeup artist. Oleg Novitsky and Pyotr Dubrov will appear in the film, with Dubrov and Mark Vande Hei assisting in the production.Shkaplerov will appear in some scenes of the movie.

The mission concluded successfully with the landing of Soyuz MS-19 on March 30, 2022, marking a historic achievement in the collaboration between space and cinema.

Expedition 65 also featured three significant spacewalks to enhance the ISS's infrastructure and power systems. On June 16, 2021, astronauts Thomas Pesquet and Shane Kimbrough performed the first spacewalk, focusing on the installation of the first Roll Out Solar Array (iROSA) on the 2B power channel and the P6 truss. This spacewalk was cut short due to a spacesuit computer issue and technical difficulties with the iROSA, lasting 7 hours and 15 minutes instead of the planned duration.

The second spacewalk, conducted on June 20, 2021, successfully deployed and connected the first iROSA to the station's power system, marking a key milestone in the ISS's upgrade. The third spacewalk, on June 25, 2021, installed and deployed the second iROSA on the 4B mast can, further advancing the station's power generation capabilities.

Another planned spacewalk on August 24, initially intended to replace a plasma measuring instrument and install a bracket for an upcoming iROSA pair, was postponed to September 12 due to minor medical issues with astronaut Mark Vande Hei. Thomas Pesquet replaced Vande Hei and completed the spacewalk, which lasted 6 hours and 45 minutes.

Russian cosmonauts Oleg Novitsky and Pyotr Dubrov also conducted three crucial spacewalks from the Russian Orbital Segment (ROS). These spacewalks were essential for installing the Nauka module, which was equipped with the European Robotic Arm, and for preparing the module for the Prichal docking node, scheduled for Expedition 66. The first spacewalk occurred on June 2, 2021, followed by two more on September 3 and 9, with durations of 7 hours and 19 minutes, 7 hours and 54 minutes, and 7 hours and 25 minutes, respectively.

Expedition 65 concluded on October 17, 2021, with the departure of Soyuz MS-18, marking the end of a period characterized by significant crew rotations and leadership milestones.

View of Earth taken during ISS Expedition 65

Expedition 66

In early 2022, a series of significant events unfolded at the International Space Station (ISS), reflecting the dynamic and continuously evolving nature of space operations.

On January 23, 2022, the SpaceX CRS-24 cargo spacecraft undocked from the ISS, marking the conclusion of its successful mission to deliver supplies and scientific experiments. This event followed a busy end to 2021. On December 22, 2021, the Progress Propulsion Module, which had been attached to the ISS, undocked. In the same timeframe, the SpaceX CRS-24 docked at the station on December 22, having launched from Earth just a day earlier on December 21. This mission was pivotal for resupplying the ISS and supporting ongoing research.

On December 8, 2021, the Soyuz MS-20 spacecraft embarked on a noteworthy mission to the International Space Station (ISS). This flight, distinct from its predecessors, was not designed to transport crew members for an ISS expedition nor to serve as a lifeboat for the station's existing occupants. Instead, Soyuz MS-20 was a pioneering venture in space tourism, commanded by a seasoned Russian cosmonaut and carrying two private space tourists.

The official portrait of the seven-member Expedition 66 crew. From left are, NASA astronauts Raja Chari and Thomas Marshburn; ESA (European Space Agency) astronaut Matthias Maurer; Roscosmos cosmonauts Anton Shkaplerov and Pyotr Dubrov; and NASA astronauts Kayla Barron and Mark Vande

The Soyuz MS-20 mission was led by Commander Alexander Misurkin, a veteran of two previous long-duration missions aboard the ISS. This flight marked a significant departure from traditional Soyuz operations, as the spacecraft was modified to be operated by a single cosmonaut. Misurkin's role was pivotal in ensuring the success of this mission, which lasted from December 8 to December 20, 2021.

The space tourists aboard Soyuz MS-20 were Japanese nationals Yusaku Maezawa and Yozo Hirano. Maezawa, a billionaire, art collector, and space enthusiast, secured both seats on this mission through the space tourism company Space Adventures. Hirano, Maezawa's production assistant, joined him on this extraordinary journey. Their spaceflight was historic, as it was the first

time two Japanese citizens had traveled to space together. The mission's duration from launch to docking with the ISS was approximately six hours.

(Dec. 11, 2021) --- The three-person Soyuz MS-20 crew (front row) participates: Japanese spaceflight participant Yusaku Maezawa, Roscosmos cosmonaut Alexander Misurkin, and Japanese spaceflight participant Yozo Hirano.

Initially, there was speculation that Austrian airline pilot Johanna Maislinger would be among the tourists. However, on May 13, 2021, Space Adventures confirmed that Maislinger was not a serious candidate and had never had the financial means to participate. The announcement also clarified that Japanese entertainer Yumi Matsutoya, previously mentioned as a potential participant, would not be joining the mission.

Soyuz MS-20 was notable for being the first tourist flight to the ISS in over twelve years. The last tourist flight, Soyuz TMA-16, had been in September 2009 with Canadian entertainer Guy Laliberté. Plans for a subsequent tourist mission aboard Soyuz TMA-18M in September 2015, featuring British singer Sarah Brightman, were canceled in May 2015.

Soyuz MS-20 docking to the ISS

This mission marked a significant shift in the approach to space tourism. Unlike previous tourist missions, which involved either a single tourist on a taxi flight or a handover period with long-duration crews, Soyuz MS-20 was dedicated entirely to space tourists. This change signaled a new era of space tourism, one that would soon be followed by similar ventures. In April 2022, American company Axiom Space, in collaboration with SpaceX, launched a mission with professional astronaut assistance and three paying tourists aboard the Crew Dragon Endeavour.

Soyuz MS-20's flight was part of a broader trend towards commercial space travel, with plans for future missions, including Soyuz MS-23, which was initially intended as a commercial flight but later adapted to a standard Russian-American expedition. The mission underscored the evolving landscape of space tourism, bridging the gap between the early 2000s era and the new wave of private space exploration.

November 2021 was marked by significant logistical movements. On November 11, the SpaceX Crew-3 mission docked with the ISS. This crewed mission, carrying NASA astronauts Raja Chari, Tom Marshburn, and Kayla Barron, as well as European Space Agency astronaut Matthias Maurer, brought new personnel to the station. Their arrival was in close succession to the departure of the SpaceX Crew-2 mission, which undocked and splashed down on November 8, 2021. This crew had spent months aboard the ISS conducting various scientific experiments and operations.

Additionally, the ISS Progress 78 spacecraft undocked on November 25, 2021, and its successor, ISS Progress 79, launched on October 27, 2021. The Progress 79 spacecraft docked with the station on October 29, continuing the vital supply missions to support the ISS's operations.

In October, the ISS saw the redocking and undocking of the ISS Progress 78 spacecraft. Specifically, it was undocked on October 20, 2021, and redocked on October 21, 2021. This maneuver was part of routine logistics and maintenance operations.

The autumn of 2021 also saw the arrival of the Prichal module, which docked with the ISS on November 26, 2021, after its launch on November 24, 2021. This module expanded the ISS's docking capabilities and supported its future missions.

Earlier in the year, on February 14, 2022, the ISS Progress 80 spacecraft launched and subsequently docked with the station on February 17, 2022. This mission continued the essential supply chain that supports the ISS's ongoing scientific and operational activities. Similarly, the Cygnus CRS-17 spacecraft, launched on February 19, 2022, was captured and docked with the ISS on February 21, 2022.

Expedition 66 marked the 66th long-duration mission aboard the International Space Station (ISS), commencing with the departure of Soyuz MS-18 on October 17, 2021. This expedition was initially led by Thomas Pesquet of the European Space Agency (ESA), who became the fourth European astronaut and the first French astronaut to command the ISS. His command lasted until November 8, 2021, when Russian cosmonaut Anton Shkaplerov, arriving aboard Soyuz MS-19, assumed command of the station.

Pesquet had been transported to the ISS on SpaceX Crew-2, which launched in April 2021. Alongside Pesquet, Crew-2 included NASA astronauts Shane Kimbrough and Megan McArthur, and Akihiko Hoshide from the Japan Aerospace Exploration Agency (JAXA). The crew from Expedition 65, who extended their stay, joined Expedition 66 with Russian cosmonaut Pyotr Dubrov and NASA astronaut Mark Vande Hei. Both Dubrov and Vande Hei had launched on Soyuz MS-18 and remained on the station until their return on Soyuz MS-19. Shkaplerov's arrival on Soyuz MS-19 also included two notable participants in a joint film project between

Roscosmos and Channel One: film director Klim Shipenko and actress Yulia Peresild.

SpaceX Crew-3 porait with NASA astronauts Raja Chari, Thomas Marshburn, Kayla Barron and ESA astronaut Matthias Maurer. Photographer: Robert Markowitz

On November 10, 2021, SpaceX Crew-3 launched, delivering NASA astronauts Raja Chari, Thomas Marshburn, Kayla Barron, and ESA astronaut Matthias Maurer to the ISS. In the evolving space exploration saga, the Crew Dragon spacecraft symbolized innovation and resilience. The first Crew Dragon capsule, christened "Endeavour," was launched as a testament to pioneering spirit. Its successor, "Resilience," followed, embodying the steadfast determination of its creators. On October 7, 2021, it was announced that the third capsule would be named "Endurance." This name pays homage to both the SpaceX and NASA teams who persevered through the challenges of a global pandemic to build the spacecraft and train its astronauts. Additionally, "Endurance" evokes the legacy of Sir Ernest Shackleton's ship, which, after being trapped by ice during the Imperial Trans-Antarctic Expedition of 1915, became a symbol of endurance in the face of adversity.

The mission's crew was carefully selected, with German ESA astronaut Matthias Maurer being the first to be chosen in September 2020. On December 14, 2020, NASA astronauts Raja Chari and Thomas Marshburn were added, completing the primary crew. Initially, the fourth seat was reserved for a Russian cosmonaut as part of a barter agreement between NASA and Roscosmos, intended to facilitate seat exchanges on Soyuz and Commercial Crew Vehicles. However, this arrangement was deferred until after the Crew-3 launch. Ultimately, in May 2021, NASA astronaut Kayla Barron was assigned to the fourth seat.

Raja Chari, who was commanding this mission, distinguished himself as the first rookie astronaut to lead a NASA space mission since Gerald Carr's command of the Skylab 4 crew in 1973. This mission also marked the inaugural spaceflight for both Matthias Maurer and Kayla Barron, adding a new chapter to their respective careers. The full lineup of the Crew-3 mission included:

Commander: Raja Chari, NASA (Expedition 66/67), making his first spaceflight.

Pilot: In his third and final spaceflight, Thomas Marshburn, NASA (Expedition 66/67).

Mission Specialist 1: Matthias Maurer, ESA (Expedition 66/67), embarking on his first spaceflight.

Mission Specialist 2: Kayla Barron, NASA (Expedition 66/67), also on her first spaceflight.

The backup crew consisted of:

Commander: Kjell N. Lindgren, NASA.

Pilot: Robert Hines, NASA.

Mission Specialist 1: Samantha Cristoforetti, ESA.

Mission Specialist 2: Stephanie Wilson, NASA.

Raja Chari and Kayla Barron took a stuffed turtle as a zero-gravity indicator on their SpaceX Crew-3 mission as a nod to their astronaut group. This turtle was named "Pfau," a German word for "peacock," in honor of Matthias Maurer and Tom Marshburn, who were part of the NASA Astronaut Group 19, known as "The Peacocks."

The Crew-3 mission was initially set to launch on October 31, 2021, but it faced delays. Unfavorable weather conditions in the Atlantic Ocean pushed the launch to November 3, 2021. Further complications, including a minor medical issue with one of the astronauts, delayed the launch to November 7, 2021. Eventually, the mission lifted off from Cape Canaveral on November 11, 2021, at 02:03:31 UTC.

The Crew-3 mission was notable not only for its ambitious goals but also for its logistical complexities. The return of Crew-3 was postponed several times, ultimately undocking on May 5, 2022, and splashing down the following day after spending 176 days in space. The mission's

European segment was aptly named "Cosmic Kiss," symbolizing the international collaboration and the unifying spirit of exploration that defines the journey to the International Space Station.

From left to right, ESA (European Space Agency) astronaut Matthias Maurer, NASA astronauts Tom Marshburn, Raja Chari, and Kayla Barron, are seen inside the SpaceX Crew Dragon Endurance spacecraft onboard the SpaceX Shannon recovery ship shortly after having landed in the Gulf of Mexico off the coast of Tampa, Florida, Friday, May 6, 2022. Maurer, Marshburn, Chari, and Barron are returning after 177 days in space as part of Expeditions 66 and 67 aboard the International Space Station. Photo Credit: (NASA/Aubrey Gemignani)

The mission faced challenges as international cooperation in space exploration was cast into uncertainty following the 2022 Russian invasion of Ukraine and the subsequent sanctions imposed on Russia. Despite these geopolitical tensions, Expedition 66 demonstrated the ongoing commitment to collaboration in space, albeit under increasingly complex circumstances.

During Expedition 66, the crew conducted a variety of scientific experiments and maintenance tasks. One key research area was the study of material science in microgravity, which has implications for future manufacturing processes and technology development. The crew also worked on experiments related to human health, including studies on the effects of long-term spaceflight on the cardiovascular system and musculoskeletal health.

A major highlight of Expedition 66 was the continued enhancement of the ISS's infrastructure. The crew installed and upgraded various systems, including new research facilities and equipment. This ongoing work ensures that the ISS remains at the cutting edge of space research and technology.

The mission underscored the ISS's role as a platform for international cooperation. The crew members' diverse backgrounds and collaborative work reflected the station's global significance and contribution to scientific progress.

Expedition 66 concluded in April 2022, with the crew returning to Earth after a successful mission. Their contributions to research and technology development aboard the ISS helped to further our understanding of space and prepared the way for future exploration missions.

Astronaut Kayla Barron checks out plants growing inside the Veggie facility.

The Soyuz MS-19 crew ship, carrying three Expedition 66 crew members, backs away from the International Space Station after undocking from the Rassvet module. The Soyuz crew ship would parachute to a landing in Kazakhstan just over four hours later with NASA astronaut Mark Vande Hei and Roscosmos cosmonauts Anton Shkaplerov and Pyotr Dubrov. At left, was one of two cymbal-shaped UltraFlex solar arrays attached to Northrop Grumman's Cygnus space freighter.

The three-person Soyuz MS-20 crew (front row) participates in a group portrait with the seven-member Expedition 66 crew. Front row, front left, are Japanese spaceflight participant Yusaku Maezawa, Roscosmos cosmonaut Alexander Misurkin, and Japanese spaceflight participant Yozo Hirano. In the middle row, from left, are Roscosmos cosmonauts Pyotr Dubrov and Anton Shkaplerov with NASA astronaut Mark Vande Hei. In the back, from left, was ESA (European Space Agency astronaut Matthias Maurer with NASA astronauts Thomas Marshburn, Raja Chari and Kayla Barron.

Expedition 66 crew members Mark Vande Hei of NASA, left, cosmonauts Anton Shkaplerov, center, and Pyotr Dubrov of Roscosmos, are seen inside their Soyuz MS-19 spacecraft after was landed in a remote area near the town of Zhezkazgan, Kazakhstan, Wednesday, March 30, 2022. Vande Hei and Dubrov are returning to Earth after logging 355 days in space as members of Expeditions 64-66 aboard the International Space Station. For Vande Hei, his mission was the longest single spaceflight by a U.S. astronaut in history. Shkaplerov was returning after 176 days in space, serving as a Flight Engineer for Expedition 65 and commander of Expedition 66.
Photo Credit: NASA/Bill Ingalls

The official portrait of the seven-member Expedition 67 crew. From left are, Flight Engineers Robert Hines of NASA; Samantha Cristoforetti of ESA (European Space Agency); Denis Matveev of Roscosmos; Commander Oleg Artemyev of Roscosmos; and Flight Engineers Sergey Korsakov of Roscosmos; Jessica Watkins of NASA; and Kjell Lindgren of NASA.

Expedition 67, the 67th long-duration mission aboard the International Space Station (ISS), commenced with the departure of the Soyuz MS-19 spacecraft on March 30, 2022. This signaled the beginning of a new phase for the ISS crew. NASA astronaut Thomas Marshburn assumed the role of ISS commander, taking over the station's leadership.

At the outset, Expedition 67 comprised Marshburn and his three colleagues from SpaceX Crew-3: Raja Chari, Kayla Barron, and Matthias Maurer. They were joined by Roscosmos cosmonauts Oleg Artemyev, Denis Matveev, and Sergey Korsakov, who had launched aboard Soyuz MS-21 on March 18, 2022. This Russian crew had transferred from Expedition 66, continuing their mission alongside the Crew-3 astronauts.

A NASA portrait of the three-member Soyuz MS-21 crew. From left are, Flight Engineer Denis Matveev; Commander Oleg Artemyev; and Flight Engineers Sergey Korsakov all of Roscosmos.

On March 18, 2022, the Soyuz MS-21 spacecraft launched from the Baikonur Cosmodrome, embarking on a historic mission to the International Space Station (ISS). Originally scheduled for March 30, Roscosmos advanced the launch date to March 18, reflecting a shift in the provisional flight manifest prepared in the summer of 2020. This mission was notable as it marked the first time a Soyuz flight to the ISS carried a crew composed solely of Roscosmos cosmonauts.

The crew for Soyuz MS-21 was announced in May 2021 and consisted of three Russian astronauts: Commander Oleg Artemyev, Flight Engineer 1 Denis Matveev, and Flight Engineer 2 Sergey Korsakov. Artemyev, a veteran of two previous spaceflights, commanded the mission. Matveev, making his first spaceflight, and Korsakov, also on his first mission, joined him. This mission was part of Expedition 66/67 on the ISS.

NASA initially considered purchasing a seat on Soyuz MS-21 for astronaut Loral O'Hara. However, NASA ultimately decided against this, deferring its seat acquisition to future missions, including Soyuz MS-22 and SpaceX Crew-5.

After a successful six-month mission, Soyuz MS-21 completed its objectives and returned to Earth on September 29, 2022. The landing occurred on the Kazakh Steppe in Kazakhstan, concluding the mission as planned.

A notable aspect of Soyuz MS-21 was the international attention garnered by the cosmonauts' arrival suits. Upon reaching the ISS, the crew wore bright yellow suits with blue accents. These colors drew considerable speculation and interpretation from international commentators, who saw them as reflecting the national colors of Ukraine amid the ongoing 2022 Russian invasion of Ukraine. The suits were perceived by some as a gesture of sympathy towards Ukraine during a time of heightened tension and strained space cooperation due to international sanctions on Russia.

In response to these interpretations, the cosmonauts clarified that the yellow suits were mandated for use and that the color scheme represented the identifying colors of Bauman Moscow State Technical University, from which all three cosmonauts had graduated. Roscosmos supported this explanation, reaffirming that the suit colors were not intended as a political statement but rather as a symbol of their alma mater.

Geopolitical tensions marked the backdrop of this expedition, notably the 2022 Russian invasion of Ukraine and the ensuing sanctions on Russia, which cast uncertainty over international cooperation in space.

During Expedition 67, the ISS also welcomed the crew of Axiom Mission 1, a notable space tourist mission. On April 9, 2022, this mission brought three private astronauts to the station, accompanied by former NASA astronaut Michael López-Alegría, who had previously commanded the ISS during Expedition 14. After completing their visit, the Axiom Mission 1 crew departed on April 25, 2022.

As Expedition 67 progressed, SpaceX Crew-3 departed on May 5, 2022. They were succeeded by SpaceX Crew-4, which delivered NASA astronauts Kjell N. Lindgren, Bob Hines, and Jessica Watkins, along with European Space Agency (ESA) astronaut Samantha Cristoforetti, to the station. Before their departure, Thomas Marshburn handed over command of the ISS to Oleg Artemyev.

A significant milestone during this expedition was the visit of Boeing's Starliner spacecraft, which docked with the ISS for the first time in May 2022 as part of its OFT-2 mission. This event underscored the ongoing advancements in commercial spaceflight.

During Expedition 67, the crew of the International Space Station (ISS) comprised a diverse and dynamic group of astronauts from various space agencies, whose roles and compositions evolved over the course of the mission.

The crews from Expedition 67 crew (top and bottom row and one on center left) with non expedition Axiom Mission-1 crew (remaining in center row).

From March 30 to April 27, 2022, the initial team included NASA astronaut Thomas Marshburn, who served as the commander. Marshburn, with his extensive experience, led the crew through the early part of the expedition. SpaceX Crew-3 members Raja Chari, Kayla Barron, and Matthias Maurer joined him, each contributing as flight engineers in their first space missions. The Russian contingent aboard comprised cosmonauts Oleg Artemyev, Denis Matveev, and Sergey Korsakov from Roscosmos. Artemyev, who was on his third spaceflight, held the position of flight engineer, while Matveev and Korsakov, on their first spaceflights, also served as flight engineers.

On April 8, 2022, the International Space Station (ISS) witnessed a landmark event with the launch of Axiom Mission 1 (Ax-1), a pioneering privately funded and operated crewed mission. This historic flight, managed by Axiom Space from their Mission Control Center in Houston, Texas, marked a significant milestone in space exploration as the first wholly commercially operated crewed mission to the ISS.

The spacecraft used for this mission was a SpaceX Crew Dragon, launched atop a Falcon 9 Block 5 rocket from Kennedy Space Center's Launch Complex 39A in Florida. The Crew Dragon Endeavour, which had previously supported Crew Dragon Demo-2 and SpaceX Crew-2 missions, embarked on this journey, carrying a diverse crew of four. The team included

Michael López-Alegría, an American born in Spain and a former NASA astronaut now with Axiom Space, alongside three space tourists: Eytan Stibbe from Israel, Larry Connor from the United States, and Mark Pathy from Canada. Peggy Whitson, another former NASA astronaut and Axiom consultant, was designated as the backup commander, with John Shoffner, an airshow pilot and entrepreneur, as the backup pilot.

Crew from Axiom Mission 1 Clockwise from left: Connor, Stibbe, Pathy and López-Alegría

The mission's primary objective was to conduct scientific research and experiments aboard the ISS. The crew spent ten days on the station, where they conducted over 25 different research experiments. Among these, the Israeli segment, named Rakia—which translates to "sky" in Hebrew—was particularly notable. This segment was named in honor of Ilan Ramon's diary fragments that survived the Space Shuttle Columbia disaster in 2003. Additionally, Connor brought artifacts from the Armstrong Air & Space Museum, including a John Glenn senatorial campaign button, a museum patch, and a piece of Kapton foil from the Apollo 11 Command Module aboard.

The mission encountered a minor delay due to adverse weather conditions affecting the planned splashdown site. The crew spent a total of 17 days in orbit, with 16 days docked to the ISS. The Crew Dragon Endeavour eventually undocked and returned to Earth, concluding the mission.

In a unique touch, the "zero-gravity indicator" for this historic flight was a toy dog named Caramel, representing the Montreal Children's Hospital Foundation. This mission demonstrated the growing role of private entities in space exploration and set the stage for future commercial endeavors in low Earth orbit.

Axiom Space, founded in 2016, has ambitious plans to expand its role in space by constructing the world's first commercial space station. By 2024, the company aims to launch the first module of the Axiom Orbital Segment, a new addition to the ISS that will support its activities in space. Axiom plans to continue offering crewed flights to the ISS, providing as many as two missions per year, aligning with opportunities made available by NASA.

From April 27 to May 5, 2022, the composition of the crew remained stable, but the mission saw a notable transition as the Axiom Mission 1 crew departed. During this period, the ISS continued to operate with the remaining astronauts from SpaceX Crew-3 and the Roscosmos team.

The crew underwent a significant change from May 5 to September 21, 2022. SpaceX Crew-3 departed the station, and the new team from SpaceX Crew-4 arrived. On April 27, 2022, SpaceX Crew-4 embarked on its mission as the fourth operational flight of the Crew Dragon spacecraft under NASA's Commercial Crew Program and the seventh overall crewed orbital flight of Crew Dragon. The mission launched at 07:52 UTC and docked with the International Space Station (ISS) at 23:37 UTC, marking another significant milestone for SpaceX and international collaboration in space exploration.

The Crew-4 mission featured the inaugural flight of the Crew Dragon spacecraft named Freedom. The name, chosen by the crew, reflects the celebration of fundamental human rights and the spirit of innovation that thrives in an environment of freedom. The mission utilized the Falcon 9 rocket booster designated B1067, which distinguished itself as the first booster to be used for a fourth flight in a Commercial Crew mission, having previously launched SpaceX Crew-3 in 2021.

The crew of Crew-4 comprised four astronauts: NASA astronauts Kjell Lindgren and Robert Hines, ESA astronaut Samantha Cristoforetti, and NASA astronaut Jessica Watkins. Kjell Lindgren, the mission commander, had previously flown to space on Expedition 44/45 and was tasked with leading the Crew-4 mission. Robert Hines served as the pilot for this mission,

marking his first spaceflight. Samantha Cristoforetti, representing the European Space Agency (ESA), was originally designated as the commander of Expedition 68 but was later reassigned to a mission specialist role due to adjustments in the mission schedule. Jessica Watkins, also a NASA astronaut, was assigned as a mission specialist, making her first spaceflight.

SpaceX Crew-4 Portrait with NASA astronauts, Robert Hines, Jessica Watkins, Kjell Lindgren and ESA astronaut Samantha Cristoforetti.

Crew-4 astronauts, from left, Jessica Watson, mission specialist; Bob Hines, pilot; Kjell Lindgren, commander and Samantha Cristoforetti, mission specialist, inside SpaceX's Crew Dragon, named Freedom by the Crew-4 crew

The European segment of the mission, named Minerva after the Roman goddess of wisdom, was a notable component of Crew-4, emphasizing the collaborative nature of the mission and the contributions of international partners. Cristoforetti's involvement marked her second mission to the ISS, reflecting her continued commitment to space exploration.

Crew-4's mission lasted a total of 170 days, 13 hours, 2 minutes, and 32 seconds, during which the crew conducted various scientific experiments and maintenance tasks aboard the ISS. The successful completion of this mission further solidified SpaceX's role in facilitating regular crewed missions to the ISS and underscored the importance of international cooperation in advancing human spaceflight.

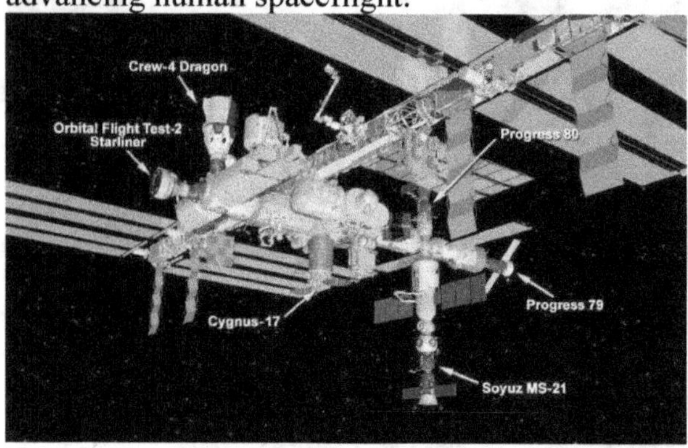

Both commercial Crew vehicles, Crew Dragon (on Crew-4 Mission) and Starliner (on OFT2 mission) docked to ports on harmony module at the same time

On September 21, 2022, the Soyuz MS-22 mission launched from Baikonur Cosmodrome, marking a significant event in the ongoing partnership between Russia and the United States aboard the International Space Station (ISS). Originally scheduled for September 13, the launch was delayed to accommodate a revised timeline, extending the mission's duration to 188 days.

The Soyuz MS-22 mission, a pivotal part of the ISS's crew rotation strategy, was initially planned to be crewed entirely by Russian astronauts. However, a notable change was introduced when American astronaut Francisco Rubio was added to the crew in place of Anna Kikina. This adjustment was part of the Soyuz-Dragon crew swap system, a strategy designed to ensure continuous space station occupancy by both American and Russian personnel. This system serves as a backup to mitigate potential disruptions caused by vehicle grounding, such as the Soyuz MS-10 launch failure, or delays in crew rotation missions, exemplified by the postponement of SpaceX Crew-3 due to adverse launch weather conditions.

The Soyuz MS-22 crew consisted of Commander Sergey Prokopyev and Flight Engineers Dmitry Petelin and Francisco Rubio. Prokopyev and Petelin were Russian cosmonauts representing Roscosmos, while Rubio was an astronaut from NASA. All three were making their first spaceflight. In preparation for this mission, a backup crew was designated, including Commander Oleg Kononenko, Flight Engineer Nikolai Chub, and Flight Engineer Loral O'Hara.

The spacecraft itself was named in honor of Konstantin Tsiolkovsky, a Russian scientist celebrated as one of the pioneers of modern rocketry and astronautics. The launch of Soyuz MS-22 coincided with the 165th anniversary of Tsiolkovsky's birth, adding a symbolic layer to the mission.

On December 15, 2022, an unexpected issue arose when a visible stream of flakes emanated from the Soyuz MS-22 spacecraft, alongside a pressure loss in the external radiator cooling loop. The cause of the leak was determined to be a micro-meteorite impact, which had created a 0.8 mm diameter hole in the external cooling radiator of the service module. This incident led to the cancellation of a scheduled spacewalk by Petelin and Prokopyev while engineers assessed the situation.

The leak caused temperatures within the spacecraft to rise significantly, with the orbital and descent modules reaching 30 °C (86 °F) and the service module peaking at 40 °C (104 °F). By January 2023, temperatures had stabilized around 30 °C. The outer surface of the Soyuz MS-22 was examined using the cameras of the European Robotic Arm and Canadarm2, helping engineers identify the damage. Progress MS-21 later encountered a similar coolant pressure leak in February 2023.

Due to the inability of Soyuz MS-22 to return its crew, an uncrewed Soyuz MS-23 was launched on February 24, 2023, to serve as a replacement vehicle. This mission aimed to bring the crew back to Earth in September 2023, extending their space stay to one year. The original mission of Soyuz MS-23 was reassigned to Soyuz MS-24, resulting in the return of dry cargo and equipment from the uncrewed Soyuz MS-22. During re-entry, temperatures likely reached 50 °C (122 °F), a scenario deemed more favorable than the worst emergency crewed landing conditions studied.

In the interim, SpaceX Crew-5 was considered as an alternative for returning the MS-22 crew in case of emergency, due to the Crew Dragon's design accommodating up to seven astronauts. To enhance safety, Francisco Rubio's Soyuz seat liner was transferred to the Crew Dragon Endurance on January 17, 2023, allowing for potential lifeboat capabilities. This was the first instance of a seat liner being transferred from a Soyuz spacecraft to a Crew Dragon. The adjustment aimed to reduce the heat load inside the MS-22 spacecraft, providing additional protection for Prokopyev and Petelin in an emergency.

Upon the arrival of Soyuz MS-23 at the ISS on February 26, 2023, Rubio's seat liner was moved to the new Soyuz on March 6, and the seat liners for Prokopyev and Petelin were transferred from MS-22 to MS-23 on March 2, in preparation for their return. The mission exemplified the collaborative efforts to ensure the safety and continuity of operations aboard the International Space Station.

The final phase of Expedition 67 saw the arrival of the Soyuz MS-22 crew from September 21 to September 29, 2022. This included Russian cosmonauts Sergey Prokopyev and Dmitry Petelin and NASA astronaut Francisco Rubio. Prokopyev, in his second spaceflight, was joined by Petelin and Rubio, both on their first spaceflights, as flight engineers. Their arrival marked the transition into Expedition 68.

As Expedition 67 drew to a close, the remaining crew continued their work on the ISS, becoming part of Expedition 68 starting in September 2022.

The four members of the Axiom-1 (Ax-1) private crew join the seven members of the International Space Station's Expedition 67 crew for a brief welcome ceremony on Saturday, April 9, 2022. From left to right: Roscosmos cosmonauts Sergey Korsakov and Denis Matveev, NASA astronauts Raja Chari (upside down) and Kayla Barron, Ax-1 mission specialists Mark Pathy of Canada and Eytan Stibbe of Israel, Roscosmos cosmonaut Oleg Artemyev, Ax-1 pilot Larry Connor, European Space Agency astronaut Matthias Maurer (upside down), Ax-1 commander Michael Lopez-Alegria and NASA astronaut Tom Marshburn. (Image credit: NASA TV)

Expedition 68

Expedition 68, a pivotal mission in the ongoing operations of the International Space Station (ISS), saw a series of significant events unfold throughout the year 2023, along with some key activities from the previous year. The mission was marked by crew changes, cargo deliveries, and the dynamic logistical operations that kept the ISS functioning smoothly.

The year began with notable activities in late 2022. On October 23, 2022, the ISS Progress 80 spacecraft undocked from the station, concluding

Both Commercial Crew Program vehicles, the SpaceX Dragon Freedom and Boeing CST-100 Spacecraft 2, docked to ISS Harmony ports.

its cargo mission. Shortly after that, on October 25, 2022, the ISS Progress 82 spacecraft was launched, followed by its docking on October 27, 2022. These maneuvers were crucial in maintaining the ISS's supply chain.

The seven-member Expedition 68 crew poses for an official portrait. From left are, NASA astronaut Frank Rubio; Roscosmos cosmonaut Dmitri Petelin; Japan Aerospace Exploration Agency (JAXA) astronaut Koichi Wakata; NASA astronauts Josh Cassada and Nicole Mann; and Roscosmos cosmonauts Sergey Prokopyev and Anna Kikina.

As the year continued, SpaceX Crew-4 made its splashdown on October 14, 2022, marking the end of its mission. The Crew-4 astronauts, including Commander Kjell N. Lindgren, Pilot Robert L. Hines, and Mission Specialists Jessica Watkins and Samantha Cristoforetti, returned to Earth after their time aboard the ISS. This was succeeded by SpaceX Crew-5's launch on October 5, 2022, which brought new crew members to the station, including Commander Nicole Aunapu Mann, Pilot Josh Cassada, and Mission Specialists Koichi Wakata and Anna McClain. The Crew-5 spacecraft docked with the ISS on October 6, 2022, continuing the station's uninterrupted crew presence.

In November 2022, Cygnus CRS-18 launched on November 7 and was captured by the ISS on November 9. This mission provided essential cargo and supplies, underscoring the collaborative efforts of various space agencies and commercial partners to support the ISS.

As the year transitioned to 2023, the SpaceX CRS-26 mission played a significant role. After its launch on November 26, the spacecraft docked with the ISS on November 27, 2022. It was later undocked on January 9, 2023, completing its resupply mission.

The beginning of 2023 continued to be active, with the undocking of the ISS Progress 81 on February 7. This was followed by the launch and docking of the ISS Progress 83 on February 9 and February 11, respectively, further ensuring the station's continued operations.

A key highlight of February 2023 was the arrival of the Soyuz MS-23 spacecraft. Launched on February 23, 2023, Soyuz MS-23 docked with the ISS on February 25, bringing new crew members to the station. The Soyuz MS-23 mission was instrumental in rotating the station's crew and maintaining the ISS's capacity to support ongoing research and operations.

In early March 2023, SpaceX Crew-6 marked a significant milestone as the sixth crewed operational flight of NASA's Commercial Crew Program and the ninth overall crewed orbital flight of a Crew Dragon spacecraft. Launched on 2 March 2023 at 05:34:14 UTC, the mission showcased the continued advancement in space travel and international cooperation. The Crew-6 spacecraft successfully docked with the International Space Station (ISS) the following day, at 06:40 UTC, after a precise journey through orbit.

Expedition 68, marking the sixty-eighth long-duration mission aboard the International Space Station (ISS), began with the departure of Soyuz MS-21 on September 29, 2022. This event transitioned the station from Expedition 67, ushering in a new team of astronauts. ESA astronaut Samantha Cristoforetti assumed command of the ISS, leading the crew through the mission until its conclusion.

SpaceX Crew 6 - Al Neyadi, Hoburg, Bowen & Fedyaev

The initial phase of Expedition 68 saw Cristoforetti and her SpaceX Crew-4 colleagues—Kjell N. Lindgren, Bob Hines, and Jessica Watkins—join the ISS. They were accompanied by Roscosmos cosmonauts Sergey Prokopyev, Dmitry Petelin, and American astronaut Francisco Rubio, who had arrived aboard Soyuz MS-22 on September 21, 2022. This marked a seamless transition from Expedition 67, setting the stage for the work ahead.

Cristoforetti's role as commander, part of the Minerva mission for ESA, was notable for her leadership throughout this period. Her tenure as commander concluded with the official handover of ISS command to Sergey Prokopyev on October 12, 2022. Following the departure of SpaceX Crew-4 on October 14, 2022, Expedition 68 continued with the arrival of SpaceX Crew-5. This new crew, consisting of NASA astronauts Nicole Mann and Josh Cassada, JAXA astronaut Koichi Wakata, and Roscosmos cosmonaut Anna Kikina, took over seamlessly in a direct handover.

Koichi Wakata's fifth spaceflight, now aboard SpaceX Crew-5, was particularly historic. Wakata, having previously flown aboard Soyuz and the Space Shuttle, became the eighth individual to travel on three different Earth-launching spacecraft. This flight was his first on the Dragon spacecraft, adding a significant milestone to his illustrious career.

Anna Kikina's inclusion in SpaceX Crew-5 was equally historic. Her flight marked the first instance of a Russian cosmonaut traveling to the ISS aboard a NASA Commercial Crew Program vehicle, making it a noteworthy event in the history of spaceflight. This was the first time in two decades that a Russian, on a mission for Russia, flew on a US spacecraft.

The crew's rotation continued with the arrival of SpaceX Crew-6 on March 3, 2023. This crew included NASA astronauts Stephen Bowen and Warren Hoburg, the United Arab Emirates Sultan Al Neyadi, and Roscosmos cosmonaut Andrey Fedyaev. As Expedition 68 drew to a close, these new crew members replaced those who had completed their mission.

On December 14, 2022, the mission encountered a significant issue when a coolant leak was detected on the Soyuz MS-22 spacecraft docked to the ISS. The leak, which involved ammonia, led to the cancellation of a planned Russian spacewalk. Subsequent investigations and imaging revealed that the MS-22 spacecraft could not safely return its crew, except in an emergency situation. Consequently, Roscosmos and NASA announced changes to the crew manifest, extending the remaining crew's stay on the ISS to 12 months and deciding that Soyuz MS-22 would return uncrewed. Soyuz MS-23, launched uncrewed in February 2023, was prepared to accommodate the crew for their extended mission. The original crew scheduled for Soyuz MS-23 was instead launched aboard Soyuz MS-24 to support Expeditions 69 and 70, making adjustments to ensure that all crew and seat liners were properly assigned.

In February, during the Progress MS-22/83P vehicle docking, another incident occurred when the Progress MS-21/82P experienced a similar coolant leak. This prompted Russian officials to delay the uncrewed launch of Soyuz MS-23 pending an investigation into the cause of the leak. Soyuz MS-23 was eventually launched on February 24, 2023, after it was deemed safe. A thermal test of the Soyuz MS-22 was conducted to assess the damage caused by the coolant loss.

Expedition 68, the sixty-eighth long-duration mission aboard the International Space Station (ISS), began with the departure of Soyuz MS-21 on September 29, 2022. This marked the official transition from Expedition 67 to the new crew's tenure. With Soyuz MS-21 leaving, Samantha Cristoforetti, the outgoing commander, handed over leadership of the ISS to Sergey Prokopyev in a formal Change of Command Ceremony held on October 12, 2022.

The arrival of SpaceX Crew-5 on October 6, 2022, signaled the start of a new phase. This crew, consisting of NASA astronauts Nicole Mann and Josh Cassada, JAXA astronaut Koichi Wakata, and Roscosmos cosmonaut Anna Kikina, joined the station's existing occupants. The same day, SpaceX Crew-4 departed, completing their mission and making way for the incoming team.

Throughout October, the ISS experienced several logistical changes. On October 23, Progress MS-19/80P undocked, while Progress MS-21/82P docked on October 27, bringing new supplies and scientific equipment. As November unfolded, the station saw the arrival of CRS Cygnus NG-18 on November 9, with its cargo contributing to the ISS's ongoing research and operational needs. The station's robotic arm carefully captured the Cygnus spacecraft before being berthed.

Two significant extravehicular activities (EVAs) were conducted during November. On November 15, EVA 1, designated US-81, lasted over seven hours and addressed maintenance and upgrades to the station's external systems. Two days later, EVA 2, known as VKD-55, focused on installing a work platform on the Nauka module, enhancing the ISS's capacity to handle large payloads.

As December began, the station continued its busy schedule. On December 3, EVA 3 (US-82) was conducted, involving the installation of an International Space Station Roll-Out Solar Array (iROSA) on Array 3A, essential for the station's power generation. However, a coolant leak was discovered on Soyuz MS-22 on December 14, leading to the cancellation of a planned Russian EVA and prompting an investigation into the leak's implications for the spacecraft's safety.

Despite these challenges, the station's work continued. On December 22, EVA 4 (US-83) was performed, installing the fourth iROSA on Array 4A. The year concluded with the undocking of CRS SpX-26 on January 9, 2023, which had delivered critical cargo to the ISS.

In mid-January, seat liners were shifted to accommodate changing crew assignments, with Francisco Rubio's seat liner moving from Soyuz MS-22 to Crew-5. This adjustment was crucial for aligning with the updated crew rotation plans. Subsequent EVAs, including EVA 5 (US-84) on January 20 and EVA 6 (US-85) on February 2, involved further maintenance and upgrades, including additional enhancements to the station's equipment.

February also brought new developments with the docking of Progress MS-22/83P on February 11, and the discovery of a leak on Progress MS-21, prompting immediate investigation. Soyuz MS-23, launched uncrewed on February 26, was prepared to accommodate the crew for future missions, following the issues faced by Soyuz MS-22.

March 2023 saw the arrival of SpaceX Crew-6 on March 3, continuing the crew rotation process. A significant moment on March 6 was the transfer of Francisco Rubio's seat liner from Crew-5 to Soyuz MS-23. Meanwhile, Soyuz MS-22 underwent a critical thermal test on March 15 to assess its emergency evacuation capability.

As Expedition 68 ended, CRS SpX-27 docked with the ISS on March 16, bringing further supplies and scientific experiments. The expedition officially concluded on March 28, 2023, with the uncrewed undocking of Soyuz MS-22, marking the transition to Expedition 69 and continuing the ISS's vital mission of scientific research and international cooperation.

In an unrelated development, SpaceX's Crew-6 mission launched on March 2, 2023, deviating from the traditional April or September launch window. This shift did not affect the customary conclusion of an expedition with the departure of a Soyuz spacecraft. Expedition 68's conclusion was marked by the uncrewed departure of Soyuz MS-22 on March 28, 2023, with all onboard crew being transitioned to Expedition 69.

Expedition 68 marked a dynamic period aboard the International Space Station (ISS), characterized by a series of arrivals and departures that facilitated both crew transitions and cargo deliveries.

(March 6, 2023) --- The 11-member crew aboard the International Space Station give thumbs up signs in this portrait. In the bottom row from left are Flight Engineers Andrey Fedyaev of Roscosmos, Sultan Alneyadi from UAE (United Arab Emirates), and Woody Hoburg from NASA. In the middle row from left are Flight Engineers Anna Kikina from Roscosmos, Koichi Wakata from JAXA (Japan Aerospace Exploration Agency), Nicole Mann from NASA, Dmitri Petelin from Roscosmos, and Frank Rubio from NASA. In the back are Flight Engineer Stephen Bowen from NASA, Commander Sergey Prokopyev from Roscosmos, and Flight Engineer Josh Cassada from NASA.

Vehicles from Previous Expeditions

The expedition inherited several key vehicles from earlier missions:

Progress MS-19/80P: This Russian cargo spacecraft, docked to the Poisk module's zenith port, had been a fixture since February 17, 2022, during Expedition 66. It was crucial in resupplying the station before undocking on October 23, 2022.

SpaceX Crew-4 "Freedom": Representing a vital part of Expedition 67 and 68, this Crew Dragon spacecraft had been docked at the Harmony module's zenith port since April 27, 2022. Its mission concluded with a departure on October 14, 2022, as it carried its crew home, paving the way for the arrival of new personnel.

Progress MS-20/81P: Another vital Russian cargo mission, this spacecraft, had been docked to the Zvezda module's aft port since June 3, 2022. It supported the station's operations until it undocked on February 7, 2023.

Soyuz MS-22/68S "Altai": Arriving at the Rassvet module's nadir port on September 21, 2022, this Soyuz spacecraft was a critical vehicle for crew transport. It marked a significant transition from Expedition 67 to Expedition 68,

and continued to serve until its undocking on March 28, 2023.

(Dec. 24, 2022) --- Expedition 68 Flight Engineers (from left) Josh Cassada of NASA, Koichi Wakata of the Japan Aerospace Exploration Agency (JAXA), and Frank Rubio of NASA, pose for a photograph while sharing a meal on Christmas Eve inside the International Space Station's Unity module.

Vehicles Docked During Expedition 68

Expedition 68 saw the arrival of several new vehicles, enhancing the station's capabilities:

SpaceX Crew-5 "Endurance": Docking at the Harmony module's forward port on October 6, 2022, Crew-5 brought a new team of astronauts, including Nicole Mann, Josh Cassada, Koichi Wakata, and Anna Kikina. This mission signaled a fresh chapter in the expedition, concluding with their undocking on March 11, 2023.

Progress MS-21/82P: This Russian cargo spacecraft arrived at the Poisk module's zenith port on October 28, 2022. It provided essential supplies and support until its undocking on February 18, 2023.

CRS Cygnus NG-18 "Sally Ride": A key cargo mission for NASA, this spacecraft docked at the Unity module's nadir port on November 9, 2022. It was integral for delivering supplies and scientific experiments until April 21, 2023, when it departed as part of Expedition 69.

CRS Dragon SpX-26: Arriving at Harmony's zenith port on November 27, 2022, this Dragon spacecraft was crucial for cargo resupply, departing on January 9, 2023, after delivering its load.

Progress MS-22/83P: Another important Russian cargo mission, it docked with the Zvezda module's aft port on February 11, 2023. This spacecraft provided additional supplies and support until August 20, 2023.

Soyuz MS-23/69S: Docking at the Poisk module's zenith port on February 26, 2023, this Soyuz spacecraft was crucial for crew transport and return. It departed on September 27, 2023, as part of the transition to Expedition 69.

SpaceX Crew-6 "Endeavour": Arriving at Harmony's zenith port on March 3, 2023, Crew-6 continued the flow of astronauts to the ISS, concluding its mission with undocking on September 3, 2023.

CRS Dragon SpX-27: Docking at Harmony's forward port on March 16, 2023, this Dragon spacecraft was pivotal for resupplying the station. It undocked on April 15, 2023, marking the end of its mission as part of Expedition 69.

The uncrewed departure of Soyuz MS-22 on March 28, 2023, marked the official end of Expedition 68. The ISS transitioned smoothly into Expedition 69, continuing its vital work in space exploration and research.

Nov. 3, 2022) --- Three Expedition 68 Flight Engineers pose for a portrait inside the International Space Station's U.S. Destiny laboratory module. From left are, astronaut Koichi Wakata of the Japan Aerospace Exploration Agency (JAXA), and NASA astronauts Frank Rubio and Nicole Mann.

The SpaceX Crew Dragon Endurance spacecraft shortly after it landed with NASA astronauts Raja Chari, Kayla Barron, Tom Marshburn, and ESA (European Space Agency) astronaut Matthias Maurer aboard, in the Gulf of Mexico.

Expedition 68 crew members Dmitri Petelin of Roscosmos, top, Frank Rubio of NASA, and Sergey Prokopyev of Roscosmos, bottom, wave farewell prior to boarding the Soyuz MS-22 spacecraft for launch, Wednesday, Sept. 21, 2022, at the Baikonur Cosmodrome in Kazakhstan. Rubio, Prokopyev and Petelin, will launch onboard the Soyuz rocket from the Baikonur Cosmodrome for a mission on the International Space Station. Photo Credit: (NASA/Bill Ingalls)

Expedition 69

The official Expedition 69 crew from left) flight engineers Frank Rubio from NASA, Dmitri Petelin from Roscosmos, Sultan Alneyadi from UAE (United Arab Emirates), Woody Hoburg from NASA, Stephen Bowen from NASA, Andrey Fedyaev from Roscosmos, and Commander Sergey Prokopyev from Roscosmos.

Expedition 69, the 69th long-duration mission to the International Space Station (ISS), commenced in March 2023 with a significant transition period marked by the uncrewed departure of the Soyuz MS-22 spacecraft. This departure set the stage for a new chapter as Russian cosmonaut Sergey Prokopyev continued his role from Expedition 68. The expedition concluded on September 27, 2023, with the departure of Prokopyev and his crewmates aboard Soyuz MS-23.

At the start of Expedition 69, the crew comprised Prokopyev, his Soyuz MS-22/23 colleagues Dmitry Petelin from Russia and American astronaut Francisco Rubio. They were complemented by American astronauts Stephen G. Bowen and Warren Hoburg, Emirati astronaut Sultan Al Neyadi, and Russian cosmonaut Andrey Fedyaev. These astronauts launched aboard SpaceX Crew-6 on March 2, 2023, and were integrated into the ISS operations, continuing from Expedition 68. However, a coolant leak in Soyuz MS-22 discovered in December 2022 necessitated a shift in the mission plan. To address this issue, Soyuz MS-22 was returned uncrewed, and Soyuz MS-23 was launched as a replacement. As a result, the crew remained aboard the ISS for over a year, extending their mission.

The timeline for crew handovers was notably altered due to these adjustments. Traditionally, U.S. crew rotations occurred during new expeditions, roughly 2 to 3 weeks after Soyuz handovers. For Expedition 69, however, the U.S. crew handover took place before Soyuz MS-22's scheduled departure on March 28. This alteration was a direct consequence of the coolant leak and the subsequent adjustments in the mission manifest.

The expedition saw further crew rotations with SpaceX Crew-7, which included NASA astronaut Jasmin Moghbeli, Danish astronaut Andreas Mogensen, Japanese astronaut Satoshi Furukawa, and Russian cosmonaut Konstantin Borisov. Additionally, Soyuz MS-24 brought Russian cosmonauts Oleg Kononenko and Nikolai Chub, along with American astronaut Loral O'Hara, to the ISS for a year-long mission. The ISS also hosted Axiom Mission 2 during Expedition 69, featuring former NASA astronaut Peggy Whitson and John Shoffner, Ali AlQarni, and Rayyanah Barnawi. Whitson's participation was particularly notable due to her previous command roles during Expeditions 16 and 51.

Axiom Mission 2 (Ax-2) represented a significant advancement in private spaceflight, building on the success of Axiom Mission 1 and continuing the trend of commercial crewed missions aboard SpaceX's Dragon spacecraft. Launched on 21 May 2023, Ax-2 took off from Launch Complex 39A at Kennedy Space Center aboard a Falcon 9 Block 5 rocket. This mission marked a notable first for crewed flights: the first stage of the Falcon 9 rocket successfully landed on land at Cape Canaveral Space Force Station's

Landing Zone 1, a departure from the usual at-sea recovery.

The Dragon crew capsule, named Freedom, docked with the International Space Station (ISS) on 22 May 2023, where it remained for eight days. During their stay, the Ax-2 crew conducted various scientific experiments, including research into the effects of microgravity on stem cells and other biological studies. They also engaged in public outreach activities, sharing their experiences with the world and highlighting the importance of private space exploration.

Axiom -2 : Whitson, Shoffner, AlQarni & Barnawi

The Axiom Mission-2 and Expedition 69 crew members pose for a portrait together during dinner time aboard the International Space Station. In the center front row, was Expedition 69 crew member and UAE (United Arab Emirates) astronaut Sultan Alneyadi flanked by (from left) Axiom Mission-2 crew members Commander Peggy Whitson, Mission Specialist Ali Alqarni, Pilot John Shoffner, and Mission Specialist Rayyanah Barnawi. In the back (from left) are, Expedition 69 crew members Roscosmos cosmonaut Dmitri Petelin, NASA astronaut Stephen Bowen, Roscosmos cosmonauts Andrey Fedyaev and Sergey Prokopyev, and NASA astronaut Woody Hoburg. Not pictured was NASA astronaut Frank Rubio.

The mission's crew was composed of a diverse group of individuals. Commanding Ax-2 was Peggy Whitson, an accomplished former NASA astronaut now with Axiom Space. Joining her was John Shoffner, a space tourist serving as the pilot, marking his first spaceflight. The other two seats on the mission were occupied by astronauts Ali AlQarni and Rayyanah Barnawi from the Saudi Space Agency. This mission was particularly historic as it included the first female Saudi astronaut, Barnawi, underscoring the growing international participation in space.

Axiom Space had initially announced a unique selection process for one of the crew members through the reality television series "Who Wants to Be an Astronaut?" produced by Discovery Channel. However, on 11 January 2022, it was revealed that Italian Air Force Colonel Walter Villadei, previously announced as a backup crew member, would serve as the pilot on Axiom Mission 3. Villadei's role in Ax-2 was as a backup commander.

On 22 September 2022, Axiom Space further solidified its commitment to international collaboration by partnering with the Saudi Space Agency to facilitate the participation of two Saudi astronauts in Ax-2. This collaboration aimed to advance research in areas such as cancer, cloud seeding, and the effects of microgravity.

After completing their mission objectives, the crew aboard Freedom undocked from the ISS and began their return journey to Earth. Twelve hours later, the spacecraft successfully splashed down in the Gulf of Mexico, off the coast of Panama City, Florida. The recovery was managed by SpaceX's recovery ship, Megan, marking a successful conclusion to Ax-2's pioneering flight.

As Expedition 69 unfolded, it underscored the ISS's role as a global platform for scientific research and international cooperation. The diverse crew, led by Commander Sergey Prokopyev and supported by Flight Engineers Oleg Kononenko, Dmitry Petelin, and Joshua Kutryk, exemplified the collaborative spirit that defines the ISS. Their work highlighted the ongoing commitment to advancing scientific knowledge and fostering international partnerships in space exploration.

One of the primary objectives of Expedition 69 was to continue the extensive scientific research programs that are central to the ISS's mission. The crew engaged in various experiments to advance our understanding of human physiology, materials science, and fundamental physics. Notably, they conducted studies on the effects of prolonged exposure to microgravity on the human body crucial for planning future deep-space missions. The research included investigations into bone density loss, muscle atrophy, and the impact of space radiation on biological systems.

(May 28, 2023) --- The SpaceX Dragon Freedom spacecraft docked to the space-facing port on the International Space Station's Harmony module. Dragon Freedom carried four Axiom Mission-2 astronauts to the orbital lab on May 22, 2023, including Commander Peggy Whitson, Pilot John Shoffner, and Mission Specialists Ali Alqarni and Rayyanah Barnawi.

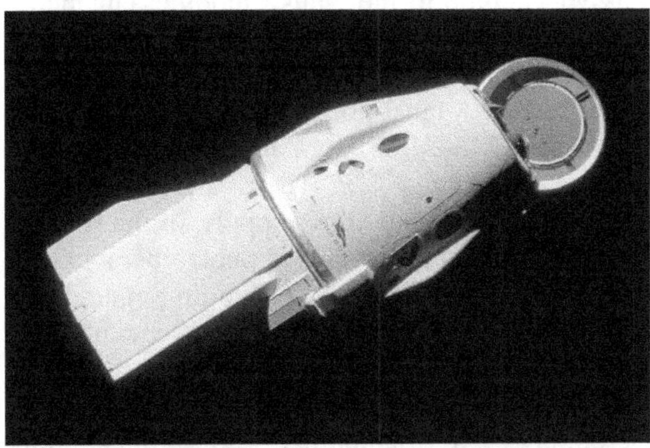

The SpaceX Dragon Freedom spacecraft with its nosecone deployed after departing the International Space Station moments earlier. Dragon Freedom returned to Earth later that day splashing down in the Gulf of Mexico with four Axiom Mission-2 astronauts including Commander Peggy Whitson, Pilot John Shoffner, and Mission Specialists Ali Alqarni and Rayyanah Barnawi.

In addition to scientific experiments, Expedition 69 performed routine maintenance and upgrades to the ISS's systems. This included addressing technical issues and ensuring the station's various modules and equipment functioned optimally. The crew's work was essential for maintaining the ISS's operational capabilities and supporting the ongoing research efforts.

A significant highlight of Expedition 69 was the successful implementation of new technology and experiments. The crew installed and tested new scientific instruments to enhance the station's research capabilities. These upgrades were crucial for expanding the range of experiments that could be conducted aboard the ISS and for improving the quality of data collected.

Throughout Expedition 69, the crew also collaborated with scientists and engineers on Earth to analyze and interpret data from their experiments. This collaboration was key to advancing our understanding of the results and applying the findings to various fields of science and technology.

(May 23, 2023) --- The Axiom Mission-2 and Expedition 69 crew members pose for a portrait together during dinner time aboard the International Space Station. In the center front row, was Expedition 69 crew member and UAE (United Arab Emirates) astronaut Sultan Alneyadi flanked by (from left) Axiom Mission-2 crew members Commander Peggy Whitson, Mission Specialist Ali Alqarni, Pilot John Shoffner, and Mission Specialist Rayyanah Barnawi. In the back (from left) are, Expedition 69 crew members Roscosmos cosmonaut Dmitri Petelin, NASA astronaut Stephen Bowen, Roscosmos cosmonauts Andrey Fedyaev and Sergey Prokopyev, and NASA astronaut Woody Hoburg. Not pictured was NASA astronaut Frank Rubio.

On March 28, 2023, the Soyuz MS-22 spacecraft, a key element of the International Space Station's operations, undocked in a notable event marking the official transition from Expedition 68 to the next phase of station activities. This event signaled the end of the crewed mission's impact and prepared the ISS for new developments.

A week later, on April 6, 2023, the Soyuz MS-23 spacecraft returned to the station, reestablishing its connection and ensuring continued crew and supply operations. This docking was a crucial step in maintaining the ISS's operational readiness.

April 15 saw the undocking of the CRS SpX-27 cargo spacecraft, which had been instrumental in delivering essential supplies and scientific experiments to the station. The following day, April 16, marked a significant achievement with EVA 1 (VKD-56). Astronauts embarked on a spacewalk lasting 7 hours and 55 minutes, performing vital maintenance and upgrades outside the ISS.

On the same day, the Nauka module underwent an important adjustment. The "RtoD" add-on heat radiator, initially attached to the Rassvet module, was relocated to the Nauka module, enhancing its thermal management capabilities.

The momentum continued on April 21 with the unberthing and release of the CRS Cygnus NG-18 cargo spacecraft. This action was part of the ongoing effort to manage cargo and maintain the station's clutter-free environment.

April closed with a significant spacewalk on April 28, EVA 2 (US-86), which lasted 7 hours and 1 minute. This spacewalk involved installing and maintaining critical station equipment.

As May unfolded, additional spacewalks and relocations were on the agenda. From May 3 to 4, EVA 3 (VKD-57) was conducted, with astronauts working outside the ISS for 7 hours and 11 minutes. On May 4, the experiment airlock "ShK" was relocated from the Rassvet module to the Nauka module's forward port, optimizing the module's operational efficiency.

The SpaceX Crew-6 spacecraft was redocked on May 6, ensuring the station remained well-staffed with a new crew. The following spacewalk, EVA 4 (VKD-58) on May 12, lasted 5 hours and 14 minutes and focused on essential maintenance tasks.

May 22 brought the arrival of Axiom Mission 2, a non-expedition crew mission, which docked with the ISS. This mission was followed by the Progress MS-23/84P docking on May 24, continuing the flow of supplies and support to the station.

By May 30, Axiom Mission 2 undocked, marking the end of its stay. This was shortly followed by the docking of SpaceX CRS-28 on June 1, bringing fresh supplies and research materials to the ISS.

June 9 and 15 saw two significant spacewalks, EVA 5 (US EVA-87) and EVA 6 (US EVA-88), respectively. EVA 5, lasting 6 hours and 3 minutes, involved the installation of the fifth iROSA (International Space Station Roll-Out Solar Array) at Array 1A. EVA 6, which lasted 5 hours and 35 minutes, completed the installation of the sixth and final iROSA at Array 1B.

June 22 featured EVA 7 (VKD-59), a spacewalk lasting 6 hours and 24 minutes, which contributed further to the station's operational and experimental capabilities. Following this, on June 29, the CRS SpX-28 spacecraft undocked, clearing the station for new arrivals.

The beginning of August was marked by the capture and berthing of CRS Cygnus NG-19 on August 4. On August 9, EVA 8 (VKD-60) was conducted, lasting 6 hours and 35 minutes. This spacewalk involved the relocation of the European Robotic Arm's Portable Workpost from the Rassvet module to the Nauka module, enhancing robotic operations on the station.

Later in August, Progress MS-22/83P undocked on August 20, followed by the docking of Progress MS-24/85P on August 25 and SpaceX Crew-7 on August 27. These arrivals ensured continued supply and crew support for the ISS.

The final notable events of September included the undocking of SpaceX Crew-6 on September 3 and the docking of Soyuz MS-24 on September 15. The ISS also celebrated a significant milestone with the Expedition 69/70 Change of Command Ceremony on September 26, transitioning leadership from Sergey Prokopyev to Andreas Mogensen. Finally, on September 27, Soyuz MS-23 undocked, officially marking the switch to Expedition 70 and setting the stage for the next chapter in the ISS's ongoing mission.

During Expedition 69, which spanned from March 2023 to September 2023, the International Space Station (ISS) hosted a variety of spacecraft that played crucial roles in its ongoing operations and international cooperation. This period was marked by significant transitions and continued collaboration among the global partners of the ISS.

The expedition began with a critical changeover. The Russian Soyuz MS-22 spacecraft, which had served as a crew transport vehicle during the previous expeditions, departed in early March 2023, making way for Soyuz MS-23. This uncrewed transition was necessitated by a coolant leak discovered in Soyuz MS-22 in December 2022, which led to its replacement by Soyuz MS-23. Despite this challenge, Soyuz MS-23, commanded by Sergey Prokopyev, continued to ensure the safe return of the crew, including Dmitry Petelin and Francisco Rubio, who had joined the ISS in the previous expedition.

The station's logistics were further supported by the arrival and departure of several key spacecraft throughout Expedition 69. The United States' Cygnus NG-18, named in honor of Sally Ride, docked at the Unity module's nadir port on November 9, 2022, as part of Expedition 68. This cargo spacecraft completed its mission with an undocking on April 21, 2023, making way for new arrivals.

The Russian Progress MS-22, a cargo vehicle, arrived at the Zvezda module's aft port on February 11, 2023, during Expedition 68. It remained docked until August 20, 2023, continuing to supply essential resources to the ISS.

SpaceX Crew-6, a mission that began under Expedition 68, docked at the Harmony module's zenith port on February 28, 2023. The crew, consisting of American astronauts Stephen Bowen and Warren Hoburg, Emirati astronaut Sultan Al Neyadi, and Russian cosmonaut Andrey Fedyaev, played a key role in the station's operations. They departed the ISS on May 6, 2023, as part of the crew handover process.

During Expedition 69, SpaceX Crew-6's role was succeeded by SpaceX Crew-7, which arrived on August 27, 2023. This mission included NASA astronaut Jasmin Moghbeli, Danish astronaut Andreas Mogensen, Japanese astronaut Satoshi Furukawa, and Russian cosmonaut Konstantin Borisov. SpaceX Crew-7 docked at Harmony's zenith port, continuing the international crew rotation and collaboration tradition.

On August 26, 2023, the SpaceX Crew-7 mission embarked on its journey as the seventh crewed operational flight under NASA's Commercial Crew Program and the eleventh overall crewed mission of a Crew Dragon spacecraft. Launched aboard the Crew Dragon spacecraft, Endurance, the mission marked a significant milestone in international collaboration and commercial space operations.

The Crew-7 team consisted of four astronauts representing a diverse range of space agencies. As the mission commander, NASA astronaut Jasmin Moghbeli became the first person of Iranian descent to command a Crew Dragon flight. Andreas Mogensen, an astronaut from the European Space Agency (ESA) and Denmark, fulfilled the pilot role. This mission was notable as Mogensen became the first non-American to serve as a Crew Dragon pilot. The crew also included Satoshi Furukawa from the Japan Aerospace Exploration Agency (JAXA) and Konstantin Borisov from Roscosmos, the Russian space agency. Both Furukawa and Borisov were embarking on their second spaceflights while Borisov was on his first.

SpaceX Crew-7 Portrait (Jasmin Moghbeli, Andreas Mogensen, Satoshi Furukawa and Konstantin Borisov) PHOTO DATE: 03-03-23 LOCATION: Building 8, Room 183 - Photo Studio PHOTOGRAPHER: Bill Stafford and Robert Markowitz

The mission, designated as Huginn for its European segment, was named after the raven of the same name in Norse mythology, symbolizing thought and wisdom. Crew-7 launched from Kennedy Space Center on a Falcon 9 Block 5 rocket, lifting off at 07:27:27 UTC. The spacecraft successfully docked with the International Space Station (ISS) at the Harmony zenith port on August 27, 2023, at 13:16 UTC.

Throughout their time aboard the ISS, the Crew-7 astronauts contributed to various scientific experiments and maintenance activities. Their mission spanned 197 days, 2 hours, and 4 minutes, during which they supported the ongoing operations of the space station and engaged in collaborative research.

The Crew-7 mission concluded with their undocking from the ISS on March 11, 2024, at 15:20 UTC. Following a successful re-entry, the Crew Dragon spacecraft splashed down in the Gulf of Mexico near Pensacola, Florida, on March 12, 2024, at 09:47 UTC. Recovery operations were conducted by the recovery vessel MV Megan, ensuring a safe return for the crew after their extensive mission.

Additional missions included the Axiom Mission-2, which docked at Harmony's zenith port on May 22, 2023. This mission, notable for former NASA astronaut Peggy Whitson and John Shoffner, Ali AlQarni, and Rayyanah Barnawi, focused on commercial objectives and scientific experiments. They undocked on May 30, 2023, marking a successful commercial visit.

Cargo resupply was managed by the arrival of Progress MS-23, which docked at the Poisk module's zenith port on May 24, 2023, and the CRS Dragon SpX-28, which arrived at Harmony Forward on June 6, 2023. Progress MS-23 was scheduled to remain until November 29, 2023, while Dragon SpX-28 departed on June 29, 2023.

In August 2023, Cygnus NG-19, named "Laurel Clark," arrived at Unity's nadir port on August 4 and was expected to remain until December 22, 2023. Progress MS-24 followed, docking at Zvezda's aft port on August 25, 2023, with plans to stay until February 13, 2024.

The arrival of Soyuz MS-24 on September 15, 2023, at the Rasssvet module's nadir port marked the continuation of the Russian crew rotation into Expedition 70. This spacecraft was scheduled to stay until April 6, 2024, ensuring a smooth transition into the next expedition phase.

On September 15, 2023, Soyuz MS-24 embarked on a significant mission to the International Space Station (ISS) from Baikonur Cosmodrome. This Russian crewed flight was launched following a series of adjustments due to an unforeseen coolant leak on the Soyuz MS-22 spacecraft, which had led to Soyuz MS-23 being launched uncrewed as a temporary replacement.

Soyuz MS-24's primary crew included Commander Oleg Kononenko, a seasoned Roscosmos astronaut undertaking his fifth spaceflight. Kononenko was joined by Nikolai Chub, also from Roscosmos, marking Chub's first spaceflight. The crew's mission was part of Expeditions 69, 70, and 71. Their journey held the potential to be historic as Kononenko, if the mission extended between 300 and 365 days, would surpass the existing spaceflight duration record of 878 days held by Gennady Padalka. This extended mission could propel Kononenko into the record books as the first astronaut to exceed 1,000 days in space.

Expedition 68 astronaut Frank Rubio of NASA, left, cosmonauts Sergey Prokopyev and Dmitri Petelin of Roscosmos, along with Expedition 68 backup crewmembers Loral O'Hara of NASA, Oleg Kononenko and Nikolai Chub of Roscosmos, right, are seen in quarantine, behind glass, at the conclusion of a press conference, Tuesday, Sept. 20, 2022, at the Cosmonaut Hotel in Baikonur, Kazakhstan. Rubio, Prokopyev, and Petelin are scheduled to launch to the International Space Station aboard the Soyuz MS-22 spacecraft on Sept. 21. Photo Credit: (NASA/Bill Ingalls)

The Soyuz MS-24 crew also included Loral O'Hara from NASA and Marina Vasilevskaya from Belarus, who were on their first spaceflight. While O'Hara participated in the Expedition 69 and 70 missions, Vasilevskaya joined as a spaceflight participant. Their presence highlighted the international collaboration on the ISS.

Upon completing her mission, Loral O'Hara and Oleg Novitsky and Marina Vasilevskaya returned to Earth on April 6, 2024, aboard the Soyuz MS-24 spacecraft. Meanwhile, Oleg Kononenko and Nikolai Chub continued their stay aboard the ISS, with plans to return to Earth on the Soyuz MS-25 spacecraft, alongside NASA astronaut Tracy Caldwell-Dyson.

Throughout Expedition 69, the ISS demonstrated its role as a hub of international collaboration and scientific research. The various spacecraft, each with its purpose and timing, contributed to the station's continued success and the advancement of its missions.

The expedition concluded in September 2024, with the crew preparing for the arrival of the next group of astronauts. The accomplishments of Expedition 69 contributed significantly to the ISS's legacy of scientific discovery and international cooperation, setting the stage for future missions and continuing the station's vital role in advancing human spaceflight.

Support teams work around the SpaceX Dragon Endurance spacecraft shortly after it landed with NASA astronaut Jasmin Moghbeli, ESA (European Space Agency) astronaut Andreas Mogensen, Japan Aerospace Exploration Agency (JAXA) astronaut Satoshi Furukawa, and Roscosmos cosmonaut Konstantin Borisov aboard in the Gulf of Mexico off the coast of Pensacola, Florida, Tuesday, March 12,2024. Moghbeli, Mogensen, Furukawa, and Borisov are returning after nearly six-months in space as part of Expedition 70 aboard the International Space Station. Photo Credit: (NASA/Joel Kowsky)

Expedition 70

The official Expedition 70 crew portrait with (top row from left) Roscosmos cosmonauts Nikolai Chub, Konstantin Borisov, and Oleg Kononenko; JAXA (Japan Aerospace Exploration Agency) astronaut Satoshi Furukawa; and NASA astronaut Loral O'Hara. In the front row are, ESA (European Space Agency) astronaut and Expedition 70 Commander Andreas Mogensen and NASA astronaut Jasmin Moghbeli.

Expedition 70 marked a period of remarkable activity on the International Space Station (ISS), spanning from late 2023 through early 2024. The expedition, which began on September 27, 2023, with the departure of Soyuz MS-23, saw a series of significant events, missions, and crew changes that highlighted the dynamic nature of life aboard the ISS.

The transition into Expedition 70 commenced with Danish astronaut Andreas Mogensen assuming command of the ISS from Russian cosmonaut Sergey Prokopyev, who had led

Expedition 69. This change in leadership marked the beginning of a new chapter in the ISS's ongoing mission to advance international collaboration in space exploration.

In the months following the expedition's start, several critical resupply missions played a crucial role in maintaining the ISS's operations. The SpaceX CRS-29 mission, launched on November 9, 2023, was part of NASA's Commercial Resupply Services program. This mission delivered essential scientific experiments, cargo, and supplies to the station, docking successfully on November 11, 2023. The arrival of SpaceX CRS-29 was instrumental in supporting ongoing research and ensuring the station remained well-stocked with necessary materials.

As November drew to a close, the ISS saw the undocking of ISS Progress 84 on November 29, 2023. This cargo spacecraft had been a vital part of the station's resupply chain, providing necessary supplies and equipment before completing its mission. Shortly thereafter, on December 1, 2023, ISS Progress 86 launched, followed by its docking on December 3, 2023. This spacecraft continued the vital task of delivering essential materials and scientific experiments, contributing to the smooth operation of the ISS.

The end of December brought further activity with the undocking of SpaceX CRS-29 on December 21, 2023, marking the completion of its mission. The following day, Cygnus CRS-19 was released after fulfilling its resupply duties, highlighting the continuous flow of cargo and equipment necessary for the station's functionality.

Holding on to each other to keep from floating away, the newcomers from the Axiom 3 mission (Ax-3) pose up front for a crowded crew picture on the International Space Station.

The new year commenced with notable events as Axiom Mission 3 launched on January 18, 2024. Conducted by Axiom Space, this private mission aimed to enhance commercial spaceflight and bring new opportunities to the ISS. The spacecraft docked with the ISS on January 20, 2024, bringing new crew members and equipment. The Axiom Mission 3 crew, which included Marcus Wandt from the European Space Agency (ESA), Walter Villadei from Italy, Alper Gezeravcı from Turkey, and Michael López-Alegría, a dual citizen of the United States and Spain, joined the existing team of astronauts aboard the ISS. This marked a historic moment as Wandt and his colleagues were greeted by Expedition 70 crew members, including his friend and colleague, Andreas Mogensen, in the first instance of two Scandinavians sharing space together.

The docking of Axiom Mission 3 was smooth, occurring at 12:16 GMT (13:16 CET). The crew settled into their new environment, with Wandt beginning his Muninn mission, which involved conducting around twenty experiments, including studies on space habitat design, cellular changes in microgravity, and public outreach programs. The mission, which began with the launch of the Falcon 9 rocket from Kennedy Space Center, concluded with the undocking of Axiom Mission 3 on February 7, 2024. After a successful 21-day stay, the SpaceX Dragon spacecraft re-entered Earth's atmosphere. They made a controlled splashdown in the Atlantic Ocean on February 9, 2024, marking a successful demonstration of Axiom Space's growing role in commercial space operations.

The expedition continued with further spacecraft arrivals and departures. On February 1, 2024, Cygnus NG-20 arrived at the ISS, delivering additional supplies and scientific payloads. The undocking of Axiom Mission 3 on February 7, 2024, was followed by the departure of Progress MS-24/85P on February 13, 2024. Progress MS-26/87P docked with the ISS on February 17, 2024, ensuring the continuity of resupply and support for ongoing research.

Throughout Expedition 70, the ISS crew undertook various maintenance tasks and spacewalks. Notably, on October 25-26, 2023, cosmonauts Oleg Kononenko and Nikolai Chub completed a spacewalk lasting 7 hours and 41 minutes. This was followed by another spacewalk on November 1, 2023, when astronauts Jasmin

Moghbeli and Loral O'Hara worked outside the station for 6 hours and 42 minutes.

One significant challenge during the expedition arose on October 9, 2023, when a malfunction in the Nauka module's RTOd heat radiator system was detected. This critical system, responsible for radiating heat from Nauka's experiments, experienced a leak, which limited its functionality. The issue, a third radiator malfunction affecting the ISS, underscored the need for reliance on the module's main launch radiator, which constrained Nauka's experimental capabilities.

SpaceX Crew-8, comprising Matthew Dominick, Michael Barratt, Jeanette Epps, and Alexander Grebenkin, arrived on March 5, 2024. A change of command ceremony occurred on March 10, 2024, with Mogensen handing over command to Kononenko. SpaceX Crew-7 undocked on March 11, 2024, and CRS SpX-30 docked on March 23, 2024. Soyuz MS-25 arrived on March 25, 2024, with crew members Novitsky, Vasilevskaya, and Caldwell-Dyson.

Between September 27, 2023, and April 6, 2024, the International Space Station (ISS) hosted a dynamic roster of astronauts and cosmonauts from various space agencies worldwide. This period saw the station's crew transition through several increments, each marked by the arrival and departure of different spacecraft and personnel.

The Soyuz MS-24 spacecraft, which docked on September 27, 2023, brought a team of three to the ISS. This crew included Oleg Kononenko from Roscosmos, embarking on his fifth spaceflight as the mission commander, and Nikolai Chub, making his inaugural spaceflight as a flight engineer. Loral O'Hara, a NASA astronaut also on her first spaceflight, joined them as another flight engineer. They were part of Increment 70a, which ran until March 5, 2024.

In mid-March, the station welcomed a new crew with the arrival of SpaceX Crew-7. This crew included Jasmin Moghbeli from NASA and Andreas Mogensen from ESA, both making their debut spaceflights as flight engineers. Also aboard were Satoshi Furukawa from JAXA and Konstantin Borisov from Roscosmos, both seasoned space travelers on their second and first flights, respectively. This team continued operations on Increment 70b until March 25, 2024.

SpaceX Crew-8 then took over from March 25 to April 6, 2024. This crew comprised Matthew Dominick and Jeanette Epps from NASA, both on their first spaceflights as flight engineers, and Michael Barratt, also from NASA, returning for his third space mission. Alexander Grebenkin from Roscosmos completed the crew, also making his debut spaceflight.

In addition to the expedition crews, Crew Dragon made notable visits to the station. The Axiom Mission 3, which docked in this period, included Michael López-Alegría of Axiom Space (formerly NASA), Walter Villadei from the Italian Ministry of Defence, Alper Gezeravcı of the Turkish Space Agency, and Marcus Wandt from the Swedish National Space Agency.

The Soyuz MS-25 mission, which arrived in this period, carried Tracy Caldwell-Dyson from NASA, embarking on her third spaceflight. Alongside her were Oleg Novitsky from Roscosmos and Marina Vasilevskaya, a flight attendant trained by the Belarus Space Agency for ISS EP-21. One week after docking, Novitsky and Vasilevskaya returned to Earth with Loral O'Hara aboard the Soyuz MS-24 spacecraft. Meanwhile, Kononenko and Chub, launched on Soyuz MS-24, returned to Earth on Soyuz MS-25, completing their mission aboard the ISS.

From March 25 to April 6, 2024, the International Space Station reached a significant milestone by occupying all seven of its currently active docking ports. This unprecedented situation reflected space operations' dynamic and intricate nature, showcasing the ISS's role as a hub for international collaboration and cutting-edge technology.

On September 27, 2023, the International Space Station (ISS) witnessed a pivotal event with the undocking of Soyuz MS-23, marking the official transition from Expedition 69. This departure was a significant milestone, setting the stage for subsequent crew rotations and mission updates aboard the station.

A week later, on October 9, 2023, the ISS encountered a noteworthy technical issue with the Nauka module. A leak was detected in the RtoD add-on heat radiator. The onboard crew immediately and meticulously investigated to address the issue and maintain the module's thermal regulation, which was crucial for its operation and longevity.

The end of October was marked by intense activity, with two critical spacewalks. On October 25 and 26, 2023, astronauts Oleg Kononenko and Sergei Chub conducted this period's first extravehicular activity (EVA), known as EVA 1 (VKD-61). This spacewalk lasted 7 hours and 41 minutes, during which the astronauts executed essential tasks outside the station, contributing to its maintenance and upgrades. Their work was instrumental in ensuring the continued functionality and safety of the ISS.

Following this, on November 1, 2023, the station hosted its second spacewalk of the season, EVA 2 (US-89), performed by astronauts Jasmin Moghbeli and Josh O'Hara. Lasting 6 hours and 42 minutes, this EVA was a routine but vital part of the station's operational maintenance, involving various repairs and adjustments to the station's systems and infrastructure.

On November 11, 2023, the CRS SpX-29 spacecraft arrived at the ISS, delivering essential supplies and equipment. This docking bolstered the station's resources and prepared it for upcoming missions. Later in November, on the 29th, the Progress MS-23/84P spacecraft undocked, vacating a docking port for future arrivals.

December saw further activity with the docking of Progress MS-25/86P on December 3, 2023, which brought additional cargo and resources. On December 21, 2023, the CRS SpX-29 spacecraft undocked, concluding its mission. The very next day, December 22, 2023, the CRS Cygnus NG-19 spacecraft was undocked and released, marking the end of its mission at the ISS.

Entering the new year, on January 20, 2024, the Axiom Mission 3 crew arrived at the ISS, a non-expedition mission contributing to various research and activities aboard the station. On February 1, 2024, the CRS Cygnus NG-20 spacecraft arrived, and its capture and berthing were completed, further enhancing the ISS's capabilities.

By February 7, 2024, the Axiom Mission 3 crew had completed their stay and undocked, returning to Earth. Shortly after that, on February 13, 2024, the Progress MS-24/85P spacecraft undocked, clearing the way for new arrivals.

On February 17, 2024, the Progress MS-26/87P spacecraft docked, delivering crucial supplies and equipment. As March approached, the ISS continued to be a hub of activity. On March 5, 2024, the SpaceX Crew-8 mission arrived, bringing a new crew for Expedition 70.

A few days later, on March 10, 2024, the ISS held a significant event, the Expedition 70 change of command ceremony. Andreas Mogensen officially handed over command to Oleg Kononenko in a formal ceremony that highlighted the collaborative spirit of the international crew.

The following day, March 11, 2024, the SpaceX Crew-7 mission undocked, concluding their mission and returning to Earth. On March 23, 2024, the CRS SpX-30 spacecraft docked, bringing essential supplies and equipment.

On March 25, 2024, Soyuz MS-25, known as "Kazbek," docked with the ISS, marking the start of Expedition 70/71 and ISS EP-21. This spacecraft, docked at the Prichal nadir port, was set to stay until September 24, 2024, and brought new crew members and visitors to the station.

During this period, the Harmony Forward port was occupied by SpaceX Crew-8, named "Endeavour," from March 5, 2024, through August 2024. This mission, part of Expedition 70/71, was crucial in maintaining the ISS's human presence and advancing scientific research. On March 23, 2024, the CRS Dragon SpX-30 docked at the Harmony Zenith port, fulfilling its cargo delivery role until April 28, 2024.

Cargo missions from various international partners further utilized the station's docking ports. The United States' CRS Cygnus NG-20, named "Patricia 'Patty' Hilliard Robertson," was docked at the Unity Nadir port from February 1, 2024, until July 12, 2024, delivering vital supplies and equipment. Russia's Progress MS-25, known as "86P," was docked at the Poisk Zenith port throughout Expedition 71. Progress MS-26, which arrived on February 17, 2024, occupied the Zvezda Aft port, serving the station's logistical needs until the end of the year.

The Prichal, also known as the Uzlovoy Module (UM), which translates to "Nodal Module Berth" in Russian, was a spherical module designed to enhance the docking capabilities of the Russian segment of the ISS. Weighing approximately 4 tonnes (8,800 pounds), Prichal was launched in November 2021 as part of a strategic upgrade to the station's docking infrastructure, allowing for greater flexibility and efficiency in managing spacecraft arrivals and departures.

Prichal was delivered to the ISS integrated with a specialized version of the Progress cargo spacecraft and launched aboard a standard Soyuz rocket. It docked to the nadir port of the Nauka module, expanding the station's docking capacity. The module features six docking ports: one active hybrid docking port and five passive hybrid ports.

The active hybrid docking port was specifically designed to be compatible with the Multipurpose Laboratory Module (MLM), facilitating its integration with the station. The remaining five passive hybrid ports support a range of docking needs, including those of Soyuz and Progress spacecraft, and accommodate heavier modules and future spacecraft equipped with modified docking systems.

Originally intended to serve as a critical component of the now-canceled Orbital Piloted Assembly and Experiment Complex (OPSE), Prichal remains a key addition to the ISS, providing essential docking capabilities and contributing to the ongoing operational flexibility and functionality of the station.

Notably, the Axiom Mission 3, designated "Freedom," occupied the Harmony Forward port from January 20 to February 7, 2024. This commercial venture mission represented the growing role of private spaceflight in supporting ISS operations.

Throughout this busy period, several ports remained vacant or were in transition as spacecraft departed or arrived. The Prichal module's aft, forward, starboard, and other ports, however, remained unused since its docking, highlighting areas of potential for future missions.

The final significant event of Expedition 70 was the swapping of seat liners between Soyuz MS-24 and MS-25 on March 29, 2024, in preparation for the landing of Soyuz MS-24. The expedition officially ended with the undocking of Soyuz MS-24 on April 6, 2024, marking the transition to Expedition 71.

Russian Spacewalkers dwarfed by the Nauka and Prichal modules iss066e121432 (Jan. 19, 2022) --- The Prichal docking and the Nauka multipurpose laboratory module figure prominently in this image taken during a spacewalk with cosmonauts Anton Shkaplerov (at bottom) and Pyotr Dubrov (partially obscured behind Prichal) who outfitted both modules to integrate with the International Space Station.

Expedition 71

329

(Oct. 4, 2023) --- The official Expedition 71 crew portrait with (bottom row from left) Roscosmos cosmonaut Alexander Grebenkin and NASA astronauts Mike Barratt, Matthew Dominick, and Jeanette Epps. In the back row (from left) are, NASA astronaut Loral O'Hara and Roscosmos cosmonauts Nikolai Chub and Oleg Kononenko.

Expedition-71, the 71st long-duration mission aboard the International Space Station (ISS), officially began with the undocking of Soyuz MS-24 on April 6, 2024. Russian cosmonaut Oleg Kononenko, who had served as the commander during Expedition 70, continued his leadership through Expedition 71. His tenure as commander was slated to conclude with the departure of Soyuz MS-25 on September 24, 2024, alongside the returning crew from both Soyuz MS-24 and Soyuz MS-25.

Initially, Expedition 71's crew included Kononenko and his Soyuz MS-24 colleague, Nikolai Chub, who had been aboard the ISS for a year-long mission beginning on September 15, 2023. American astronaut Tracy Caldwell-Dyson joined the ISS on March 23, 2024, arriving aboard Soyuz MS-25 and joining the Expedition 71 team. In addition to these astronauts, SpaceX Crew-8, which had launched on March 4, 2024, contributed to the mission. This crew comprised American astronauts Matthew Dominick, Michael Barratt, Jeanette Epps, and Russian cosmonaut Alexander Grebenkin. The SpaceX Crew-8 members were integrated into Expedition 71 following the departure of Soyuz MS-24.

On March 4, 2024, SpaceX's Crew-8 mission marked a significant milestone as the eighth crewed operational flight under NASA's Commercial Crew Program and the thirteenth overall crewed orbital flight of a Crew Dragon spacecraft. This mission, known officially as USCV-8, carried four astronauts to the International Space Station (ISS), continuing the legacy of collaborative space exploration between the United States and Russia.

(Oct. 4, 2023) --- Official SpaceX Crew-8 portrait with Roscosmos cosmonaut and Mission Specialist Aleksandr Grebenkin, and Pilot Michael Barratt, Commander Matthew Dominick, and Mission Specialist Jeanette Epps, all three NASA astronauts.
Credit: NASA/Bill Stafford

The Crew-8 mission featured a distinguished crew of four: NASA astronauts Matthew Dominick, Michael Barratt, and Jeanette Epps, alongside Roscosmos cosmonaut Alexander Grebenkin. Matthew Dominick, who was making his first spaceflight, was the mission commander. Michael Barratt, a veteran of two previous spaceflights, took on the pilot role. Jeanette Epps, also embarking on her maiden space journey, was a mission specialist. Alexander Grebenkin, making his debut flight, joined as the second mission specialist. Notably, Epps had previously been assigned to Boeing Starliner missions but joined Crew-8 for this SpaceX flight. The mission launched at 03:53 UTC from Launch Complex 39A at the Kennedy Space Center aboard a Falcon 9 Block 5 rocket. Crew Dragon Endeavour, the spacecraft used for this mission, marked its 50th astronaut launch. The Crew-8 crewed spacecraft docked with the ISS at the Harmony module's forward port on March 5, 2024, at 08:00 UTC. This initial docking phase was a crucial step in integrating the crew into the ISS's ongoing operations.

(June 18, 2024) -- A SpaceX Dragon Endeavour spacecraft docked to the zenith port of the International Space Station's Harmony module as the orbiting complex soared 265 miles above the Indian Ocean.

To accommodate an upcoming Boeing Starliner mission, Crew Dragon Endeavour was relocated to the zenith port of the Harmony module on May 2, 2024. This relocation was part of the preparation for the Boeing Crew Flight Test, scheduled to dock at the forward port of Harmony. Boeing Starliner Calypso successfully launched on June 5, 2024, and docked with the forward port of Harmony on June 6.

The Crew-8 mission's duration was planned for 176 days, with the crew spending 58 days at the forward port before relocating to the zenith port. As of the latest update, Crew Dragon Endeavour remained docked at the zenith port, continuing its mission with an expected end in September 2024. The spacecraft's return will be marked by its landing in either the Atlantic Ocean or the Gulf of Mexico, where it will be recovered by either the MV Megan or MV Shannon.

Crew rotations were a key feature throughout the expedition. On June 6, 2024, the Boeing Crew Flight Test docked with the ISS. This significant event brought NASA astronauts Barry Wilmore and Sunita Williams aboard. The flight marked the first crewed test flight of the Starliner. Crew members were scheduled to stay aboard the station for approximately 6 days, but the mission was extended due to issues with the spacecraft. It was the first launch of humans from Cape Canaveral since Apollo 7 in October 1968, and first launch of humans from SLC-41

As the expedition progressed, further crew changes were scheduled. From April 6 to August 19, 2024, Expedition 71 included the initial crew members from Soyuz MS-24 and Soyuz MS-25. The team would transition to the new arrivals from August 19 to August 26, 2024. SpaceX Crew-9 was expected to arrive during this period, featuring astronauts Zena Cardman, Nick Hague, Stephanie Wilson, and cosmonaut Aleksandr Gorbunov. Cardman, Hague, and Wilson were making their respective first, third, and fourth spaceflights, while Gorbunov would be on his first mission.

Following this, from August 26 to September 11, 2024, the crew would consist of the members from SpaceX Crew-9, who would operate on the ISS before the transition to the next crew. The final phase of Expedition 71, from September 11 to September 24, 2024, would include the crew from Soyuz MS-26. This crew, which would include cosmonauts Aleksey Ovchinin and Ivan Vagner and astronaut Donald Pettit, would complete their missions before the official changeover to Expedition 72.

In April 2024, the International Space Station (ISS) witnessed significant activities marking the transition between expeditions and the arrival and departure of various spacecraft. The month began with the undocking of Soyuz MS-24 on April 25th, including the Visiting Expedition 21. This event marked the official conclusion of Expedition 70. Following this, the first extravehicular activity (EVA) of the month, EVA-1, occurred on April 25th, performed by cosmonauts Oleg Kononenko and Aleksandr Chub. This spacewalk lasted four hours and thirty-six minutes and was part of ongoing maintenance and upgrades to the ISS's external systems.

Expedition 71 NASA astronaut Tracy Dyson, Roscosmos cosmonaut Oleg Novitskiy, and Belarus spaceflight participant Marina Vasilevskaya are seen in quarantine, behind glass, during a press conference, Wednesday, March 20, 2024 a the Cosmonaut Hotel in Baikonur, Kazakhstan. Dyson, Novitskiy, and Belarus spaceflight participant Marina Vasilevskaya are scheduled to launch aboard their Soyuz MS-25 spacecraft on March 21. Photo Credit: (NASA/Bill Ingalls)

Three days later, on April 28th, the CRS SpX-30 spacecraft undocked, having completed its resupply mission. The redocking of SpaceX Crew-8 soon followed this on May 2nd, crucial for maintaining crew rotation and support on the station.

May continued with significant logistics events, including the Progress MS-25 (86P) undocking on May 28th. The following month, June, Progress MS-27 (88P) docked with the ISS on June 1st, delivering necessary supplies and equipment. Additionally, Boeing's Crew Flight Test, an important milestone in developing commercial crew vehicles, docked with the ISS on June 6th.

Soyuz MS-25 represents a notable mission in the history of crewed spaceflight, conducted by Roscosmos from the Baikonur Cosmodrome. Launched on March 23, 2024, following a brief delay due to a voltage issue in one of the spacecraft's power generators, this mission carries significant milestones and historical importance.

The primary crew of Soyuz MS-25 includes Russian cosmonaut Oleg Novitsky, who serves as the mission commander. Novitsky, hailing from Chervyen in Belarus, was embarking on his fourth spaceflight. He was joined by Oleg Kononenko, another Russian cosmonaut on his fifth spaceflight, and Nikolai Chub, a cosmonaut making his debut. Notably, Maryna Vasileuskaya from Belarus was also aboard as a spaceflight participant, marking her first spaceflight. From the United States, Tracy Caldwell-Dyson, a NASA astronaut with two prior spaceflights, rounds out the crew as a flight engineer.

The Soyuz rocket shortly after having been rolled out to launch pad at Site 31, Monday, March 18, 2024, at the Baikonur Cosmodrome in Kazakhstan. Expedition 71 NASA astronaut Tracy Dyson, Roscosmos cosmonaut Oleg Novitskiy, and Belarus spaceflight participant Marina Vasilevskaya are scheduled to launch aboard their Soyuz MS-25 spacecraft on March 21. Photo Credit: (NASA/Bill Ingalls)

Soyuz MS-25's mission was particularly historic as it features the first launch of two women—Tracy Caldwell-Dyson and Maryna Vasileuskaya—aboard a Soyuz spacecraft. This mission highlights an important step in gender representation within the realm of space exploration.

The mission was originally slated for March 21, 2024, but was rescheduled to March 23 due to technical difficulties. Upon successful launch, the crew will spend approximately six months aboard the International Space Station (ISS). Roscosmos cosmonaut Oleg Novitsky and Maryna Vasileuskaya will participate in the 21st ISS visiting expedition, which entails a brief stay of around 13 days on the station.

Following the completion of her mission, Tracy Caldwell-Dyson was scheduled to return to Earth on September 24, 2024. She will journey back aboard the Soyuz MS-25 spacecraft with Roscosmos cosmonauts Oleg Kononenko and

Nikolai Chub. Kononenko, who arrived at the ISS with NASA astronaut Loral O'Hara on Soyuz MS-24, will remain on the station for an extended period. If his mission extends to 300–365 days, he will set a new record for cumulative spaceflight duration, surpassing the previous record of 878 days held by Gennady Padalka, and potentially achieving a total of 1,036–1,101 days in space.

On June 5, 2024, Boeing's Starliner capsule embarked on its inaugural crewed mission, known as Boeing Crew Flight Test (Boe-CFT). This landmark flight, which marked the debut of the Boeing Starliner with astronauts aboard, was launched from Cape Canaveral Space Force Station. The mission carried NASA astronauts Barry E. Wilmore and Sunita Williams, who were set to spend eight days aboard the International Space Station (ISS) before returning to Earth on June 14.

NASA's Boeing Crew Flight Test. Left: Suni Williams, the pilot, and to the right: Barry "Butch" Wilmore, spacecraft commander.

However, the mission faced significant challenges. Shortly after docking with the ISS, the Starliner experienced thruster malfunctions and a helium leak, leading NASA to delay the spacecraft's return until these issues could be resolved or better understood. As a result, Wilmore and Williams remain aboard the ISS, extending their stay beyond the planned duration. NASA was evaluating their return options and plans to make a decision by mid-August, considering whether to return them on a SpaceX capsule while the Starliner remains in orbit uncrewed.

A United Launch Alliance Atlas V rocket with Boeing's CST-100 Starliner spacecraft aboard on the launch pad at Space Launch Complex 41 ahead of the NASA's Boeing Crew Flight Test, Saturday, May 4, 2024 at Cape Canaveral Space Force Station in Florida. NASA's Boeing Crew Flight Test was the first launch with astronauts of the Boeing CFT-100 spacecraft and United Launch Alliance Atlas V rocket to the International Space Station as part of the agency's Commercial Crew Program. The flight test, targeted for launch at 10:34 p.m. EDT on Monday, May 6, serves as an end-to-end demonstration of Boeing's crew transportation system and will carry NASA astronauts Butch Wilmore and Suni Williams to and from the orbiting laboratory. Photo Credit: (NASA/Joel Kowsky)

The Boeing Crew Flight Test had been scheduled to occur much earlier, with its initial timeline set for 2017. However, various delays pushed back the launch. The Starliner underwent two uncrewed orbital flight tests prior to this mission: Boe-OFT in 2019 and Boe-OFT-2 in 2022. Both tests were crucial in preparing the spacecraft for crewed missions.

The spacecraft was integrated with the Atlas V launch vehicle on April 16, 2024, in preparation for its flight. The launch, initially set for May 7, 2024, was delayed due to an oxygen valve issue on the Atlas V rocket, a part of United Launch Alliance's (ULA) fleet. Further complications arose from a helium leak, which was critical for pressurizing the reaction control system thrusters. A second attempt on June 1 also faced problems when the ground launch sequencer computer detected a loss of redundancy due to a faulty power supply. Finally, the third attempt on June 5 was successful, with liftoff occurring at 14:52:15 UTC.

During the flight, additional helium leaks were detected, and as the Starliner approached the ISS, five out of its 28 thrusters failed. After adjustments, four of the five thrusters were brought back online, allowing the Starliner to dock with the ISS, albeit with a delay.

The mission was initially intended to include Nicole Aunapu Mann as the commander, making her the first woman to command the maiden crewed flight of an orbital spacecraft. However, Mann was reassigned to SpaceX Crew-5, becoming the first female commander of a NASA Commercial Crew Program launch. Eric Boe, originally assigned as the pilot, was replaced for medical reasons by Michael Fincke. Chris Ferguson, initially the mission commander, was replaced by Barry E. Wilmore for personal reasons. Wilmore and Sunita Williams, also on her third spaceflight, were confirmed as the primary crew for the Boe-CFT.

The Starliner Calypso capsule used for this mission originated in the first Orbital Flight Test. Boeing prepared the vehicle for the CFT mission by modifying its docking system and updating its parachute and airbag systems. This mission was also notable as the first crewed launch from Cape Canaveral Space Force Station since Apollo 7 in 1968 and the first crewed spacecraft launch using the Atlas V since Mercury-Atlas 9 in 1963.

As a short-duration mission, the Boe-CFT was designed to meet NASA and Boeing's test objectives, demonstrating Starliner's capability for operational crewed missions. Upon its eventual return, the Starliner will make a historic ground landing in the Western United States—a first for a crewed capsule launched from the U.S., as previous missions have all concluded with ocean splashdowns.

The pace of operations remained brisk throughout the summer. On June 24th, NASA astronauts Dyson and Barratt conducted EVA-2. This extravehicular activity was notably brief, lasting just thirty-one minutes, and was part of the ongoing efforts to enhance the station's capabilities. The Cygnus NG-20 spacecraft was then unberthed and released on July 12th, following its successful mission.

Looking ahead, August 2024 was set to be a busy month with several planned EVAs. These include EVA-3 (US-91), EVA-4 (VKD-63), EVA-5 (VKD-64), and EVA-6 (US-92), each designed further to advance the station's scientific and technological objectives. Additionally, Boeing's Crew Flight Test was scheduled for undocking later in the month. On August 5th, CRS Cygnus NG-21 was anticipated to capture and berth, bringing a new supply of equipment and experiments to the ISS.

Further logistical maneuvers are expected as the summer progresses, including the undocking of Progress MS-26 (87P) on August 13th and the docking of Progress MS-28 (89P) on August 17th. SpaceX Crew-9 was also slated to dock on August 19th, while SpaceX Crew-8 will undock on August 26th, concluding their mission.

In September, the ISS will welcome Soyuz MS-26 on the 11th, marking the beginning of Expedition 71. On September 23rd, a significant change of command ceremony will be held, transitioning from Oleg Kononenko to a yet-to-be-named successor for Expedition 71/72. The month will conclude with the undocking of Soyuz MS-25 on September 24th, officially marking the end of Expedition 72 and the beginning of the new expedition cycle.

A series of notable activities and mission milestones marked the timeline of events for Expedition 71. Following the departure of Soyuz MS-24, the expedition witnessed several key events: an extravehicular activity (EVA) conducted by Kononenko and Chub on April 25, 2024, lasting 4 hours and 36 minutes; the undocking of CRS SpX-30 on April 28, 2024; and the redocking of SpaceX Crew-8 on May 2, 2024.

In the following months, the station saw continued logistical and scientific operations. Progress MS-25/86P undocked on May 28, 2024, followed by the docking of Progress MS-27/88P on June 1, 2024. The Boeing Crew Flight Test's June 6, 2024 docking was another significant event, providing further resources and support to the ISS.

The expedition also included multiple extravehicular activities. EVA-2, performed by Dyson and Barratt on June 24, 2024, lasted 31 minutes. Subsequent planned activities for July included several more EVAs, namely EVA-3 through EVA-6 , expected to address various scientific and maintenance tasks.

The station's logistical operations continued with the unberthing and release of CRS Cygnus NG-20 on July 12, 2024, followed by the capture and berthing of CRS Cygnus NG-21 on August 5, 2024. Progress MS-26/87P undocked on August 13, 2024, and Progress MS-28/89P docked on August 17, 2024. SpaceX Crew-9 was anticipated to dock in August, with SpaceX Crew-8 scheduled to undock on August 26, 2024.

The seven Expedition 71 crew with the two Crew Flight Test members aboard the space station. Front from left: Suni Williams, Oleg Kononenko, and Butch Wilmore. Second row from left: Alexander Grebenkin, Tracy C. Dyson, and Mike Barratt. Back are, Nikolai Chub, Jeanette Epps, and Matthew Dominick.

A United Launch Alliance Atlas V rocket with Boeing's CST-100 Starliner spacecraft aboard launches from Space Launch Complex 41 at Cape Canaveral Space Force Station, Wednesday, June 5, 2024, in Florida. NASA's Boeing Crew Flight Test was the first launch with astronauts of the Boeing CFT-100 spacecraft and United Launch Alliance Atlas V rocket to the International Space Station as part of the agency's Commercial Crew Program. The flight test, which launched at 10:52 a.m. EDT, serves as an end-to-end demonstration of Boeing's crew transportation system and will carry NASA astronauts Butch Wilmore and Suni Williams to and from the orbiting laboratory. Photo Credit: (NASA/Joel Kowsky)

As Expedition 71 neared its conclusion, Soyuz MS-26 was set to dock on September 11, 2024, leading up to Expedition 71/72 Change of Command Ceremony, during which Oleg Kononenko would transfer command to his successor. The official end of Expedition 71 was marked by the undocking of Soyuz MS-25 on September 24, 2024, signifying the transition to Expedition 72 and the continuation of the ISS's scientific and international collaboration mission.

Incidents in Space

Since its inception, the International Space Station (ISS) has witnessed a series of incidents and accidents, reflecting the inherent challenges of operating and maintaining a human outpost in space. This narrative provides a detailed account of notable mishaps and technical failures, focusing on the significant events that have impacted the station's operations.

In November 2001, the Progress M1-7 cargo spacecraft experienced difficulties in docking due to debris obstructing the docking ring, remnants from a previous Progress mission. An unplanned spacewalk on December 3 allowed astronauts to inspect and remove the foreign object, eventually enabling successful docking.

2003 saw a critical issue during the descent of Soyuz TMA-1 on May 4. The spacecraft switched to a ballistic reentry, causing it to land approximately 460 kilometers from the intended site. This malfunction was traced back to a fault in the descent control system, which caused the spacecraft to exceed its yaw limits and trigger the ballistic mode. Subsequent simulations struggled to replicate the exact failure scenario, though evaluations suggested a rare combination of input signals led to the fault.

On February 27, 2004, a malfunction during the VKD-9 spacewalk cut the mission short. Aleksandr Kaleri's spacesuit cooling system failed due to a pinched coolant tube. This disruption in the cooling water flow led to the premature termination of the spacewalk, with the remaining activities rescheduled. Later that year, on June 24, a similar issue occurred during VKD-9a when Mike Fincke's spacesuit lost oxygen pressure rapidly. The spacewalk, lasting just over 14 minutes, was halted due to an improperly seated oxygen flow switch, prompting updated procedures to prevent future occurrences.

August 3, 2004, brought about another challenge during Expedition 9's EVA-3 spacewalk. The ISS lost its attitude control and drifted 80 degrees off its nominal position, leading to a temporary loss of primary communication. The crew had to maneuver away from the Service Module to allow its thrusters to correct the station's orientation. This incident led to revised attitude control protocols and power conservation measures for subsequent spacewalks.

In 2005, a significant issue arose when gap fillers between heat shield tiles on the STS-114 orbiter were identified during a rendezvous pitch maneuver. Such protrusions had previously caused heating and tile damage. NASA addressed the problem during EVA-3 on August 3, where the gap fillers were removed. An additional spacewalk was considered but later deemed unnecessary.

A more complex problem occurred on October 26, 2006, when the Progress M-58 spacecraft's orientation antenna failed to retract fully during docking. Although the spacecraft connected to the ISS, mission control delayed hard docking by over four hours to mitigate risks of antenna interference with the Zvezda service module. Expedition 14's crew partially retracted the antenna during subsequent spacewalks.

2007 saw a series of incidents, starting with the June 8 launch of STS-117/13A, where part of the thermal insulation blanket detached. This issue was resolved during EVA-3 on June 15. On August 15, EVA-3 of STS-118/13A.1 was cut short after a hole was discovered in Rick Mastracchio's spacesuit glove. This hole, caused by a sharp edge, was non-leaking but led to an early termination of the spacewalk as a precaution.

In October 2007, the Soyuz TMA-10 experienced a failure during its descent, resulting in ballistic reentry and increased stress on the crew. This issue was anticipated to recur with the next Soyuz landing, prompting further investigation.

2008 brought its own set of challenges. On April 19, Soyuz TMA-11's landing was performed in ballistic mode, resulting in a touchdown 420 kilometers from the nominal site. The separation failure between the Instrumentation and Propulsion Module and the Descent Module was traced to faulty pyro bolts. The crew was initially reported to be healthy but Yi So-Yeon was later hospitalized due to vertebrae bruising. To address similar issues, an unplanned EVA was conducted on July 10 during Soyuz TMA-12's stay on the ISS.

November 20, 2008, saw a high CO_2 level in Shane Kimbrough's spacesuit during EVA-2 of STS-126/ULF2. The spacewalk was terminated as a precaution due to the CO_2 buildup, later controlled through periodic rests.

In 2009, a collision risk with debris on March 12 led to the crew taking shelter in their Soyuz spacecraft, as avoidance maneuvers were impossible. On July 22, EVA-3 of STS-127/21A faced a CO2 removal system failure in Chris Cassidy's spacesuit, which led to early termination of the spacewalk.

The year 2010 saw its own set of incidents, including the July 2 abort of Progress M-06M's docking due to an automated system issue. This problem was resolved with a successful docking on July 4. On September 23, Soyuz TMA-18's undocking was delayed due to docking system issues , later resolved through electrical jumper installation.

2011 experienced a close call on June 28, when debris passed within 260 meters of the ISS. Although no collision occurred, the crew took precautionary measures by boarding their Soyuz spacecraft. Later, on August 24, Progress M-12M failed to reach orbit due to an engine shutdown, temporarily grounding Soyuz-FG flights.

In 2012, a piece of debris from a previous collision led to another shelter-in-place event on March 24. The crew resumed regular activities after the debris passed safely.

In 2013, March 1 saw the CRS-2 Dragon cargo spacecraft experiencing thruster failures, which delayed docking by a day. On July 16, ESA astronaut Luca Parmitano's EVA-23 spacewalk was interrupted by water accumulation in his helmet, caused by a faulty Fan Pump Separator.

The year 2014 saw significant incidents, including the aborted docking attempt of Soyuz TMA-12M on March 25, due to a failed burn. The docking was rescheduled. On October 28, the Cygnus Orb-3 cargo spacecraft failed to reach orbit following an Antares rocket explosion, destroying both the rocket and spacecraft but causing no injuries.

In 2015, Progress M-27M lost contact shortly after launch on April 28, with the spacecraft spinning uncontrollably due to a flawed design. Similarly, on June 28, SpaceX's CRS-7 Dragon failed to orbit due to a rocket disintegration caused by a defective bolt. The crew had to shelter in their Soyuz TMA-16M spacecraft on July 16 due to a collision risk with Soviet satellite debris, which passed without incident.

2016 faced a significant issue on January 15 when a water bubble formed in Tim Kopra's helmet during EVA-35, attributed to a blockage in the spacesuit's water management system. In December, Progress MS-04 failed to orbit due to a third-stage failure, leading to a delay in Progress MS-05's launch.

2017 saw a parachute system issue during the descent of Soyuz MS-02 on April 10, causing partial depressurization but no harm to the crew. On June 14, debris from the launch of Progress MS-06 started a fire at the impact site, resulting in fatalities during recovery efforts.

Lastly, in 2018, an air leak was discovered in the Soyuz MS-09 spacecraft on August 29. The leak was initially patched with Kapton tape and later sealed with an epoxy patch, demonstrating the ISS crew's ability to respond to such emergencies effectively.

Epilogue

Expedition 72, scheduled to commence in February 2025, will feature a diverse and internationally representative crew, continuing the International Space Station's tradition of global cooperation in space exploration. The mission was set to bring a blend of seasoned astronauts and promising newcomers to the orbital laboratory.

The initial crew for Expedition 72 will include American astronauts Zena Cardman, Nick Hague, and Stephanie Wilson. Cardman, known for her background in astrobiology, will bring her expertise to the mission. Hague, a spaceflight veteran, was returning to space to continue his research and contributions. Wilson, an accomplished astronaut with previous ISS experience, will provide valuable knowledge and skills to the team. Joining the American astronauts will be Russian cosmonauts Aleksandr Gorbunov, who will be transferred from Expedition 71, and Aleksey Ovchinin. Ovchinin, having a solid history with previous space missions, will continue his contributions to the station's research and operations. Additionally, Ivan Vagner, another experienced Russian cosmonaut, will join the team, contributing to the mission's scientific and technical objectives. The American crew will be joined by Donald Pettit, whose previous work on the ISS has significantly contributed to space research and technology.

In March 2025, Expedition 72 will transition to Expedition 73, with SpaceX Crew-10 scheduled to arrive. This new crew will consist of representatives from the United States, Europe, and Japan, including Takuya Onishi from Japan, who will bring his expertise in various scientific disciplines. The new Russian crew members for this expedition will include Kirill Peskov. Their arrival will be marked by the expected transfer from Expedition 72, ensuring a seamless handover of responsibilities and continuity in the ISS's operations and research activities.

Further details about the Expedition 73 crew will be confirmed as the mission date approaches. However, it was anticipated that the team will include astronauts from the United States and Europe, further emphasizing the collaborative nature of the ISS missions.

Expedition 73 will continue until March 2025, at which point the new Expedition 74 team will succeed the crew. This transition will be facilitated by Soyuz MS-27, which will carry the incoming crew members to the ISS. The new crew for Expedition 74 will include Russian cosmonauts Sergey Ryzhikov and Sergey Mikajew, and American astronaut Jonny Kim. Ryzhikov and Mikajew bring extensive experience from previous missions, while Kim, a newcomer, will contribute fresh perspectives and skills to the ongoing scientific endeavors aboard the ISS.

This rotation of crew members ensures that the International Space Station remains a hub of continuous scientific research, international cooperation, and technological advancement. Each expedition builds upon the achievements of its predecessors, pushing the boundaries of human space exploration and maintaining the ISS as a pivotal platform for scientific discovery.

End of Mission

Originally envisioned as a fifteen-year project, the International Space Station (ISS) has far exceeded its anticipated lifespan, now operating for over twenty years. This extension reflects the ISS's remarkable success and the robust international collaboration that has sustained its mission. Some of the ISS's initial modules have been orbiting Earth for more than two decades as a testament to its longevity. However, the aging infrastructure has sparked concerns about the reliability of these early components, leading to discussions about potentially redirecting resources toward new space exploration endeavors, including renewed lunar missions.

The ISS was one of the most costly single structures ever built, with an estimated total expenditure of \$150 billion as of 2010. This substantial figure includes NASA's contribution of \$58.7 billion, adjusted to \$89.73 billion in 2021, and significant financial inputs from Russia, Europe, Japan, and Canada. The construction costs were further inflated by the 36 shuttle flights required for assembly, each with an estimated price tag of \$1.4 billion. Despite its high price, the ISS has demonstrated its value. Operational costs from 2000 to 2015 averaged \$7.5 million per person-day, less than half the inflation-adjusted cost of its predecessor, Skylab.

The original plan was to deorbit the ISS by 2016, but this timeline has been revised in light of evolving needs and capabilities. In September 2015, NASA extended Boeing's contract as the ISS's primary contractor through September 2020, with provisions to maintain the station's core structural elements beyond this date. This extension extended the ISS's operational period until the end of 2028. Efforts to further secure the station's future included the Space Frontier Act of 2018, which sought to prolong operations until 2030. Although the Senate approved this bill unanimously, it did not pass the House of Representatives. In December 2018, the Leading Human Spaceflight Act, with similar objectives, was confirmed and later integrated into the CHIPS and Science Act, signed into law by President Joe Biden in August 2022.

According to the Outer Space Treaty, which governs international space activities, each participating nation remains legally responsible for the spacecraft and modules it launches. The ISS could pose significant risks without proper maintenance, including dangers from orbital debris and uncontrolled re-entry. Russia has announced its intention to withdraw from the ISS program after 2025 to address these concerns. However, Russian modules will continue providing vital orbital station-keeping services until 2028.

In 2024, Russia's announcement to withdraw from the ISS program marks a significant shift in the global space exploration landscape. This decision, driven by financial constraints, the desire to develop new space infrastructure, and shifting national priorities, will involve a phased reduction of Russia's involvement. The plan includes scaling back crew assignments and scientific contributions and gradually decommissioning the Russian segment of the ISS while continuing to support station operations through 2028.

In January 2020, NASA awarded Axiom Space a contract under the NextSTEP2 program to develop a commercial module for the ISS. This contract, based on a firm fixed-price agreement, involves constructing a module to attach to the forward port of the Harmony (Node 2) module. Although initially commissioned for a single module, Axiom Space has ambitious plans for a full segment of five modules, including a node module, an orbital research and manufacturing facility, a crew habitat, and an Earth observatory with large windows. The addition of this segment was expected to enhance the ISS's capabilities significantly, allowing for larger crews and expanding private spaceflight opportunities. Axiom plans to transition this segment into a standalone space station after the ISS's decommissioning, with the Canadarm2 playing a crucial role in its assembly and future operations. Axiom Space aims to launch the first module, Hab One, by the end of 2026.

Looking beyond the ISS, NASA was contemplating extending the station's operations beyond 2031 if commercial low-Earth orbit (LEO) destinations are not yet ready to accommodate its research needs. This potential extension underscores the ISS's continued importance as a hub for scientific research and international cooperation.

As the ISS approaches the end of its operational life, NASA has evaluated several disposal strategies. Early options included natural orbital decay, boosting the station to a higher orbit, and controlling de-orbit to a remote oceanic area. By late 2010, NASA favored using a modified Progress spacecraft for de-orbiting, but this approach was later deemed insufficient. Instead, NASA decided to develop a specialized spacecraft for this task. Both raising the ISS to a stable orbit and uncontrolled de-orbiting presented too many risks, including potential hazardous debris from space collisions or atmospheric re-entry.

In January 2022, NASA announced plans to deorbit the ISS by January 2031 using a specially designed vehicle. This vehicle, scheduled for launch in 2030, will dock at the Harmony forward port or another designated location after removing the Axiom Orbital Segment. It will remain attached while the ISS's orbit naturally decays to approximately 220 kilometers, then perform a series of burns to lower the orbit further before conducting the final deorbiting maneuver. To mitigate risks associated with reliance on Russian modules, NASA has secured special funding to develop its deorbiting module, awarding a contract worth up to $843 million to SpaceX in June 2024. This vehicle, based on the Cargo Dragon spacecraft, will be equipped with a lengthened trunk module, 46 Draco thrusters, and a substantial propellant load.

As the ISS nears retirement, NASA was intensifying efforts to transition operations to future private space stations in low-Earth orbit.

This shift aims to maintain continuity in microgravity research and technological development, with private sector initiatives such as Axiom Space's ambitious plans to construct its own space station playing a central role. NASA officials emphasize the importance of avoiding disruptions in space-based research, with plans to ensure a viable successor was operational by 2028. This initiative aligns with the strategy outlined by the White House Office of Science and Technology Policy to maintain an uninterrupted American presence in low-Earth orbit and foster a robust commercial space industry.

Despite these efforts, transitioning to commercial space stations involves significant challenges, including technical complexities and budgetary considerations. NASA plans to support private space station providers through its Commercial Low Earth Orbit Destinations (CLD) program while gradually reducing its involvement. Post-2030, NASA envisions operating a national laboratory, the LEO National Lab, designed to facilitate government-sponsored research across multiple commercial platforms, complementing rather than competing with private sector interests.

International partners such as Japan, Canada, and the European Space Agency remain committed to supporting the ISS until its planned retirement. Meanwhile, Russia plans to focus on its own orbital space station by 2028, reflecting human space exploration's dynamic and evolving nature. As NASA works to ensure a seamless transition to future space stations, the legacy of the ISS as a pioneering platform for scientific research and international cooperation in space will continue to influence the future of space exploration.

About the Author

Thornton D. "TD" Barnes is a distinguished author, entrepreneur, and former military intelligence specialist. Born in Dalhart, Texas and raised on a ranch near Clayton, New Mexico and Dalhart, Texas, he cultivated a passion for exploration. After high school in Oklahoma, Barnes embarked on a ten-year military journey, initially serving in Korea as an intelligence specialist. While in the Army, he also specialized in missile and radar electronics, defending against Soviet threats and later attending the Artillery Officer Candidate School. An injury ended his military career, but Barnes soon transitioned to aerospace endeavors. He worked on pivotal projects at NASA's High Range in Nevada, including the X-15, the NASA NERVA nuclear rocket project, and atomic bomb testing at the Nevada Test Site. Furthermore, he participated in the CIA's Mach 3 A-12 Project OXCART and stealth projects at Area 51.

Barnes founded and led an oil and gas exploration company outside the aerospace sphere for over 40 years, delving into uranium and gold mining ventures. In retirement, he's dedicated to preserving Area 51's history, serving as president of Roadrunners Internationale and the Nevada Aerospace Hall of Fame Director Emeritus. His contributions have been spotlighted in documentaries on National Geographic, the History Channel, and other major networks. Barnes has authored several books, including "The Secret Genesis of Area 51" and "The CIA Area 51 Chronicles." He currently resides in Henderson, Nevada, continuing to influence aerospace, exploration, and literature, focusing on the formally highly classified of the CIA's era at Area 51.

Bibliography

NASA Archives
https://en.wikipedia.org/wiki/List_of_human_spaceflights_to_the_International_Space_Station
https://en.wikipedia.org/wiki/List_of_human_spaceflights_to_the_International_Space_Station#/medi a/File:STS-96_crew.jpg
https://www.nasa.gov/international-space-station/space-station-visiting-vehicles/
Memories of the author, T.D. Barnes during his time with NASA